Advanced Power Electronics Converters for Future Renewable Energy Systems

This book narrates an assessment of numerous advanced power converters employed on primitive phase to enhance the efficiency of power translation pertaining to renewable energy systems. It presents the mathematical modelling, analysis, and control of recent power converter topologies, namely, AC/DC, DC/DC, and DC/AC converters. Numerous advanced DC-DC converters, namely, multi-input DC-DC converter, Cuk, SEPIC, Zeta, and so forth have been assessed mathematically using state space analysis applied with an aim to enhance power efficiency of renewable energy systems.

The book:

- Explains various power electronics converters for different types of renewable energy sources
- Provides a review of the major power conversion topologies in one book
- Focuses on experimental analysis rather than simulation work
- Recommends usage of MATLAB, PSCAD, and PSIM simulation software for detailed analysis
- Includes DC-DC converters with reasonable peculiar power rating

This book is aimed at researchers and graduate students in electric power engineering, power and industrial electronics, and renewable energy.

Advanced Power Electronics Converters for Future Renewable Energy Systems

Edited by
Neeraj Priyadarshi, P. Sanjeevikumar, Farooque Azam,
C. Bharatiraja and Rajesh Singh

CRC Press
Taylor & Francis Group
Boca Raton London

CRC Press is an imprint of the
Taylor & Francis Group, an **informa** business

First edition published 2023
by CRC Press
6000 Broken Sound Parkway NW, Suite 300, Boca Raton, FL 33487–2742

and by CRC Press
4 Park Square, Milton Park, Abingdon, Oxon, OX14 4RN

CRC Press is an imprint of Taylor & Francis Group, LLC

ISBN: 978-1-032-34714-1 (hbk)
ISBN: 978-1-032-34715-8 (pbk)
ISBN: 978-1-003-32347-1 (ebk)

DOI: 10.1201/9781003323471

Typeset in Times
by Apex CoVantage, LLC

Contents

Editor Biographies

Dr Neeraj Priyadarshi is associated with the Department of Electrical Engineering, JIS College of Engineering, Kolkata, India. He received an MTech degree in power electronics and drives in 2010 from the Vellore Institute of Technology (VIT), in Vellore, India, and a PhD from Government College of Technology and Engineering, in Udaipur, Rajasthan, India. Dr Priyadarshi has a post-doc from the Department of Energy Technology, at Aalborg University, in Esbjerg, Denmark. His current research interests include power electronics, control systems, power quality, and solar power generation. Currently, he is associated with the Department of Electrical Engineering, JIS College of Engineering, Kolkata, 741235, India. He has published over 115 papers in journals and conferences of high repute like IEEE Systems Journal, IET Electric Power Applications, IET Power Electronics IEEE Access Journal, International Transaction of Electrical Energy System, Wiley, Energies MDPI, International Journal of Renewable Energy Research, and so on. Dr Priyadarshi is a reviewer of IEEE Systems Journal, Electric Power Components and Systems, Taylor and Francis, IEEE Access, IET Renewable Power Generation, International Journal of Modeling and Simulation, International Journal of Renewable Energy Research, International Journal of Power Electronics and Drives and Reputed Elsevier SCI indexed Journals, among others. Under his guidance, 25 DST, government of Rajasthan projects have been sanctioned for financial support.

Padmanaban Sanjeevikumar, PhD, is a Full Professor with the Department of Electrical Engineering, IT and Cybernetic, University of South-Eastern Norway, Porsgrunn, Norway. He serves as an Editor/Associate Editor/Editorial Board of refereed journals, in particular, the IEEE Transactions on Industry Applications, the Deputy Editor/Subject Editor of IET Renewable Power Generation, and IET Generation, Transmission and Distribution Journal, Subject Editor of FACETS and Energies MDPI Journal.

Farooque Azam received his bachelor's degree in Computer Science and Engineering from Visvesvaraya Technological University, in India, in 2011, master's degree in Computer Science and Engineering from Jawaharlal Nehru Technological University, India, in 2013 and PhD in Computer Science and Engineering in the year 2021.

He is currently working as an Associate Professor in the School of Computer Science and Engineering, at REVA University, in Bengaluru, India. He has authored 60 scientific papers of International repute. His research interests include renewable energy, vehicular communication, and V2G communication. He is the reviewer for IEEE Access, SN Journal of Applied Science, and Journal of Photovoltaics. He has received a best paper award at the First IEEE International Conference on Advances in Information Technology, organised by Adichunchunagiri Institute of Technology (AIT), in Karnataka, India, in July 2019. He has published 11 Indian Patent and got a grant of three international patent. He is Academic Editor for ITEES, Hindawi, JCNC, Hindawi. He has also received Best paper awards in Conferences held at SJBIT, Bangalore in 2021, IACIT Conference at REVA University, Bangalore in 2021 and 2022 respectively. He is also the recipient of Best Researcher Award from REVA University, Bangalore in 2021.

Dr C. Bharatiraja received a bachelor of engineering degree in electrical and electronics engineering from Kumaraguru College of Engineering, in Coimbatore, India, in 2002, and a master of engineering degree in power electronics engineering from Government College of Technology, in Coimbatore, India, in 2006. He received his PhD in 2015. He completed his first postdoctoral fellowship at the Centre for Energy and Electric Power, Faculty of Engineering and the Built Environment, at Tshwane University of Technology, in South Africa, in 2016, with National Research Foundation funding. He was the award recipient of DST, Indo-US Bhaskara Advanced Solar Energy in 2017 and through this, he completed his second postdoctoral fellowship in the Department of Electrical and Computer Engineering, at Northeastern University, USA. He is a visiting researcher scientist at Northeastern University, Boston, USA. He is a visiting researcher at University of South Africa. He was also an award recipient of Young Scientists Fellowship, Tamil Nadu State Council for Science and Technology, in 2018. He was collaborated with leading Indian overseas universities for both teaching and research. He has completed six sponsored projects from various government and private agencies. He also has signed MoUs with various industries. Currently, he is running two DST funded projects as project worth of 67 Lakhs. He is a senior member of IEEE, IEI, and IET. He is currently working as an associate professor in the Department of Electrical and Electronics Engineering, at SRM Institute of Science and Technology, Kattankulathur Campus, in Chennai, India. His research interests include power electronics converter topologies and controls for PV and EV applications, PWM techniques for power converters and adjustable speed drives, wireless power transfer, and smart grid. He has authored more than 100 research papers, which are published in international journals, including various IEEE Transaction.

Dr Rajesh Singh is currently associated with Lovely Professional University as Professor with more than 17 years of experience in academics. He has been awarded as gold medalist in M. Tech from RGPV, in Bhopal (M.P.), India, and honors in his BE from Dr B.R. Ambedkar University, in Agra (U.P.), India. His areas of expertise include embedded systems, robotics, wireless sensor networks, and Internet of Things. He has been honored as keynote speaker and session chair at international/national conferences, faculty development programs, and workshops. He has 152 patents in his account. He has published more than a hundred research papers in referred journals/conferences and 24 books in the areas of embedded systems and Internet of Things with reputed publishers like CRC/Taylor & Francis, Narosa, NIPA, River Publishers, Bentham Science, and RI publication. He is editor to five books, including a special issue published by AISC book series, Springer, in 2017 and 2018, and IGI global, in 2019. Under his mentorship, students have participated in national/international competitions including "Innovative Design Challenge competition" by Texas and DST and Laureate award of excellence in robotics engineering, Madrid, Spain, in 2014 and 2015. His team has been the winner of "Smart India Hackathon-2019" hardware version conducted by MHRD, government of India, for the problem statement of Mahindra and Mahindra. With his mentorship, student team got "InSc award 2019" for students projects program. He has been awarded the Gandhian Young Technological Innovation (GYTI) award, as mentor to "On Board Diagnostic Data Analysis System-OBDAS", appreciated under "Cutting Edge Innovation" during the Festival of Innovation and Entrepreneurship at Rashtrapati Bahawan, India, in 2018. He has been honored with a Certificate of Excellence from the third faculty branding awards-15, organised by EET CRS research wing for excellence in professional education and industry, for the category "Award for Excellence in Research", 2015, and young investigator award at the International Conference on Science and Information, in 2012.

Contributors

T. Abhilash
Yeungnam University
South Korea

Ankit Anand
National Institute of Technology
Rourkela, India

Ravi Anand
National Institute of Technology
Patna, India

C. R. Ananthraj
National Institute of Technology
Rourkela, India

T. Sudhakar Babu
Chaitanya Bharati Institute of Technology
India

M. Baranidharan
Vellore Institute of Technology
Vellore, India

Arnab Ghosh
National Institute of Technology
Rourkela, India

Vikash Gurugubelli
National Institute of Technology Rourkela
India

Vanshika Jindal
Delhi Technological University
Delhi, India

Dheeraj Joshi
Delhi Technological University
Delhi, India

N. Kalaiarasi
SRM Institute of Science and Technology
India

R. Rajesh Kanna
Vellore Institute of Technology
India

A. Kirubakaran
National Institute of Technology
Warangal, India

Jia Shun Koh
UCSI University
Malaysia

Deepak Kumar
National Institute of Technology
Delhi, India

Kuncham Sateesh Kumar
SRM Institute of Science
   and Technology
India

Jordan S. Z. Lee
UCSI University
Malaysia

Wei Hong Lim
UCSI University
Malaysia

Rajib Kumar Mandal
National Institute of Technology
Patna, India

Balaji Mendi
National Institute of Technology
Rourkela, India

S. V. K. Naresh
National Institute of Technology Andhra
   Pradesh
India

Jeeban Kumar Nayak
National Institute of Technology
   Rourkela
Odisha, India

Monalisa Pattnaik
National Institute of Technology
   Rourkela
Odisha, India

Sankar Peddapati
National Institute of Technology Andhra
    Pradesh
India

A. Sivapriya
SRM Institute of Science and Technology
India

Jammy Ramesh Rahul
NIT Andhra
Pradesh, India

Anmol Ratna Saxena
National Institute of Technology
Delhi, India

Ashutosh Kumar Singh
National Institute of Technology
Patna, India

R. Raja Singh
Vellore Institute of Technology
India

V. T. Somasekhar
National Institute of Technology
Warangal, India

Gopalakrishna Srungavarapu
National Institute of Technology
    Rourkela
Odisha, India

Nadia M. L. Tan
University of Nottingham
Ningbo, China

Rodney H. G. Tan
UCSI University
Malaysia

Preface

This book demonstrates the employment of high gain power electronics converters for future solar and wind renewable energy systems. Chapter 1 explains the evolution and analysis of a novel three-port battery-integrated tapped-inductor dc-dc boost converter (BITIC) for the power management of LVPDS. A State of Charge calculation using Coulomb counting method employed along with the battery model is explained in Chapter 2. Chapter 3 deals with SEPIC/Zeta converter because, in comparison to other converters, SEPIC and Zeta converters have reduced ripple in output voltage. Chapter 4 illustrates a solar-powered gearless elevator drive control using a four quadrant DC to DC converter system. Non-isolated quadratic bidirectional DC-DC converters for renewable energy systems are discussed in Chapter 5. Chapter 6 explains modeling and MPPT control of a PMSG-based wind turbine system. Hybrid multilevel inverter topologies for medium-voltage applications are presented in Chapter 7. Chapter 8 presents a streamlined approach on maximum power point tracking (MPPT) control of a single-stage grid-connected solar inverter. A review of cascaded H-bridge and modular multilevel converter topologies, modulation technique, and comparative analysis is explained in Chapter 9. Chapter 10 presents a review analysis of cascaded H-bridge and modular multilevel inverters. Chapter 11 presents the development of battery energy storage system (BESS) with photovoltaic (PV) emulated output characteristics for direct PV string DC coupling. Investigations on the effect of SMC parameters of STATCOM–ES on small signal stability of the power system is discussed in Chapter 12. Chapter 13 deals with a new hybrid islanding detection technique for microgrid using a virtual synchronous machine. Behavioral analysis of multi-source DC to DC converter for integrating renewable energy and sourcing to residential loads is presented in Chapter 14.

1 Three-Port DC-DC Converters for Integration of Solar PV and Energy Storage Devices with LVPDS Distribution System

Anmol Ratna Saxena and Deepak Kumar

CONTENTS

1.1 INTRODUCTION

The energy sector has undergone massive transformations during the last few decades due to the penetration of renewable energy sources (RES). The introduction of RES makes it possible to use electricity in remote regions that are still deprived of grid connectivity [1]–[3]. The systems formed without grid connectivity are called standalone off-grid AC or DC systems. The standalone off-grid AC or DC systems are generally designed to have relatively lower voltage levels as compared to conventional AC grid-connected systems [1], [2]. Therefore, these systems when designed for DC are generally called a low voltage low power residential dc distribution system (LVPDS) and are

DOI: 10.1201/9781003323471-1

preferred over the AC system, as the LVPDS system does not require the extra power-conversion stages (ac-dc or dc-ac) [4], as most of the loads and RES are DC in nature. These systems can be installed near the consumer and are thus best suited for the distributed generation. Owing to the DC nature of renewable energy sources, DC distribution systems are gaining popularity and are most suited for applications like residential and commercial buildings, PV-based irrigation pumps, aerospace applications, marine power systems, data centers, and so on.

In LVPDS, multiple dc-dc converters are required to interconnect the different elements (a power source, a load, and a battery) and achieve objectives such as maximum power point tracking (MPPT), load voltage regulation (LVR), and charging/discharging of batteries [5], [6]. This requires multiple dc-dc converters in series and parallel configurations [3]. In this system, conventional second-order dc-dc boost converters are commonly used to interconnect the low voltage renewable energy sources with loads. The voltage gain of a traditional boost converter decreases as electrical loads increase, and thus efficiency drops significantly [6]. Another issue with this system is that the power demand for these loads may suddenly increase/decrease [7]–[9]. During the period of increasing power consumption, this places an added burden on the primary source. During light load conditions, on the other hand, the source has extra power [10], [11].

Imbalance in source and load power has a substantial impact on system stability and results in system performance problems such as the risk of overloading of the converter [12], high ripples, stresses, and so on [13]. In the case of fuel cells, it causes fuel starvation during high load and fuel avalanche during light load [14]. This imbalance in the power is usually dealt with by an additional converter called a bidirectional converter [15], which is used for the integration of the battery with the system. A bidirectional converter increases the size of the system and requires an additional control scheme. Moreover, a conventional bidirectional buck-boost converter needs to be operated in buck or boost mode and has low voltage gain [16].

As a result, for balancing the power between the source and the load, a single power electronic converter with high voltage boosting and the ability to interface battery storage is required. Another necessity that arises as a result of integrating the battery into the converter is to charge the battery and achieve the MPPT at the same time [1]–[3], [17]. Three-port converters (TPC) are reported in the literature to integrate the battery with the source converter [18]. The battery inside such converters can be charged by different methods such as constant voltage, constant current, two-stage (constant current-constant voltage charging), and pulse charging. Pulse charging charges the battery using a series of current pulses, resulting in a high capacity retention rate as compared to other techniques [19].

Formulation, operation, design, and analysis of battery integrated converters [14], [18], [20] are required to show its salient features such as high voltage gains, the bidirectional power flow path for battery, and pulse charging/discharging of the battery. Battery integrated converters can integrate the battery within the converter, however, such converters necessitate the use of multiple switches for different power flow paths.

1.2   RELATED WORKS

The majority of researchers have focused on dc/dc converters with multiple inputs and outputs that integrate multiple sources with multiple loads. Most of these converters cannot include the bidirectional device, that is, the battery, and thus require an additional bidirectional converter. By knowing the necessity of power flow management for a DC Nano-grid, a literature survey is made to identify the suitable non-isolated dc/dc converters which help to integrate the source, load, and battery. Thus, this study mainly focuses on the feasibility of the converter to integrate source, load, and battery at the same time while maintaining the continuous source current.

Integration of the battery with the converter imposes several constraints on the voltage level and power management. Different range of load resistance is used with different dc/dc converters to achieve maximum power point tracking (MPPT). The equivalent load resistance for a boost converter should be lower than the actual load resistance. If the battery is used as a load, as shown

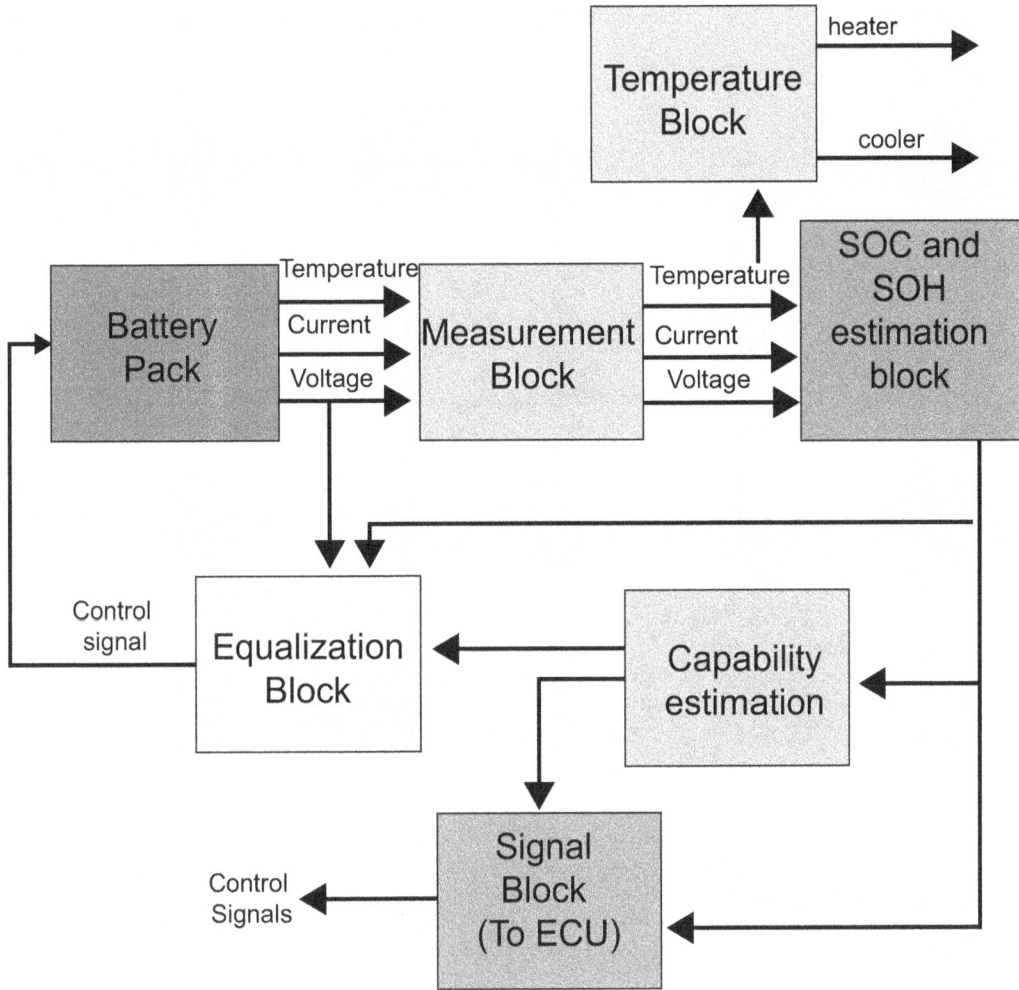

FIGURE 1.1   Battery at the load side of the boost converter [21].

in Figure 1.1, the battery voltage should be higher than the source voltage to achieve maximum power point tracking (MPPT) [21]. Interfacing of battery with load requires another converter with the objective to achieve voltage regulation across the load and maintain MPPT. In such a case, the MPPT function is affected by the change in load power and source power.

Integration of other boost converters at the load side with the battery causes the interaction among the converters, which causes the increase of battery ripples as shown in Figure 1.2 [22]. Moreover, there is no direct connection for the source to load in this topological configuration. As a result, the battery responds to any load demand. As we know, the MPPT function is affected by the change in load and source power. To reduce this effect, battery-integrated dc/dc converter modules have been developed in which the battery is connected between the source and load as shown in Figure 1.3 [4]. This allows the converter to handle the fluctuating source and load power and thus increase the reliability of the system. This will also help to overcome the partial shading of a particular module. However, MPPT is disrupted if the converter's output voltage falls below the MPPT reference value.

Thus, the voltage across the load must be regulated at a high voltage level (greater than the sum of the source and battery voltage). Also, the battery cannot deliver the power alone in this configuration. In literature, there are several three-port converters to manage the power among the source, load, and battery. One of the three-port converters reported in the literature is shown in Figure 1.4

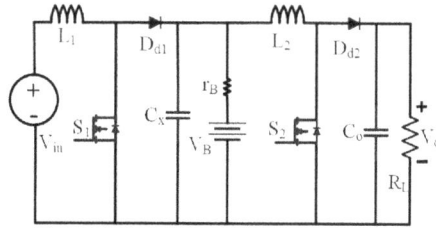

FIGURE 1.2   Two boost converter in a cascade [22].

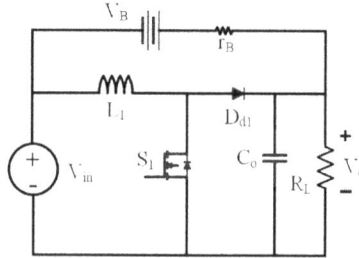

FIGURE 1.3   Battery within source and load side of converter [4].

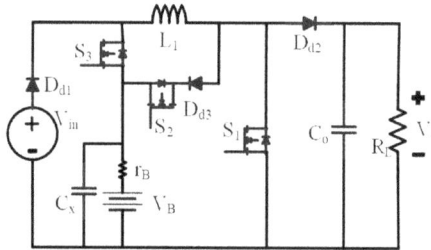

FIGURE 1.4   Non-isolated three-port DC/DC converter [23].

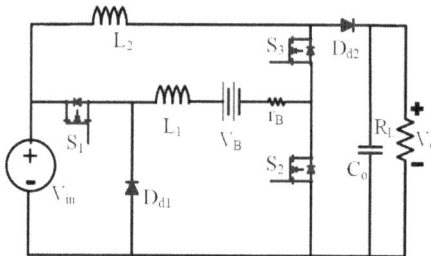

FIGURE 1.5   A different three-port converter [24].

[23]. The problem with this configuration is the discontinuous source current during the battery discharging operation which causes the loss of the appreciable amount of source power.

A different three-port converter reported, shown in Figure 1.5, has the configuration to deliver power from source to load/battery or battery to load [24]. Thus, the battery can be charged or discharged in this configuration. In this configuration, the battery can alone supply power to the load. But, source voltage has to be kept higher than the battery voltage during battery charging, which cannot be ensured during low irradiance for the photovoltaic panel, and thus maximum power cannot be tracked in such a case. Figure 1.6 depicts a non-isolated three-port converter that integrates source, load, and battery as an extension of the parallel combination of the boost converter and bidirectional buck-boost converter [25]. In this configuration, the number of components used is comparatively high and there is no direct connection between source and battery. Another converter, BI2C (Battery integrated second-order boost converter), to integrate the source, load, and battery using the embedded battery within the converter is shown in Figure 1.7. MPPT and load voltage regulation can be achieved simultaneously [26]. The battery can be charged or discharged without depending on the source and battery voltage level. However, the operation of the converter is not possible without a source, that is, the battery alone cannot regulate the load voltage.

A modified version of the BI2C, where all the power flow is possible using three switches is shown in Figure 1.8 [27]. In this configuration, the battery voltage is kept high in comparison to the load voltage to provide a power flow path for the battery to load in the event of a power outage. However, the use of a switch at the input side of the source reduces the efficiency to track the maximum power. In another configuration, source, battery, and load are also integrated using a

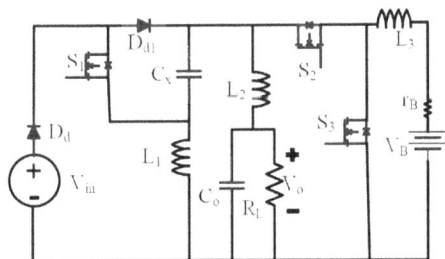

FIGURE 1.6   Non-isolated three-port converter with integrated bidirectional converter [25].

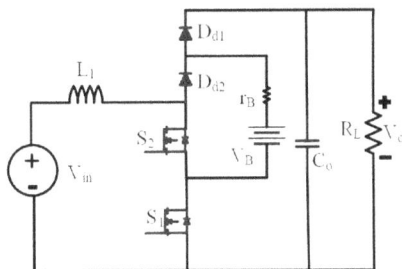

FIGURE 1.7   Second-order battery integrated boost converter [26].

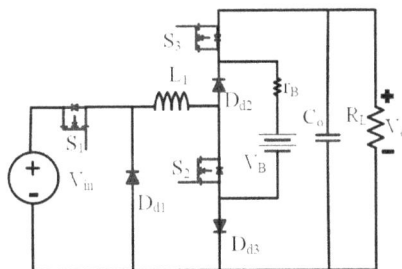

FIGURE 1.8   Three-switch battery integrated converter [27].

single switch with variable frequency and variable duty ratio operation as shown in Figure 1.9 [28]. However, this makes the control complex and increases the stresses across the switches. This causes the restricted duty ratio control and operates the inductor L_2 in discontinuous conduction mode (DCM). It has an inverted output voltage.

1.3   THREE-PORT BATTERY INTEGRATED TAPPED-INDUCTOR BOOST CONVERTER

A new battery integrated tapped-inductor boost converter (BITIC) based on a combination of the multi-port converter and tapped-inductor dc-dc boost converter is presented. Due to the integrated battery and use of the tapped-inductor, the voltage transfer ratio is increased. Moreover, the proposed converter compared to similar converters have a low number of components counts for all the power flow path required among source, battery, and load.

The proposed battery integrated tapped-inductor boost converter (BITIC) is a three-port converter that offers high-voltage gain, thereby making it a viable option for PV/fuel-cell applications that require integrating energy storage devices. Figure 1.10 shows the proposed BITIC circuit configuration. The converter has three switches, two diodes, one tapped-inductor, and one filter capacitor. The inductor having 'N' turns is tapped in the ratio $N_1:N_2$ having inductance in the ratio $L_1:L_2$. The salient features of the proposed converter are (i) compact design due to low component count; (ii) can operate as a dual-input single-output (DISO), single-input dual-output (SIDO) converter, single-input single-output (SISO); (iii) additional control parameter offers a wide range of voltage gain even at low source voltage; (iv) eliminates the requirement of multiple converters required for MPPT, and load voltage regulation; and (v) exhibits pulse charging scenario and pulse discharging scenario for battery which enhances its lifetime [19].

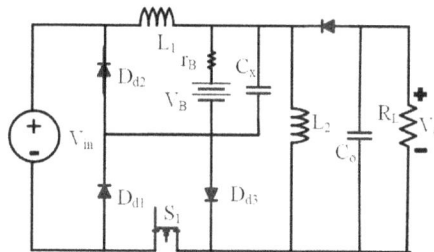

FIGURE 1.9   Single-switch-based battery integrated converter [28].

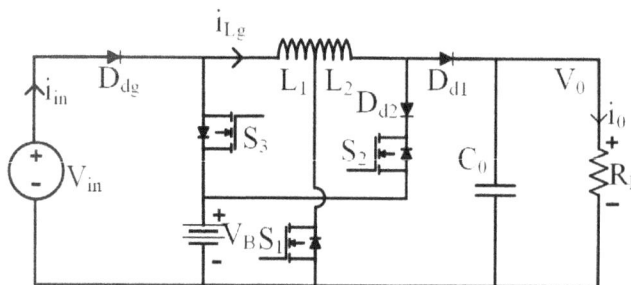

FIGURE 1.10   Circuit configuration of three-port battery integrated tapped-inductor boost converter (BITIC).

1.3.1 CONVERTER OPERATION

This section covers the functioning of the converter and its analysis in the continuous conduction mode (CCM). Depending upon the switching sequence of MOSFETs the converter operates either as a DISO converter, SIDO converter, or SISO converter. The switching state of each switching element (ON/OFF state) is given in Table 1.1 where '0' represents the OFF state while '1' represents the ON state of the switch.

The equivalent circuit of BITIC under battery discharging (DISO), battery charging (SIDO), source to load (SISO), and battery to load (SISO) modes are presented in Figures 1.11 (a) to (d) respectively. The switch S_2 is inactive during the battery discharging operation (DISO) while switch

TABLE 1.1
Switching Sequence Table for Battery Charging and Discharging Modes

Switching states	S_1	S_2	S_3	D_{d1}	D_{d2}
Battery discharging operating mode ($D_1 > D_3$), DISO					
S_{1d1}	1	0	1	0	0
S_{2di}	1	0	0	0	0
S_{3di}			NA		
S_{4di}	0	0	0	1	0
Battery discharging operating mode ($D_1 < D_3$), DISO					
S_{1d1}	1	0	1	0	0
S_{2di}			NA		
S_{3di}	0	0	1	1	0
S_{4di}	0	0	0	1	0
Battery dharging operating mode ($D_1 < D_2$), SIDO					
S_{1do}	1	0	0	0	0
S_{2do}	0	1	0	0	1
S_{3do}	0	0	0	1	0
Isolated battery operation (PV to load), SISO					
S_{1si}	1	0	0	0	0
S_{2si}	0	0	0	1	0
Isolated source operation (battery to load), SISO					
S_{1si}	1	0	1	0	0
S_{2si}	0	0	1	1	0

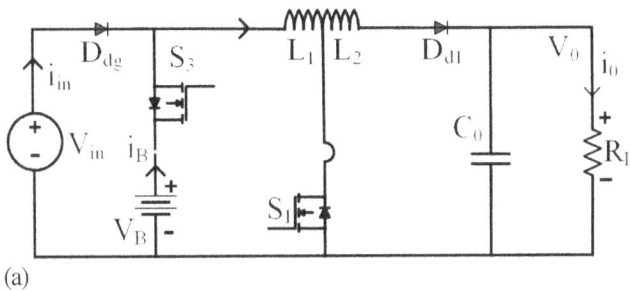

(a)

FIGURE 1.11   Equivalent circuit of BITIC under (a) battery pulse discharging scenario (DISO mode), (b) battery pulse charging scenario (SIDO mode), (c) source to load (SISO mode), and (d) battery to load (SISO mode).

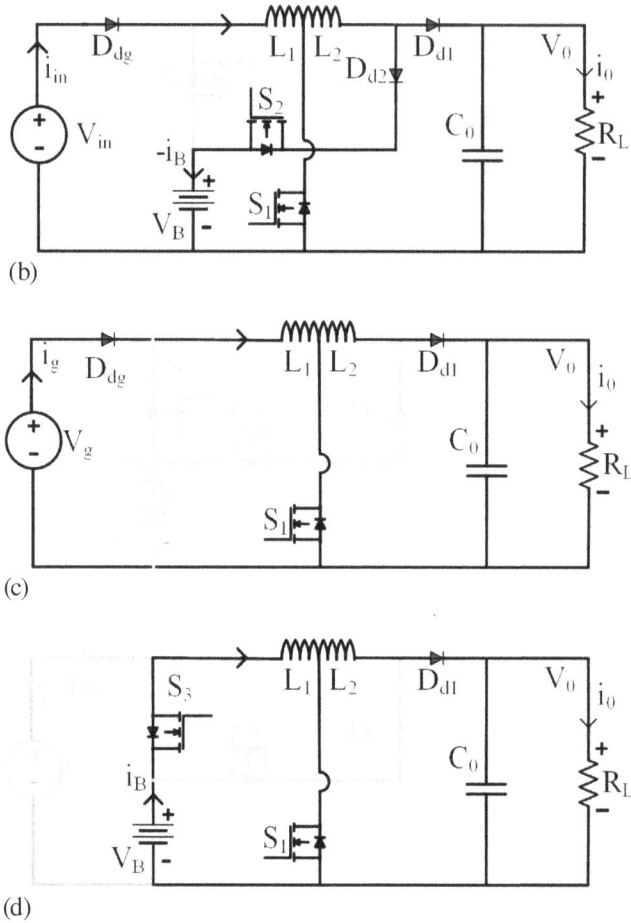

(b)

(c)

(d)

FIGURE 1.11 Continued

S_3 is inactive during the battery charging operation (SIDO). D_1, D_3, and D_3 are the duty ratios of switch S_1, S_2, and S_3, respectively.

1.3.1.1 Battery Pulse Charging Scenario (BPCS)

This is a single-input dual-output (SIDO) operation of the converter. When the power demanded by the load is less than the source power, the switching sequence makes the converter operate as a SIDO converter wherein the battery charges to store the excess power. In the case of the battery charging operation ($D_2 > D_1$), the switching modes are M_{1do}, M_{2do}, and M_{3do}. Figure 1.13 shows the converter equivalent circuits during these states. The key converter waveform, for battery charging operating mode, is shown in Figure 1.12(c), wherein the battery current is shown negative (below the x-axis) to show the charging of the battery. The switching sequence consists of the following modes of operation.

Mode-M_{1do}: This state of the converter in this mode is similar to Mode-M_{2di} of discharging operating mode. Switch S_1 is 'ON' in this state, hence switch S_2 is reverse biased and cannot be set 'ON'. Also, diodes D_{d1} and D_{d2} are reverse biased and do not conduct. Figure 1.13(a) shows

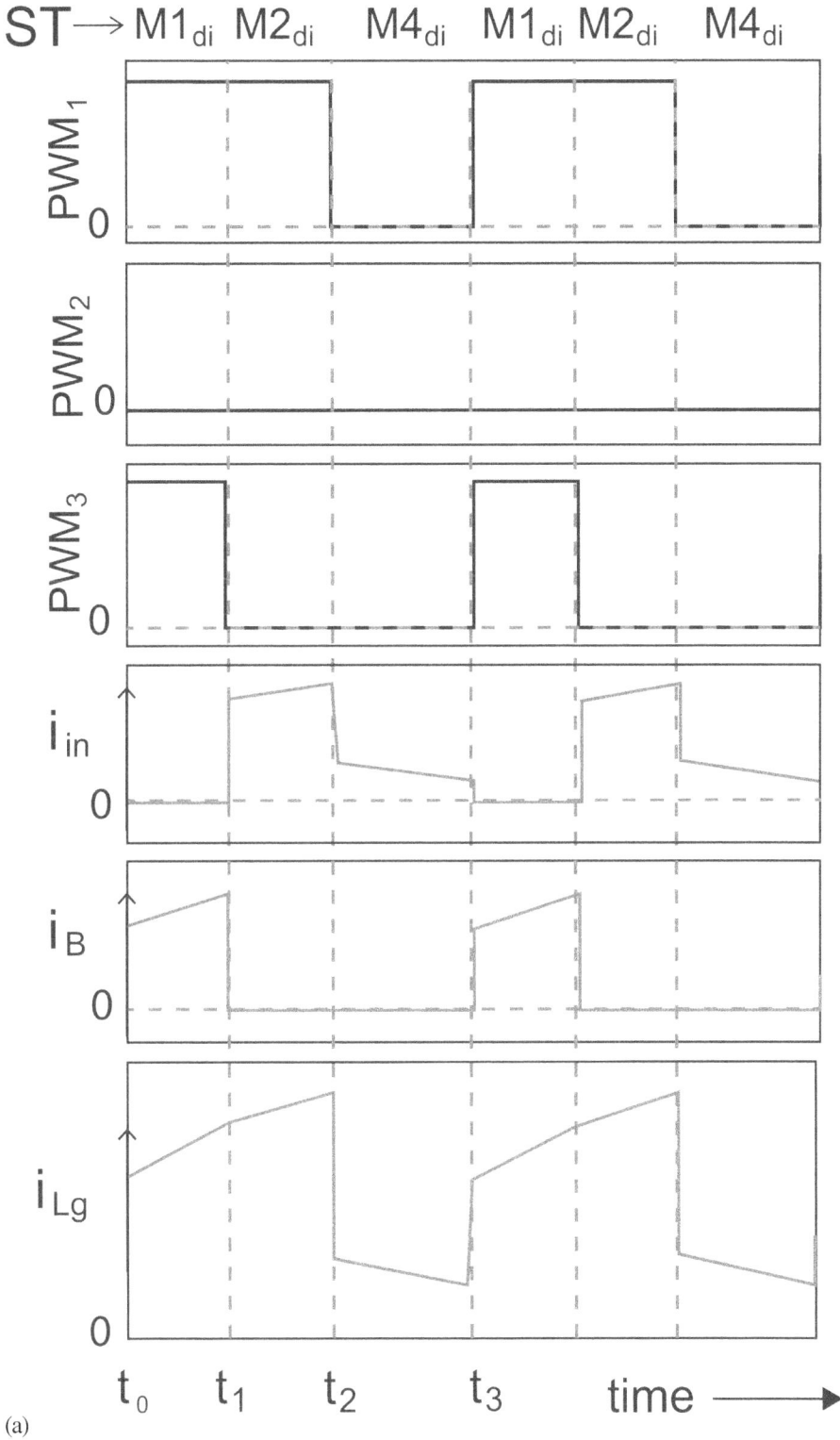

FIGURE 1.12 Three modes of the converter: (a) dual-input mode ($D_1 > D_3$), (b) dual-input mode ($D_1 < D_3$), (c) dual-output mode (top to bottom waveforms: PWM_1, PWM_2, PWM_3, i_{in}, i_B, and i_{Lg}).

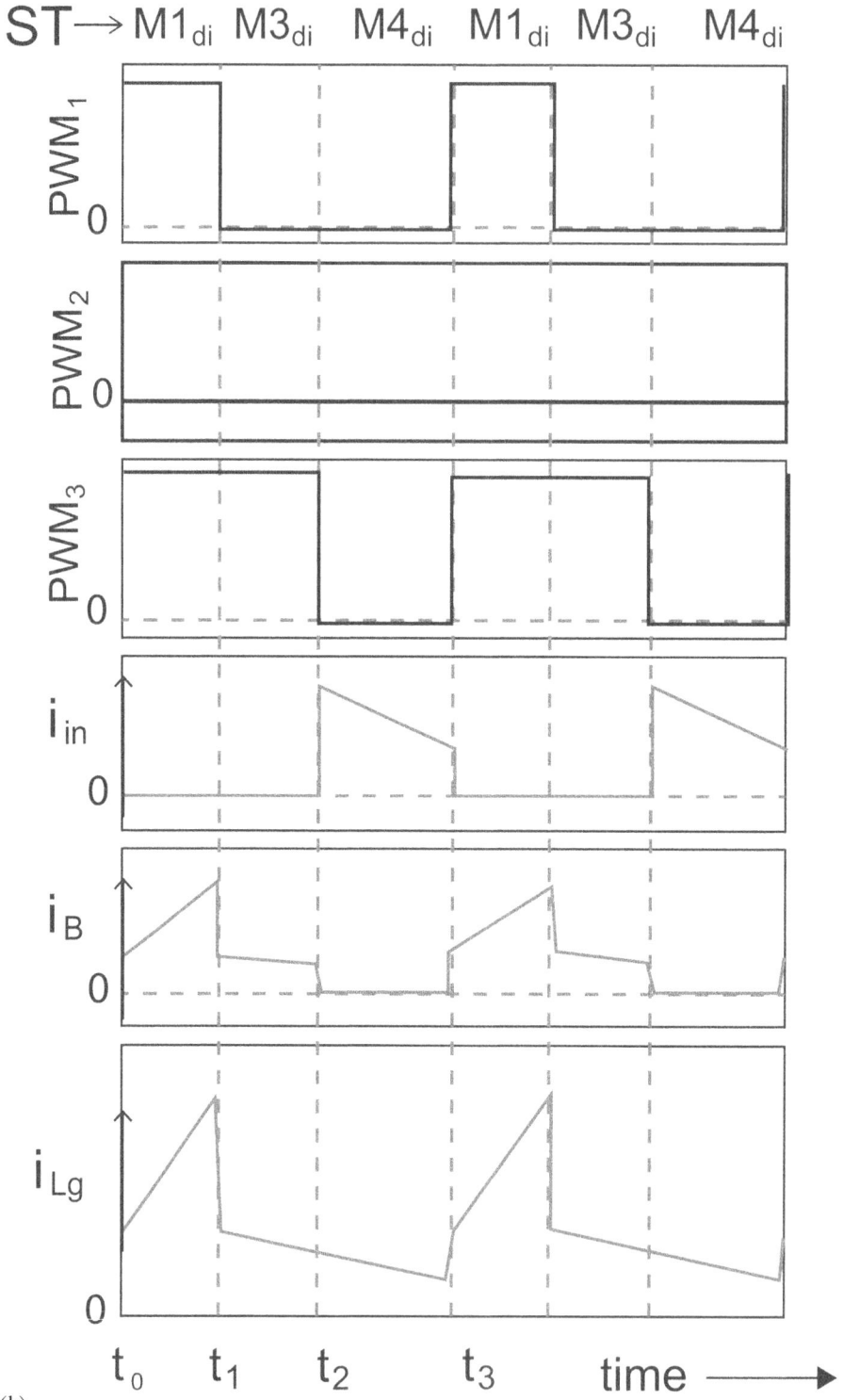

(b)

FIGURE 1.12 Continued

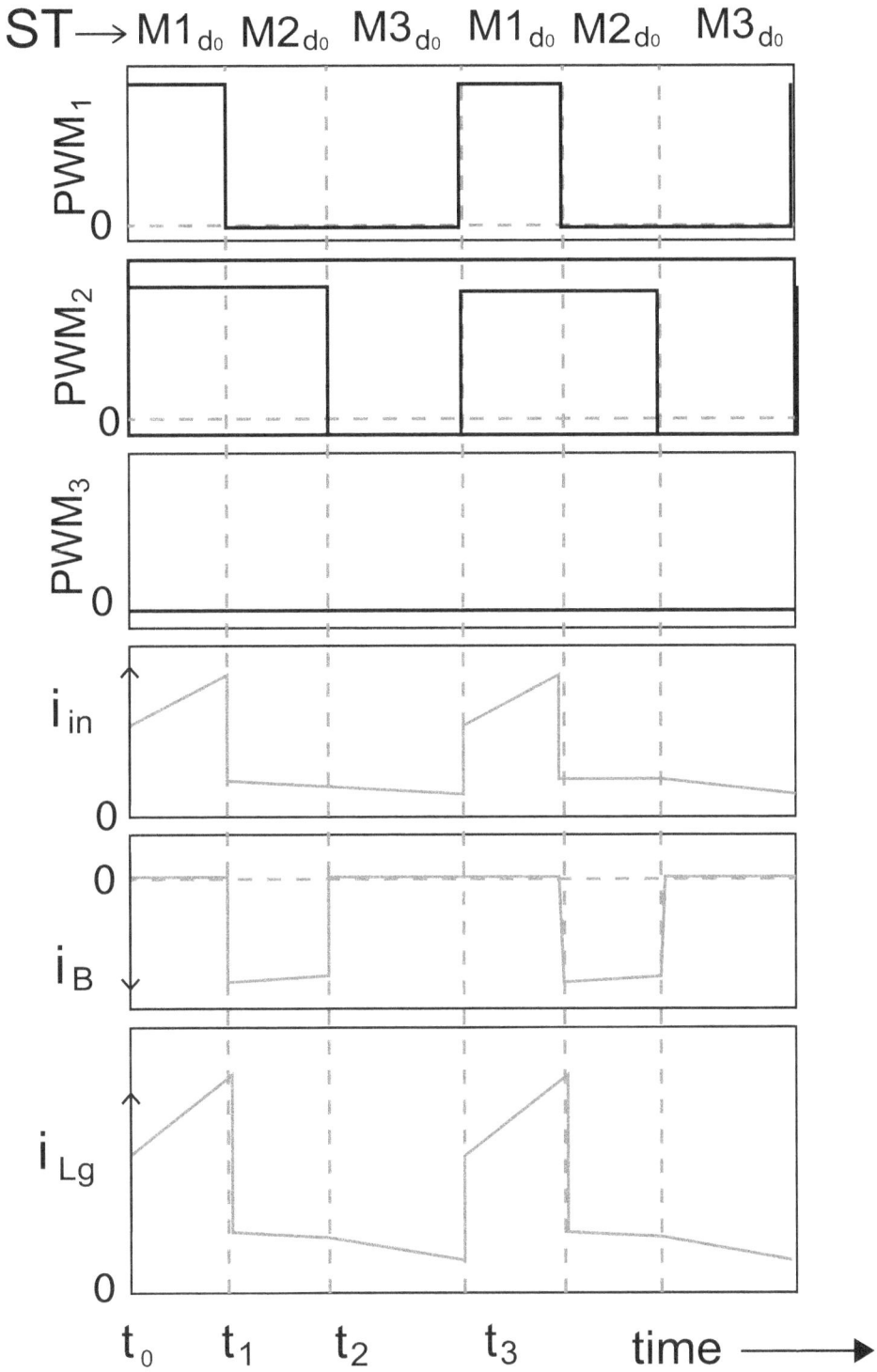

$ST \rightarrow M1_{d_0} \quad M2_{d_0} \quad M3_{d_0} \quad M1_{d_0} \quad M2_{d_0} \quad M3_{d_0}$

(c)

FIGURE 1.12 Continued

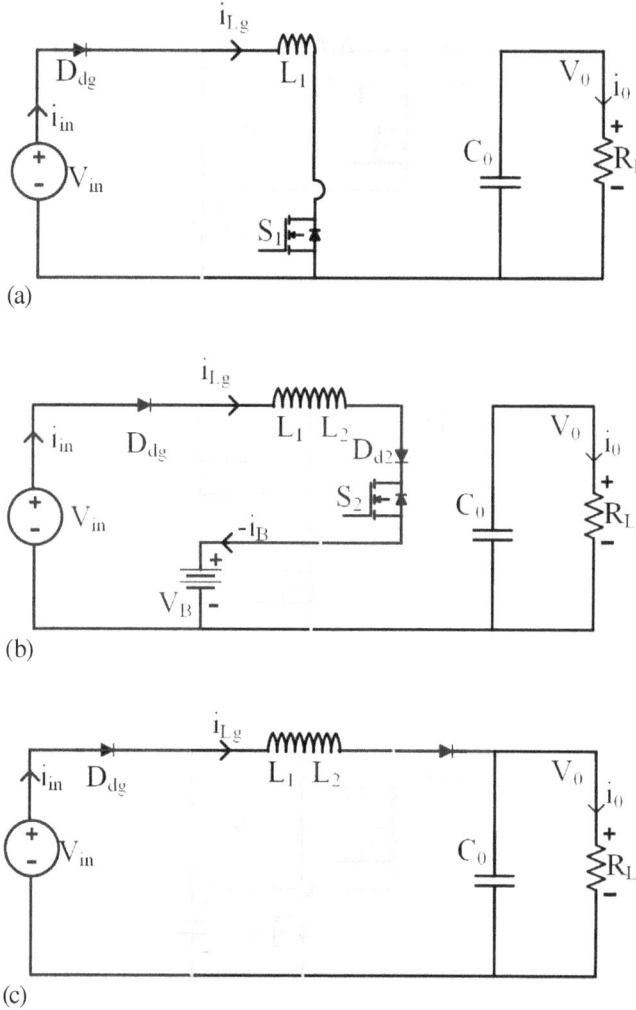

FIGURE 1.13 Equivalent circuit for the switching state under battery charging mode (DO mode): (a) M_{1do}, (b) M_{2do}, and (c) M_{3do}.

the equivalent circuit for this switching condition. The source is connected to the inductor in this state, and the capacitor is supplying power to load. Inductor current i_L in the N_1 part of the tapped-inductor rises.

Mode-M_{2do}: This mode happens as the switch S_1 is turned 'OFF'. As for $V_0 < V_B$, therefore the inductor current decreases, and its stored energy is used to charge the battery. As seen in Figure 1.13(b), capacitor C_0 is still discharging to deliver power to the load.

Mode-M_{3do}: As illustrated in Figure 1.13(c), this mode is similar to mode-M4di of the battery discharge operating mode. In case $D_1 > D_2$ with switch S_2 inactive, the converter operates as a tapped-inductor boost converter with a single-input single-output (SISO) power flow between source and load.

1.3.1.2 Battery Pulse Discharging Scenario (BPDS)

The battery can be discharged under two cases: (i) Dual-Input Single-Output case occurs when the source power is complemented by the battery to compensate for the deficient amount of load power, and (ii) SISO occurs when the battery alone supplies the load power.

1.3.1.2.1   Dual-Input Single-Output Operation

When the load's power demand exceeds the source's power, the switching sequence makes the converter operate as a DISO converter wherein the battery discharges to supply the deficit power. Both the primary power source and battery supply power to the load. The converter may exhibit two switching sequences (consisting of multiple switching states as shown in Figure 1.13) depending upon whether $D_1 > D_3$ or $D_1 < D_3$. The modes M_{1di}, M_{2di}, and M_{4di} occur, sequentially in one switching sequence, when $D_1 > D_3$, while switching sequence S_{1di}, S_{3di}, and S_{4di} occurs sequentially for $D_1 < D_3$. The converter operation for $D_1 > D_3$ is as follows while the key converter waveforms are shown in Figure 1.12(a), wherein the battery discharging current is shown positive (above x-axis).

Mode M_{1di}: This switching state starts with the turning 'ON' of the switches S_1 and S_3 simultaneously at 't = t_0'. The battery discharges and pumps in energy into the N_1 turn of the tapped-inductor for the interval D_3T_s. Capacitor C_0 supplies power to the load. The equivalent circuit for this mode is given in Figure 1.13(a).

Mode M_{2di}: This switching mode starts with turning 'OFF' of the switch S_3 at 't = t_1 = D_3T_s'. As a result, the source is now connected and pumps in more energy into the N_1 turn of the tapped-inductor for the interval '(D_3–D_1) Ts'. Capacitor C_0 continues to supply power to load. The equivalent circuit for this mode is given in Figure 1.13(b).

Mode M_{4di}: This switching mode occurs with the turning 'OFF' of the switch S_1 at 't = t_2 = D_1T_s'. As all the switches are 'OFF', the source is directly connected to the load and the inductor releases the stored energy to the load for the interval '(1–D_1) T_s'. The equivalent circuit for this mode is given in Figure 1.13(d). The converter equivalent circuit operation will be in sequence for $D_1 < D_3$ are shown in Figure 1.13(a), Figure 1.13(c), and Figure 1.13(d). During this case, switching states S_{1di} and S_{4di} are the same as for the case of $D_1 > D_3$ except that this state exists for interval D_1T_s and (D_3–D_1)T_s. The operation of switching-state S_{3di} is discussed next with the key waveforms shown in Figure 1.12(b).

Mode M_{3di}: This mode occurs as the switch S_1 stops conducting and thus connects the battery to load through the complete tapped-inductor. The battery discharges for the duration '(D_3–D_1) T_s'. This mode shows that the converter exerts more power burden on the battery instead of the source compared to the earlier case of $D_1 > D_3$.

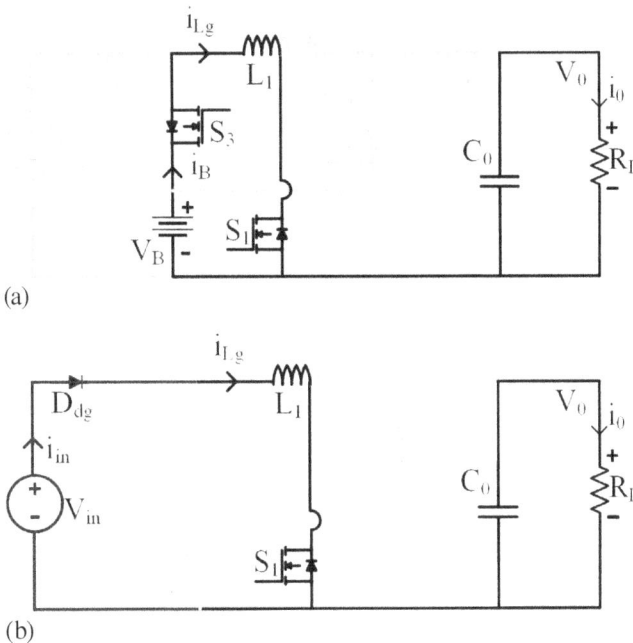

(a)

(b)

FIGURE 1.14 Equivalent circuit for each switching state in DI mode: (a) M_{1di}, (b) M_{2di}, (c) M_{3di}, and (d) M_{4di}.

(c)

(d)

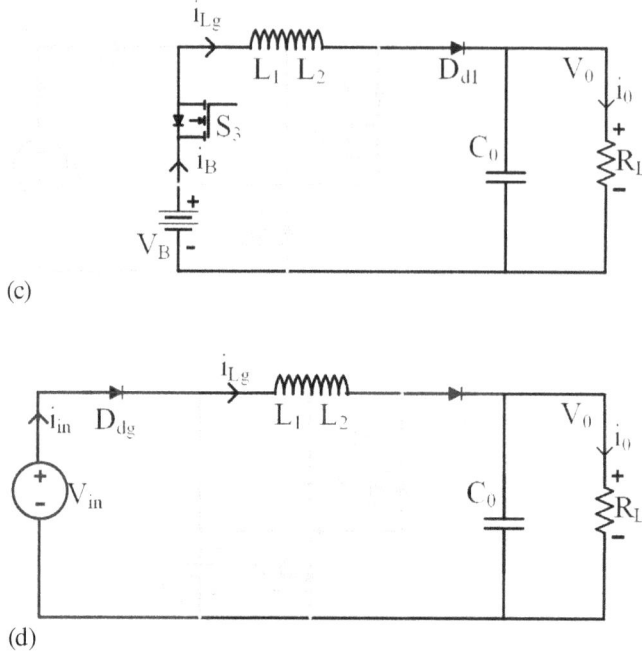

FIGURE 1.14   Continued

1.3.1.2.2   Single-Input Single-Output (SISO) Operation

If the source power is not available, the battery can discharge to supply the power to the load by controlling the switch S_1. The M_{1si} (equivalent circuit same as M_{1di}) and M_{2si} (equivalent circuit same as M_{3di}) are the switching state shown in Figure 1.15(a) and (b) which occurs in one switching cycle for this operation.

1.3.1.3   Isolated Battery Scenario (PV to Load)

The operation of the converter in an isolated battery scenario is the same as a conventional second-order boost converter. The converter is in single-input single-output (SISO) mode as shown in Table 1.1. During this operation, the battery is isolated from the converter circuit and power to load is supplied from the source only.

Under such a scenario, two switching modes occur, M_{1si} (equivalent circuit same as of M_{2di}) and M_{2si} (equivalent circuit same as of M_{4di}), as shown in Figure 1.14(b) and Figure 1.14(d) respectively. In the other switching operation, the battery supplies the power to load if the source power is not available.

1.3.2   STEADY-STATE ANALYSIS

The tapped-inductor has a turn ratio of $N_1:N_2$. The corresponding inductance for the primary and secondary sides of the tapped-inductor is denoted as L_1 and L_2 respectively. Defining the factor, tapping ratio, $K = N_1/(N_1 + N_2)$, the relationship between the inductance of the primary and secondary side to its turns ratio is given by (1). The value of the equivalent inductance of the tapped-inductor is given by (2).

$$\left. \begin{array}{l} L_1 = KN_1^2 \\ L_2 = KN_2^2 \end{array} \right]$$

(1)

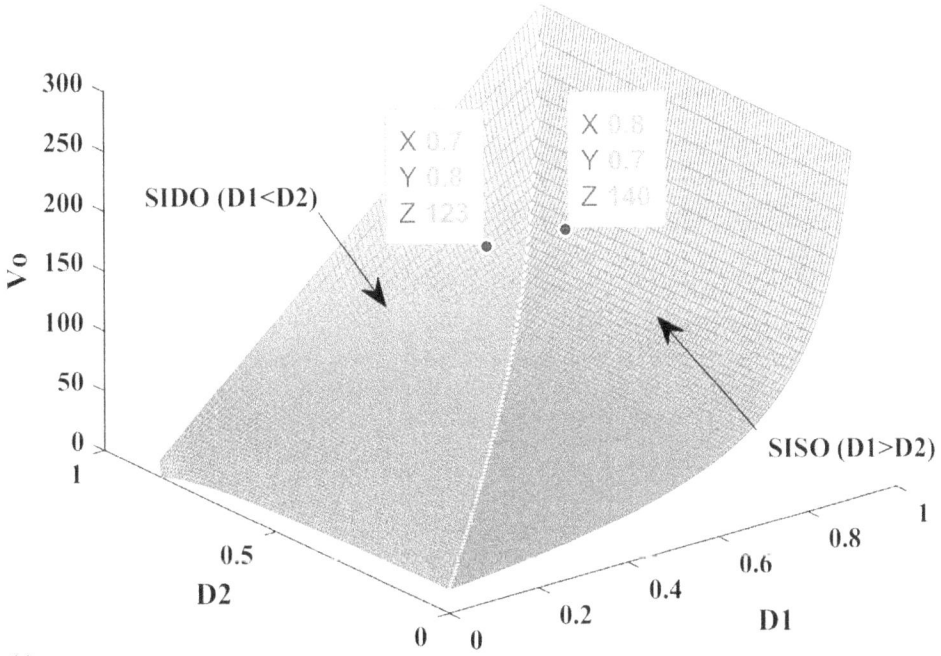

(a)

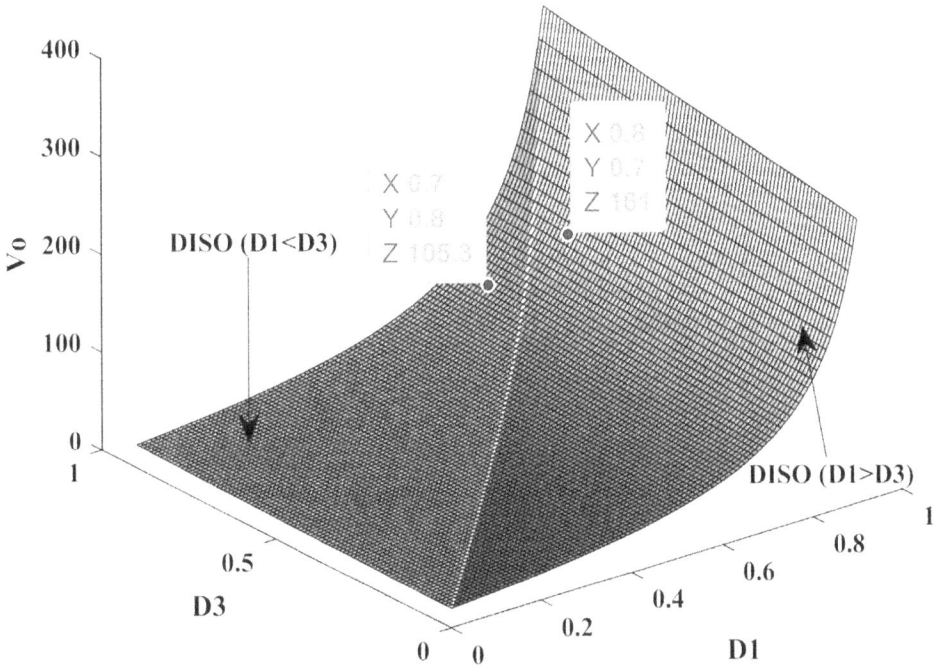

(b)

FIGURE 1.15   Output voltage (V_0) of the converter for V_{in} = 20 V, V_B = 24 V; K = 2/3 w.r.t. duty ratios for (a) SISO and SIDO mode, and (b) DISO mode.

$$L_{eq} = K \left(N_1 + N_2 \right)^2 = \frac{L_1}{K^2} \qquad (2)$$

Now, steady-state expressions for various voltages and currents are derived under different power flow operating modes as given in the following subsections.

1.3.2.1 Dual-Input Single-Output (DISO) Mode

1.3.2.1.1 Case-I ($D_1 > D_3$)

There are three operating states during this mode: M_{1di}, M_{2di}, and M_{4di}, given in Figure 1.14. The corresponding input currents at the primary of the tapped-inductor are denoted as i_{L1}, i_{L2}, and i_{L4} respectively. The voltages across the tapped-inductor are $V_{L1} = V_B$, $V_{L1} = V_{in}$, and $V_{Leq} = V_{in} - V_B$ respectively for S_{1di}, S_{2di}, and S_{4di}, and current through the output capacitor during this switching state is $I_{c0} = -V_0/R_L$, $I_{c0} = -V_0/R_L$, and $I_{c0} = i_{L4} - V_0/R_L$. Here, the average value of input tapped-inductor currents i_{L1} and i_{L2} is equal, that is, $i_{L1} \approx i_{L2}$ as inductance value is not changed for S_{1di} and S_{2di}. The input current to primary of tapped-inductor is i_{L1} and i_{L4} under S_{1di} and S_{4di} respectively and its relationship is given by

$$\frac{i_{L1}}{i_{L4}} = \frac{N_1 + N_2}{N_1} = \frac{1}{K} \qquad (3)$$

Using (3), the relationship obtained is

$$\frac{L_{eq}}{L_1} = \frac{1}{K^2} \qquad (4)$$

Taking the derivative of (4), it is found that under the switching mode M_{4di}, the following relationship holds for S_{4di}:

$$\frac{d}{dt} i_{L1} = \frac{K}{L_1} \left(V_{in} - V_0 \right) \qquad (5)$$

Using (5), for M_{4di}, volt-sec balance across the tapped-inductor gives the following relationship:

$$V_B D_3 + V_0 \left(D_1 - D_3 \right) + K \left(V_{in} - V_0 \right) = 0$$

Rearranging this equation yields the output voltage expression as given by

$$V_0 = \frac{\left(D_1 - D_3 \right) + K \left(1 - D_1 \right)}{K \left(1 - D_1 \right)} V_{in} + \frac{D_3}{K \left(1 - D_1 \right)} V_B \qquad (6)$$

Now, the expression for source current and battery current is obtained based on the charge-balance principle. Applying the charge balance across the output capacitor C_0 gives

$$-I_0 D_3 - I_0 \left(D_1 - D_3 \right) + \left(I_{L4} - I_0 \right)\left(1 - D_1 \right) = 0$$

Rearranging this expression gives

$$i_{L4} = \frac{i_0}{1 - D_1} \tag{7}$$

Using (7),

$$i_{L1} = \frac{i_0}{K\left(1 - D_1 \right)} \tag{8}$$

Now, the steady-state value of source current i_{in} is given by

$$i_{in} = 0 D_3 + i_{L2} \left(D_1 - D_3 \right) + i_{L4} \left(1 - D_1 \right) \tag{9}$$

As $i_{L1} \approx i_{L2}$, and using (8) and (9), the steady-state source current expression for DO mode for $D_1 > D_3$ is

$$i_{in} = i_0 \left(1 + \frac{D_1 - D_3}{K\left(1 - D_1 \right)} \right) \tag{10}$$

The average battery current is given by

$$i_B = D_3 i_{L1} = \frac{D_3 I_0}{K\left(1 - D_1 \right)} \tag{11}$$

1.3.2.1.2 Case-II ($D_1 < D_3$)

There are three operating states during this mode: M_{1di}, M_{3di}, and M_{4di}. Consider the input current at the primary of tapped-inductor as i_{L1}, i_{L3}, and i_{L4} for these three switching states respectively. The voltages across the inductor are $V_{L1} = V_B$, $V_{Leq} = V_B{-}V_0$, and $V_{Leq} = V_{in}{-}V_0$, and the current through the capacitor is $I_{c0} = -V_0/R_L$, $I_{c0} = i_{L3}{-}V_0/R_L$, and $I_{c0} = i_{L4}{-}V_0/R_L$ for M_{1di}, M_{3di}, and M_{4di} respectively. The steady-state relationships obtained using the same procedure as of case ($D_1 > D_3$) are given here:

$$V_0 = \frac{\left(1 - D_3\right)}{\left(1 - D_1\right)} V_{in} + \frac{D_1 - KD_1 + KD_3}{K\left(1 - D_1\right)} V_B \tag{12}$$

$$i_{in} = \frac{\left(1 - D_3\right)}{\left(1 - D_1\right)} I_0 \tag{13}$$

$$i_B = \frac{I_0\left((1-K)D_1 + KD_3\right)}{K\left(1-D_1\right)} \tag{14}$$

1.3.2.2 Single-Input Dual-Output (SIDO) Mode

The switching states for SIDO mode are M_{1do}, M_{2do}, and M_{3do} as given in Figure 1.13. Considering the input current flowing through the primary side of the inductor is i_{L1}, i_{L2}, and i_{L3} respectively for three switching states, the voltages across the tapped-inductor are $V_{L1} = V_{in}$, $V_{Leq} = V_{in} - V_B$, and $V_{Leq} = V_{in} - V_0$, and currents through the capacitor are $I_{c0} = -V_0/R_L$, $I_{c0} = -V_0/R_L$, and $I_{c0} = i_{L3} - V_0/R_L$ respectively. The steady-state relationships obtained for DO mode are given here:

$$V_0 = \frac{D_1 + K\left(1-D_1\right)}{K\left(1-D_2\right)} V_{in} + \frac{\left(D_1 - D_2\right)}{\left(1-D_2\right)} V_B \tag{15}$$

$$i_{in} = i_0 \left(1 + \frac{D_1}{K\left(1-D_2\right)} + \frac{D_2 - D_1}{1-D_2}\right) \tag{16}$$

$$i_B = \frac{D_2 - D_1}{1-D_2} i_0 \tag{17}$$

1.3.2.3 Single-Input Single-Output (SISO) Mode

If source power is unavailable, the battery can be used as a source, with the converter providing the same high voltage gain as a tapped-inductor boost converter. Similarly, without the use of a battery, the source can directly supply power to the load. The load voltage (V_0) of the converter under SISO mode is controlled by switch S_1 and its expression is given by (18). Switch S_3 to keep 'ON' keeping the gate to source voltage more than the threshold voltage.

$$V_0 = \left(1 + \frac{D}{K\left(1-D\right)}\right) V_B \tag{18}$$

According to the expression obtained in (6), (12), (15), and (18), the load voltage of the BITIC converter under the different modes for a 20 V source voltage, 24 V battery voltage, and tapping ratio, K = 2/3 as shown in Figure 1.15.

For the source power without the battery, the output voltage expression is

$$V_0 = \left(1 + \frac{D}{K\left(1-D\right)}\right) V_{in} \tag{19}$$

The current through the tapped-inductor is given by

$$i_{Lg} = \frac{V_0}{R_L} \left(1 + \frac{D_1}{K\left(1-D_1\right)}\right) \tag{20}$$

The average inductor current in Mode-I is given by $<i_{Lp1}> = \dfrac{V_0}{R_L K (1-D)}$ and the average inductor current in Mode-II is given by

$$< i_{Lp2} > = K < i_{Lp1} >$$

(21)

1.3.3 DESIGN OF CONVERTER

1.3.3.1 Inductor Design

The inductor value is chosen based on the SISO configuration, which uses a conventional tapped inductor boost converter.

The inductor current ripple (Δi_{Lp1}) for $0 < t < D_1 T_s$ and (Δi_{Lp2}) for $D_1 T_s < t < T_s$, shown in Figure 1.16(a), is

$$\Delta i_{Lp1} = \frac{V_{in} D_1}{L_1 f_s}$$

(22)

$$\Delta i_{Lp2} = \frac{K (V_{in} - V_0)(1 - D_1)}{L_1 f_s}$$

(23)

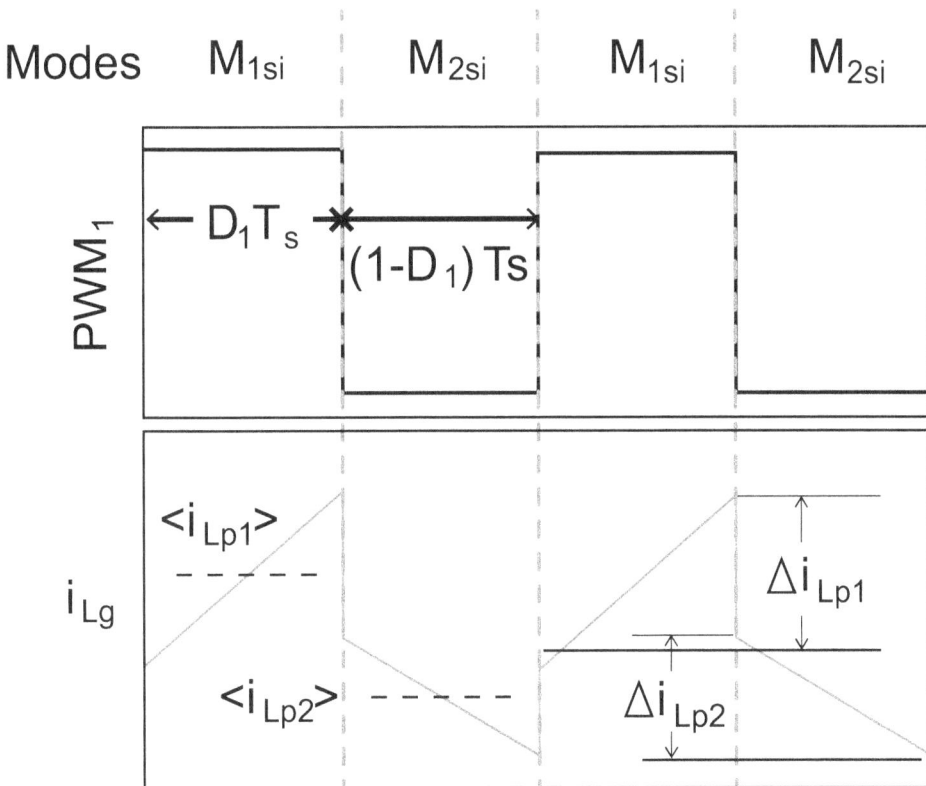

(a)

FIGURE 1.16 Primary current of tapped inductor for (a) SISO mode, and (b) dual-input mode ($D_1 > D_3$).

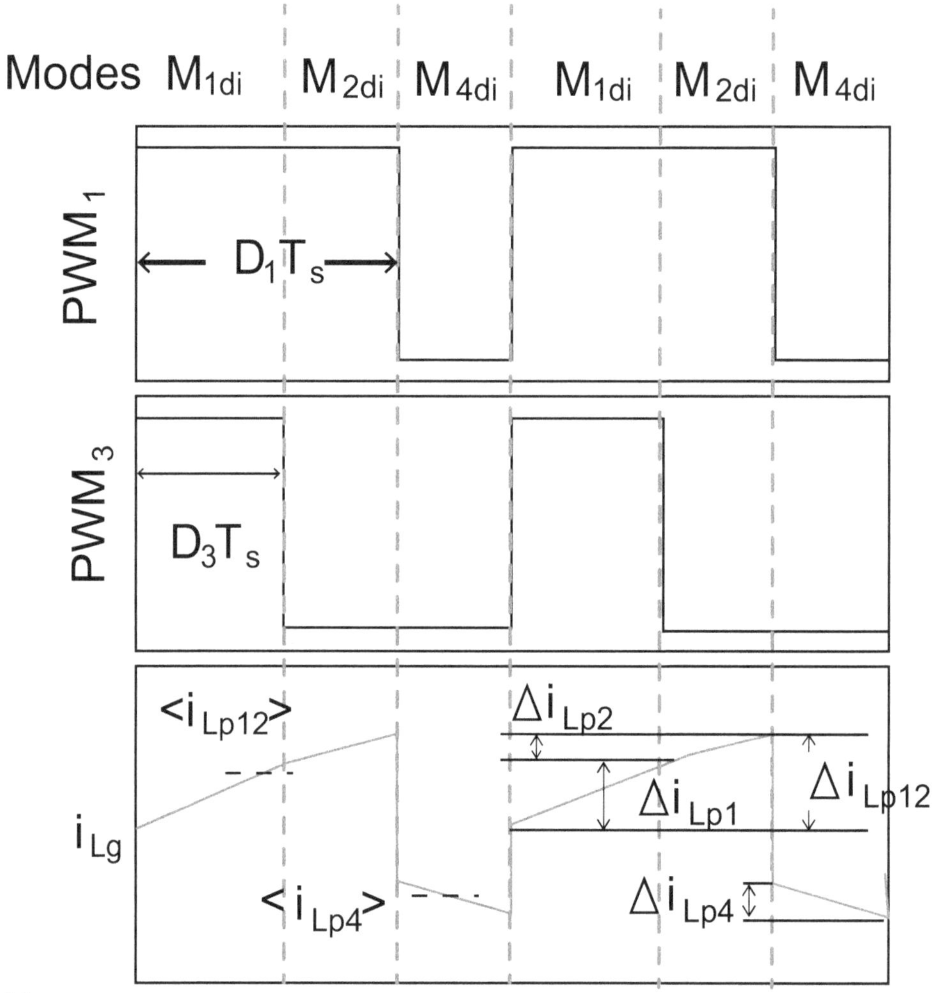

(b)

FIGURE 1.16 Continued

The maximum value and minimum value of the source current are given by $I_{max} = I_{Lp1} + \left|\frac{\Delta I_{Lp1}}{2}\right|$, $I_{min} = I_{Lp2} - \left|\frac{\Delta I_{Lp2}}{2}\right|$. Therefore, the current ripple is $\Delta I_r = I_{max} - I_{min}$.

The value of inductance required for the given value of the inductor ripple is given by

$$L_1 = \frac{D_1}{f_s\left(1 - \frac{1-D_1}{K(1-D_1)}\right)\left(\frac{1-K}{KR_L(1-D_1)} + \left(\frac{"I_r}{I_L}\right)\frac{1 + \frac{D_1}{K(1-D_1)}}{R_L}\right)} \tag{24}$$

and $L_{eq} = \dfrac{L_1}{K^2}$. The converter's time-domain analysis for the other modes is carried out to determine the inductor's critical value to maintain continuous conduction mode (CCM).

1.3.3.1.1 Inductor Current Ripple for Dual-Input Mode

1.3.3.1.1.1　Case-I ($D_1 > D_3$)　The inductor current waveform for the dual-input mode under the condition $D_1 > D_3$ is shown in Figure 1.16(b).

Using Figure 1.16(b), peak to peak ripple for the dual-input case with $D_1 > D_3$ is given by

$$\Delta i_L \simeq\, <i_{Lp12}> - <i_{Lp4}> + \frac{\Delta i_{Lp12}}{2} + \frac{\Delta i_{Lp4}}{2} \tag{25}$$

$$\text{where } <i_{Lp12}> \approx \frac{I_0}{K(1-D_1)}, \ <i_{Lp4}> = \frac{I_g}{1-D_1}$$

$$\Delta i_{Lp1} = \frac{V_B D_3}{L_1 f_S}, \ \Delta i_{Lp2} = \frac{v_0(D_1 - D_3)}{L_1 f_s}, \ \Delta i_{Lp12} = \frac{V_B D_3 + V_g(D_1 - D_3)}{L_1 f_S},$$

$$\Delta i_{Lp4} = \frac{(1-K)(V_{in} - V_0)(1-D_1)}{L_2 f_S}$$

The critical value of inductance is obtained by the condition $\Delta i_{Lp4} = 2I_g$, which gives

$$L_{cr} = \frac{(1-K)(1-D_1)(V_0 - V_{in})}{2I_g f_s} \tag{26}$$

1.3.3.1.1.2　Case-II ($D_1 < D_3$)　Peak to peak ripple for the dual-input case with $D_1 < D_3$ is given by

$$\Delta i_L \simeq\, <i_{Lp1}> - <i_{Lp34}> + \frac{\Delta i_{Lp1}}{2} + \frac{\Delta i_{Lp34}}{2} \tag{27}$$

$$\text{where } <i_{Lp1}> = \frac{I_0}{K(1-D_1)}, \ <i_{Lp34}> = \frac{I_0}{(1-D_1)}, \ \Delta i_{Lp1} = \frac{V_B D_1}{L_1 f_s},$$

$$\Delta i_{Lp3} = \frac{V_B(D_3 - D_1)}{L_1 f_s}, \ \Delta i_{Lp4} = \frac{(1-K)(1-D_3)(V_{in} - V_0)}{L_2 f_s}$$

The critical value of inductance L_2 can be known by the condition $\Delta i_{34} = 2I_0$.

Dual-output mode tapped when $D_2 > D_1$, $D_3 = 0$. The primary side of the tapped-inductor current ripple is given by

$$\Delta i_L \simeq\, <i_{Lp1}> - <i_{Lp23}> + \frac{\Delta i_{Lp1}}{2} + \frac{\Delta i_{Lp23}}{2} \tag{28}$$

$$\text{where, } <i_{Lp1}> = \frac{I_g}{K(1-D_1)}, \quad <i_{Lp23}> = \frac{I_g}{(1-D_1)}, \quad <\Delta i_{Lp1}> = \frac{V_{in}D_1}{L_1f_s},$$

$$<\Delta i_{Lp2}> = \frac{(V_{in}-V_B)(1-K)(D_2-D_1)}{L_2f_s}$$

$$<\Delta i_{Lp3}> = (1-K)\frac{(V_{in}-V_B)(D_2-D_1)+(V_{in}-V_0)(1-D_2)}{L_2f_s}$$

The critical value of inductance L_2 can be known by the condition $\Delta i_{23} = 2I_0$.

1.3.3.2 Design of Capacitor

The following expression derived for the SISO mode calculates the value of the output capacitor for a given value of output voltage ripple (ΔV_0). This converter acts like a conventional tapped boost converter [29] in SISO mode, hence the design remains the same.

$$C_0 = \frac{D_1}{\left(\frac{\Delta V_0}{V_0}\right)R_{Lmin}f_s} \tag{29}$$

1.3.3.3 Selection of Switches

The voltage stresses of the switching elements for the BITIC are shown in Table 1.2. The average and RMS value of currents flowing in the different elements for BITIC are shown in Tables 1.2 through Table 1.6 and the maximum current across the elements is also presented in Table 1.7 under the different modes.

1.4 RESULTS AND DISCUSSIONS

Figure 1.17 shows the steady-state results for all of the scenarios. The voltage considered for simulation of battery integrated tapped-inductor boost converter (BITIC) are: source voltage (V_{in} = 24 V) battery voltage V_B = 36 V (> source voltage, V_{in}). For IBS, the BITIC converter operates in SISO mode as a tapped-inductor boost converter with output voltage V_0 regulated at 48 V at the duty ratio

TABLE 1.2

Voltage Stresses of the Switching Components

SISO mode	Voltage stress
S_1	$V_0+(1-K)(V_{in}-V_0)$
S_2	$V_{in}-V_B$
S_3	V_B-V_{in}
D_{d1}	$V_0+\left(\frac{1}{K}-1\right)V_{in}$
D_{d2}	$V_B+\left(\frac{1}{K}-1\right)V_{in}$

TABLE 1.3
Average and RMS Value of Current for DI Mode ($D_1 < D_3$)

DI mode ($D_1 < D_3$)	Average current	RMS current
L	$\dfrac{I_0\left(D_1 + K\left(1 - D_1\right)\right)}{K\left(1 - D_1\right)}$	$\dfrac{I_0\sqrt{D_1 + K^2\left(1 - D_1\right)}}{K\left(1 - D_1\right)}$
S_1	$\dfrac{D_1}{K\left(1 - D_1\right)}I_0$	$\dfrac{\sqrt{D}}{K\left(1 - D_1\right)}I_0$
S_2	0	0
S_3 or battery	$\dfrac{D_1 + K\left(D_3 - D_1\right)}{K\left(1 - D_1\right)}I_0$	$\dfrac{\sqrt{D_1 + K^2\left(D_3 - D_1\right)}}{K\left(1 - D_1\right)}I_0$
D_{d1}	I_0	$\dfrac{I_0}{\sqrt{1 - D_1}}$
D_{d2}	0	0
C_0	0	$\sqrt{\dfrac{D_1}{1 - D_1}}I_0$

TABLE 1.4
Average and RMS Value of Current for DI Mode ($D_1 > D_3$)

DI mode ($D_1 > D_3$)	Average current	RMS current
L	$\dfrac{I_0\left(D_1 + K\left(1 - D_1\right)\right)}{K\left(1 - D_1\right)}$	$\dfrac{I_0\sqrt{D_1 + K^2\left(1 - D_1\right)}}{K\left(1 - D_1\right)}$
S_1	$\dfrac{D_1}{K\left(1 - D_1\right)}I_0$	$\dfrac{\sqrt{D}}{K\left(1 - D_1\right)}I_0$
S_2	0	0
S_3 or battery	$\dfrac{D_3}{K\left(1 - D_1\right)}I_0$	$\dfrac{\sqrt{D_3}}{K\left(1 - D_1\right)}I_0$
D_{d1}	I_0	$\dfrac{I_0}{\sqrt{1 - D_1}}$
D_{d2}	0	0
C_0	0	$\sqrt{\dfrac{D_1}{1 - D_1}}I_0$

$D_1 = 0.444$. The input current ripple for the tapping ratio $K = 4/5$ and $L_{eq} = 2.22$ mH is 30% for IBS, as shown in Figure 1.17. For $C_0 = 20$ uF, the output voltage ripple is 1%. During the BPCS (DO), the simulated waveform is shown for $D_1 = 0.4$, $D_2 = 0.6$. The inductor current ripple is ≤30% for IBS and BPCS for the selected value of tapping ratio and inductor value. However, the inductor of the BITIC converter operates in discontinuous conduction mode (DCM) for BPDS, and thus input current ripple is large enough as shown for the duty ratio $D_1 = 0.4$, $D_3 = 0.3$ (Case-I: $D_1 > D_3$), and

TABLE 1.5
Average and RMS Value of Current for DO Mode

DO mode ($D_1 < D_2$)	Average current	RMS current
L	$\dfrac{I_0\left(D_1 + K(1-D_1)\right)}{K(1-D_2)}$	$\dfrac{I_0\sqrt{D_1 + K^2(1-D_1)}}{K(1-D_2)}$
S_1	$\dfrac{D_1}{K(1-D_2)}I_0$	$\dfrac{\sqrt{D_1}}{K(1-D_2)}I_0$
S_2 or battery	$\dfrac{(D_2-D_1)}{(1-D_2)}I_0$	$\dfrac{\sqrt{D_1-D_2}}{(1-D_2)}I_0$
S_3	0	0
D_{d1}	I_0	$\dfrac{I_0}{\sqrt{1-D_2}}$
D_{d2}	$\dfrac{(D_2-D_1)}{(1-D_2)}I_0$	$\dfrac{\sqrt{D_1-D_2}}{(1-D_2)}I_0$
C_o	0	$\sqrt{\dfrac{D_1}{1-D_1}}I_0$

TABLE 1.6
Average and RMS Value of Current for SISO Mode

SISO mode	Average current	RMS current
L	$\dfrac{(D_1 + K(1-D_1))}{K(1-D_1)}I_0$	$\dfrac{\sqrt{D_1 + K^2(1-D_1)}}{K(1-D_1)}I_0$
S_1	$\dfrac{D_1}{K(1-D_1)}I_0$	$\dfrac{\sqrt{D}}{K(1-D_1)}I_0$
S_2	0	0
S_3	0	0
D_{d1}	I_0	$\dfrac{I_0}{\sqrt{1-D_1}}$
D_{d2}	0	0
C_o	0	$\sqrt{\dfrac{D_1}{1-D_1}}I_0$

$D_1 = 0.3$, $D_3 = 0.4$ (Case-II: $D_1 < D_3$) in Figure 1.17. The voltage stresses appearing across the different elements for IBS are shown in Figure 1.18(a). The maximum value of voltage stress (appearing across the diode D_{d1}) across the switching elements is compared for conventional boost converter (CBC), SOBIC, and BITIC converter as shown in Figure 1.18(b). The maximum value of voltage stress of the BITIC converter increases with the increase in the turns ratio of N_2 (tapping ratio,

$K = \dfrac{N_1}{N_1 + N_2}$ decreases). The voltage stresses are shown for battery voltage $V_B = 24$ V, regulated

TABLE 1.7

Current Stresses for Different Modes of BITIC

	Dual input ($D_1 > D_3$)	Dual input ($D_1 < D_3$)	Dual output ($D_1 < D_2$)	SISO
S_1	$<i_{Lp12}> + \dfrac{\Delta i_{Lp12}}{2}$	$<i_{Lp1}> + \dfrac{\Delta i_{Lp1}}{2}$	$<i_{Lp1}> + \dfrac{\Delta i_{Lp1}}{2}$	$<i_{Lp1}> + \dfrac{\Delta i_{Lp1}}{2}$
S_2	0	0	$<i_{Lp23}> + \dfrac{\Delta i_{Lp23}}{2}$	0
S_3	$<i_{Lp12}> + \dfrac{\Delta i_{Lp12}}{2}$	$<i_{Lp1}> + \dfrac{\Delta i_{Lp1}}{2}$	0	0
D_{d1}	$<i_{Lp2}> + \dfrac{\Delta i_{Lp4}}{2}$	$<i_{Lp34}> + \dfrac{\Delta i_{Lp34}}{2}$	$<i_{Lp23}> + \dfrac{\Delta i_{Lp23}}{2}$	$<i_{Lp2}> + \dfrac{\Delta i_{Lp2}}{2}$
D_{d2}	0	0	$<i_{Lp23}> + \dfrac{\Delta i_{Lp23}}{2}$	0

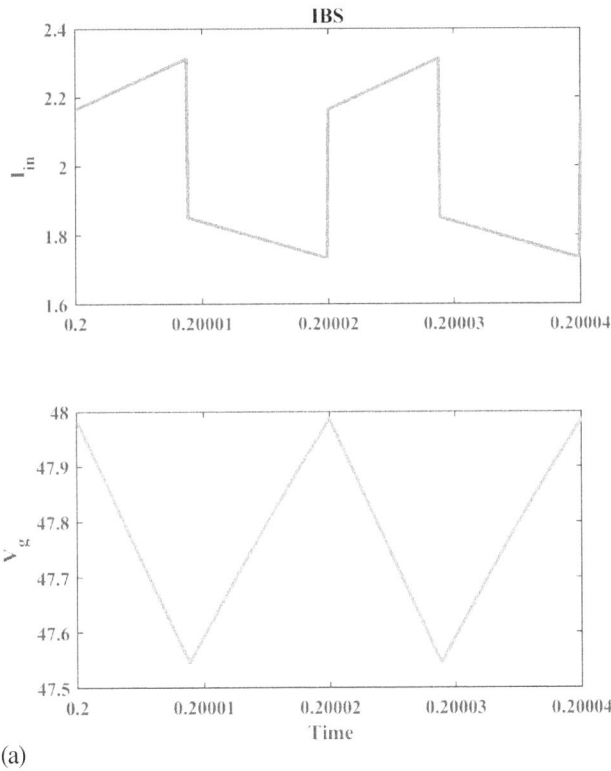

FIGURE 1.17 Steady-state waveforms of BITIC under (a) IBS, (b) BPCS, (c) BPDS ($D_1 > D_3$), and (d) BPDS ($D_1 < D_3$).

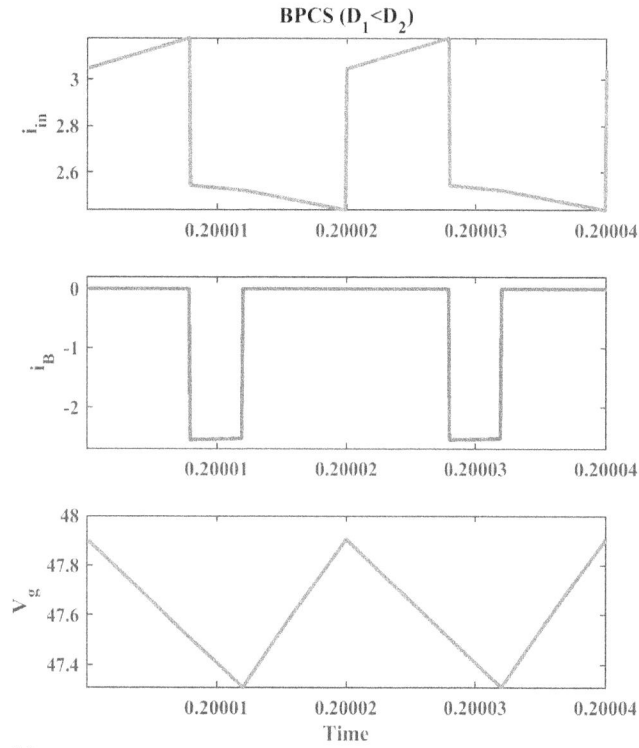

(b)

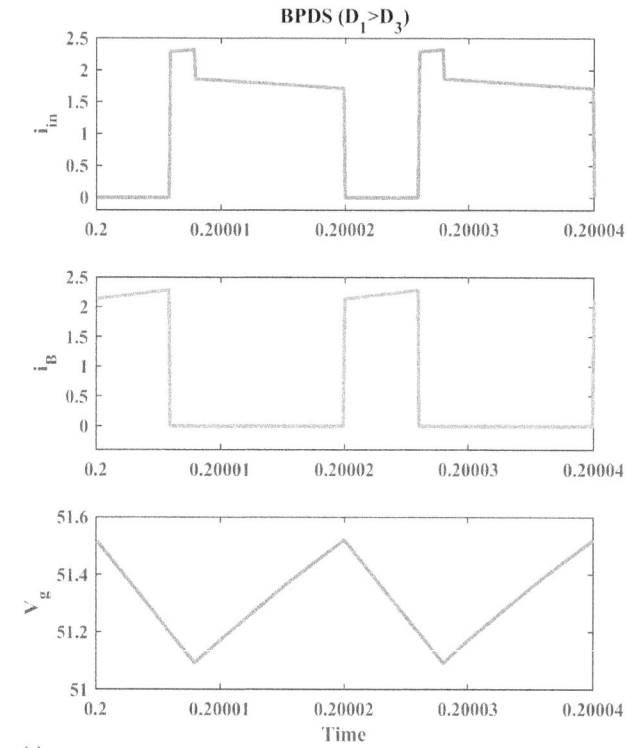

(c)

FIGURE 1.17   Continued

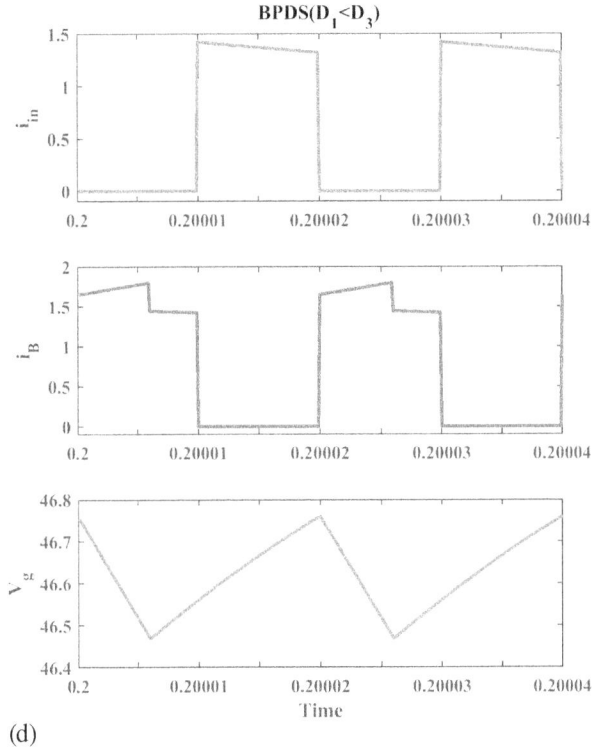

(d)

FIGURE 1.17 Continued

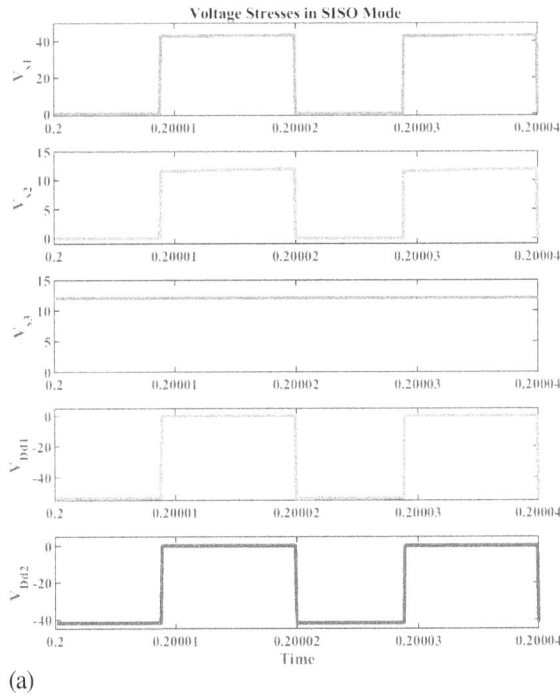

(a)

FIGURE 1.18 (a) Voltage stress of the switching elements for BITIC converter under IBS; (b) maximum voltage stresses for switching elements for CBC, SOBIC, and BITIC with $V_0 = 48$ V, $V_B = 24$ V.

Maximum Voltage Stresses

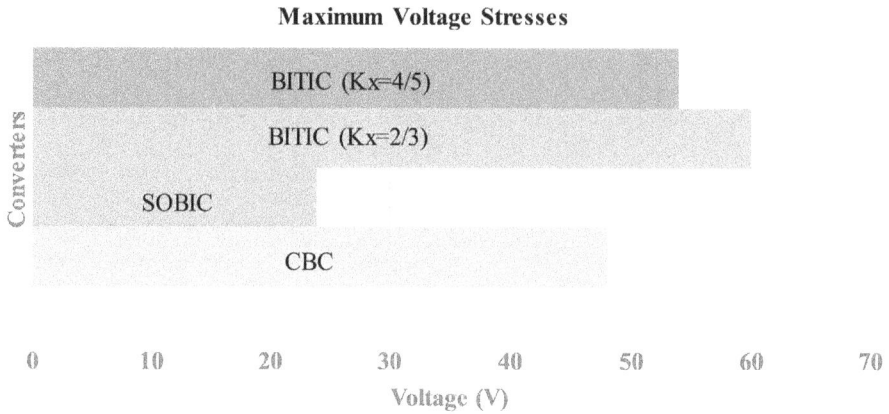

(b)

FIGURE 1.18 Continued

load voltage $V_0 = 48$ V and tapping ratios K = 2/3 and 4/5. Figure 1.19 shows the experimental setup and laboratory prototype of the BITIC, as well as the tapped-inductor. The experimental steady-state waveforms for the BITIC during BPDS (DI) are shown in Figures 1.20(a) and (b), during BPCS (DO) is shown in Figure 1.20(c), and during IBS is shown in Figure 1.20(d). In all of the scenarios, the converter maintains a high value of load voltage (V_0) while keeping the battery in charging/discharging/isolated state depending upon the power requirement.

Considering the parasitic resistances of the elements of the converter components as follows: $r_L = 0.1$ Ω, $r_{Co} = 0.1$ Ω, $r_{Dd} = 0.06$ Ω, $r_{sw} = 0.04$ Ω, $V_F = 0.4$ V. The efficiency of the converter under SISO, DI, and DO mode for 100 V regulated load is shown in Figure 1.21. For DI mode, input power from the source is considered fixed at 50 W and the rest power for the load is delivered by the battery. With the increase of load power, the power losses P_L increases, and hence drop in efficiency occurs. For DO mode, input power is 120 W, which is more than the required load power. As a result, the battery absorbs the surplus power. With the increase of load in DO mode, the power for charging the battery is reduced and therefore P_L decreases. Therefore, BITIC converter shows an improvement in the efficiency with the increase of load power in DO mode up to 120 W.

Table 1.8 gives a quick comparison of the BITIC converter with the other converter in terms of the number of passive energy storage devices and the voltage-gain expression. The voltage gains are shown for the isolated battery scenario (IBS) or single-input single-output operation (SISO). The converter offers the advantage of high voltage gain with a low number of component counts.

1.5 CONCLUSION

Many three-port battery-integrated converters are reviewed for the study of their feasibility for the low voltage direct current (LVPDS) distributed system. The new compact three-port battery-integrated tapped-inductor boost converter was presented and analyzed in this chapter. Due to its low component counts and high voltage-gain features, the battery-integrated tapped-inductor boost converter was found suitable for standalone renewable energy systems. The power flows among the source, load, and battery make it suitable for power management for renewable energy applications

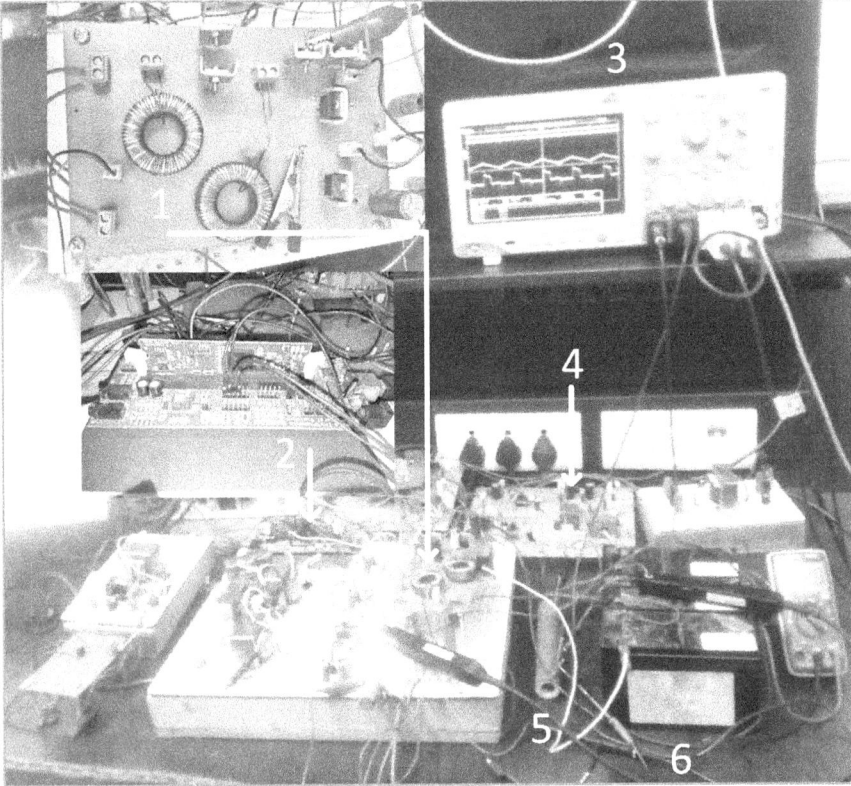

(a)

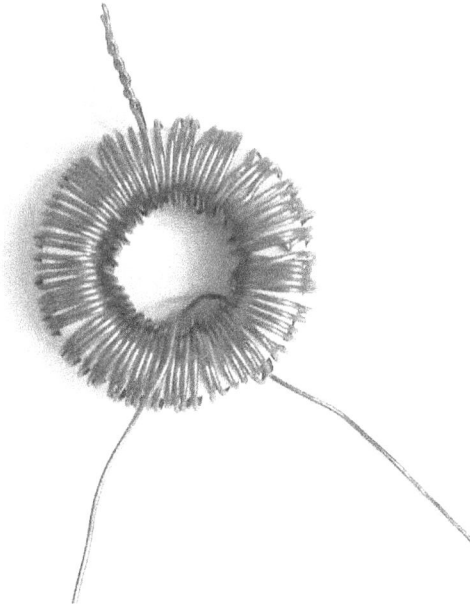

(b)

FIGURE 1.19 (a) Experimental setup: (1) BISIBC, (2) TMS320F28335 peripheral board, (3) digital storage oscilloscope (DSO), (4) sensors (LA55P current sensor, LV25P voltage sensor), (5) electrical load, and (6) battery; (b) tapped inductor used for the experiment.

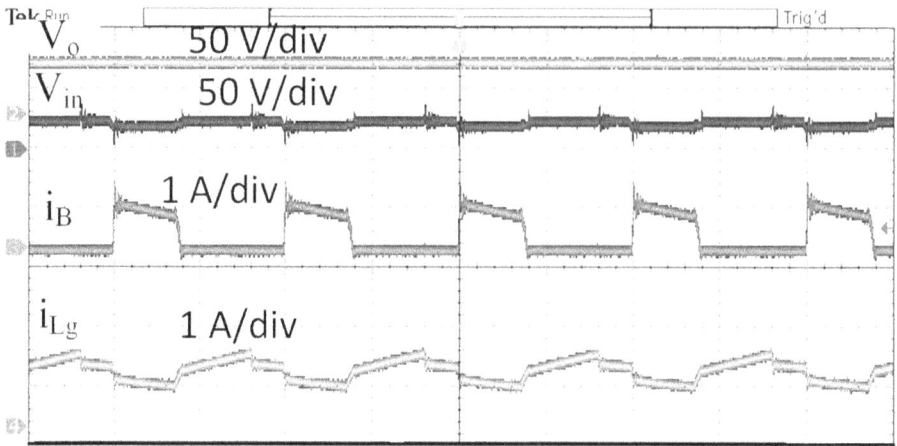

(a)

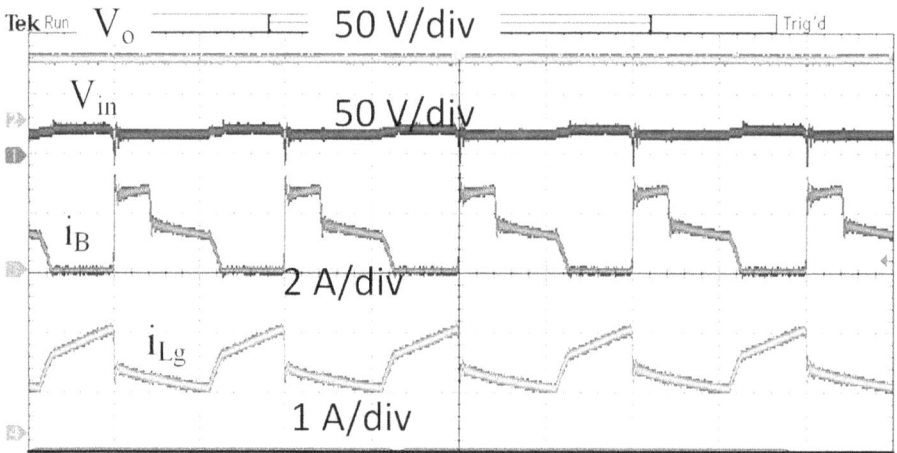

(b)

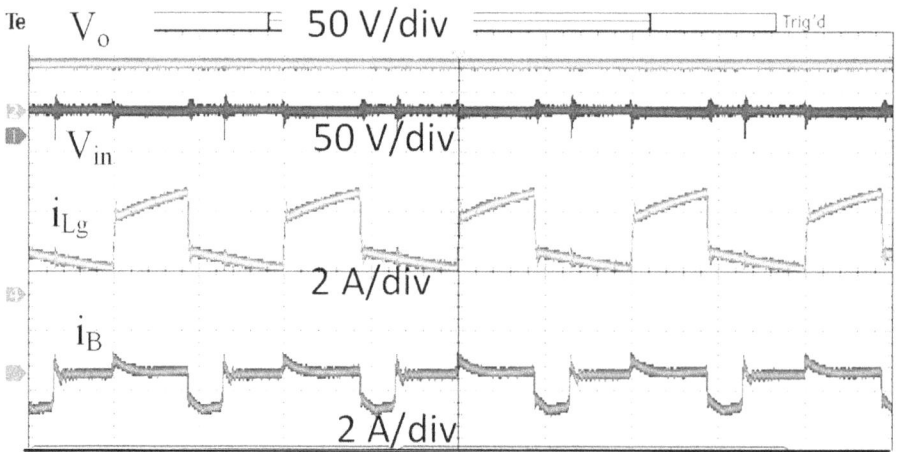

(c)

FIGURE 1.20 Steady-state waveform for (a) DI mode: $D_1 > D_3$, $K = 2/3$, $R_L = 100\ \Omega$; (b) DI mode: $D_1 < D_3$, $K = 2/3$, $R_L = 100\ \Omega$; (c) DO mode: $D_1 < D_2$, $K = 2/3$, $R_L = 100\ \Omega$; (d) SISO mode: $D_2 = D_3 = 0$.

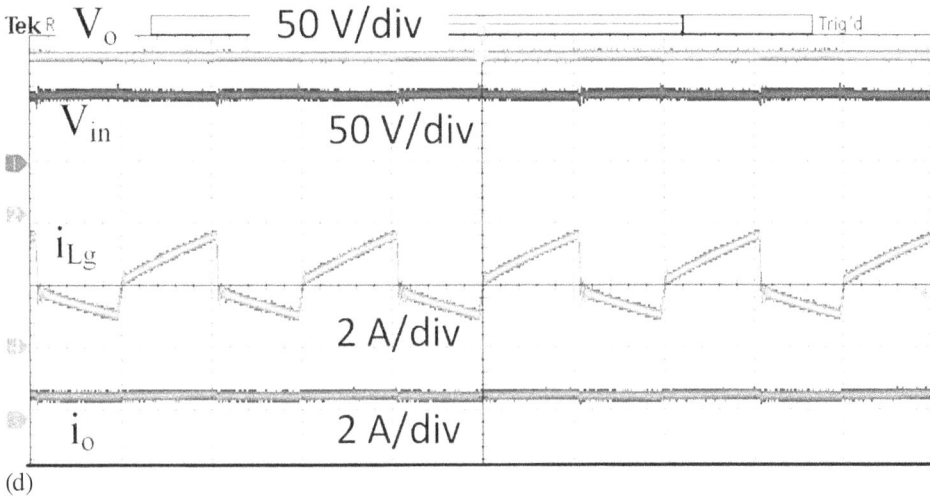

(d)

FIGURE 1.20 Continued

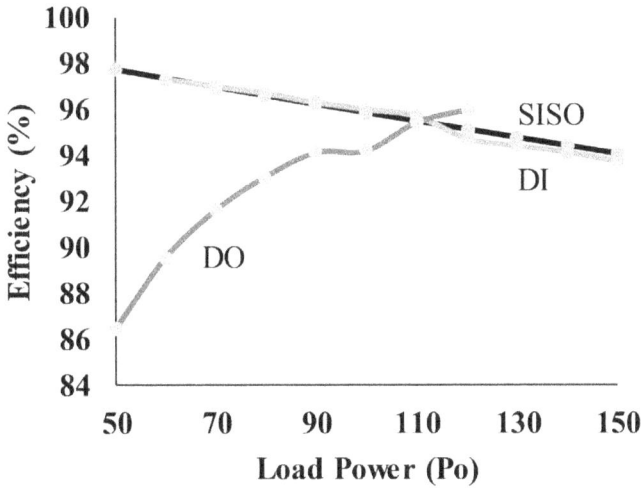

FIGURE 1.21 Efficiency of BITIC converter under different scenarios (DO(BPCS), DI(BPDS), SISO (IBS)).

TABLE 1.8
Comparison of Three-Port Converters

Ref	Switching devices		Energy storage devices		Voltage gain without battery
	Sw	Dd	L	C	
[23]	3	3	1	2	$\dfrac{V_O}{V_{in}} = \dfrac{1}{1-D}$
[27]	3	4	1	1	$\dfrac{V_O}{V_{in}} = \dfrac{1}{1-D}$
[30]	4	5	1	2	$\dfrac{V_O}{V_{in}} = \dfrac{1}{1-D}$

(Continued)

TABLE 1.8
Continued

Ref	Switching devices		Energy storage devices		Voltage gain without battery
	Sw	Dd	L	C	
[31]	3	3	2	1	$\dfrac{V_O}{V_{in}} = \dfrac{1}{1-D}$
[20]	2	4	2	1	$\dfrac{V_O}{V_{in}} = \dfrac{1}{1-D}$
[32]	2	5	2	2	$\dfrac{V_O}{V_{in}} = \left(\dfrac{1}{1-D}\right)^2$
[14]	2	4	2	2	$\dfrac{V_O}{V_{in}} = \left(\dfrac{1}{1-D}\right)^2$
[33]	3	5	2	4	$\dfrac{V_O}{V_{in}} = \dfrac{1+(N_2/N_1)}{1-D}$
BITIC	3	2	1	2	$\dfrac{V_O}{V_{in}} = \left(1 + \dfrac{(N_1+N_2)D}{N_1(1-D)}\right)$

under varying source and load power. The power can be maintained to the load even in the absence of a source or battery.

REFERENCES

[1] N. Priyadarshi, S. Padmanaban, M. S. Bhaskar, F. Blaabjerg, and A. Sharma. Fuzzy SVPWM-based inverter control realisation of grid integrated photovoltaicwind system with fuzzy particle swarm optimisation maximum power point tracking algorithm for a grid-connected PV/wind power generation system: Hardware implementation. *IET Electric Power Applications*, vol. 12, no. 7, pp. 962–971, 2018.

[2] S. Padmanaban et al. A novel modified sine-cosine optimized MPPT algorithm for grid integrated PV system under real operating conditions. *IEEE Access*, vol. 7, no. c, pp. 10467–10477, 2019.

[3] N. Priyadarshi, A. K. Sharma, and F. Azam. A hybrid firefly-asymmetrical fuzzy logic controller based MPPT for PV -wind-fuel grid integration. *International Journal of Renewable Energy Research*, vol. 7, no. 4, pp. 1546–1560, 2017.

[4] K. Kamalapathi et al. A hybrid moth-flame fuzzy logic controller based integrated cuk converter fed brushless DC motor for power factor correction. *Electronics (Switzerland)*, vol. 7, no. 11, 2018.

[5] N. Priyadarshi, S. Padmanaban, J. B. Holm-Nielsen, M. S. Bhaskar, and F. Azam. Internet of things augmented a novel PSO-employed modified zeta converter-based photovoltaic maximum power tracking system: Hardware realisation. *IET Power Electronics*, vol. 13, no. 13, pp. 2775–2781, 2020.

[6] N. Priyadarshi, M. S. Bhaskar, S. Padmanaban, F. Blaabjerg, and F. Azam. New CUK-SEPIC converter based photovoltaic power system with hybrid GSA-PSO algorithm employing MPPT for water pumping applications. *IET Power Electronics*, vol. 13, no. 13, pp. 2824–2830, 2020.

[7] Y. Zhang, H. Liu, J. Li, M. Sumner, and C. Xia. DC-DC boost converter with a wide input range and high voltage gain for fuel cell vehicles. *IEEE Transactions on Power Electronics*, vol. 34, no. 5, pp. 4100–4111, 2019.

[8] S. Yousefizadeh, J. Di. Bendtsen, N. Vafamand, M. H. Khooban, T. Dragicevic, and F. Blaabjerg. EKF-based predictive stabilization of shipboard DC microgrids with uncertain time-varying load. *IEEE Journal of Emerging and Selected Topics in Power Electronics*, vol. 7, no. 2, pp. 901–909, 2019.

[9] M. Veerachary, and A. R. Saxena. Design of robust digital stabilizing controller for fourth-order boost DC-DC converter: A quantitative feedback theory approach. *IEEE Transactions on Industrial Electronics*, vol. 59, no. 2, pp. 952–963, 2012.

[10] N. Priyadarshi, A. Anand, A. K. Sharma, F. Azam, V. K. Singh, and R. K. Sinha. An experimental implementation and testing of GA based maximum power point tracking for PV system under varying ambient conditions using dSPACE DS 1104 controller. *International Journal of Renewable Energy Research*, vol. 7, no. 1, pp. 255–265, 2017.

[11] N. Priyadarshi, S. Padmanaban, J. B. Holm-Nielsen, F. Blaabjerg, and M. S. Bhaskar. An experimental estimation of hybrid ANFIS-PSO-based MPPT for PV grid integration under fluctuating sun irradiance. *IEEE Systems Journal*, vol. 14, no. 1, pp. 1218–1229, 2020.

[12] A. V. J. S. Praneeth, N. Yalla, and S. S. Williamson. DC-DC converter with reduced circulating current in on-board battery chargers for electric transportation. *2019 IEEE Transportation Electrification Conference, ITEC-India 2019*, 2019.

[13] J. Su, T. T. Lie, and R. Zamora. Integration of electric vehicles in distribution network considering dynamic power imbalance issue. *IEEE Transactions on Industry Applications*, vol. 56, no. 5, pp. 5913–5923, 2020.

[14] A. R. Saxena, and D. Kumar. Transformerless high-gain battery-integrated DC-DC boost converter for fuel-cell stacks: Design, analysis, and control. *International Transactions on Electrical Energy Systems*, pp. 1–20, 2020.

[15] B. M. Reddy, and P. Samuel. A comparative analysis of non-isolated bi-directional dc-dc converters. *1st IEEE International Conference on Power Electronics, Intelligent Control and Energy Systems, ICPEICES 2016*, 2017.

[16] Z. Wang, P. Wang, B. Li, X. Ma, and P. Wang. A bidirectional DC-DC converter with high voltage conversion ratio and zero ripple current for battery energy storage system. *IEEE Transactions on Power Electronics*, vol. 36, no. 7, pp. 8012–8027, 2021.

[17] N. Priyadarshi, V. K. Ramachandaramurthy, S. Padmanaban, and F. Azam. An ant colony optimized mppt for standalone hybrid pv-wind power system with single cuk converter. *Energies*, vol. 12, no. 1, 2019.

[18] D. Kumar, and A. R. Saxena. A battery integrated three-port bidirectional charger/discharger for light electric vehicles with G2V and V2G power flow capability. *International Journal of Circuit Theory and Applications*, 2021.

[19] H. Lv, X. Huang, and Y. Liu. Analysis on pulse charging–discharging strategies for improving capacity retention rates of lithium-ion batteries. *Ionics*, vol. 26, no. 4, pp. 1749–1770, 2020.

[20] A. R. Saxena, and D. Kumar. Design and control of a reconfigurable high-gain battery integrated dc-dc boost converter for time-varying loads. *International Journal of Circuit Theory and Applications*, pp. 1–21, 2020.

[21] B. Nayak, A. Mohapatra, and K. B. Mohanty. Selection criteria of dc-dc converter and control variable for MPPT of PV system utilized in heating and cooking applications. *Cogent Engineering*, vol. 4, no. 1, pp. 1–16, 2017.

[22] M. Knowles, A. Morris, D. Baglee, and D. Kok. Battery ripple effects in cascaded and parallel connected converters. *IET Power Electronics*, vol. 8, no. 5, pp. 841–849, 2015.

[23] H. Wu, K. Sun, S. Ding, and Y. Xing. Topology derivation of nonisolated three-port DC-DC converters from DIC and DOC. *IEEE Transactions on Power Electronics*, vol. 28, no. 7, pp. 3297–3307, 2013.

[24] N. Vázquez, C. M. Sanchez, C. Hernández, E. Vázquez, L. D. C. García, and J. Arau. A different three-port DC/DC converter for standalone PV system. *International Journal of Photoenergy*, vol. 2014, 2014.

[25] H. Zhu, D. Zhang, B. Zhang, and Z. Zhou. A nonisolated three-port DC-DC converter and three-domain control method for PV-battery power systems. *IEEE Transactions on Industrial Electronics*, vol. 62, no. 8, pp. 4937–4947, 2015.

[26] A. Deihimi, and M. E. S. Mahmoodieh. Analysis and control of battery-integrated dc/dc converters for renewable energy applications. *IET Power Electronics*, vol. 10, no. 14, pp. 1819–1831, 2017.

[27] P. Zhang, Y. Chen, and Y. Kang. Nonisolated wide operation range three-port converters with variable structures. *IEEE Journal of Emerging and Selected Topics in Power Electronics*, vol. 5, no. 2, pp. 854–869, 2017.

[28] L. An, and D. D. C. Lu. Design of a single-switch DC/DC converter for a PV-battery-powered pump system with PFM+PWM control. *IEEE Transactions on Industrial Electronics*, vol. 62, no. 2, pp. 910–921, 2015.

[29] E. A. Rahimi, T. Md Rabiul Islam, H. Gholizadeh, S. Mahdizadeh. Design and implementation of a high step-up DC-DC converter based on the conventional boost and buck-boost converters with high value of the efficiency suitable for renewable application. *Sustainability*, vol. 13, no. 19, p. 10699, 2021.

[30] T. Cheng, D. D. C. Lu, and L. Qin. Non-isolated single-inductor DC/DC converter with fully reconfigurable structure for renewable energy applications. *IEEE Transactions on Circuits and Systems II: Express Briefs*, vol. 65, no. 3, pp. 351–355, 2018.

[31] N. Vázquez, C. M. Sanchez, C. Hernández, E. Vázquez, L. D. C. García, and J. Arau. A different three-port DC/DC converter for standalone PV system. *International Journal of Photoenergy*, vol. 2014, 2014.

[32] S. Rostami, V. Abbasi, N. Talebi, and T. Kerekes. Three-port DC–DC converter based on quadratic boost converter for stand-alone PV/battery systems. *IET Power Electronics*, 2020.

[33] L. J. Chien, C. C. Chen, J. F. Chen, and Y. P. Hsieh. Novel three-port converter with high-voltage gain. *IEEE Transactions on Power Electronics*, vol. 29, no. 9, pp. 4693–4703, 2014.

2 Modelling a Battery and SOC Calculation Using Coulomb Counting Technique and Simulation of Equalising Techniques Used in Battery Management Systems

C. R. Ananthraj, Arnab Ghosh, and Vikash Gurugubelli

CONTENTS

DOI: 10.1201/9781003323471-2

2.1 INTRODUCTION

2.1.1 MOTIVATION

Electric vehicles (EVs) are set to form the most prominent form of transport in the coming centuries. As it can be witnessed from the developing market of EVs and deteriorating petroleum products, there is a need to invest in more advanced and efficient storage systems for the EVs, as the storage is the most important part of an electric vehicle. Also, it is necessary to take into account the various risk factors associated with every storage system, as in thermal runaway, overheating, overcharging, over discharging and so on. All these risk factors can be eliminated only through proper monitoring and an efficient battery management system (BMS). Comprehensive and mature BMS are now found in laptops and mobile systems but have not been fully employed in EVs and hybrid EVs. This is because of the fact that the number of cells in an EV storage is hundreds of times larger than that used in a laptop or mobile device. And EV storage not only has to be long lasting but should also handle higher power demands; hence an efficient management of the storage system becomes necessary. These make EV storage and their BMS a complicated system as compared to portable electronics. With EVs being the future of transport as predicted now, advanced BMS technologies form the cornerstone of the reliability of such vehicles. Such shows the need to push forward to new and efficient technologies in BMS and battery storage.

2.1.2 LITERATURE REVIEW

A battery management system is a very complex circuitry which provides all the necessary protection and monitor battery conditions and operations to extract the maximum output of a battery. State of Charge (SOC) forms a very important measure of battery characteristic. It is the current capacity of the battery expressed in terms of its rated capacity. It gives us the charge left within a battery and thus allows the battery to be safely charged and discharged without damaging the battery. BMS is usually a separate entity with hardware and algorithms; it is connected on a battery to optimise its operation rather than integrated to the charger [1]–[4]. BMS is also composed of several sensory devices for measuring, calculating and monitoring battery state and parameters. These data will be used in the SOC calculation algorithm and other control decisions taken by the BMS. The most important block in any BMS model is the battery model block. To form this block requires a detailed understanding of battery characteristics and behaviour. This model is usually formed using the discharge and charge characteristics of the battery. The block diagram of the BMS is shown in Figure 2.1.

A BMS must monitor and operate on the battery charging and discharging operations at the same time. Also, it must control the rate of charging and other features. At the same time the BMS should keep on monitoring the SOC of the battery to avoid any undesirable operating condition to the battery [5]–[8]. So, the primary objective of a BMS is thus to find the operating conditions and the mode in which the battery is to be operated. The importance of BMS in the current and future

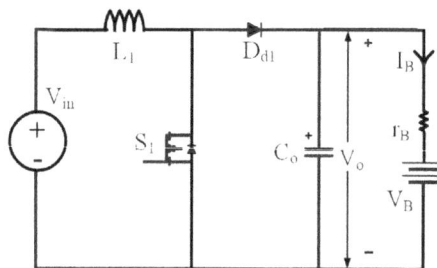

FIGURE 2.1 Block diagram of BMS.

scope comes with the fact that electrical vehicles are said to be the forthcoming means of transport. The increasing market of electrical vehicles and diminishing petroleum fuel resources made it necessary to conduct research not only in the battery storage system but also for an efficient battery management system to increase the efficiency of battery. And as a result, the reliability of an EV is increased. It is very clear that the success of an EV is directly dependent on the battery management system used [9–14].

To efficiently model a BMS, it is first necessary to properly understand the behaviour of a battery storage system. Hence the first challenge faced while designing a battery management system is proper modelling of a battery. The proper behaviour of a battery can only be modelled using battery charge and discharge characteristics. Also, each battery behaviour will be different. And then after proper understanding of battery behaviour, the storage system is modelled, and then the battery management system is designed efficiently.

The control strategy of any BMS is based on the data collected by constant monitoring of the cells by the sensors. As a result, the BMS helps to meet the EV power demands.

Some of the important features of a battery management system are as follows:

   i. Maintenance of State of Charge using the energy management system algorithm
   ii. Constant monitoring of current, voltage and temperature conditions of individual cells
   iii. Protection against overcharge and overdischarge
   iv. Monitoring the State of Health (SOH) of a battery
   v. Monitoring of currents, voltages and temperatures of the battery pack
   vi. Internal fault analysis
   vii. Determining the safe levels of charge and discharge currents of the battery storage system

2.1.2.1 Measurement Block

It captures cell currents, voltages and temperatures. Measured values are processed as digital signals. Although measuring individual cell voltages is costly and requires more hardware, the advantage of this technique is that we can achieve cell optimisation and protection at a cellular level.

2.1.2.2 Battery Algorithm Block

The primary objective is SOC and SOH estimation using the measured battery parameters, which are the battery currents, voltages and temperatures. A battery's SOC may be calculated by dividing the actual capacity of the battery by its rated capacity. SOC is similar to the fuel gauge of a motor vehicle [15–20]. It shows the amount of usable charge remaining in a battery and this in turn can be used to estimate the travelling distance of the EV. Also, SOC is dependent on the temperature and charging and discharging cycle. Therefore, the algorithm block should also take into effect the impact of these factors on SOC estimation. The respective sensors provide the inputs to calculation units, which are mainly voltage, current and temperature. The digitalisation of the analog inputs is achieved by using an analog-digital converter. The SOC of the battery is useful not only to calculate the remaining running distance of an EV but also to keep the batteries at a safe SOC to avoid the risk of over-discharge or overcharge of the battery storage. Charge dumping is another factor to be cautious about in EVs. It occurs due to regenerative braking of EVs. This can lead to the cells being overcharged, particularly when cells are already at a higher State of Charge. If such an event rises, the BMS should control the charging of the battery in order to ensure the safety of the battery and the driver. Thus, SOC estimation remains the most important part of any BMS [21–26]. The most common and easiest method is direct measurement. In this method, the open circuit voltage (OCV)-SOC characteristic of the battery is already saved and SOC is deduced from this characteristic. But a major disadvantage of this technique is that it does not take into account the temperature effect of the battery. So, it is important to use a method that does take into account the temperature effect of the cell/battery.

2.1.2.3   Capability Estimation Block

After SOH and SOC are calculated, the next objective of a BMS is to determine the maximum discharge and charge currents at any instant maintaining the safety as the primary objective. This block provides the necessary control signals for these currents. Also it makes sure that the battery is not charged or discharged outside indicated limits.

2.1.2.4   Cell Equalisation Block

Manufactured cells are not always identical. There usually exist a variation on cell capacity of up to 15%. Other than that, there might exist other deviations such as variations in internal resistance and charge and discharge characteristics. This module compares cell voltages and calculates the difference between the cell with the largest voltage and the cell with the lower voltage. If the difference is greater than a pre-determined threshold, the charging of battery storage is interrupted, and cells are equalised using either dissipative or active equalisation procedures, depending on the system's needs. Active cell balancing or active cell equalising is an advanced technique as compared to dissipative cell equalising in terms of efficiency and performance. The only factor restricting the usage is the cost and the complexity of the circuit [27–29].

2.1.2.5   Thermal Management Block

Thermal management is defined as the monitoring and controlling of the battery temperature so as to ensure its safe operation and to ensure that the battery does not explode and harm the user due to very high or low temperature. Usually, the outputs of a thermal management block control a cooling system and a heating system. The primary objective of the cooling and heating system being maintaining the temperature of the battery storage within safe limits. In the event of an abnormal temperature situation, it also transmits a control signal to the Electronic Control Unit (ECU) [30]–[51].

2.1.3   CONTRIBUTION AND CHAPTER ORGANISATION

In this chapter an equivalent battery model is formed using real-time test data. State of charge calculation by coulomb counting method is adopted and the battery behaviour is observed for various discharge currents. Using this battery simulation model, an equalising battery circuit is formed. The dissipative circuit modelled comprises load resistances to a battery to discharge the extra charge in a battery. Then an active equalising circuit model is presented. This circuit, where there is an extra charge in the battery, is used to charge the under-charged battery. The next section consists of the simulation results and discussions on the battery model formed. The behaviour of a battery for both equalising circuits is justified and the last part includes the conclusions of the forementioned simulation results.

2.2   BATTERY MODEL AND FORMATION OF EQUALISING TECHNIQUES

2.2.1   BATTERY MODEL FORMULATION

The equivalent circuit model of a battery, as shown in Figure 2.2, is used as reference to model the battery in MATLAB. The presented model has small differences when compared to a traditional equivalent circuit model. Unlike in a traditional model, internal resistance R_d of a battery is shown as a nonlinear resistance which is dependant on SOC of the battery. This concept of nonlinear transfer resistance in the circuit will increase the enactment of the model formed as compared to the traditional model. Traditional battery models usually make use of constant RC networks. These RC networks only function at specific voltages and temperatures, which is a disadvantage of traditional battery types. As a result, the notion of nonlinear networks is proposed in order to improve model accuracy. Temperature and self-discharge effects that exist in batteries are neglected for simplicity and on the assumption that a battery will be operated at a constant temperature. A second parallel

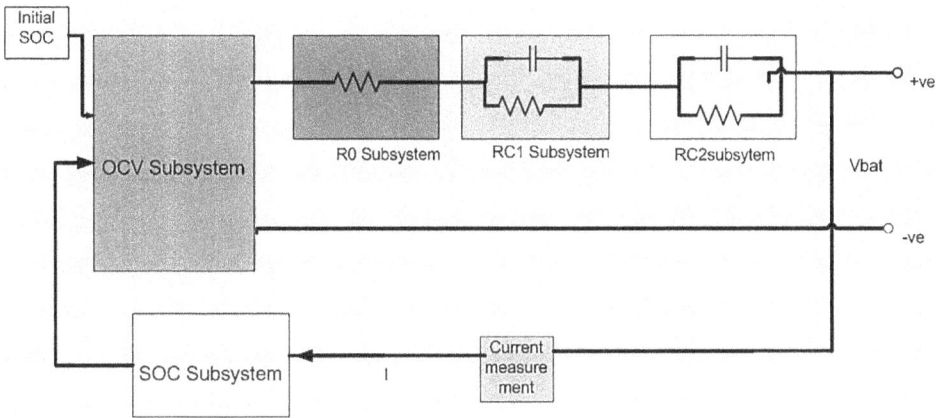

FIGURE 2.2   The battery's equivalent circuit model.

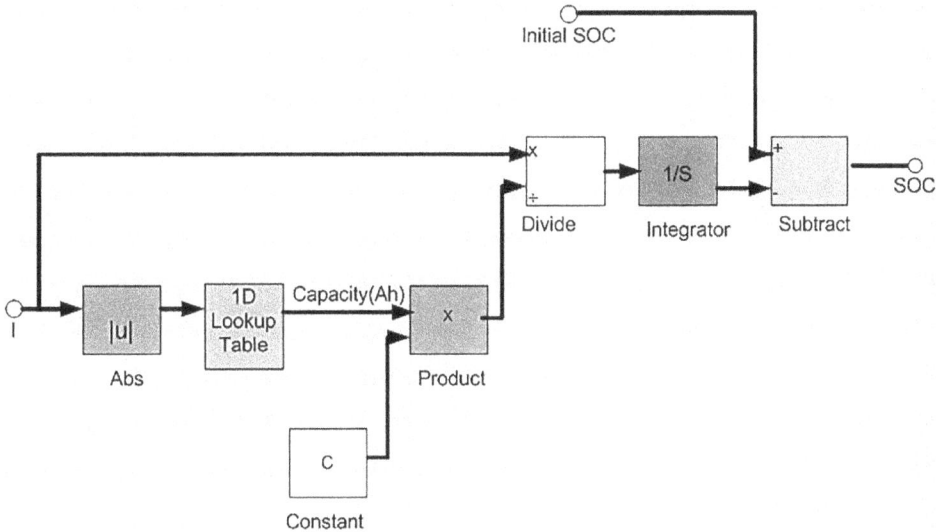

FIGURE 2.3   SOC subsystem.

RC network in series with the present one can be used to mimic the relaxation effect that exists in a battery to improve the battery model behaviour. OCV stands for the system's open circuit voltage. Internal resistance is abbreviated as R_i. The RC parallel circuits R_d and C_d are utilised to explain the battery's transient reaction. The double layer capacitance is represented by R_d and C_d. Figure 2.3 shows the SOC subsystem [2].

2.2.2   BATTERY MODEL FORMULATION

2.2.2.1   Dissipative Equalising

Considering the importance of battery equalising, an equalising circuit using passive equalisation was formed (Figure 2.4). The battery model developed and the SOC calculation module formed earlier were implemented in this circuitry. The circuit is made up of control switches and a battery algorithm block. The measured value from the battery is used to calculate the SOC of each battery of a battery pack using the coulomb counting algorithm as mentioned earlier. The SOC of each battery

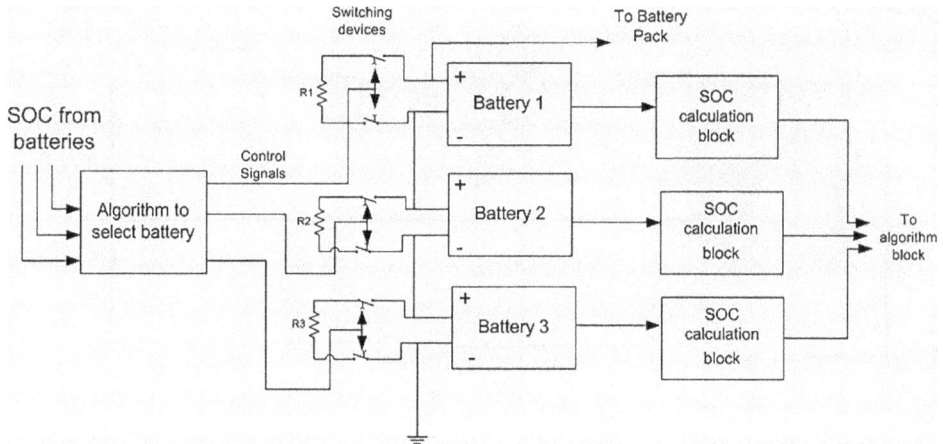

FIGURE 2.4 Dissipative equalising.

is compared in the algorithm block. A threshold value is set in the algorithm block. The battery SOC is compared, and minimum and maximum values are found out from these. If the difference in minimum and maximum value is found to be more than the threshold value, then the algorithm will start its operation. It then finds the maximum and minimum values of the aforementioned batteries in the battery pack. Then it calculates the maximum and minimum values in the battery pack. The difference between the maximum and minimum value, if found to be more than a threshold value the control switches, usually MOSFET or IGBT of those batteries which has higher SOC are turned on. These batteries are then discharged to a resistor, until their values fall within the threshold limit.

In the simulation, to test this circuitry, the SOC1 = SOC of battery 1 was set as 92%. SOC2 = SOC of battery 2 was set as 94% and SOC3 = State of Charge of battery 3 was set as 83%. Maximum was found to be SOC2 = 94% and minimum as SOC3 = 83%. The threshold value was set as 5%. Clearly the difference was more than 5%. So the algorithm will start its function. It will switch on those batteries whose difference in SOC values with minimum values more than 5%. So here switch 1 and 2 will be turned on. The two batteries will be discharged until the difference between the threshold value falls between the 5%.

2.2.3 ACTIVE EQUALISING

Battery charge equalisation techniques are necessary to equalise the charge levels of overcharged and undercharged cells in a battery pack. Usually, this module is given for each battery module in a battery pack. The battery modules consist of cells in series to reach the desired voltage level. These modules are then connected in parallel and series combinations to reach the desired voltage and current levels. If there are unequal charge levels in batteries in a module, it may lead to internal current flows and the battery may heat up, which may eventually lead to explosions. The proposed battery charge equalisation algorithm is based on the concept of active equalising technique. Figure 2.5 shows the active equalising circuit, and this technique involves charging an undercharged battery from a separate charger or from the other batteries with high charge. Thus, this adds to the complexity of the circuit and makes it more expensive. The system is composed of measurement and calculation units to measure the SOC of the cells and to determine which needs to be charged or discharged. A bidirectional SEPIC converter was used in place of the DC-DC converter block. The microcontroller unit will form the necessary algorithm to select the control switches. The battery units used in the simulation were the battery units formed in the earlier section. Control switches were realised using MOSFETs.

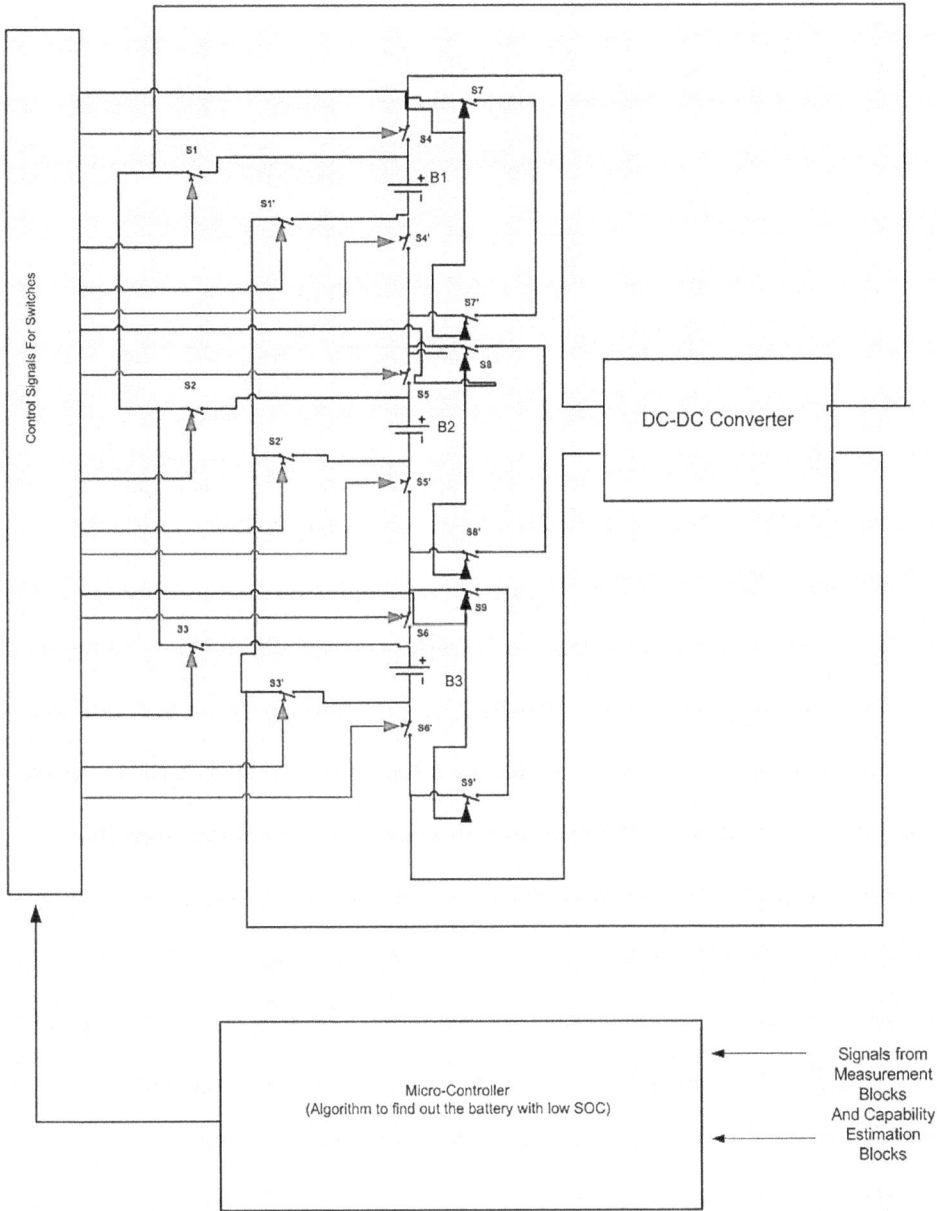

FIGURE 2.5 Active equalising circuit.

The system simulated is a simple system in which the microcontroller chooses those batteries which are undercharged and then they are connected at output part of DC-DC converter. In the example, simulated battery 1 is undercharged. The algorithm block compares the SOC of all three batteries and finds battery 1 to be lesser than the permissible difference in batteries. Thus, battery 1 is connected to the output end. In order to do that, it should be disconnected from the input end. Switches 4–4', 5–5' and 6–6' are usually on producing the required voltage level of battery module. Now switch 1–1' will be turned on, thus battery 1 will appear across the output end of DC-DC converter. Switch 4–4' will be turned off, thus disconnecting it from the module, and switch 7–7' will be turned on to complete the circuit. This whole operation is done while the module is disconnected

TABLE 2.1

Switching Sequences (Where NoE Means No Effect)

	S1& S1'	S2& S2'	S3& S3'	S4& S4'	S5& S5'	S6& S6'	S7& S7'	S8& S8'	S9& S9'
Battery 1	ON	NoE	NoE	OFF	NoE	NoE	ON	NoE	NoE
Battery 2	NoE	ON	NoE	NoE	OFF	NoE	NoE	ON	NoE
Battery 3	NoE	NoE	ON	NoE	NoE	OFF	NoE	No	ON

from the whole pack, otherwise it may create further complications and the necessary gate pulses are given to the DC-DC converter and it will then charge battery 1 using the extra charge in battery2 and battery3. If more than one cell has an undercharge condition it will be connected parallel across the output and charging them simultaneously. In case of overcharged cells, they remain at input, and undercharged cells go to output side and the same operation continues until the charges are equalised. The way in which the battery can be connected to the output side is shown in Table 2.1. How it is achieved is reviewed in the following section.

2.2.3.1 Algorithm and Flowchart

The algorithm keeps monitoring the cell status of each pack. It uses the measured value and calculates the SOC of the cells in a battery pack. The algorithm then checks to see if all the cells are in normal operational conditions or not. If they are overcharged or undercharged their cells are brought to normal charged conditions. The SOC of cells is calculated and the cell with highest and lowest SOC is determined. Their difference is found out in a pack. If the difference is more than the allowable limit, then the battery pack starts the cell equalisation algorithm. A DC-DC converter is utilised to charge or discharge the cells as and when necessary. The cell (cells) with the lowest SOC is linked to the output side of the DC-DC converter, and higher SOC cells are then discharged into this cell.

Under normal operation, there is no need for equalising the control signals and the clock to the converter is zero. In this case, switches are as follows. 1–1', 2–2', 3–3' will be off. 4–4', 5–5', 6–6' will be the switches that connect the cells in series. They will be always on to keep the voltage at prescribed limits. Switches 7–7', 8–8', 9–9' are the bypass switches. They will be turned off by default. In this case, simulated the active equalising one of the batteries was put as overcharged and the other two were kept at a lesser charge. The difference is such that it will cross the threshold limit of the batteries. Then active equalising algorithm in the circuit model formed works in such a manner that the battery which is overcharged is used to transfer the charge that is extra to other batteries of same battery pack. Figure 2.6 shows the flowchart of the active equalizing.

2.3 BIDIRECTIONAL SEPIC CONVERTER AS BATTERY CHARGER

SEPIC is an acronym for single-ended primary-inductor converter. The SEPIC converter can have an output voltage which can be equal to, larger than, or lower than input voltage. The duty cycle given to the control switches regulates output. A unidirectional SEPIC converter is shown in Figure 2.7.

A SEPIC converter can be broken down into two parts: a boost converter and an inverted buck-boost converter. As a result, it is like a commercial buck-boost converter which gives a non-inverting performance, which is an advantage over a traditional buck-boost converter. It accomplishes this by connecting energy from the input to the output through a series capacitor.

Another feature is that it is able to truly shut down when switch Q1 is turned off. At this point output will be zero. The working of a unidirectional SEPIC is shown in Figure 2.8.

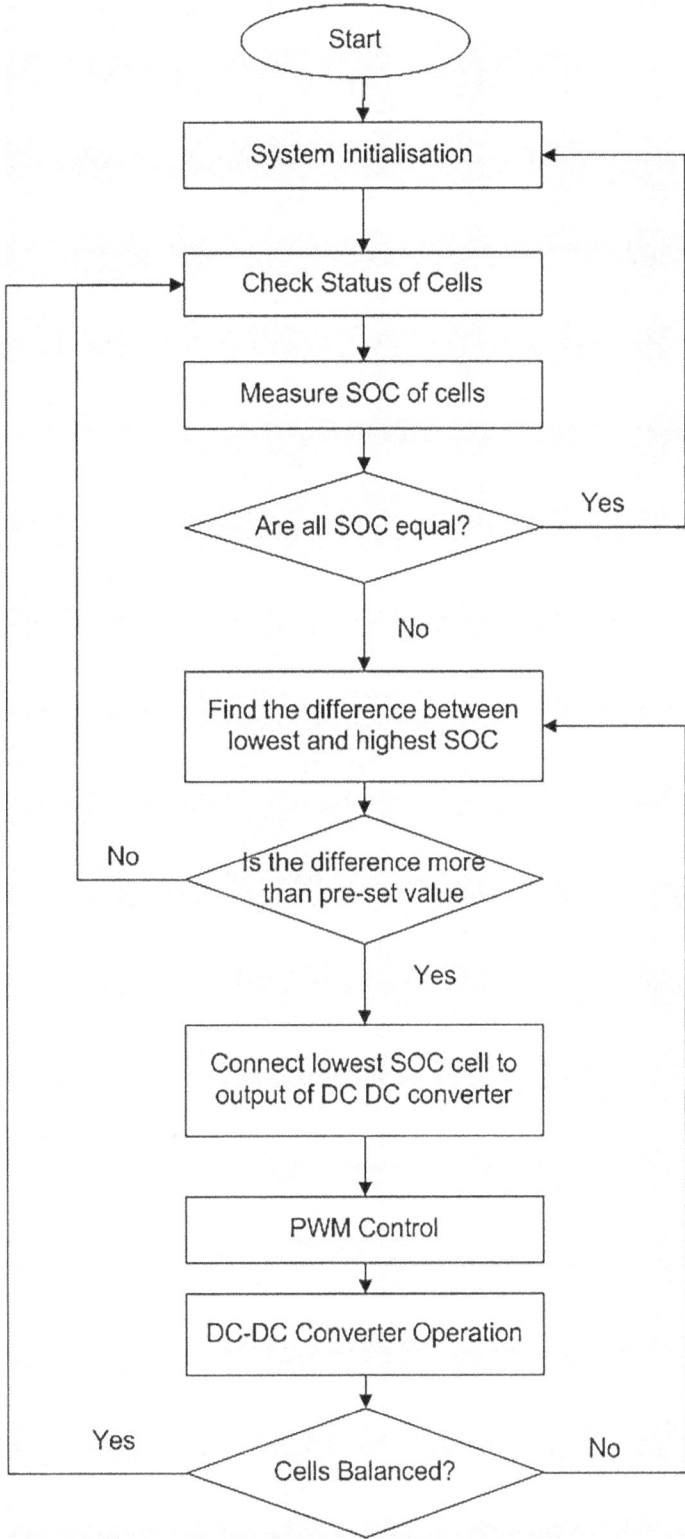

FIGURE 2.6 Flowchart of active equalising.

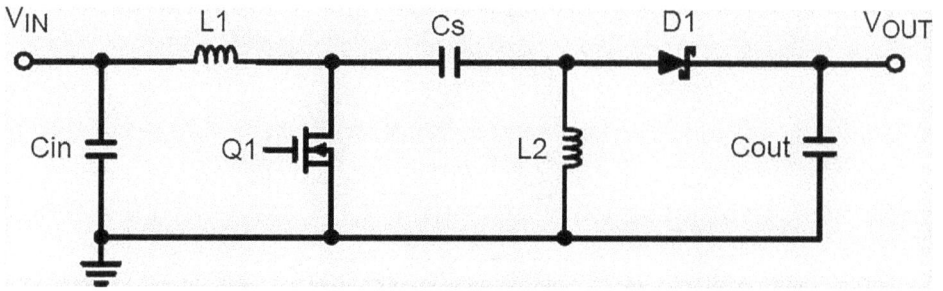

FIGURE 2.7   SEPIC converter circuit diagram.

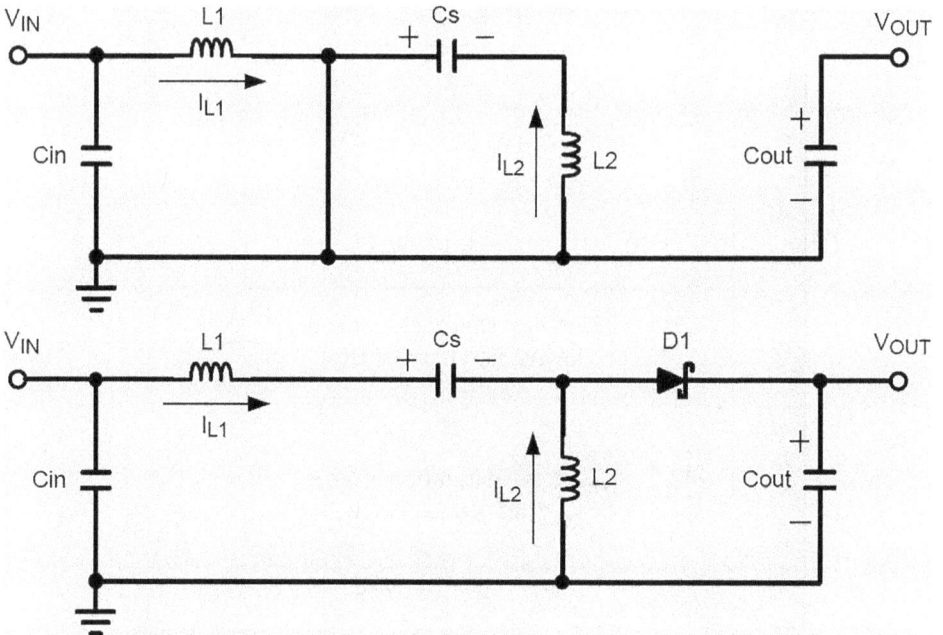

FIGURE 2.8   Switching in SEPIC converter.

2.3.1   SEPIC CONVERTER DESIGN

Duty cycle for a SEPIC converter, taking into account continuous conduction mode, is calculated using the following equations.

$$D = \frac{V_{out} + V_D}{V_{out} + V_D + V_{in}} \tag{1}$$

Peak to peak ripple current set at percentage of peak input voltage is a reasonable rule of thumb when choosing inductor values. As a result, the following equation can be used to choose inductor values. The hardware model of SEPIC converter is shown in Figure 2.9.

$$L_1 = L_2 = L = \frac{V_{in}(\min)}{\Delta I_L \times f_{sw}} \times D_{\max} \tag{2}$$

When the power switch Q1 is switched on in a SEPIC converter, the inductor charges and output capacitor supplies the current at output terminals. As a consequence, massive ripple currents pass through the output capacitor. Therefore, the output capacitor chosen must be able to deal with the full RMS current. Output capacitor's RMS current is

$$I_{Cout} = I_{out} \times \frac{V_{out} + V_D}{\sqrt{V_{in}(\min)}} \tag{3}$$

And output capacitance is given by

$$C_{out} \geq \frac{I_{out} \times V_D}{V_{ripple} \times 0.5 \times f_{sw}} \tag{4}$$

Selection of coupling capacitor: Most of the time, a capacitor that fulfils the RMS current requirement can create a tiny ripple voltage on C_s. As a result, input voltage is also taken as the peak voltage. Output voltage ripple is given by

$$\Delta V_{C_S} = \frac{I_{out} \times D_{\min}}{C_S \times f_{sw}} \tag{5}$$

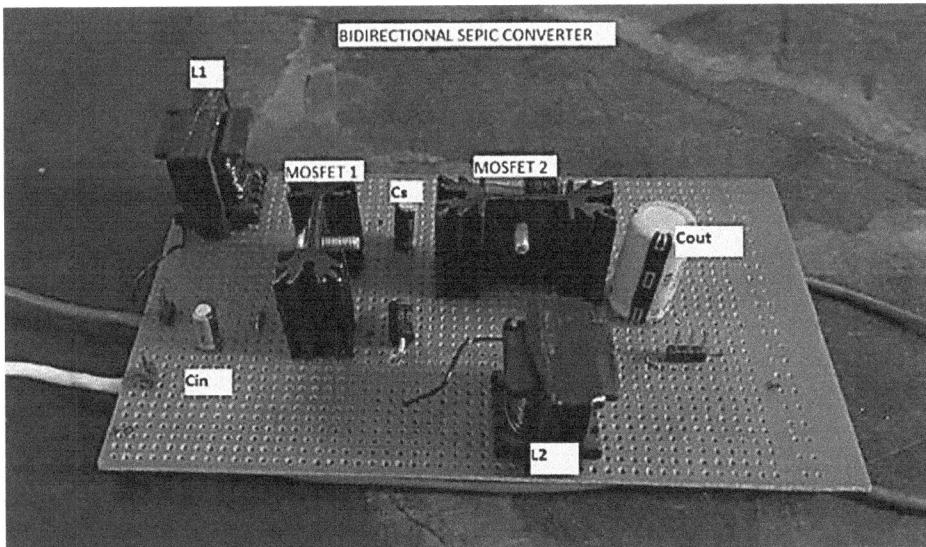

FIGURE 2.9   Hardware model of SEPIC converter.

2.3.2 Bidirectional SEPIC Converter

Based on the design constraints the values on inductor and capacitor were chosen. The value of inductance and capacitances are as follows: $L_1 = 1.3$ mH, $L_2 = 1.3$ mH, $C_{in} = 10$ μF, $C_{out} = 470$ μF, $C_s = 10$ μF. For the value of C_{in}, any value above 10 μF is considered apt.

2.3.2.1 Gate Driver Circuit for SEPIC Converter and DC Power Supply

To provide the necessary gate pulse to the DC-DC converters a gate driver circuit using NE555 timer was designed. The necessary control signal for the NE555 was given from the Arduino UNO microcontroller. The Arduino microcontroller will act as the closed loop control circuit. TLP350 opto-isolator was used to isolate both the control and power circuit of the converter. The NE555 timer is operated in monostable operation. The duty cycle is varied using the input square wave duty cycle. The resistor and capacitor values are chosen according to the design constraints of the NE555 timer operation.

As input a square pulse is given to the gate driver circuit. The circuit diagram of gate driver is shown in Figure 2.10. The duty cycle of the gate driver can thus be varied using the help of this square pulse input. An Arduino UNO microcontroller board is used to achieve this task. The input output relation of the gate driver circuit was designed to follow an inverted pattern. For an input duty cycle of 25% we will get output from gate driver as 75%. And hence the input duty cycle is programmed accordingly. The input voltage was 5 V and the input square pulse was having the same voltage. The pin diagrams of the TLP350 and 555 timer are shown Figures 2.11 and 2.12,

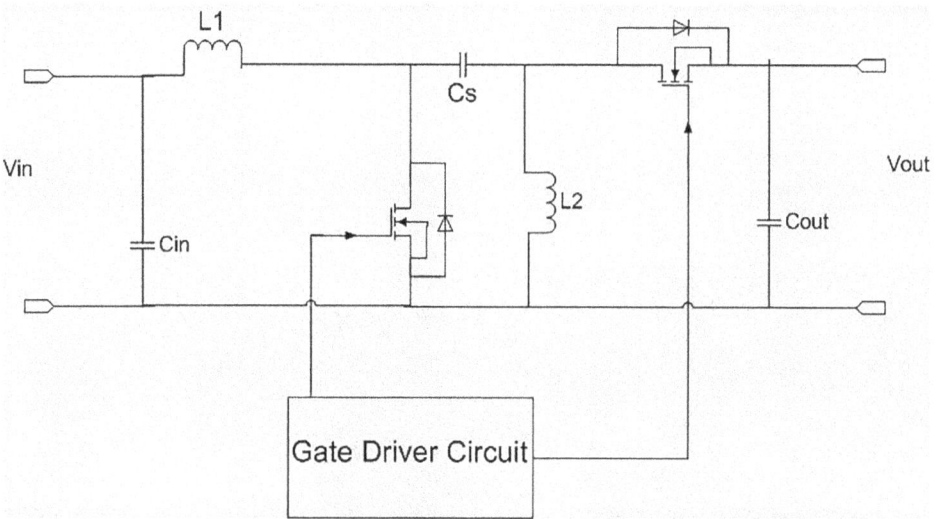

FIGURE 2.10   Circuit diagram of gate driver circuit.

Truth Table

Input	LED	Tr1	Tr2	Output
H	ON	ON	OFF	H
L	OFF	OFF	ON	L

Pin Configuration (top view)

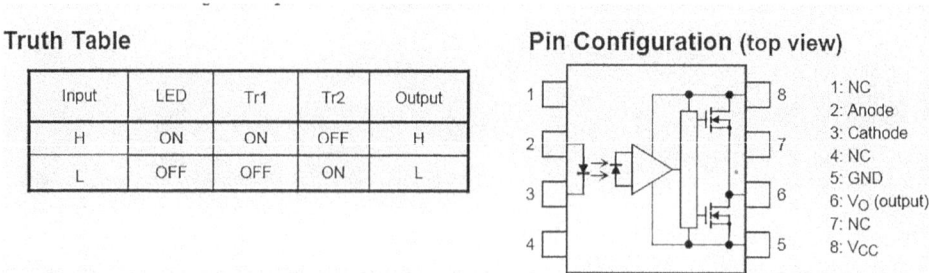

1: NC
2: Anode
3: Cathode
4: NC
5: GND
6: V_O (output)
7: NC
8: V_{CC}

FIGURE 2.11   Pin diagram of TLP350.

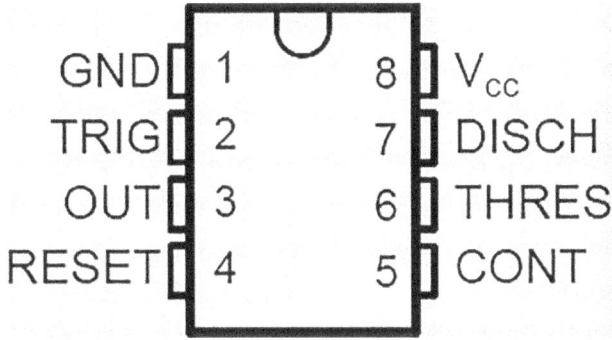

FIGURE 2.12 Pin diagram of 555 timer.

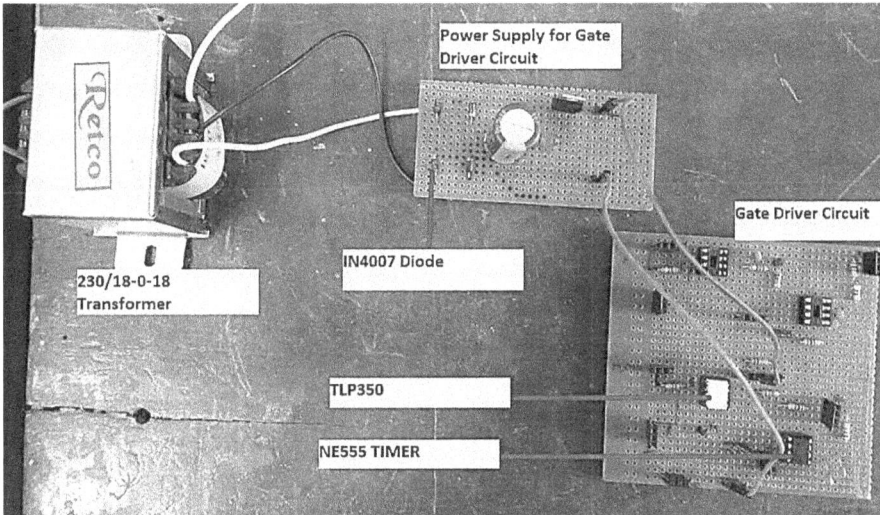

FIGURE 2.13 Hardware model of power supply and gate driver circuit.

respectively. The hardware model of the power supply and gate driver circuit is shown in Figure 2.13. The output voltage of the gate driver square pulse was the same as the bias voltage of NE555 timer. In the circuit the bias voltage was given as 15V. This gate pulse was then clipped to 10 V using a Zener protection device across the Gate-Source terminals of the SEPIC converter. The connection diagram of the gate driver is shown in Figure 2.14.

TLP350 is an opto-isolator. The device converts electrical signal at input to light and this light energy is then used to close a switch and in this way it provides an output using the bias voltage Vcc. TLP350 is used to separate and isolate optically the control circuit and power circuit of the system. In this was even if there is maloperation at one side it doesn't affect the other part of the circuit. And it also protects the measuring and microcontroller blocks. The system bus is operating at 40 V, and this voltage is too much for the operation of small voltage devices. And hence the optical isolator is of utmost importance to the driver circuit. The results of gate driver circuit are discussed in the next section. To provide biasing to the gate driver circuit a 15 V DC supply was made. It was made using a 230/18–0–18 V transformer followed by a rectifier circuit. The 18 V rms voltage was transformed to 15 V DC by using a bridge diode rectifier and a voltage regulator followed by that. The circuit produced a constant 15.2 V DC output. The circuit diagram is shown in Figure 2.15.

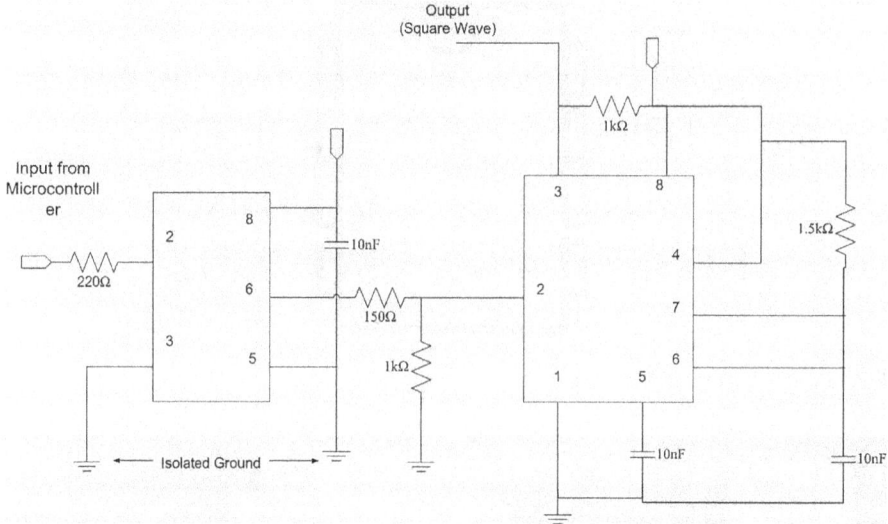

FIGURE 2.14   Connection diagram of gate driver.

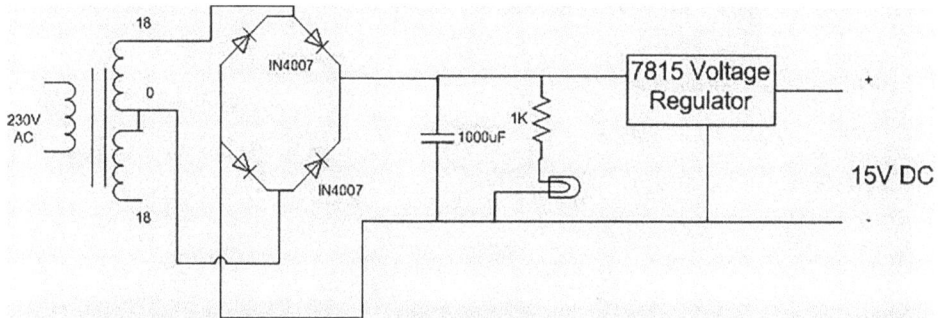

FIGURE 2.15   Circuit diagram of power supply.

2.4   RESULTS AND DISCUSSION

2.4.1   BATTERY MODEL AND SOC CALCULATION

The battery model was formed in Simulink and the battery characteristics were compared to that of Li-ion battery. The SOC computation module was formed with the battery circuit and the relationship between various battery characteristics was determined. The SOC computation module was implemented using the coulomb counting method, and hence the initial SOC of the battery was provided to the Simulink model. Various relationships of battery discharge characteristics were plotted as shown in Figure 2.16. The variation of voltage vs SOC was determined. It was found out that the terminal voltage was a maximum when the SOC was also at its maximum. As the battery is discharged and the SOC is decreased, the terminal voltage also decreased following a nonlinear pattern. And it reached its minimum value when its SOC was found to be 0. As we can see, the minimum value is not 0 but a slightly lesser value than the initial value. As in the case of a real physical value, the final value of terminal voltage also did not fall to zero. And the simulated formulated battery was found to be similar to a physical battery.

The voltage vs time graph (Figure 2.17) for the battery was plotted at different discharge currents. It was discharged at 0.25C, 0.5C and 1C. It was found that the battery decreases to minimum voltage value faster when it is discharged at a higher current. The characteristic is given, and it was found to be similar to the real physical battery system. The coulomb counting method was used to calculate the SOC in the simulated battery system. It is a widely used method of SOC calculation. Although it has a disadvantage in that the initial value of SOC needs to be assumed, still it provides

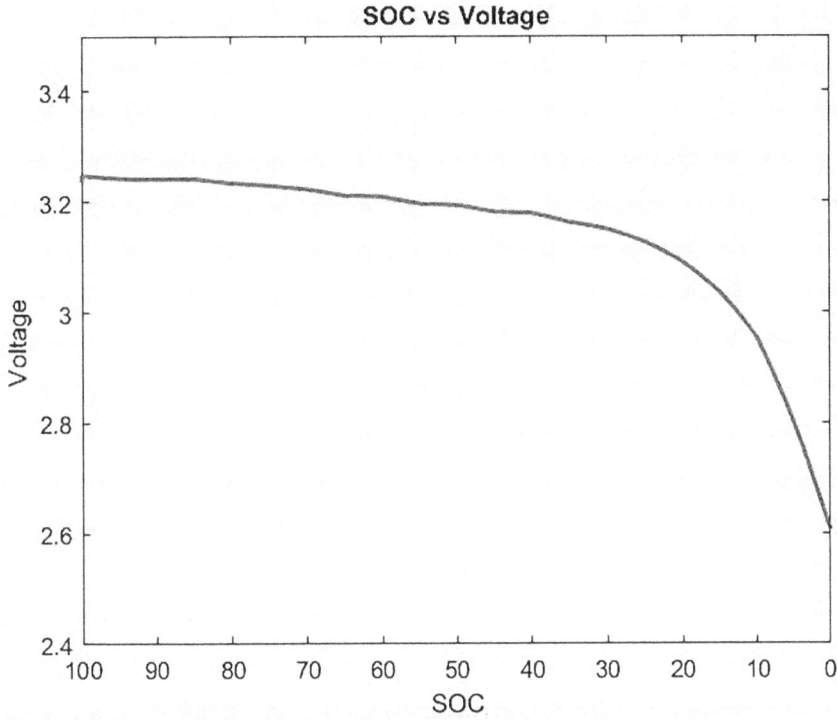

FIGURE 2.16 Voltage-SOC response.

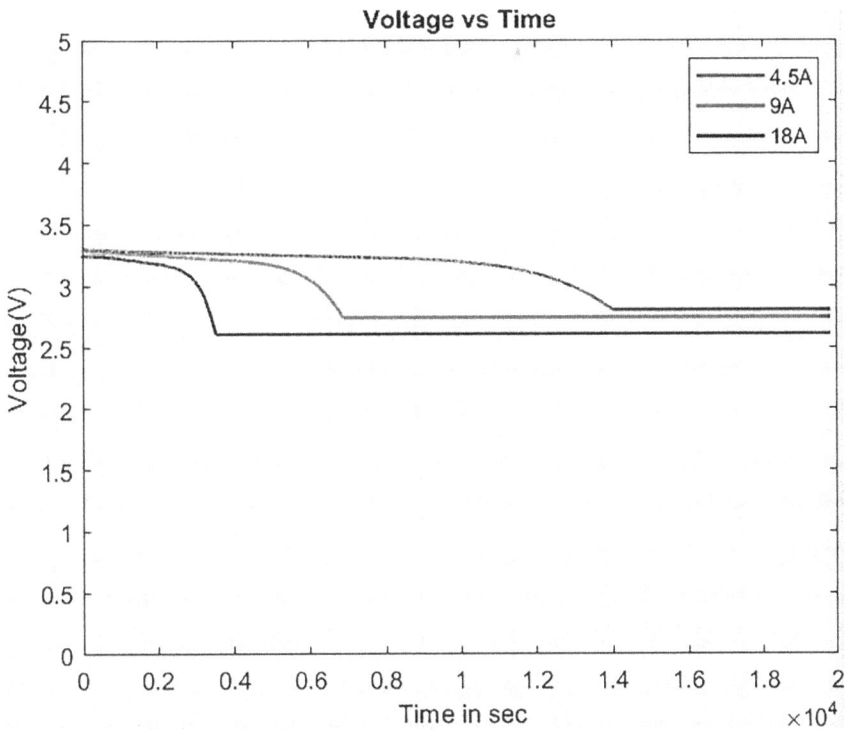

FIGURE 2.17 Voltage versus time response.

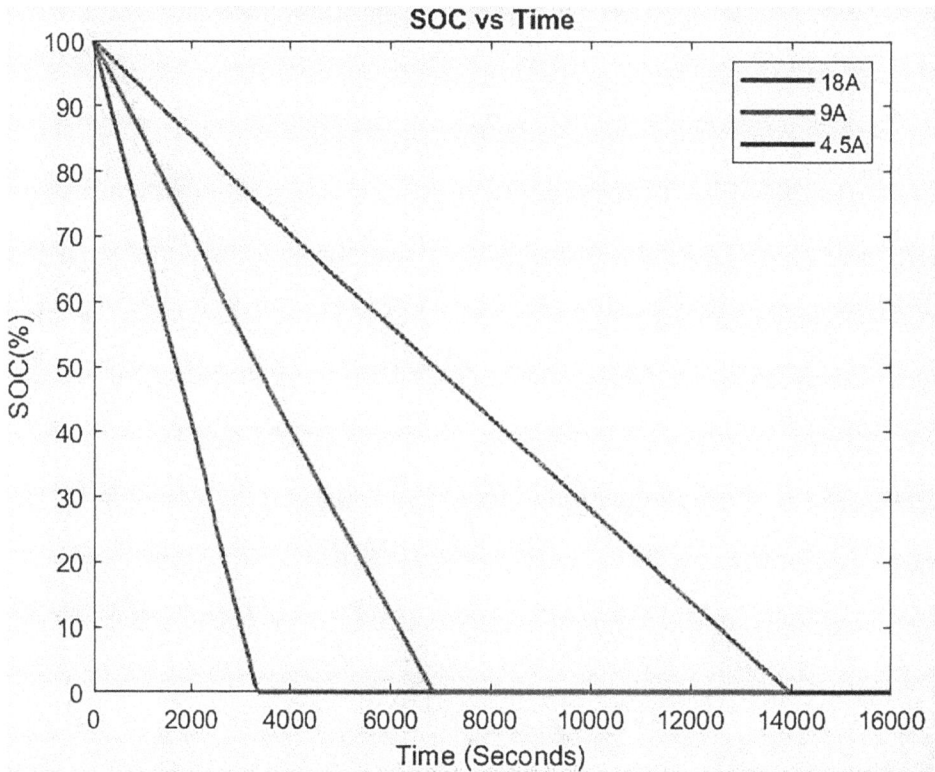

FIGURE 2.18   SOC-time response (Discharging of all batteries).

accurate SOC calculation of the battery system. The SOC vs time graph was plotted for various discharge currents at 0.25C, 0.5C and 1C and it was found that the SOC decrease to a lower value much faster when the discharge current value is higher. The graph is given in Figure 2.18.

2.4.2   DISSIPATIVE EQUALISATION

In Figure 2.19 we can see that battery 1 has SOC of 85%, battery 2 has SOC of 94% and battery 3 has SOC of 83%. Battery 3 has minimum SOC of 83%. The threshold value is set at 5%, the same as the earlier case. But this time only battery 2 will be discharged, as the SOC of battery 2 and minimum SOC has a difference of more than 5%. Hence from the graph we can see battery 2 being discharged till the difference is reduced to the threshold value. The voltage-time response (discharging of battery 2) is shown in Figure 2.20. We can also observe that battery 1 is being maintained at the same SOC as the value is within the threshold value.

From the graph shown in Figure 2.21, one can see that switch 1 and 2 turns on to discharge batteries 1 and 2, which was at SOC 92% and 94% respectively. Then the difference between minimum and the corresponding SOC values are calculated in the algorithm block. If the difference is more than the threshold value, the algorithm block generates control signals. The threshold value here is set as 5%. So, the batteries 1 and 2 will be more than the threshold values, as the minimum value among the three batteries in the battery pack is 83%. So, the switches of batteries 1 and 2 will be turned on. The batteries 1 and 2 will be discharged to the resistor. The voltage-Time Response (Discharging of batteries 1 and 2) is shown Figure 2.22.

The discharge continues until 88%, at which point both battery 1 and battery 2 will fall within the threshold range. Now we can see the difference in SOC values are now 5% with the minimum

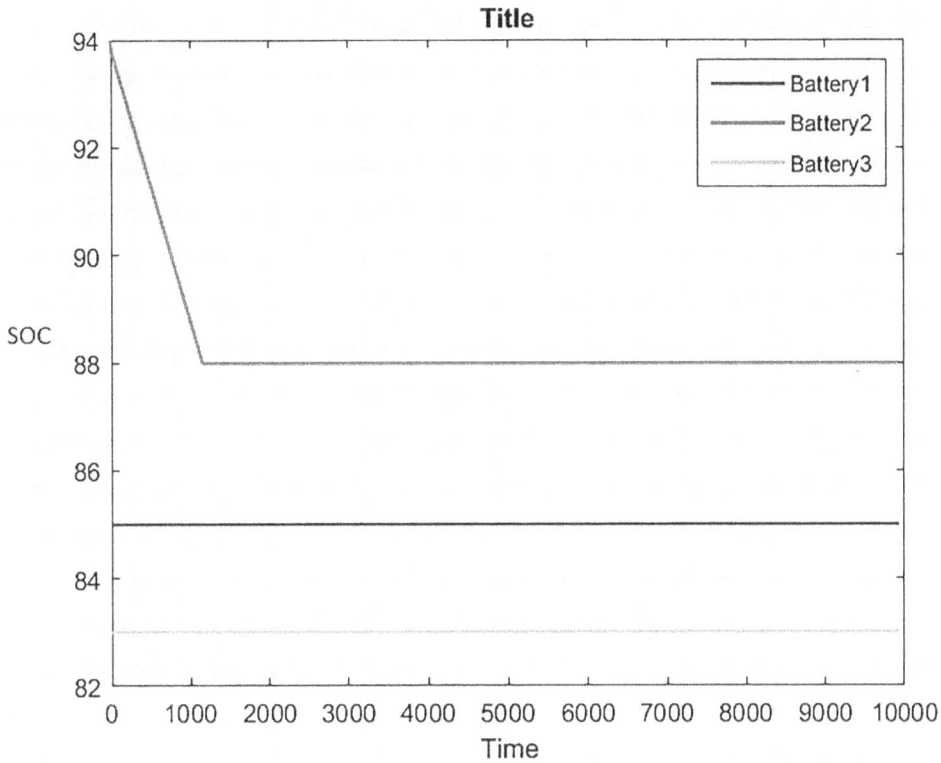

FIGURE 2.19   SOC-time response (Discharging of battery 2).

value of SOC. This algorithm can be applied to any number of cells in a battery pack. The point at which the discharging of the battery stops the SOC of batteries would have reached permissible limits. This is how dissipative equalisation works. Although it is a common and easy method of cell equalisation, and also less costly, this method has a disadvantage in that it generates heat which must be removed from the system. Hence it can be concluded that the dissipative cell equalising technique chooses that battery which is overcharged as compared to the other batteries, and the extra charge is being dissipated in a resistor.

In this simulation, one battery SOC was initialised to 89% (battery 1) and other two 95% (battery 2 and battery 3). Battery 2 and battery 3 were initialised with an SOC of 95%. This gives a situation in which batteries 2 and 3 were overcharged as compared to battery 1. Or the reverse is also true, that is, battery 1 was undercharged as compared to batteries 2 and 3. In both cases, battery 1 needs to be charged using the extra charge from batteries 2 and 3. The control algorithm first measures the SOC of all batteries and then finds the maximum and minimum values. The threshold limit is set as 5%. And the difference is found between maximum and minimum limits. Algorithm then check if it is more than the threshold limit. In this case the threshold limit is exceeded. The Algorithm block now have to connect battery 1 to output side of DC-DC converter. By default, the position of switches is that 1–1', 2–2', 3–3' are in off position, 4–4', 5–5', 6–6' are in on position, 7–7', 8–8', 9–9' are in off position. These connections are to make the series circuit in battery pack. In order to connect battery1 at output side we need to switch on 1–1' and 7–7' and at the same time switch of 4–4'. The algorithm block sends in control signals to the switches. The algorithm block then also determines the value of the duty cycle of the pulse to be given according to the charging voltage required and type of converter used. This simulation made use of a SEPIC converter. More about the same will be explained in coming sections. The undercharged battery is thus charged from overcharged battery using DC-DC converter.

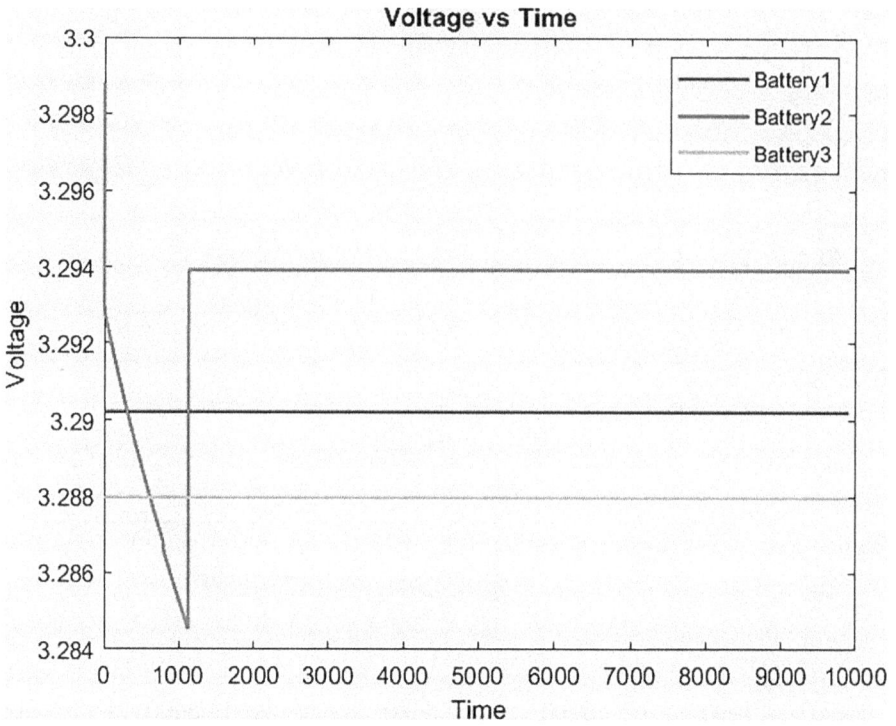

FIGURE 2.20   Voltage-time response (Discharging of battery 2).

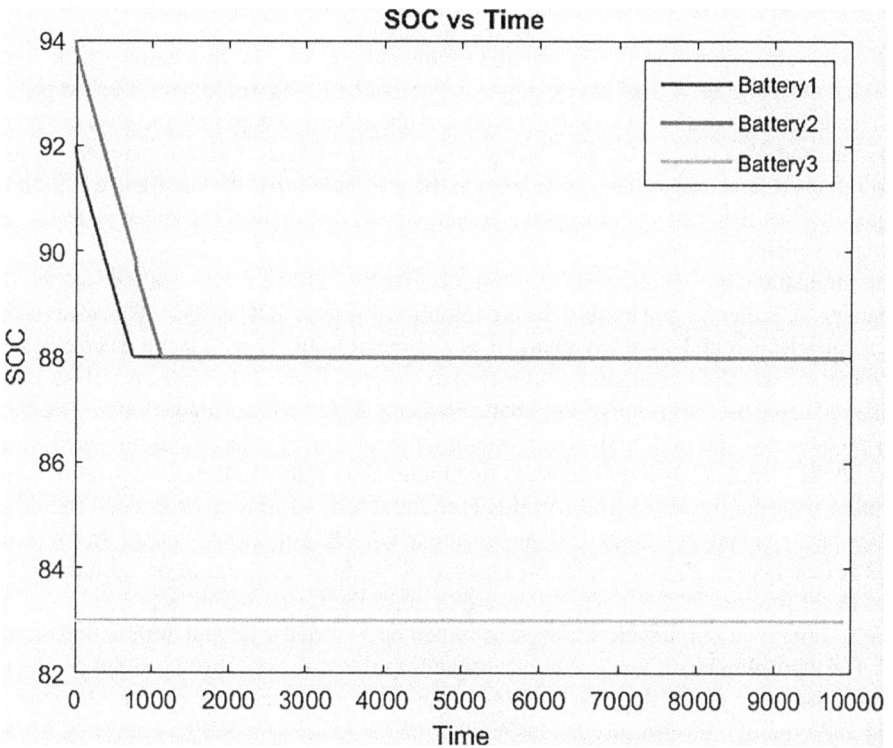

FIGURE 2.21   SOC-time response (Discharging of batteries 1 and 2).

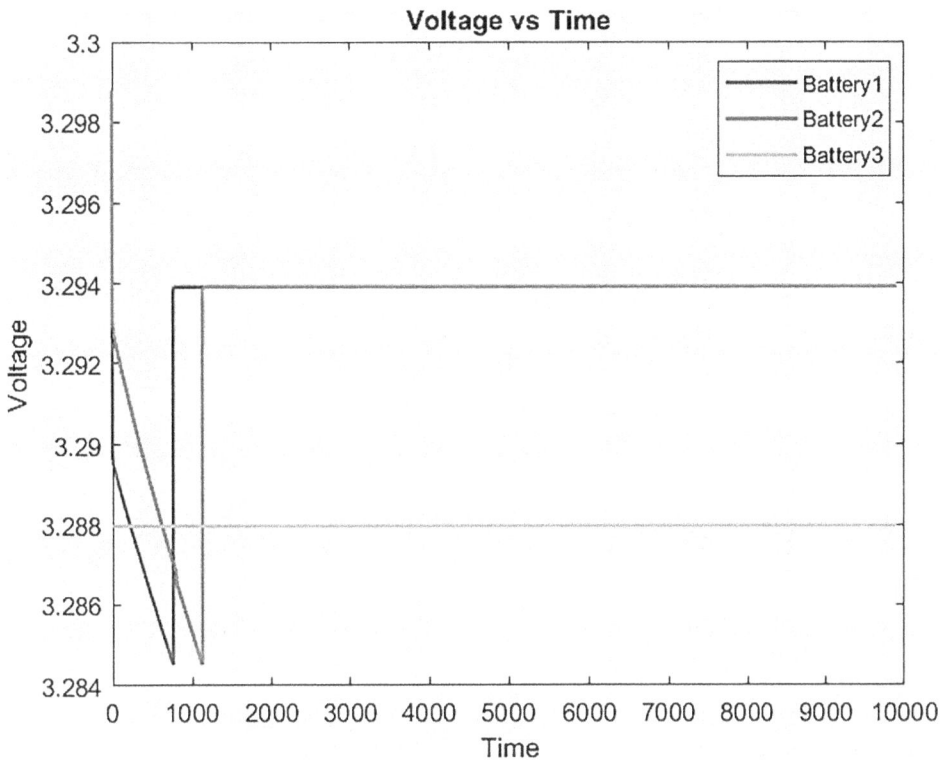

FIGURE 2.22   Voltage-Time Response (Discharging of batteries 1 and 2).

2.4.3 ACTIVE EQUALISING

The simulation of an active equalising circuit using DC-DC converter was completed. The battery SOCs were initialised at battery1 = 89%, battery 2 = 95% and battery 3 = 95%. This created a scenario in which battery 1 was undercharged as compared to battery 2 and 3. Batteries 2 and 3 had extra charge, which can be transferred to battery 1. This was done using the DC-DC SEPIC converter. The output response of battery1 was shown to decrease at first but then the SOC was found to rise. Similarly, the SOC of battery 2 and battery 3 was found to decrease. Since battery 1 is the output and batteries 2 and 3 form the output the charge in the input side which is being discharged is being used to charge battery 1. A drawback to this technique is the increased number of power electronic switches and complexity of the circuit. Also, this needs an extra microcontroller and algorithm to achieve equalising.

2.4.4 TEST RESULT OF BIDIRECTIONAL DC-DC CONVERTER

Simulation of the DC-DC converter for charging the battery was tested and the results are shown. The DC-DC converter was found to maintain almost a constant value of output voltage for a constant input voltage. Then the simulation of the same for a constant voltage method of charging was tested using a battery storage. The variation of SOC was plotted. The SOC response of battery 1, 2, and 3 are shown in Figures 2.23, 2.24, and 2.25, respectively. The SOC was found to increase for the same and the figures show the results.

The output of DC-DC converter simulation (SEPIC) is shown in Figure 2.26. The output voltage was found to be a constant value for a constant input DC voltage. For a 40 V input the output voltage was found to hold steady at 27 V for a duty cycle of 40%. The same design was used to charge

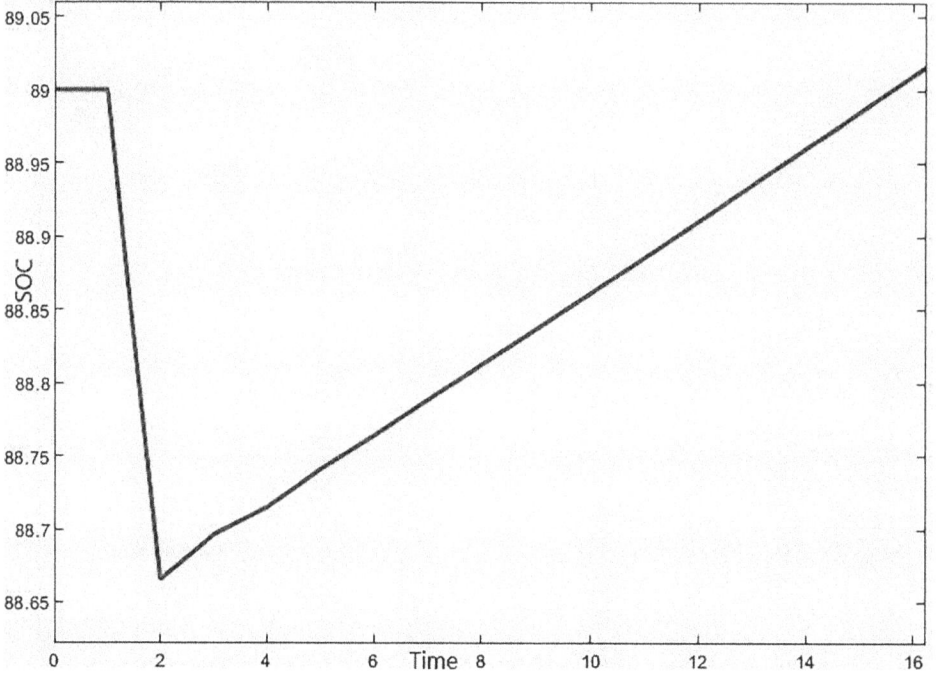

FIGURE 2.23 SOC response of battery 1.

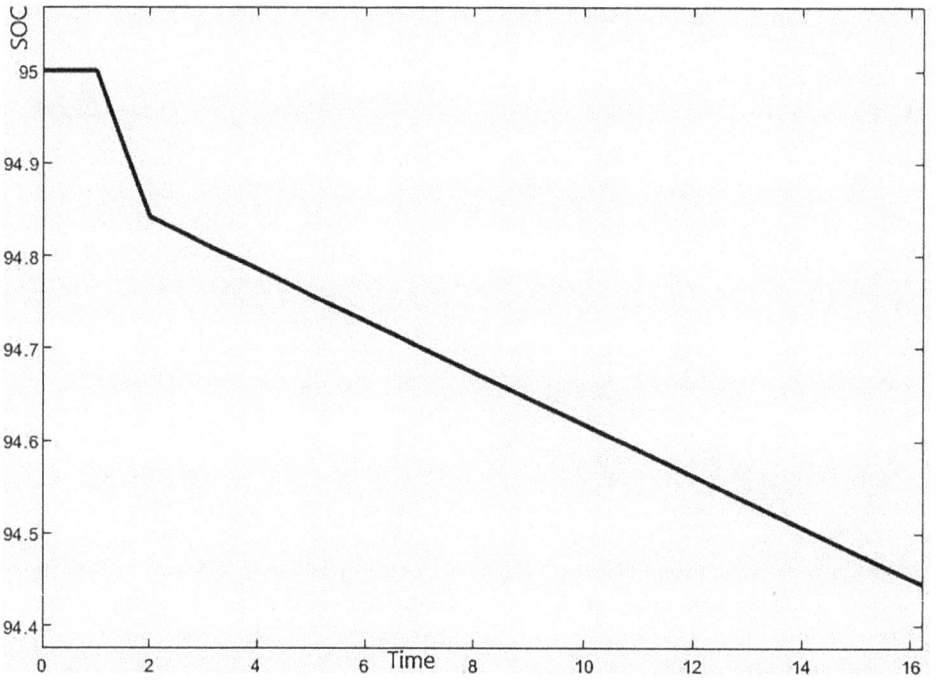

FIGURE 2.24 SOC response of battery 2.

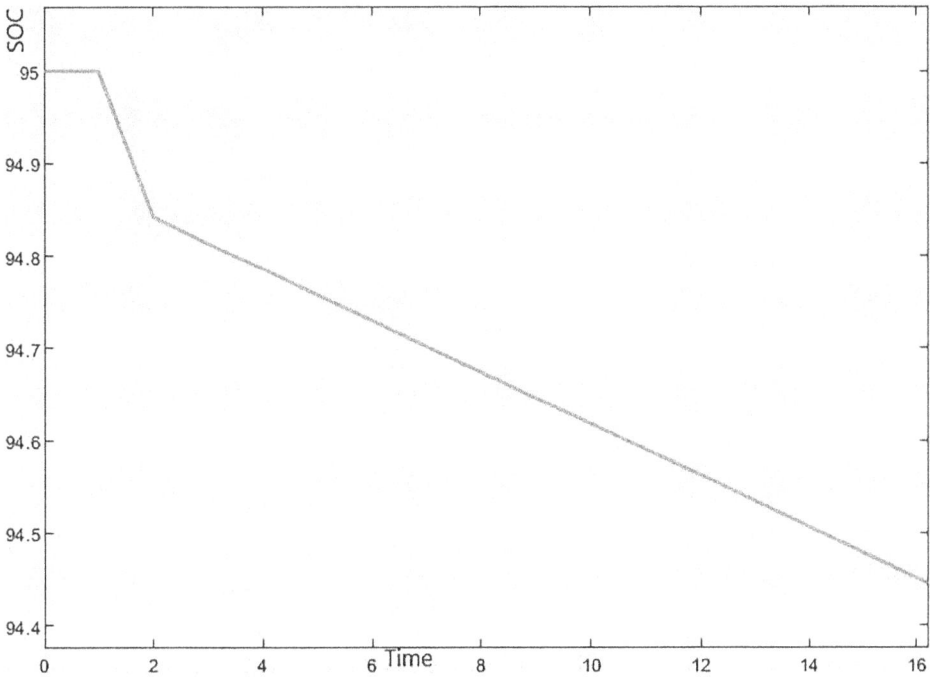

FIGURE 2.25   SOC response of battery 3.

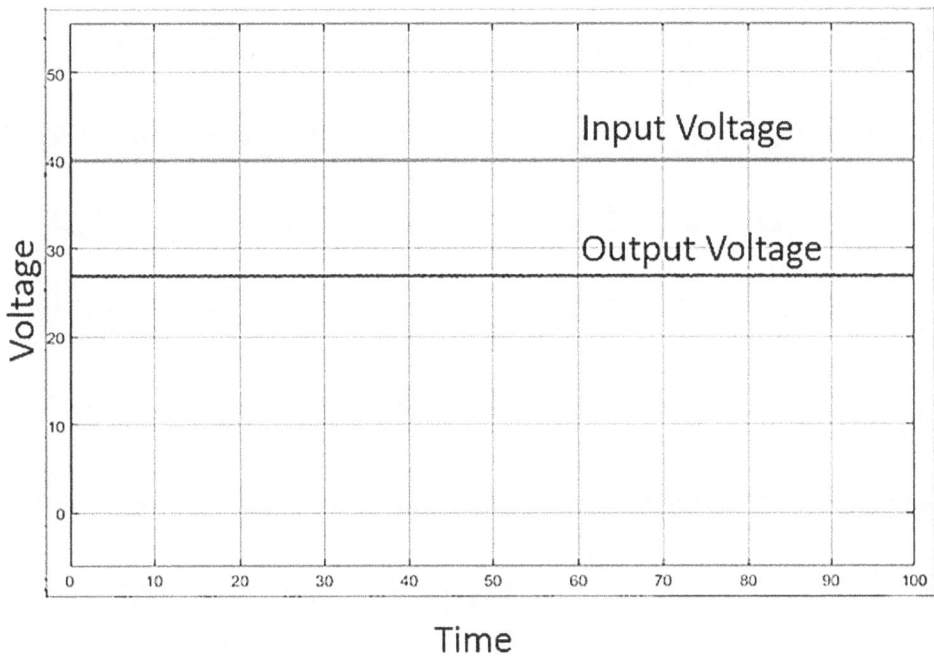

FIGURE 2.26   Simulation result of SEPIC converter.

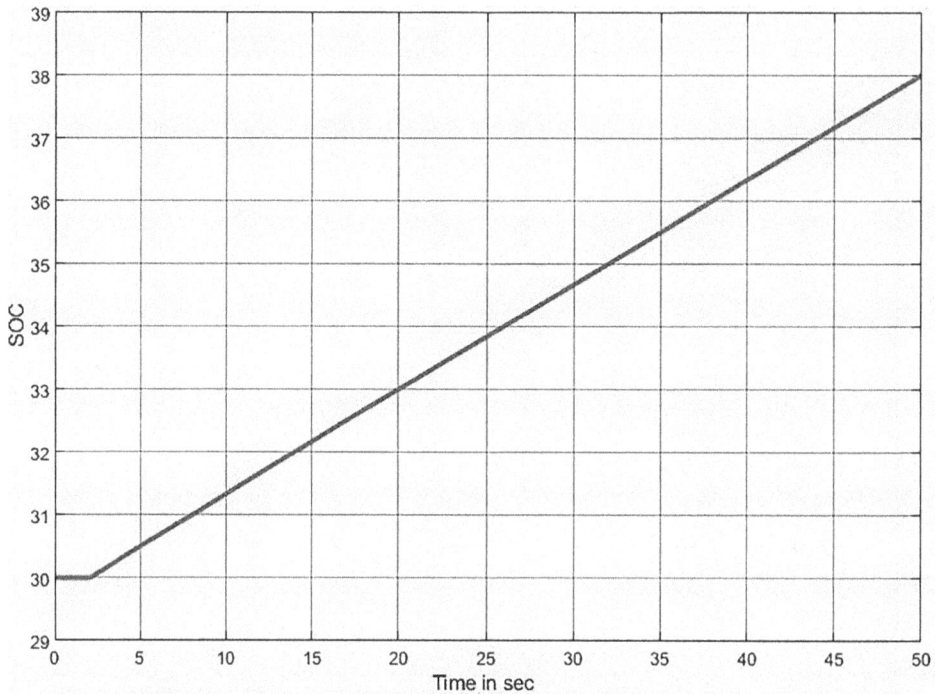

FIGURE 2.27   Variation of SOC while charging from SEPIC converter.

a battery and the output variation of SOC seemed to increase. The variation of SOC while charging from SEPIC converter is shown in Figure 2.27. This shows that the circuit can be used as a charger for constant voltage method of battery charging.

2.4.5   THE EXPERIMENTAL TEST RESULTS OF DC-DC CONVERTER

The bidirectional converter was tested for input voltages of 10 V and 15 V. The duty cycle of the gate driver circuit was set at 37.5%. The output produced a voltage of approximately 4 V, as displayed in Figure 2.28. The given outcome demonstrates the satisfactory operation, and the circuit can sustain a steady DC output if given a constant DC input voltage. Buck mode operation of SEPIC converter was successfully tested.

Bidirectional SEPIC converter was then tested for a different value of input voltage. This time the input was set at 15 V and the duty cycle was maintained the same. The SEPIC converter was found to change the output for any variation in input voltage. Buck mode operation for a different value of input voltage was found. The converter maintained a steady output voltage for a constant input in the second case also. This result is given in Figure 2.29.

The test data from the bidirectional converter was taken using a digital storage oscilloscope (DSO). The bidirectional converter should work in buck mode while charging. The bus voltage is the input voltage. The output voltage should be maintained at 27.6 for 24 V battery pack to achieve constant voltage charging. And while discharging the battery pack of 24 volt should be able to maintain the bus at a constant voltage of 40 V. The test result shows the working of SEPIC converter in buck mode as satisfactory.

2.5   CONCLUSIONS

BMS is still a growing field. The research in the field has shown the importance of battery management not only in electric vehicles but also as a storage device in renewable energy systems. The growing demand for electric vehicles is a result of the depleting petroleum product in the world. So,

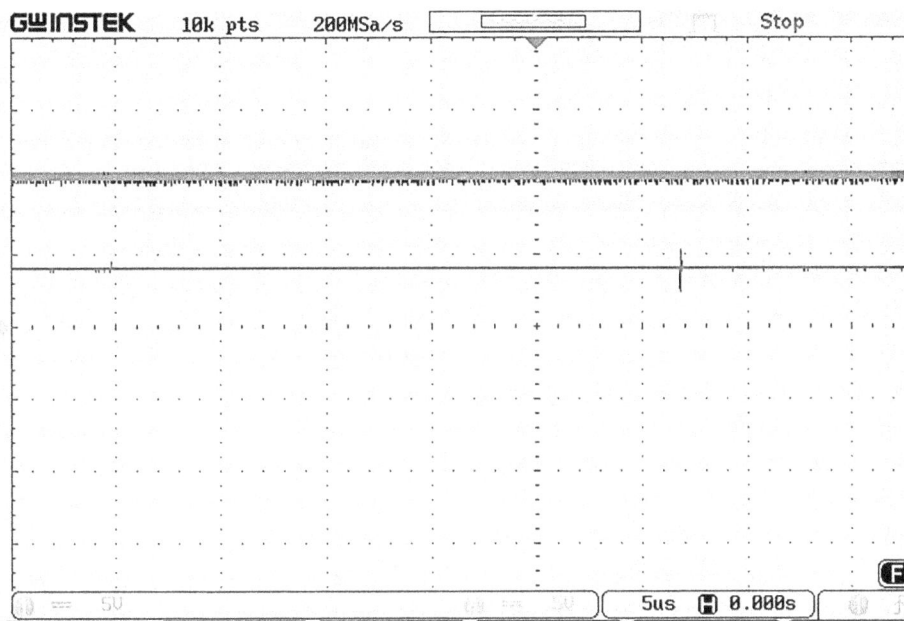

FIGURE 2.28   Bidirectional SEPIC converter test result (input voltage = 10 V).

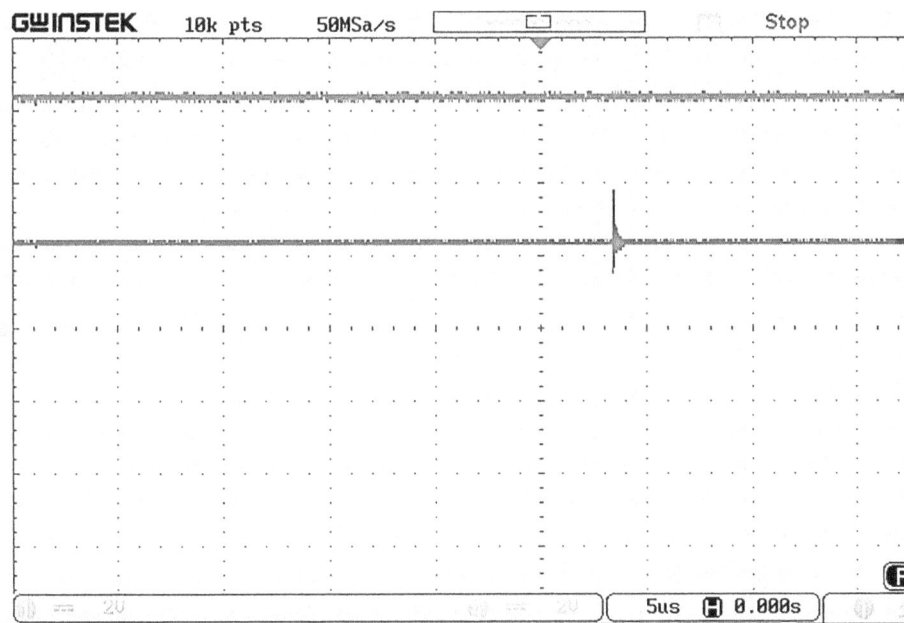

FIGURE 2.29   Bidirectional SEPIC converter test result (input voltage = 15 V).

in order to increase the reliability on EVs, more and more research is being done in the field of BMS. The preceding simulation the model was formed using mathematical components. The Simulink model of Li-ion battery was formed using the data from the experimental parameter extractions. The model formed was used in SOC calculation as well as the dissipative equalising developed. The characteristic of the battery formed was compared to the real physical battery characteristic and the characteristics showed closeness to the physical battery system. SOC calculation using coulomb counting was employed rather than the direct OCV measurement. In direct OCV measurement the OCV-SOC characteristics are found and SOC is directly measured from the preceding characteristic. But, in the coulomb counting method the SOC is calculated using a formula which is dependent on the current and initial SOC.

The dependencies of SOC on battery voltage and time were plotted for varying values of current. The graphs plotted showed that SOC is dependent on OCV of the battery but follows a nonlinear relationship. Also, it was found that if the discharge current is higher, the battery SOC will decrease at a much higher rate. In that case the terminal voltage of the battery will also fall to a certain value. The importance of these data lies in the fact that the SOC determines the running distance of the EV and also using SOC we can determine how much the battery needs to be charged and discharged so as to avoid any unforeseen circumstances. The simulation result showed the efficiency of the coulomb counting method of SOC calculation and also the battery response to different discharge currents. The dissipative cell equalising technique is the conventional technique used. It is simple and less costly than the active equalising techniques which make use of complicated electronic circuitry and complex algorithms. This technique uses a resistor connected to the battery to help the overcharged battery to discharge. The graph plotted for SOC vs time of batteries of a battery pack showed that only the battery whose SOC is more than the threshold difference is discharged, and other batteries remain the same. In the graph, batteries 1 and 2 had SOC 92% and 94% respectively. The threshold value was set as 5% and minimum value was 83%. Since the difference with minimum value of SOC is more than 5% threshold the control switches of those batteries will be turned on and the excess charge will be discharged to the resistor connected. In the second simulation graph, it is seen that battery 1 is at 85% and battery 2 at 94%. Minimum value is of battery 3, that is, 83%. But the difference between battery 1 and battery 3 does not fall above threshold value, hence this time the control switch of battery 1 will not turn on as selected by the algorithm. So SOC of battery 1 will remain the same. But battery 2 will be discharged as the difference is more than the threshold value in this circuit. Hence the dissipative cell equalising works to equalise batteries in a battery pack for the safe operations and to ensure the long life of battery storage devices. But since the dissipative cell equalising has a disadvantage in that the extra energy is being lost and generation of heat, this type of equalisation must be replaced with active equalising circuit.

A more advanced equalising technique using active equalising was formed. In this the extra charge that lies within a battery is used to charge the undercharged cell. This method is an advanced technique as compared to passive equalising, as the energy is not lost as heat in this circuit. Further research involves developing a more advanced active equalising circuit, thus reducing the cost and complexity of the circuit. Although the various switches used increase the complexity of the circuit, it can increase the efficiency and reliability of the battery management system. The simulation of SEIC converter found that the battery can be charged using constant voltage method of charging and the hardware test result showed that the bidirectional converter was able to maintain a constant DC output voltage if a constant DC input voltage was maintained. This showed that the operation of DC-DC converter was as expected with the simulation, and hence it can be used to charge a battery for constant voltage method of charging.

REFERENCES

[1] A. Merabet, K. T. Ahmed, H. Ibrahim, R. Beguenane, and A. M. Ghias. Energy management and control system for laboratory scale micro grid based wind-PV-battery. *IEEE Transactions on Sustainable Energy*, vol. 8, no. 1, pp. 145–154, 2016.

[2] C. R. Ananthraj, and A. Ghosh. Battery management system in electric vehicle. *2021 4th Biennial International Conference on Nascent Technologies in Engineering* (ICNTE), 2021, pp. 1–6.

[3] M. S. S. Chandra, L. V. Kumar, and S. Mohapatro. Voltage control and energy management of solar PV fed stand-alone low voltage DC microgrid for rural electrification. *IEEE National Power Systems Conference (NPSC)*, 2020, pp. 1–6.

[4] F. Azam, S. Kumar, and N. Priyadarshi. A framework for secured dissemination of messages in internet of vehicle using blockchain approach. *Innovations in Electronics and Communication Engineering: Proceedings of the 9th ICIECE 2021*, p. 401.

[5] K. W. E. Cheng, B. P. Divakar, H. Wu, K. Ding, and H. F. Ho. Battery Management System (BMS) and SOC development for electrical vehicles. *IEEE Transactions on Vehicular Technology*, vol. 60, no. 1, pp. 76–88, 2011.

[6] V. Gurugubelli, A. Ghosh, and A. K. Panda. Droop controlled voltage source converter with different classical controllers in voltage control loop. *2022 IEEE International Conference on Power Electronics, Smart Grid, and Renewable Energy (PESGRE)*, 2022, pp. 1–6, IEEE.

[7] M. A. Hannan, M. M. Hoque, A. Hussain, Y. Yusof, and P. J. Ker. State-of-the-art and energy management system of lithium-ion batteries in electric vehicle applications: Issues and recommendations. *IEEE Access*, vol. 6, pp. 19362–19378, 2018.

[8] A. Ghosh, S. Banerjee, M. K. Sarkar, and P. Dutta. Design and implementation of type-II and type-III controller for DC–DC switched mode boost converter by using K-factor approach and optimisation techniques. *IET Power Electronics*, vol. 9, no. 5, pp. 938–950, 2016.

[9] H. Tiwari, and A. Ghosh. Power flow control in solar PV fed DC microgrid with storage. *2020 IEEE 9th Power India International Conference (PIICON)*, 2020, pp. 1–6.

[10] J. Meher, and A. Ghosh. Comparative study of DC/DC bidirectional SEPIC converter with different controllers. *2018 IEEE 8th Power India International Conference (PIICON)*, 2018, pp. 1–6.

[11] G. Vikash, D. Funde, and A. Ghosh. Implementation of the virtual synchronous machine in grid-connected and stand-alone mode. *DC—DC Converters for Future Renewable Energy Systems*, Springer, Singapore, 2022, pp. 335–353.

[12] V. Gurugubelli, A. Ghosh, and A. K. Panda. Different oscillator-controlled parallel three-phase inverters in stand-alone microgrid. *Sustainable Energy and Technological Advancements*, Springer, Singapore, 2022, pp. 67–79.

[13] H. Zhang, Y. Wang, H. Qi, and J. Zhang. Active battery equalization method based on redundant battery for electric vehicles. *IEEE Transactions on Vehicular Technology*, vol. 68, no. 8, pp. 7531–7543, 2019.

[14] S. Patel, A. Ghosh, and P. K. Ray. Adaptive power management in PV/Battery integrated hybrid microgrid system. *2022 IEEE International Conference on Power Electronics, Smart Grid, and Renewable Energy (PESGRE)*, 2022, pp. 1–6.

[15] V. Gurugubelli, A. Ghosh, A. K. Panda, and S. Rudra. Implementation and comparison of droop control, virtual synchronous machine, and virtual oscillator control for parallel inverters in standalone microgrid. *International Transactions on Electrical Energy Systems*, vol. 31, no. 5, p. e12859, 2021.

[16] V. Gurugubelli, and A. Ghosh. Control of inverters in standalone andgrid-connected microgrid using different control strategies. *World Journal of Engineering*, 2021.

[17] G. Vikash, A. Ghosh, and S. Rudra. Integration of distributed generation to microgrid with virtual inertia. *2020 IEEE 17th India Council International Conference (INDICON)*, 2020, pp. 1–6, IEEE.

[18] G. Vikash, and A. Ghosh. Parallel inverters control in standalone microgrid using different droop control methodologies and virtual oscillator control. *Journal of the Institution of Engineers (India): Series B*, pp. 1–9, 2021.

[19] V. Gurugubelli, A. Ghosh, and A. K. Panda. Comparison of deadzone and vanderpol oscillator controlled voltage source inverters in islanded microgrid. *2021 IEEE 2nd International Conference on Smart Technologies for Power, Energy and Control (STPEC)*, 2021, pp. 1–6, IEEE.

[20] V. Gurugubelli, A. Ghosh, and A. K. Panda. Design of different classical controllers in the voltage control loop of a virtual synchronous machine in standalone mode. *2022 IEEE International Conference on Power Electronics, Smart Grid, and Renewable Energy (PESGRE)*, 2022, pp. 1–6, IEEE.

[21] N. Priyadarshi, S. Padmanaban, J. B. Holm-Nielsen, F. Blaabjerg, and M. S. Bhaskar. An experimental estimation of hybrid ANFIS–PSO-based MPPT for PV grid integration under fluctuating sun irradiance. *IEEE Systems Journal*, vol. 14, no. 1, pp. 1218–1229, 2019.

[22] S. Padmanaban, N. Priyadarshi, J. B. Holm-Nielsen, M. S. Bhaskar, F. Azam, A. K. Sharma, and E. Hossain. A novel modified sine-cosine optimized MPPT algorithm for grid integrated PV system under real operating conditions. *IEEE Access*, vol. 7, pp. 10467–10477, 2019.

[23] S. Padmanaban, N. Priyadarshi, M. S. Bhaskar, J. B. Holm-Nielsen, E. Hossain, and F. Azam. A hybrid photovoltaic-fuel cell for grid integration with jaya-based maximum power point tracking: Experimental performance evaluation. *IEEE Access*, vol. 7, pp. 82978–82990, 2019.

[24] N. Priyadarshi, V. K. Ramachandaramurthy, S. Padmanaban, and f. Azam. An ant colony optimized MPPT for standalone hybrid PV-wind power system with single Cuk converter. *Energies*, vol. 12, no. 1, p. 167, 2019.

[25] N. Priyadarshi, S. Padmanaban, L. Mihet-Popa, F. Blaabjerg, and F. Azam. Maximum power point tracking for brushless DC motor-driven photovoltaic pumping systems using a hybrid ANFIS-FLOWER pollination optimization algorithm. *Energies*, vol. 11, no. 5, p. 1067, 2018.

[26] N. Priyadarshi, A. K. Sharma, and F. Azam. A hybrid firefly-asymmetrical fuzzy logic controller based MPPT for PV-wind-fuel grid integration. *International Journal of Renewable Energy Research (IJRER)*, vol. 7, no. 4, pp. 1546–1560, 2017.

[27] N. Priyadarshi, A. Anand, A. Sharma, F. Azam, V. Singh, and R. Sinha. An experimental implementation and testing of GA based maximum power point tracking for PV system under varying ambient conditions using dSPACE DS 1104 controller. *International Journal of Renewable Energy Research (IJRER)*, vol. 7, no. 1, pp. 255–265, 2017.

[28] N. Priyadarshi, M. S. Bhaskar, S. Padmanaban, F. Blaabjerg, and F. Azam. New CUK–SEPIC converter based photovoltaic power system with hybrid GSA–PSO algorithm employing MPPT for water pumping applications. *IET Power Electronics*, vol. 13, no. 13, pp. 2824–2830, 2020.

[29] N. Priyadarshi, S. Padmanaban, M. S. Bhaskar, F. Blaabjerg, and A. Sharma. Fuzzy SVPWM-based inverter control realisation of grid integrated photovoltaic-wind system with fuzzy particle swarm optimisation maximum power point tracking algorithm for a grid-connected PV/wind power generation system: Hardware implementation. *IET Electric Power Applications*, vol. 12, no. 7, pp. 962–971, 2018.

[30] K. Kamalapathi, N. Priyadarshi, S. Padmanaban, J. B. Holm-Nielsen, F. Azam, C. Umayal, and V. K. Ramachandaramurthy. A hybrid moth-flame fuzzy logic controller based integrated cuk converter fed brushless DC motor for power factor correction. *Electronics*, vol. 7, no. 11, p. 288, 2018.

[31] N. Priyadarshi, S. Padmanaban, J. B. Holm-Nielsen, M. S. Bhaskar, and F. Azam. Internet of things augmented a novel PSO-employed modified zeta converter-based photovoltaic maximum power tracking system: hardware realisation. *IET Power Electronics*, vol. 13, no. 13, pp. 2775–2781, 2020.

[32] N. Priyadarshi, S. Padmanaban, D. M. Ionel, L. Mihet-Popa, and F. Azam. Hybrid PV-wind, micro-grid development using quasi-Z-source inverter modeling and control—experimental investigation. *Energies*, vol. 11, no. 9, p. 2277, 2018.

[33] F. Azam, S. K. Yadav, N. Priyadarshi, S. Padmanaban, and R. C. Bansal. A comprehensive review of authentication schemes in vehicular ad-hoc network. *IEEE Access*, vol. 9, pp. 31309–31321, 2021.

[34] F. Azam, N. Priyadarshi, H. Nagar, S. Kumar, and A. K. Bhoi. An overview of solar-powered electric vehicle charging in vehicular adhoc network. *Electric Vehicles*, pp. 95–102, 2021.

[35] N. Priyadarshi, F. Azam, A. K. Sharma, P. Chhawchharia, and P. R. Thakura. An interleaved ZCS supplied switched power converter for fuel cell-based electric vehicle propulsion system. *Advances in Smart Grid Automation and Industry 4.0*, pp. 355–362. Springer, Singapore, 2021.

[36] F. Azam, A. Biradar, N. Priyadarshi, S. Kumari, D. Almakhles, and S. Tangade. A framework for secured dissemination of messages in Internet of Vehicle (IoV) using blockchain approach. *2021 IEEE International Conference on Mobile Networks and Wireless Communications (ICMNWC)*, 2021, pp. 1–6.

[37] F. Azam, A. Biradar, N. Priyadarshi, S. Kumari, and S. Tangade. A review of blockchain based approach for secured communication in Internet of Vehicle (IoV) scenario. *2021 Second International Conference on Smart Technologies in Computing, Electrical and Electronics (ICSTCEE)*, 2021, pp. 1–6.

[38] N. Priyadarshi, M. S. Bhaskar, P. Sanjeevikumar, F. Azam, and B. Khan. High-power DC-DC converter with proposed HSFNA MPPT for photovoltaic based ultra-fast charging system of electric vehicles. *IET Renewable Power Generation*, 2022.

[39] V. Gurugubelli, A. Ghosh, and A. K. Panda. A new virtual oscillator control for synchronization of single-phase parallel inverters in islanded microgrid. *Energy Sources, Part A: Recovery, Utilization, and Environmental Effects*, vol. 44, no. 4, pp. 8842–8859, 2022.

[40] N. Priyadarshi, S. Padmanaban, M. S. Bhaskar, F. Azam, B. Khan, and M. G. Hussien. A novel hybrid grey wolf optimized fuzzy logic control based photovoltaic water pumping system. *IET Renewable Power Generation*, 2022.

[41] V. Gurugubelli, A. Ghosh, and A. K. Panda. Parallel inverter control using different conventional control methods and an improved virtual oscillator control method in a standalone microgrid. *Protection and Control of Modern Power Systems*, vol. 7, no. 1, pp. 1–13, 2022.

[42] N. Priyadarshi, P. Sanjeevikumar, M. S. Bhaskar, F. Azam, I. B. Taha, and M. G. Hussien. An adaptive TS-fuzzy model based RBF neural network learning for grid integrated photovoltaic applications. *IET Renewable Power Generation*, 2022.

[43] B. Sujith, A. Ghosh, and V. Gurugubelli. Design of PFC boost converter with stand-alone inverter for microgrid applications. In *2022 IEEE Delhi Section Conference (DELCON)*, IEEE, Februay 2022, pp. 1–5.

[44] J. K. Nayak, H. Thalla, and A. Ghosh. Efficient maximum power point tracking algorithms for photovoltaic systems with reduced number of sensors. *Process Integration and Optimization for Sustainability*, pp. 1–23, 2022.

[45] S. Patel, A. Ghosh, and P. K. Ray. Improved power flow management with proposed fuzzy integrated hybrid optimized fractional order cascaded proportional derivative filter (1+ proportional integral) controller in hybrid microgrid systems. *ISA Transactions*, 2022.

[46] D. Ravi, and A. Ghosh. Voltage mode control of buck converter using practical PID controller. In *2022 International Conference on Intelligent Controller and Computing for Smart Power (ICICCSP)*, IEEE, July 2022, pp. 1–6.

[47] S. Saurav, and A. Ghosh. Fourth order interleaved boost converter with PID, type II and type III controllers for smart grid applications. *Cyber-Physical Systems: Foundations and Techniques*, pp. 179–207, 2022.

[48] S. Joarder, and A. Ghosh. Design and implementation of dual active bridge converter for DC microgrid application. In *2022 IEEE Delhi Section Conference (DELCON)*, IEEE, February 2022, pp. 1–6.

[49] T. Barker, and A. Ghosh. Neural network-based PV powered electric vehicle charging station. In *2022 IEEE Delhi Section Conference (DELCON)*, IEEE, February 2022, pp. 1–6.

[50] A. Singh, and A. Ghosh. Comparison of quantitative feedback theory dependent controller with conventional PID and sliding mode controllers on DC-DC boost converter for microgrid applications. *Technology and Economics of Smart Grids and Sustainable Energy*, vol. 7, no. 1, pp. 1–12, 2022.

[51] H. Tiwari, A. Ghosh, P. K. Ray, B. Subudhi, G. Putrus, and M. Marzband. Direct power control of a three-phase AC-DC converter for grid-connected solar photovoltaic system. In *2021 International Symposium of Asian Control Association on Intelligent Robotics and Industrial Automation (IRIA)*, IEEE, September 2021, pp. 125–130.

3 Design and Control of Bidirectional SEPIC/ Zeta Converter

Jeeban Kumar Nayak, Vikash Gurugubelli, and Arnab Ghosh

CONTENTS

3.1 INTRODUCTION

In the present scenario, the demand of microgrid (MG) and nano grid (NG) is increasing instead of conventional power transmission system because it is inexpensive to maintain, less harmful to the environment, the power loss can be decreased in transmission and distribution systems and a good voltage regulation profile can be maintained. So, if we can control the cost of power generation and storage systems as minimum as possible, there will be no need for a conventional power system. As a result, MGs and NGs will quickly take over the entire power system. The main characteristics of these grids are that they integrate local generation, which increases reliability (It can operate in islanded mode of operation as well as in grid-connected mode.). The local generations may be solar energy, diesel generators (DGs), wind energy and so on. Other than DGs, all local remaining generations are non-polluting to the environment.

DOI: 10.1201/9781003323471-3

A bidirectional converter (BDC) is a device that transfers the power between the DC bus system and the energy storage system in either direction, depending upon the application. Applications of BDCs are hybrid electric vehicles, electric vehicles, and battery backup systems. So many researchers are working in this area to improve the efficiency of the BDC. BDC has two modes of operation, depending on the load demand. If the DC bus requires excess power to meet load demand, then BDC can allow flow of power from the energy storage element to load and the DC bus having excess power then the BDC can allow power flow from the dc bus to the storage system. In space applications, DC buses are backed up with rechargeable batteries or supercapacitors. These storage systems can connect to DC buses through a BDC. Depending on the application, one can use either isolated BDC or non-isolated BDC. BDC with soft switching will give more efficiency compared with normal switching.

The controllers are very important for the stable operation of MG. There are two modes of operation in the MG; one is islanded mode and another is grid-connected mode. Similarly, the BDC is also having two modes of operation. To satisfy all the conditions in the system we need several controllers. In an islanded mode, suppose the load demand is more than the local generation, then the BDC acts like a boost converter to transfer energy from the storage system to s local load. So, we need a control algorithm to decide which control can come into the picture, depending upon the load conditions and MG operation mode. Therefore, efficient control techniques are required to get the desired MG operation.

3.2   LITERATURE REVIEW

The SEPIC/Zeta converter is a dc-dc converter that operates in either BUCK mode or BOOST mode [1]. The irregular nature of these resources raises concerns about system reliability and stability. To address such issues, energy storage systems are required. These energy storage systems (ESSs) should be able to store energy in both directions, the surplus energy produced from solar energy, wind energy and so on, and when the generated energy is insufficient, let it go. A bidirectional dc to dc converter (BDC) is essential for ESS [2]. BDCs have been classified into two types in previous studies: non-isolated and isolated converters. By adjusting the isolated converter's turns ratio, high voltage conversion ratios can be obtained from isolated converters [3]. The dual active bridge (DAB) is replaced by a series resonant converter (SRCs). Regrettably, these converters employ a transformer for their operation, resulting in an increase in size and price [4]. The conversion ratio in a conventional buck-boost BDC is significantly reduced by parasitic capacitance and inductance, making it ineffective and appropriate for non-isolated applications [5].

In non-isolated BDC, buck-boost, multilevel converters, SEPIC/Zeta and so on are included. Multilevel converters require more components; the control circuit becomes more complicated [6]. Other research studies shows that BDC based on a coupled inductor is the best option for reducing stress across switches, because of the presence of storage energy in elements [7]. Bidirectional dc-dc converters are used in battery chargers to control the charging and discharging of the battery, and also to control the voltage regulate output voltage of the discharger when the stored energy is used again at a predetermined value of the battery [8]. The simulation yields a linear dynamic model with small-signal analysis of the multiple-input converter, from which the transfer functions are derived. It is possible to determine the type of feedback control design that will be used [9]. The method is basically on a small-signal pulse width modulation switched model in DCM combined with a pulse width modulation-based Zeta converter. This derived Zeta converter topology is the same as the original Zeta converter topology, in terms of its operational behavior, and serves solely for the modelling procedure [10]. Using the same power components to achieve bidirectional power flow provides simple and galvanically isolated topology in an efficient manner that is especially appealing for use in battery charging circuits in uninterruptible power supply (UPS) [11]. The steady-state and transient analysis for ideal and non-ideal components using average equations modelled in SSR, as well as continuous and conduction modes are discussed in [12]. A BDC is used to connect various storage devices, such as batteries, that meet during peak demand of power [13]. An interleaved BDC based on switching capacitor that combines a 3-ø interleaved circuit design with switched mode ultracapacitor is described in [14], [15]. That converter has characteristics of wide voltage gain, reduced ripples in current on the LV side, less power consumption and low voltage stresses in power switches [16]–[20].

3.3 METHODOLOGY

3.3.1 Bidirectional SEPIC/Zeta Converter

The bidirectional SEPIC/Zeta dc to dc (BSZDC) converter is shown in Figure 3.1. The proposed converter can transfer power in both directions depending on demand of the load. One side of the converter is battery and other side is DC bus. The converter has one pair of inductors, one pair of capacitors and one pair of switches. The switches are bidirectional switches; it can allow current in both directions, so the converter names itself as a bidirectional converter. MOSFET with body diode can act like a bidirectional switch. The applications of this converter are hybrid electric vehicles, electric vehicles and battery backup systems [21]–[55].

3.3.1.1 Converter Operation

The BSZDC topology operates in two modes. One is SEPIC mode or boost mode of operation and the second one is Zeta mode or buck mode of operation. In SEPIC mode the power feed is from battery to DC bus. In Zeta mode the power flows from DC bus to battery. In SEPIC mode the switch S_1 acts like a MOSFET and the switch S_2 acts like a diode as shown in Figure 3.2. In Zeta mode the switch S_2 acts like a MOSFET and the switch S_1 acts as a diode, as shown in the Figure 3.3. Both modes are same as conventional SEPIC or Zeta converters. In SEPIC mode the duty ratio should be greater than 0.5, because it can operate in a boost mode. Similarly, in Zeta mode, the duty ratio should be less than 0.5, because it can operate in a buck mode.

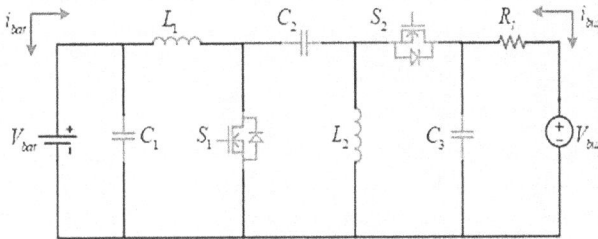

FIGURE 3.1 Bidirectional SEPIC/Zeta configuration.

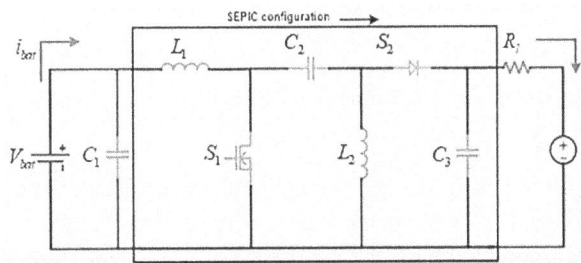

FIGURE 3.2 Bidirectional converter in SEPIC mode.

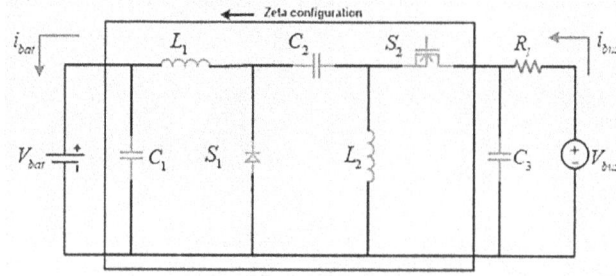

FIGURE 3.3 The bidirectional converter in Zeta mode of operation.

3.3.1.2 Circuit Parameter Design

Design of circuit parameter for conventional SEPIC or Zeta converter is known to us. In case of bidirectional converter, the circuit parameters should satisfy both modes of operation. Actually, the BSZDC converter acts like a SEPIC converter from left to right, as shown in Figure 3.2; similarly, it acts like a Zeta converter from right to left, as shown in Figure 3.3. The equations (1) to (4) should satisfy both modes of the BSZDC converter.

$$L_1 = \frac{V_{in}}{\Delta i_{L_1}} * DT_S \tag{1}$$

$$L_2 = \frac{V_{in}}{\Delta i_{L_2}} * DT_S \tag{2}$$

$$C_1 = \frac{1}{\Delta v_{C_1}}\left[I_{L_1} * D^1 T_s\right] \tag{3}$$

$$C_2 = \frac{I_0}{\Delta v_{C_2}} * DT_S \tag{4}$$

Where V_{in} is input voltage, if the BSZDC converter in SEPIC mode then V_{in} is replace with battery voltage (V_{bat}). D is the duty ratio, T_s is the switching period, I_{L1} is the average current in inductor L_1, and I_0 the load current, Δi_{L1} and Δi_{L2} are the maximum current ripples in the inductors L_1 and L_2 respectively. Δv_{C1} and Δv_{C2} are the maximum ripple voltage in the capacitors C_1 and C_2 respectively. Inductor values are designed by considering the maximum ripple is less than 30%, and the capacitor values are designed by taking the assumption that the maximum ripple is less than 2%. The designed circuit parameters as shown in Table 3.1.

In this study, the modelling of proposed converter was done by using the state space averaging technique.

3.3.2 SEPIC CONVERTER STATE SPACE AVERAGING MODEL

3.3.2.1 Mode-I (Switch S_1 On)

In this mode, switch (MOSFET) is on and diode is off due to short circuit across the circuit, as shown in Figure 3.4. Here the battery is used as a source so the battery voltage is considered as input, that is, $V_{bat} = V_{in}$, assuming the voltage across C_2 is the same as input voltage, that is, initially charged.

TABLE 3.1

Circuit Parameters of BSZDC Converter

Circuit parameters value

Battery voltage (V_{bat}) 30 V

Bus voltage (V_{bus}) 45 V

Inductor (L_1) 2 mH

Inductor (L_2) 2 mH

Capacitor(C_1) 200 μF

Capacitor(C_2) 150 μF

Capacitor(C_3) 200μF

Load resistance (R) 15 Ω

Switching frequency (f_s) 25 kHz

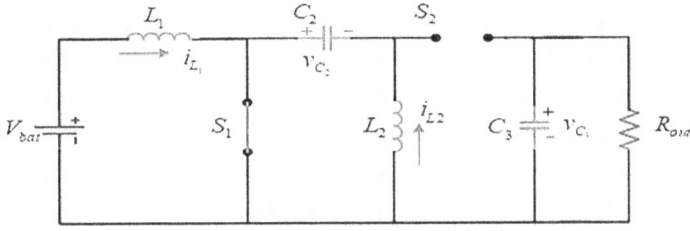

FIGURE 3.4   SEPIC mode during turn-on.

To get the state space matrix, all the state equations are to be writen in terms of state variables and input voltage. In this case, there are four state variables: i_{L1}, i_{L2}, v_{C2}, v_{C3}.

By applying KVL, the voltage across inductor L_1 and L_2 can be found.

$$V_{L1} = V_{in}$$

$$\frac{di_{L1}}{dt} = \frac{V_{in}}{L_1} \tag{i}$$

and $V_{L2} = v_{C2}$

$$\frac{di_{L2}}{dt} = \frac{v_{C2}}{L_2} \tag{ii}$$

By applying KCL, current through capacitor C_2 that is,

$$i_{C2} = -i_{L2}$$

$$\frac{dv_{c2}}{dt} = -\frac{i_{L2}}{C_2} \tag{iii}$$

And current through capacitor C_3 is

$$i_{C3} = \frac{-v_{C3}}{R_{out}C_3}$$

$$\frac{dv_{c3}}{dt} = -\frac{1}{R_{out}C_3} \qquad \qquad \text{(iv)}$$

From equations (i), (ii), (iii) and (iv), we can write the state matrix as

$$\begin{bmatrix} \dot{i}_{L_1} \\ \dot{i}_{L_2} \\ \dot{v}_{C_2} \\ \dot{v}_{C_3} \end{bmatrix} = \begin{bmatrix} 0 & 0 & 0 & 0 \\ 0 & 0 & \dfrac{1}{L_2} & 0 \\ 0 & \dfrac{-1}{C_2} & 0 & 0 \\ 0 & 0 & 0 & \dfrac{-1}{R_{out}C_3} \end{bmatrix} \begin{bmatrix} i_{L_1} \\ i_{L_2} \\ v_{C_2} \\ v_{C_3} \end{bmatrix} + \begin{bmatrix} \dfrac{1}{L_1} \\ 0 \\ 0 \\ 0 \end{bmatrix} V_{bat}$$

3.3.2.2 Mode-II (Switch Off)

From Figure 3.5, by applying KVL, we can have the voltage across inductor L_1 is

$$v_{L1} = -\left[v_{C2} + v_{C3}\right] + V_{bat}$$

$$\text{and } \frac{di_{L1}}{dt} = -\frac{1}{L_1}\left[v_{C2} + v_{C3}\right] + \frac{1}{L1}\left[V_{bat}\right] \qquad \qquad \text{(v)}$$

The voltage across inductor L_2 is

$$v_{L2} = v_{C3}$$

$$\text{and } \frac{di_{L2}}{dt} = -\frac{v_{C3}}{L_2} \qquad \qquad \text{(vi)}$$

By applying KCL, the current through capacitor C_2 is

$$i_{C2} = i_{L1}$$

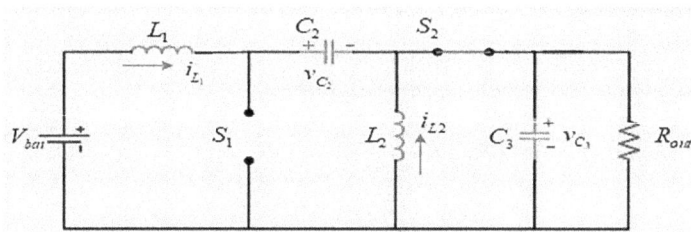

FIGURE 3.5 SEPIC mode during turn-off.

$$\text{and } \frac{dv_{C2}}{dt} = \frac{i_{L1}}{C_2} \qquad \text{(vii)}$$

the current through capacitor C_3 is

$$i_{C3} = \left(i_{L1} + i_{L2}\right) - \frac{v_{C3}}{R_{out}}$$

$$\text{and } \frac{dv_{C3}}{dt} = \frac{1}{C_3}\left[i_{L1} + i_{L2}\right] - \frac{v_{C3}}{R_{out}C_3} \qquad \text{(viii)}$$

Now, the state matrix can be written from the preceding equations (v), (vi), (vii) and (viii) as follows:

$$
\begin{bmatrix} \dot{i}_{L_1} \\ \dot{i}_{L_2} \\ \dot{v}_{C_2} \\ \dot{v}_{C_3} \end{bmatrix}
=
\begin{bmatrix}
0 & 0 & \frac{-1}{L_1} & \frac{-1}{L_1} \\
0 & 0 & 0 & \frac{-1}{L_2} \\
\frac{1}{C_2} & 0 & 0 & 0 \\
\frac{1}{C_3} & \frac{1}{C_3} & 0 & \frac{-1}{R_{out}C_3}
\end{bmatrix}
\begin{bmatrix} i_{L_1} \\ i_{L_2} \\ v_{C_2} \\ v_{C_3} \end{bmatrix}
+
\begin{bmatrix} \frac{1}{L_1} \\ 0 \\ 0 \\ 0 \end{bmatrix} V_{bat}
$$

Applying state space averaging method,

$$[A] = [A_1] \times D + [A_2] \times (1-D) \qquad \text{(a)}$$

$$[B] = [B_1] \times D + [B_2] \times (1-D) \qquad \text{(b)}$$

Where, $[A_1]$ - matrix A during ON period
$[A_2]$ - matrix A during OFF period
$[B_1]$ - matrix B during ON period
$[B_2]$ - matrix B during OFF period
D- duty cycle

The state space matrix after applying averaging method is given as follows:

$$
\begin{bmatrix} \dot{i}_{L_1} \\ \dot{i}_{L_2} \\ \dot{v}_{C_2} \\ \dot{v}_{C_3} \end{bmatrix}
=
\begin{bmatrix}
0 & 0 & \frac{-(1-D)}{L_1} & \frac{-(1-D)}{L_1} \\
0 & 0 & \frac{D}{L_2} & \frac{-(1-D)}{L_2} \\
\frac{(1-D)}{C_2} & \frac{-D}{C_2} & 0 & 0 \\
\frac{(1-D)}{C_3} & \frac{(1-D)}{C_3} & 0 & \frac{-1}{R_{out}C_3}
\end{bmatrix}
\begin{bmatrix} i_{L_1} \\ i_{L_2} \\ v_{C_2} \\ v_{C_3} \end{bmatrix}
+
\begin{bmatrix} \frac{1}{L_1} \\ 0 \\ 0 \\ 0 \end{bmatrix} V_{bat}
$$

3.3.2.3 Transfer Function of SEPIC Converter

The transfer function (TF) of the aforementioned SEPIC converter can be obtained with the help of MATLAB code. The analysis of the system is done by RL plot and step response plot. Transfer function is denoted by $G_1(s)$ and is given as follows:

$$G_1(s) = \frac{1.504 \times 10^6 s^2 - 4.274 \times 10^{-8} s + 2.256 \times 10^{12}}{s^4 + 501.3 s^3 + 2.503 \times 10^6 s^2 + 6.516 \times 10^8 s + 1.504 \times 10^{12}}$$

3.3.3 ZETA CONVERTER STATE SPACE AVERAGING MODEL

3.3.3.1 Mode-I (Switch On)

In this mode of operation, switch S_2 is turn-on and switch S_1 (diode) is off as shown in Figure 3.6. The bus voltage acts as input voltage to charge the battery in buck mode when bus power is more than load power. To get the state space matrix, all the state equations are to be written in terms of state variables and input voltage, that is, i_{L1}, i_{L2}, v_{C1}, v_{C2} and v_{bus}.

By applying KVL, the voltage across inductor L_1 is

$$v_{L1} = V_{bus} + v_{C2} - v_{C1}$$

$$\frac{di_{L1}}{dt} = \frac{1}{L1}\left(v_{C2} - v_{C1}\right) + \frac{1}{L1}V_{bus} \tag{a}$$

The voltage across inductor L_2 is

$$v_{L2} = V_{bus}$$

$$\text{and } \frac{di_{L2}}{dt} = \frac{V_{bus}}{L_2} \tag{b}$$

By applying KCL, current through capacitor C_1 is

$$i_{C1} = i_{L1} - \frac{v_{C1}}{R_{in}}$$

$$\text{and } \frac{dv_{C1}}{dt} = \frac{1}{C_1}\left[i_{L1} - \frac{v_{C1}}{R_{in}}\right] \tag{c}$$

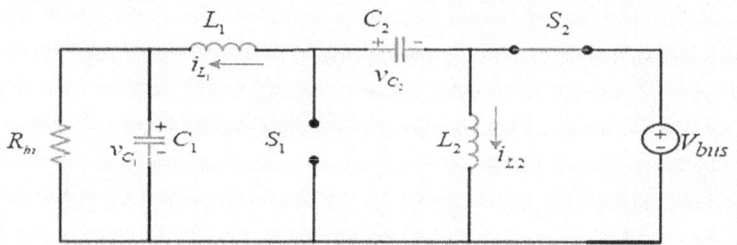

FIGURE 3.6 Zeta mode of operation during turn-on.

The current through capacitor C_2 is

$$i_{C2} = -i_{L1}$$

$$\text{and} \quad \frac{dv_{C2}}{dt} = -\frac{i_{L1}}{C_2} \tag{d}$$

From the preceding state equation (a), (b), (c) and (d), the SS matrix can be written as

$$
\begin{bmatrix} \dot{i}_{L_2} \\ \dot{i}_{L_1} \\ \dot{v}_{C_2} \\ \dot{v}_{C_1} \end{bmatrix} =
\begin{bmatrix}
0 & 0 & 0 & 0 \\
0 & 0 & \dfrac{1}{L_1} & \dfrac{-1}{L_1} \\
0 & \dfrac{-1}{C_2} & 0 & 0 \\
0 & \dfrac{1}{C_1} & 0 & \dfrac{-1}{R_{in}C_1}
\end{bmatrix}
\begin{bmatrix} i_{L_2} \\ i_{L_1} \\ v_{C_2} \\ v_{C_1} \end{bmatrix} +
\begin{bmatrix} \dfrac{1}{L_2} \\ \dfrac{1}{L_1} \\ 0 \\ 0 \end{bmatrix} V_{bus}
$$

3.3.3.2 Mode-II (Switch Off)

In this mode of operation, switch S_2 is off and switch S_1 (diode) is on, as depicted in Figure 3.7. The voltage across inductor L_1 is

$$v_{L1} = -\frac{v_{C1}}{L_1}$$

And, voltage across L_2 is

$$v_{L2} = -v_{C2}$$

By applying KCL, the current through capacitor C_1 is

$$i_{C1} = \frac{v_{C1}}{R_{in}}$$

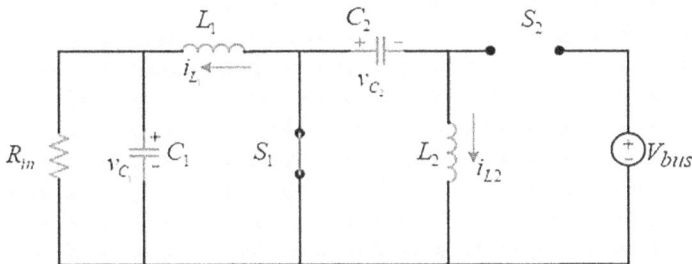

FIGURE 3.7　Zeta mode of operation during turn-off.

and current through capacitor C_2 is

$$i_{C2} = i_{L2}$$

State equations can be written with the help of the preceding equations

$$\frac{di_{L1}}{dt} = -\frac{v_{C1}}{L_1} \tag{P}$$

$$\frac{di_{L2}}{dt} = -\frac{v_{C2}}{L_2} \tag{Q}$$

$$\frac{dv_{C1}}{dt} = \frac{v_{C1}}{C_1 R_1} \tag{R}$$

$$\frac{dv_{C2}}{dt} = \frac{i_{L2}}{C_2} \tag{S}$$

The required state space matrix can be written from state equations (P), (Q), (R) and (S) as follows:

$$
\begin{bmatrix} \dot{i}_{L_2} \\ \dot{i}_{L_1} \\ \dot{v}_{C_2} \\ \dot{v}_{C_1} \end{bmatrix} =
\begin{bmatrix}
0 & 0 & \dfrac{-1}{L_2} & 0 \\
0 & 0 & 0 & \dfrac{-1}{L_1} \\
\dfrac{1}{C_2} & 0 & 0 & 0 \\
0 & \dfrac{1}{C_1} & 0 & \dfrac{-1}{R_{in}C_1}
\end{bmatrix}
\begin{bmatrix} i_{L_2} \\ i_{L_1} \\ v_{C_2} \\ v_{C_1} \end{bmatrix} +
\begin{bmatrix} 0 \\ 0 \\ 0 \\ 0 \end{bmatrix} V_{bus}
$$

Now, the averaged state space equation can be obtained by using the formula as mentioned in equations (a) and (b).

$$
\begin{bmatrix} \dot{i}_{L_2} \\ \dot{i}_{L_1} \\ \dot{v}_{C_2} \\ \dot{v}_{C_1} \end{bmatrix} =
\begin{bmatrix}
0 & 0 & \dfrac{-(1-D)}{L_2} & 0 \\
0 & 0 & \dfrac{D}{L_1} & \dfrac{-1}{L_1} \\
\dfrac{(1-D)}{C_2} & \dfrac{-D}{C_2} & 0 & 0 \\
0 & \dfrac{1}{C_1} & 0 & \dfrac{-1}{R_{in}C_1}
\end{bmatrix}
\begin{bmatrix} i_{L_2} \\ i_{L_1} \\ v_{C_2} \\ v_{C_1} \end{bmatrix} +
\begin{bmatrix} \dfrac{D}{L_2} \\ \dfrac{D}{L_1} \\ 0 \\ 0 \end{bmatrix} V_{bus}
$$

3.3.3.3 Transfer Function of Zeta Converter

The TF of the preceding Zeta converter can be found with the help of MATLAB code. The analysis of this system is done by RL plot and step response plot. The TF is denoted by $G_2(s)$ and given as follows:

$$G_2(s) = \frac{1.5 \times 10^6 s^2 + 2 \times 10^{12}}{s^4 + 333.3s^3 + 4.233 \times 10^6 s^2 + 5.778 \times 10^8 s + 1.333 \times 10^{12}}$$

3.4 RESULTS AND DISCUSSIONS

In this section, the characteristics of both SEPIC and Zeta topology are discussed through the root-locus diagram and step response. The transient behavior of the system is also studied by suddenly changing loads.

3.4.1 SEPIC MODE

The root-locus plot and step response for the transfer function mentioned in subsection 3.3.2.3 is given in Figure 3.8 and Figure 3.9 correspondingly. From the root-locus plot it is noticed that there are two closed loop poles (black colour squares) present in right half of S-plane the step response (SR) of the system increasing exponentially and getting unbounded, which makes the system unstable.

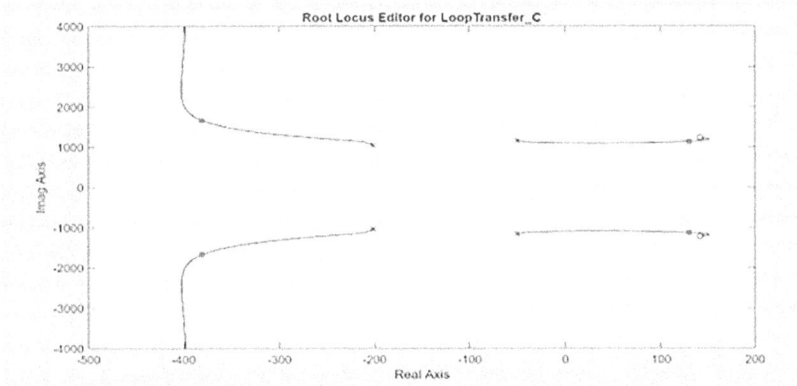

FIGURE 3.8   Root-locus diagram without controller in SEPIC mode.

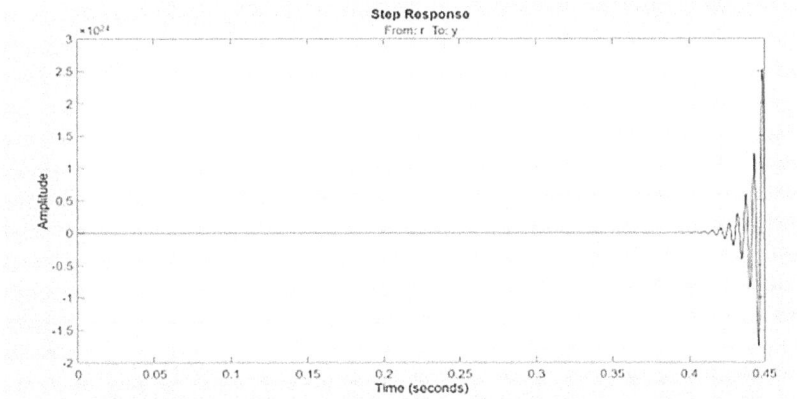

FIGURE 3.9   Step response without controller in SEPIC mode.

To make the system stable and robust, the IMC (Internal Model Controller) of second degree is used. The output of the controller is fed to a PWM generator and then to the gate terminal of the switch (here MOSFET). Duty ratio 'D' will change as per output voltage by the controller to maintain constant output voltage. The TF of the controller is denoted by $C_1(s)$ and given as follows:

$$C_1(s) = \frac{0.048131(s^2 + 243.4s + 1.028 \times 10^6)}{s^2 + 575.4s}$$

After using the controller with the preceding transfer function, the root-locus plot and step response is shown in Figure 3.10 and Figure 3.11, respectively. From the root-locus plot, it is clearly seen that all closed-loop poles (shown in black colour squares) are present on the negative side of S-plane and nearer to the origin, which implies the steady state error is much less and stability of the system is improved.

From step response characteristics, it is noticed that the rise time (T_r) and settling time (T_s) are improved, that is, rise time is 0.0125 sec or 12.5 millisec and settling time is 0.0213 sec or 21.3 millisec.

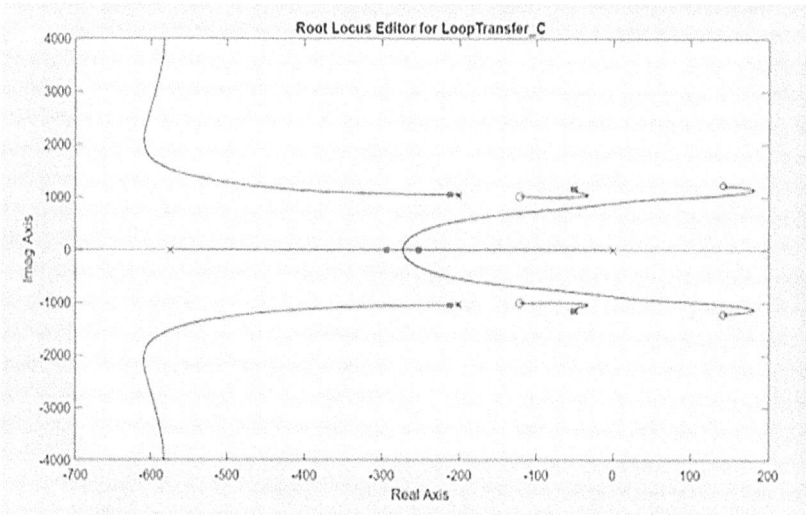

FIGURE 3.10   Root-locus diagram with controller in SEPIC mode.

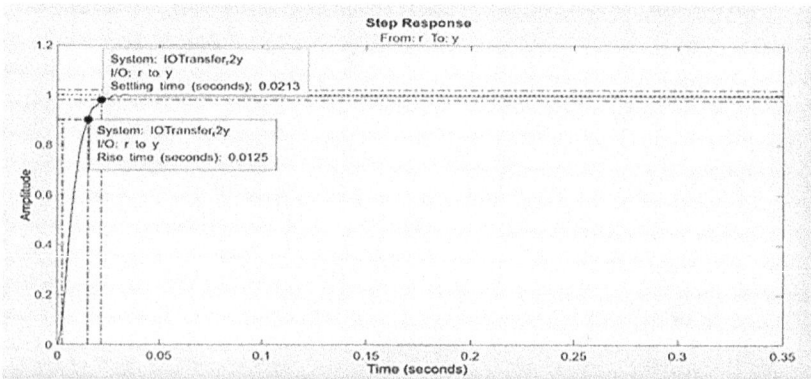

FIGURE 3.11   Step response with controller in SEPIC mode.

3.4.1.1 Transient Behavior/Response of the System with Variation of Load

With the implementation of IMC controller, wide range of load variation without affecting the stability can be achieved. The load variation is done by connecting different resistances in parallel and with the help of switching action, increasing and decreasing of load is achieved in different time interval, as shown in Figure 3.12. To verify this, three loads are taken: R1, R2 and R3. The values and states of load are given in Table 3.2. Here in the figure, '0' means load is OFF and '1' means load is ON. The load R1 is in for whole time period, R2 is ON at 0.4 sec and OFF at 0.5 sec, R3 is on at 0.8 sec and off at 0.9 sec. The values or magnitude of loads is given in Table 3.2.

From Figure 3.13, it is observed that at steady state, when the load is 15 ohm, the load voltage is 45 volt and load current is 3 amp. But when the load R2 = 20 ohm is suddenly on at 0.4 sec, there is a sudden drop in voltage and load current suddenly increases to balance the constant power, and within 0.1 sec it again comes to steady state with the load current of 5.25 amp. At 0.5 sec, the 20-ohm load is cut off, so the load current decreases suddenly and load voltage increases to maintain the constant power.

At 0.8 sec, the load R1 is already in ON condition and R2 is in OFF condition, load R3 is suddenly on; as a result the equivalent load resistance became 10 ohm. After addition of extra load, there is a spike in voltage and within 0.1 sec it comes to steady state: 45 v.

The base load for the propose converter is 15 ohm. However, the load variation of about 33% can be achieved without losing the stability and steady state error. From the preceding discussion it is observed that during decrease in load (increase in load resistance), the transient time is slightly more compared to increase in load (decrease in load resistance).

3.4.2 ZETA MODE

This is the stepdown mode of converter. In this mode, batteries can be charged, or other loads with rated output voltage can be used. The stability and steady state error (SSE) analysis is observed by analyzing the RL (root-locus) diagram and step response (SR) of the system.

The RL plot and SR of the system mentioned in subsection 3.3.3.3 is shown in Figure 3.14 and Figure 3.15 respectively. It is observed that all closed-loop poles lie in the left half of S-plane and

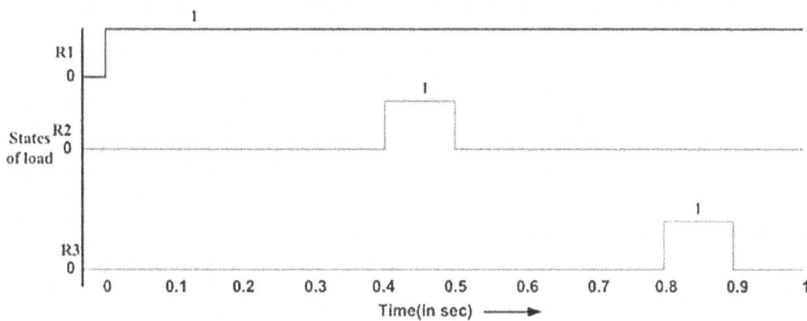

FIGURE 3.12 States of load during different time intervals.

TABLE 3.2
Turn On and Turn Off Time of Load

Loads	Values (ohm)	ON time (sec)	OFF time (sec)
R1	15	0	–
R2	20	0.4	0.5
R3	30	0.8	0.9

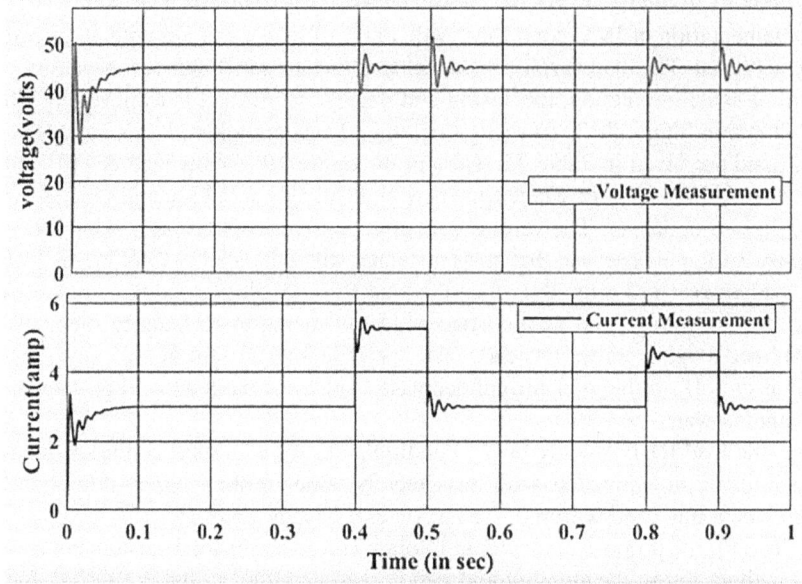

FIGURE 3.13   Load voltage and current waveform with variation of load in SEPIC mode.

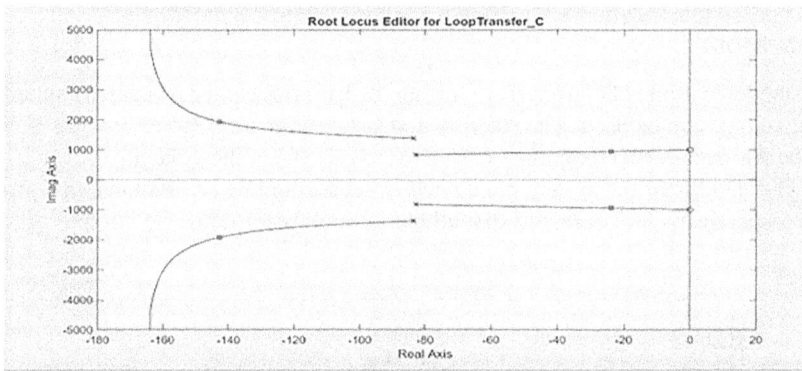

FIGURE 3.14   Root-locus diagram without controller in Zeta mode.

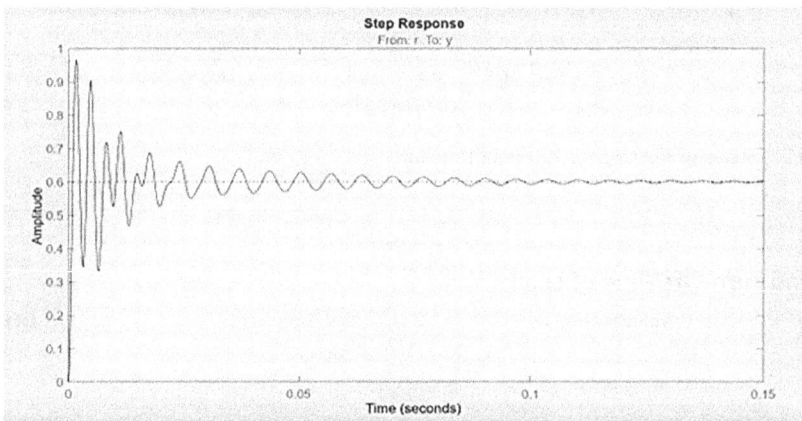

FIGURE 3.15   Step response without controller in Zeta mode.

the system is in stable condition, but the SSE is present at the output and settling time is more than 100 millisec, as shown in Figure 3.16.

To improve the transient behavior and SSE of the system, a PID controller, with proper tuning, is used in the proposed model. The value of K_p = 160, K_i = 360 and K_d = 150.

Where, K_p = Gain of proportionality
K_i = Gain of integral controller
K_d = Gain of differential controller

After introducing the PID controller into the system, the RL plot and SR of the system (see Figure 3.16 and Figure 3.17), it is clearly observed that closed-loop poles (black colour squares) are introduced in negative real axis and SSE is also zero. The rise time and settling time is 0.00927 sec or 9.27 millisec and 0.0356 sec or 35 millisec.

3.4.2.1 Transient Behavior/Response Analysis

Here also the transient response of the system is studied by varying the load in the same time interval as in case SEPIC the values of load resistance are different. The load R1 is ON for the whole time period, R2 turns ON at 0.4 sec and turns OFF at 0.5 sec, R3 turns ON at 0.8 sec and turns OFF at 0.9 sec. The values of load resistance, turn ON time and turn OFF time are given in Table 3.3 and Figure 3.18.

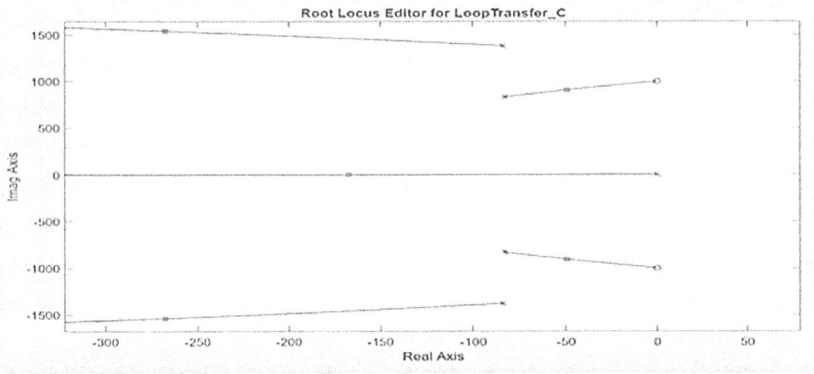

FIGURE 3.16   Root-locus diagram with controller in Zeta mode.

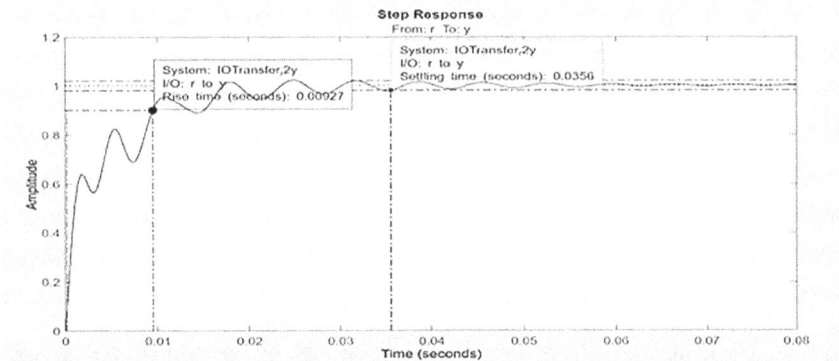

FIGURE 3.17   Step response with controller in Zeta mode.

TABLE 3.3

Values of Load Resistance, Turn-On and Turn-Off Time of Loads in Zeta Mode

Loads	Values (ohm)	ON time (in sec)	OFF time (in sec)
R1	60	0	–
R2	50	0.4	0.5
R3	30	0.8	0.9

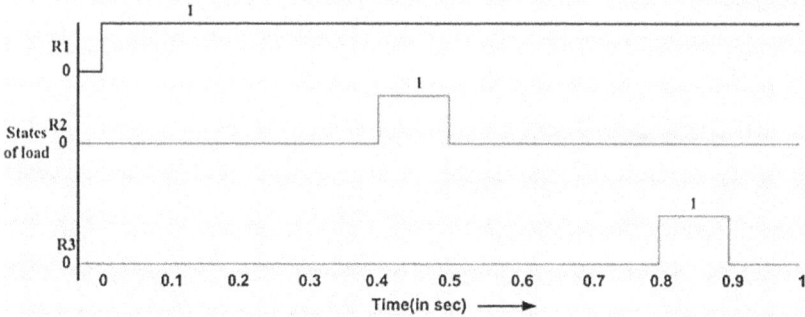

FIGURE 3.18   Load variation in different time interval.

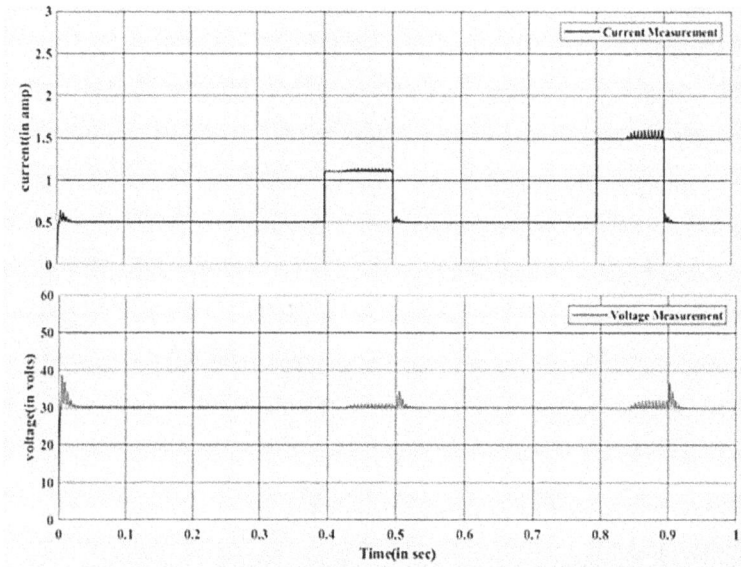

FIGURE 3.19   Load voltage and current waveform with variation of load in Zeta topology.

From Figure 3.19, it is observed that, initially, the steady states come within 0.5 sec, that is, load voltage is 30 V and load current is 0.5 amp for 60 ohms of load. When load R2 = 50 ohm is connected at time 0.4 sec, the equivalent resistance becomes 27.27 ohm and voltage is 30 V, the load current increases to 1.1 amp. Momentarily, there are some ripples present in output voltage during increase in load. The load current decreases to 0.5 amp at time 0.5 sec as load is decreased but the average load terminal voltage remains constant at 30 V.

When the load R3 = 30 ohm is connected at time 0.8 sec the equivalent load becomes 20 ohm and the load current increases to 1.5 amp. The transient during switching operation dies out within 0.5 sec and the system comes to a stable condition.

From the preceding transient behaviour of the Zeta topology with PID controller it is observed that the load variation of 33% can be achieved without affecting the terminal voltage and stability of the system.

3.5 CONCLUSION

For the bidirectional SEPIC/Zeta converter, the stability analysis, transient analysis and steady state analysis are studied through root-locus diagram and step response of the system. It is analyzed that the transient characteristics, that is, rise time and settling time with 2% error tolerance of the system is improved. The behaviour of the converter, when load is varying, is studied in both cases and it is observed that due to load variation, a transient in the circuit occurs for a few milliseconds, but the stability and steady state response remain unaffected.

The operational frequency of the proposed topology is 25 kHz. Future work will include study on the efficiency of the system by measuring the losses with high operational frequency and how losses will reduce to increase efficiency of the system.

ACKNOWLEDGEMENT

The idea of this work is supported by DST project (SP/YO/2019/1349).

REFERENCES

[1] L. S. Yang, H. Liang, and T. Liang. Analysis and implementation of a novel bidirectional DC–DC converter. *IEEE Transactions on Industrial Electronics*, vol. 59, no.1, pp. 422–434, 2011.

[2] M. Shahin. Analysis of bidirectional SEPIC/Zeta converter with coupled inductor. *proceedings of International Conference on Technological Advancements in Power and Energy (TAP Energy)*, pp. 103–108, 2015.

[3] G. Chen, Y. S. Lee, S. Y. R. Hui, and D. Xu. Actively clamped bidirectional flyback converter. *IEEE Transactions on Industrial Electronics*, vol. 47, no. 4, pp. 770–779, 2000.

[4] F. M. Lbanez, J. M. Echeverria, J. Valido, and L. Fontan. A step-up bidirectional series resonant DC/DC converter using a continuous current mode. *IEEE Transactions on Power Electronics*, vol. 30, no. 3, pp. 1393–1402, 2014.

[5] H. Y. Lee, T. J. Liang, J. F. Chen, and K. H. Chen. Design and implementation of a bidirectional SEPIC-Zeta DC-DC Converter. *IEEE International Symposium on Circuits and Systems (ISCAS)*, pp. 101–104, 2014.

[6] K. Filsoof, and P. W. Lehn. A bidirectional modular multilevel DC-DC converter of triangular structure. *IEEE Transactions on Power Electronics*, vol. 30, no. 1, pp. 54–64, 2015.

[7] I. D. Kim, S. H. Paeng, J. W. Ahn, E. C. Nho, and J. S. Ko. New bidirectional ZVS PWM SEPIC/Zeta DC-DC converter. *Proceedings IEEE Industrial Electronics, Vigo*, 2007, pp. 555–560.

[8] I. D. Kim, Y. H. Lee, B. H. Min, E. C. Nho, and J. W. Ahn. Design of bidirectional PWM SEPIC/Zeta DC-DC converter. *Proceeding of International Conference on Power Electronics*, 2007, pp. 614–619.

[9] J. S. Salenga, and E. R. Magsino. Dynamic analysis of two input Zeta converter topology for modular hybrid PV-wind microgrid system. *Proceeding of TECON IEEE Region 10 Conference*, 2015, pp. 1–6.

[10] V. Gurugubelli, and A. Ghosh. Control of inverters in standalone and grid-connected microgrid using different control strategies. *World Journal of Engineering*, July 2021.

[11] E. Niculescu, M. C. Niculescu, and D. M. Purcaru. Modeling the PWM Zeta converter in discontinuous conduction mode. *Proceeding of MELECON 14th IEEE Mediterranean Electrotechnical Conference*, 2008, pp. 651–657.

[12] M Jain, M. Daniele, and P. K. Jain. A bidirectional DC-DC Converter topology for low power application. *IEEE Transactions on Power Electronics*, vol. 15, pp. 595–606, 2000.

[13] D. Ravi, S. S. Letha, P. Samuel, and B. M. Reddy. An overview of various DC-DC converter techniques used for fuel cell-based applications. *International conference on Power Energy, Environment and Intelligent control*, pp. 16–21, 2018.

[14] V. Gurugubelli, A. Ghosh, A. K. Panda, and S. Rudra. Implementation and comparison of droop control, virtual synchronous machine, and virtual oscillator control for parallel inverters in standalone microgrid. *International Transactions on Electrical Energy Systems*, vol. 31, no. 5, p. e12859, 2021.

[15] M. Venmathi, and R. Ramaprabha. Implementation of SEPIC/Zeta three-port bidirectional DC-DC converter for renewable energy applications.

[16] Y. Zhang, W. Zhang, F. Gao, S. Gao, and D. J. Rogers. A switched capacitor interleaved bidirectional converter with wide voltage-gain range for super capacitors in EVs. *IEEE Transaction on Power Electronics*, vol. 35, no. 2, pp. 1536–1547, 2019.

[17] G. Vikash, and A. Ghosh. Parallel inverters control in standalone microgrid using different droop control methodologies and virtual oscillator control. *Journal of the Institution of Engineers (India): Series B*, pp. 1–9, 2021.

[18] G. Vikash, D. Funde, and A. Ghosh. Implementation of the virtual synchronous machine in grid-connected and stand-alone mode. In *DC—DC Converters for Future Renewable Energy Systems*, Springer, Singapore, 2022, pp. 335–353.

[19] N. Priyadarshi, S. Padmanaban, P. K. Maroti, and A. Sharma. An extensive practical investigation of FPSO-based MPPT for grid integrated PV system under variable operating conditions with anti-islanding protection. *IEEE Systems Journal*, vol. 13, no. 2, pp. 1861–1871, 2018.

[20] N. Priyadarshi, S. Padmanaban, J. B. Holm-Nielsen, F. Blaabjerg, and M. S. Bhaskar. An experimental estimation of hybrid ANFIS–PSO-based MPPT for PV grid integration under fluctuating sun irradiance. *IEEE Systems Journal*, vol. 14, no. 1, pp. 1218–1229, 2019.

[21] S. Padmanaban, N. Priyadarshi, J. B. Holm-Nielsen, M. S. Bhaskar, F. Azam, A. K. Sharma, and E. Hossain. A novel modified sine-cosine optimized MPPT algorithm for grid integrated PV system under real operating conditions. *IEEE Access*, vol. 7, pp. 10467–10477, 2019.

[22] S. Padmanaban, N. Priyadarshi, M. S. Bhaskar, J. B. Holm-Nielsen, E. Hossain, and F. Azam. A hybrid photovoltaic-fuel cell for grid integration with jaya-based maximum power point tracking: Experimental performance evaluation. *IEEE Access*, vol. 7, pp. 82978–82990, 2019.

[23] N. Priyadarshi, V. K. Ramachandaramurthy, S. Padmanaban, and F. Azam. An ant colony optimized MPPT for standalone hybrid PV-wind power system with single Cuk converter. *Energies*, vol. 12, no. 1, p. 167, 2019.

[24] N. Priyadarshi, S. Padmanaban, L. Mihet-Popa, F. Blaabjerg, and F. Azam. Maximum power point tracking for brushless DC motor-driven photovoltaic pumping systems using a hybrid ANFIS-FLOWER pollination optimization algorithm. *Energies*, vol. 11, no. 5, p. 1067, 2018.

[25] N. Priyadarshi, A. K. Sharma, and F. Azam. A hybrid firefly-asymmetrical fuzzy logic controller based MPPT for PV-wind-fuel grid integration. *International Journal of Renewable Energy Research (IJRER)*, vol. 7, no. 4, pp. 1546–1560, 2017.

[26] F. Azam, S. Kumar, and N. Priyadarshi. A framework for secured dissemination of messages in internet of vehicle using blockchain approach. *Innovations in Electronics and Communication Engineering: Proceedings of the 9th ICIECE 2021*, p. 401.

[27] N. Priyadarshi, A. Anand, A. Sharma, F. Azam, V. Singh, and R. Sinha. An experimental implementation and testing of GA based maximum power point tracking for PV system under varying ambient conditions using dSPACE DS 1104 controller. *International Journal of Renewable Energy Research (IJRER)*, vol. 7, no. 1, pp. 255–265, 2017.

[28] N. Priyadarshi, M. S. Bhaskar, S. Padmanaban, F. Blaabjerg, and F. Azam. New CUK–SEPIC converter based photovoltaic power system with hybrid GSA–PSO algorithm employing MPPT for water pumping applications. *IET Power Electronics*, vol. 13, no. 13, pp. 2824–2830, 2020.

[29] F. Azam, A. Biradar, N. Priyadarshi, S. Kumari, and S. Tangade. A review of blockchain based approach for secured communication in Internet of Vehicle (IoV) scenario. *2021 Second International Conference on Smart Technologies in Computing, Electrical and Electronics (ICSTCEE)*, 2021, pp. 1–6.

[30] N. Priyadarshi, S. Padmanaban, M. S. Bhaskar, F. Blaabjerg, and A. Sharma. Fuzzy SVPWM-based inverter control realisation of grid integrated photovoltaic-wind system with fuzzy particle swarm optimisation maximum power point tracking algorithm for a grid-connected PV/wind power generation system: Hardware implementation. *IET Electric Power Applications*, vol. 12, no. 7, pp. 962–971, 2018.

[31] K. Kamalapathi, N. Priyadarshi, S. Padmanaban, J. B. Holm-Nielsen, F. Azam, C. Umayal, and V. K. Ramachandaramurthy. A hybrid moth-flame fuzzy logic controller based integrated cuk converter fed brushless DC motor for power factor correction. *Electronics*, vol. 7, no. 11, p. 288, 2018.

[32] N. Priyadarshi, S. Padmanaban, J. B. Holm-Nielsen, M. S. Bhaskar, and F. Azam. Internet of things augmented a novel PSO-employed modified zeta converter-based photovoltaic maximum power tracking system: Hardware realisation. *IET Power Electronics*, vol. 13, no. 13, pp. 2775–2781, 2020.

[33] N. Priyadarshi, S. Padmanaban, D. M. Ionel, L. Mihet-Popa, and F. Azam. Hybrid PV-wind, micro-grid development using quasi-Z-source inverter modeling and control—experimental investigation. *Energies*, vol. 11, no. 9, p. 2277, 2018.

[34] F. Azam, S. K. Yadav, N. Priyadarshi, S. Padmanaban, and R. C. Bansal. A comprehensive review of authentication schemes in vehicular ad-hoc network. *IEEE Access*, vol. 9, pp. 31309–31321, 2021.

[35] F. Azam, N. Priyadarshi, H. Nagar, S. Kumar, and A. K. Bhoi. An overview of solar-powered electric vehicle charging in vehicular adhoc network. *Electric Vehicles*, pp. 95–102, 2021.

[36] N. Priyadarshi, F. Azam, A. K. Sharma, P. Chhawchharia, and P. R. Thakura. An interleaved ZCS supplied switched power converter for fuel cell-based electric vehicle propulsion system. In *Advances in Smart Grid Automation and Industry 4.0*, pp. 355–362. Springer, Singapore, 2021.

[37] F. Azam, A. Biradar, N. Priyadarshi, S. Kumari, D. Almakhles, and S. Tangade. A framework for secured dissemination of messages in Internet of Vehicle (IoV) using blockchain approach. *2021 IEEE International Conference on Mobile Networks and Wireless Communications (ICMNWC)*, 2021, pp. 1–6.

[38] B. Sujith, A. Ghosh, and V. Gurugubelli. Design of PFC boost converter with stand-alone inverter for microgrid applications. In *2022 IEEE Delhi Section Conference (DELCON)*, IEEE, February 2022, pp. 1–5.

[39] V. Gurugubelli, A. Ghosh, and A. K. Panda. Parallel inverter control using different conventional control methods and an improved virtual oscillator control method in a standalone microgrid. *Protection and Control of Modern Power Systems*, vol. 7, no. 1, pp. 1–13, 2022.

[40] V. Gurugubelli, A. Ghosh, and A. K. Panda. A new virtual oscillator control for synchronization of single-phase parallel inverters in islanded microgrid. *Energy Sources, Part A: Recovery, Utilization, and Environmental Effects*, vol. 44, no. 4, pp. 8842–8859, 2022.

[41] N. Priyadarshi, S. Padmanaban, M. S. Bhaskar, F. Azam, B. Khan, and M. G. Hussien. A novel hybrid grey wolf optimized fuzzy logic control based photovoltaic water pumping system. *IET Renewable Power Generation*, 2022.

[42] N. Priyadarshi, P. Sanjeevikumar, M. S. Bhaskar, F. Azam, I. B. Taha, and M. G. Hussien. An adaptive TS-fuzzy model based RBF neural network learning for grid integrated photovoltaic applications. *IET Renewable Power Generation*, 2022.

[43] G. Vikash, A. Ghosh, and S. Rudra. Integration of distributed generation to microgrid with virtual inertia. In *2020 IEEE 17th India Council International Conference (INDICON)*, IEEE, July 2020, pp. 1–6.

[44] V. Gurugubelli, A. Ghosh, and A. K. Panda. Comparison of deadzone and vanderpol oscillator controlled voltage source inverters in islanded microgrid. In *2021 IEEE 2nd International Conference on Smart Technologies for Power, Energy and Control (STPEC)*, IEEE, December 2021, pp. 1–6.

[45] V. Gurugubelli, A. Ghosh, and A. K. Panda. Droop controlled voltage source converter with different classical controllers in voltage control loop. In *2022 IEEE International Conference on Power Electronics, Smart Grid, and Renewable Energy (PESGRE)*, IEEE, January 2022, pp. 1–6.

[46] V. Gurugubelli, A. Ghosh, and A. K. Panda. Design of different classical controllers in the voltage control loop of a virtual synchronous machine in standalone mode. In *2022 IEEE International Conference on Power Electronics, Smart Grid, and Renewable Energy (PESGRE)*, IEEE, January 2022, pp. 1–6.

[47] V. Gurugubelli, A. Ghosh, and A. K. Panda. Different oscillator controlled parallel three-phase inverters in standalone microgrid. In *1st International Symposium on Sustainable Energy and Technological Advancements (ISSETA 2021)*, September 2021.

[48] J. K. Nayak, H. Thalla, and A. Ghosh. Efficient maximum power point tracking algorithms for photovoltaic systems with reduced number of sensors. *Process Integration and Optimization for Sustainability*, pp. 1–23, 2022.

[49] S. Patel, A. Ghosh, and P. K. Ray. Improved power flow management with proposed fuzzy integrated hybrid optimized fractional order cascaded proportional derivative filter (1+ proportional integral) controller in hybrid microgrid systems. *ISA Transactions*, 2022.

[50] D. Ravi, and A. Ghosh. Voltage mode control of buck converter using practical PID controller. In *2022 International Conference on Intelligent Controller and Computing for Smart Power (ICICCSP)*, IEEE, July 2022, pp. 1–6.

[51] S. Saurav, and A. Ghosh. Fourth order interleaved boost converter with PID, type II and type III controllers for smart grid applications. *Cyber-Physical Systems: Foundations and Techniques*, pp. 179–207, 2022.

[52] S. Joarder, and A. Ghosh. Design and implementation of dual active bridge converter for DC microgrid application. In *2022 IEEE Delhi Section Conference (DELCON),* IEEE, February 2022, pp. 1–6.

[53] T. Barker, and A. Ghosh. Neural network-based PV powered electric vehicle charging station. In *2022 IEEE Delhi Section Conference (DELCON),* IEEE, February 2022, pp. 1–6.

[54] A. Singh, and A. Ghosh. Comparison of quantitative feedback theory dependent controller with conventional PID and sliding mode controllers on DC-DC boost converter for microgrid applications. *Technology and Economics of Smart Grids and Sustainable Energy,* vol. 7, no. 1, pp. 1–12, 2022.

[55] H. Tiwari, A. Ghosh, P. K. Ray, B. Subudhi, G. Putrus, and M. Marzband. Direct power control of a three-phase AC-DC converter for grid-connected solar photovoltaic system. In *2021 International Symposium of Asian Control Association on Intelligent Robotics and Industrial Automation (IRIA),* IEEE, September 2021, pp. 125–130.

4 Solar-Powered Gearless Elevator Drive Control Using Four Quadrant DC to DC Converter System

M. Baranidharan and R. Raja Singh

CONTENTS

DOI: 10.1201/9781003323471-4

Nomenclature

I_{ph}	Current in photovoltaic cell (A)	F_e	Active force
I_{sc}	Short circuit current (A)	F_b	Friction caused by the electromagnetic brake
K_1	Equivalent of I_{sc} when cell at 25°C	k_1, k_2	Stiffness
T	Operating temperature (K)	b_1, b_2	Damping coefficients
I_{sr}	Solar irradiation (W/m²)	x_1, x_2	Rope's altered extension while sliding
I_{rs-pv}	PV module reverse saturation current	M_c	Mass of the automobile
μ_{sc}	Short circuit co-efficient	M_w	Counterweight
Tc	Temperature at the operating condition	T_m	Electromechanical time constants
T_{ref}	Reference temperature	T_a	Electrical time constants
Gtc	Irradiance for test conditions	I_f	Field current
Gref	Reference irradiance	D_f	Boost converter duty ratio
T_r	Nominal temperature (°C)	C_{dc}	DC link capacitance (F)
E_b	Motor back emf	V_{dc}	DC link voltage (V)
I_a	Armature current (A)	I_d	Diode current (A)
V_a	Armature voltage (V)	K	Back emf constant
R_s	Resistance connected series (Ω)	Irs	Reverse current saturation
R_{sh}	Resistance connected parallel (Ω)	R_a	Armature resistance (Ω)
I_{sh}	Shunt current (A)	L_a	Armature inductance
V_t	Diode thermal voltage (V)	β	Pitch angle (θ)
I_d	Diode current (A)	ω_r	Rotor speed
V_s	Supply voltage (V)	T_e	Electromagnetic torque
Irs	Reverse current saturation	R_a	Armature resistance (Ω)
CW	Counterweight	UM	Upward motion
EC	Elevator cabin	DM	Downward motion
$\omega*m$	Reference speed	PI	Proportional integral controller
$\mathbf{E_c}$	Error signal	V_o	Average voltage
δ	Duty cycle	f	Chopping frequency

4.1 INTRODUCTION

DC motors and drives are utilized in many claims, particularly in adjustable speed drive and position control applications, because of its comfort, efficacy, budget, reliability, and ease of application. DC motors are typically measured by adjusting the field flux or the armature voltage, depending on their kind. This armature voltage regulate technique is commonly employed in practice by regulate dc motor speed. DC motors are operated in these circumstances using power electronics components such as a PWM chopper and a regulated rectifier. Due to its ability to transfer bidirectional power both through motor and regenerative braking action, DC choppers may be used to drive dc motors [1]. A high execution motor drive system has various features, similar dynamic speed command tracking and load regulation reaction, making it a key component of manufacturing applications today [2]. In this chapter, the four-quadrant chopper is used as all the quadrants and it's regenerating the power from the load source for elevator application. Basically, a chopper is a static power electronics device that converts fixed dc input voltage to a variable dc output voltage. A chopper should be used to stage down or stage up the fixed input volts like a transformer. It is generally preferred because of its even control ability, high effectiveness and fast response. This chopper works in all quadrants for the P can flow either from supply to load or load to supply. A Class-E chopper works as a stage-down chopper in the I mode and as a stage-up chopper in the II mode. Type-E chopper is another name for this type of chopper. With the use of a circuit schematic and a Simulink model, this chapter explains the working principle and functioning of a Class E chopper. In industry and in our daily lives, DC machines play a critical role, though DC machines have the distinct benefit of having easily adjustable properties. This chapter will show you how to use a PI controller to create a four-quadrant operation for a DC motor [3]. CW, CCW, FB and RB are the four quadrants in which the motor can be controlled.

The Class E chopper is accomplished at working in all modes of the V-I plane. An T_L in the first quadrant works in the opposite direction of rotation [4]. As a result, in order to lift the loaded elevator, the motor's produced torque must be in the rotational direction, or positive. Because the P is positive, this mode is known as the forward motoring. The unloaded cage is lifted up in the II mode. Because the counter load is heavier than the empty cage, the elevator's N may reach very upper levels. So, the T_M necessity to rotate in the other way or be negative. Because the P also negative despite the positive speed, this quadrant is called the forward regenerating quadrant [5]. The empty cage's downward motion is represented by the third quadrant. Torque from the counterweight and friction at the transmitting sections will resist the downward travel; to move the cage downwards, the T_M must be in the rotational direction. When compared to the first quadrant, an electric machine works in the opposite way [6]. The T is negative as the N is raised in the negative way, but the P is positive, hence this quadrant is called Reverse motoring. The weighted cage moves downward in the IV quadrant. Because the loaded cage is heavier than the balanced cage, the motor torque must be polarized in the differing direction of rotation in order to operate as a brake. The T_M is positive, because the N and P is negative, hence this quadrant is called reverse regenerating. Elevators or lifts, it is a type of vertical transportation utilized mostly to go among several floorings in high-rises for buildings. They're utilized to move people and products from floor to floor. Lifts may be realized on huge boats with numerous decks as well. Indeed, elevators consume developed a required quality in high-rises through further great buildings to category it modest for those to mobility impairments to direct those in wheelchairs. From the several types of lifts or elevators are building lifts, capsule lifts, hydraulic elevators, pneumatic elevators, passenger lifts, freight elevators, traction elevators/ cable driven, residential elevators, machine room-less elevators and so on. The elevator sector is a significant part of world technology and economics, and its importance symbolizes one of the most essential characteristics of contemporary civilization. The primary function of all passenger elevators is to provide vertical movement in buildings and structures for various reasons. Elevators not only make it easier for individuals to move around on a daily basis, but they are sometimes the only way to do it [7]. The breakthroughs of mechanics, electromechanics, power- and microelectronics, and mechatronics boosted the quality of elevator construction, as well as almost all other fields of technology, at the turn of the previous two centuries. In recent years, the use of renewable energy sources in high-rise buildings has been encouraged.

An independently activated motor's regenerative braking is rather straightforward. The load's energy is recycled back into the supply system in regenerative mode. During this method, DCM works as generator. The motor, now operating as a generator, then rises the stored magnetic energy in the armature circuit as long as the chopper is turned on [8]. When the chopper is turned off, a huge voltage develops across the motor terminals, which is bigger than the supply voltage V, and the energy stored in the inductance as well as the energy given by the machine is returned to the supply system. After the motor's voltage drops below V, the diodes in the line stop current flow, preventing the load from shorting out. Motor braking is quite effective up to very low speeds. Industrial and commercial applications both demand an IV-quadrant operation. These applications need the capacity to drive and brake, that is, motoring and producing capabilities. Electric adhesion systems, crane and elevators and so on. The distinct quadrant operations operate the motor with normal and reverse voltage and currents to run and break the motor in forward and reverse way [9]. In this chapter, an independently excited DC motor is employed to implement regenerative mode of operation in an elevator. For a four-quadrant chopper supplied DC motor driver system, current, speed, and torque are suggested through Matlab simulation [10]. Mathematical modeling also derived in next section of this chapter. System architecture of the IV-quadrant chopper is addressed in the third section. The fourth portion also includes a simulation model environment. In the chapter's last part, the acquired simulation results are reviewed.

4.2   MATHEMATICAL MODELING

A review of mathematics applied to DC motors, on the other way, will be an extra benefit in modeling accuracy and will assist to know the fundamental working of DC machines.

4.2.1   MODELING OF A SOLAR PHOTOVOLTAIC (PV)

In essence, PV units' concepts are either a 1-D type or 2-D type. This current source, which fluctuates linearly with sun irradiation, is used to represent the light-generated current. This is the easiest and most often used model because it strikes a good mix between simple and precision [11], [12].

In the single-diode model, KCL can be used:

$$I = I_{ph} - I_d - I_p \tag{1}$$

According to the equation, the expression for diode current follows,

$$I_d = \left[\exp \frac{V + I \cdot R_s}{\eta \cdot V_{t \cdot N_s}} - 1 \right] \times I \tag{2}$$

I_{sh} is assumed for

$$I_p = \frac{V + I \cdot R_s}{R_p} \tag{3}$$

Photovoltaic current is given as

$$I_{ph} = \left[\mu_{sc} \left(T_c - T_{ref} \right) + I_{sc} \right] \times G_{pu} \tag{4}$$

$$G_{pu} = \frac{G_{tc}}{G_{ref}} \tag{5}$$

$$I_{rs} = I_{sc} / \left[\exp\left(\frac{q \times V_{oc}}{\eta \times k \times T_c \times N_s} - 1 \right) \right] \tag{6}$$

The characteristic equation in a single diode model is

$$I = I_{ph} - I_0 \left[\exp\left(\frac{V + R_{se}I}{V_t a} \right) - 1 \right] - \frac{V + R_{se}I}{R_{pa}} \tag{7}$$

In this design, a supplemental diode is linked in parallel to the single diode circuit. It's utilized to produce an additional precise I-V typical bend so takes into account fluctuation popular present stream by short current levels caused by charge recombination [13] in the semiconductor's reduction zone [14]. Figures 4.1 and 4.2 indicate that the model is an additional exact PV array with DC-DC converter, perturbation and observation algorithm respectively, however the one diode model is desirable due to the equation's difficulty [15].

This equation for a common double diode model is

$$I = I_{ph} - I_{01}\left\{\exp\left[\frac{(V + IR_{sc})}{V_t}\right] - 1\right\} - I_{o2}\left\{\exp\left[\frac{(V + IR_{se})}{V_t}\right] - 1\right\} - \frac{V + R_{se}}{R_{pa}} \tag{8}$$

The current source represents the photocurrent in the cell.

The essential of Iph is Rsh with Rse. Because Rsh's value is usually extremely high and Rse is usually very little, they can be neglected to maximize the experiment's efficiency for single diode PV cell circuit at Figure 4.3. PV components are grouped together to form PV modules, which are

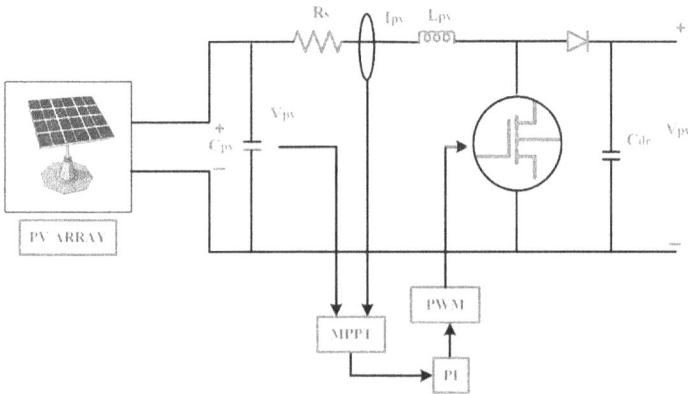

FIGURE 4.1 PV array with DC-DC converter.

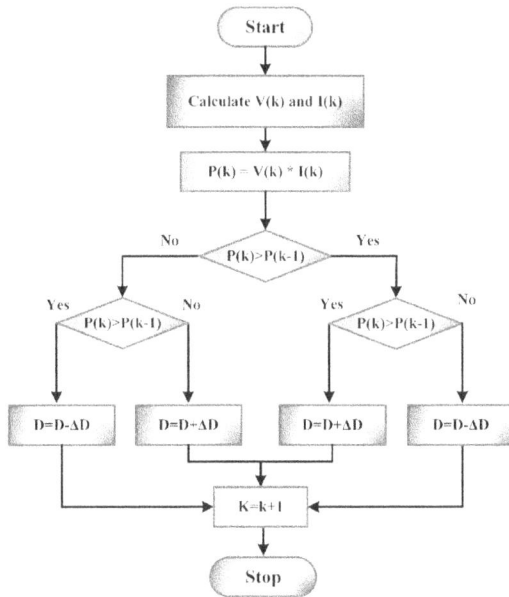

FIGURE 4.2 Perturbation and observation (P&O) algorithm.

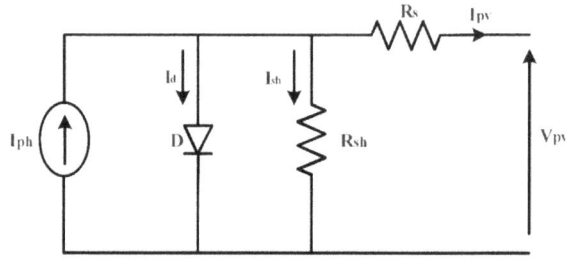

FIGURE 4.3 Equivalent circuit for a single diode PV cell.

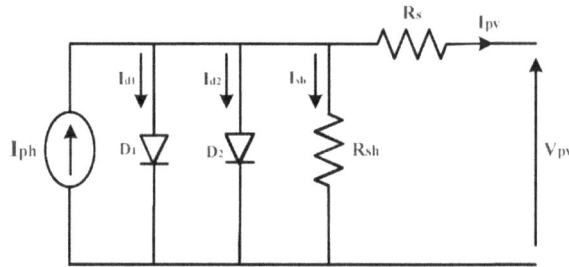

FIGURE 4.4 Double diode PV cell equivalent circuit.

then connected in a parallel series manner to make PV arrays [16]. The solar panel may be quantitatively designed, as demonstrated in the photo-current formula module [17].

$$I_{ph} = \left[I_{SCr} + K_i (T - 298) \right] * \frac{\lambda}{1000} \tag{9}$$

I_{rs} is the reverse current saturation module

$$I_{rs} = I_{scr} / \left[\exp\left(\frac{q \times V_{oc}}{N_s \times kAT} - 1 \right) \right] \tag{10}$$

It is expected that the module overload current (I_0) varies with PV cell temperature.

$$I_o = I_{rs} \left[\frac{T}{T_r} \right] \exp\left[\frac{q * E_{go}}{Bk} \left\{ \frac{1}{T_r} - \frac{1}{T} \right\} \right] \tag{11}$$

The PV module I_{PV} is

$$I_{PV} = N_p * I_{ph} - N_p I_o \left[\exp\left\{ \frac{q^* (V_{PV} + I_{PV} R_s)}{N_s AkT} \right\} - 1 \right] \tag{12}$$

$V_{pv} = V_{oc}$, Np = 1 and Ns = 36

Figure 4.4 depicts a two-diode model for matching a real bend, through the additional diode
supplying an optimist problem by the two in the exponential period disagreement denominator.

Resistances in series (Rs) and parallel (Rp) may be found in a PV cell or circuit, creating a model characteristic [18].

4.2.1.1 PV Array Planning

Photovoltaic cells heating energy into electricity, and a collection of photovoltaic cells are combined to make panels to adjust for the energy consumption. To satisfy the V and I requirements, the panels are coupled in a series-parallel configuration [19]. To use the reported array V and I, the PV array power and variation in power are determined. The switching frequency is raised if the varying the power and voltage are both more than zero; otherwise, the duty ratio is lowered if P is higher than 0 and V is below zero. The switching frequency is raised if P and V are both lesser than zero; otherwise, the switching frequency is lowered if P and V are both lesser than zero [20]. Receiving the Vpv and Ipv data completes the loop. This P&O method was chosen because it is easy to implement and provides improved efficacy [21]. For such MATLAB simulation, a SunPower SPR-E19–315 model PV array with four series modules and 40 parallel strings was chosen. Figure 4.5 depicts the V, I and P response at an irradiance of 800–1000 W/m2 and a temperature of 35°C. The highest power is attained for 1000 W/m2, and the boost converter's gate signal is tweaked to obtain optimum power by using MPPT method [22].

4.2.2 DC Motor

In a DC machine excited in its own right, both field and armature windings are the most important parts of the motor and each one has its own power source. To excite the flux, the motor's field windings are employed. With the aid of the brush and commutator section, the rotor pulls armature current. A field current (I_f) excites a separately activated DC motor, and current in armature movements through the circuit as an outcome. The motor produces a torque and back EMF in order to balance the T_L. An equivalent circuit for a separately excited DC motor is shown in Figure 4.6, and equivalent equations also defined.

The present state of the field it doesn't matter what the armature current I_a is also the field current is unaffected by changes in the armature current [23]. In most cases, the I_f is significantly lower than

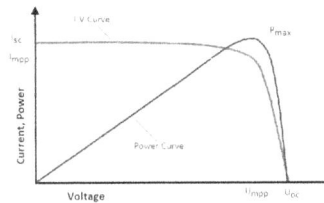

FIGURE 4.5 Characteristics of PV cell [22].

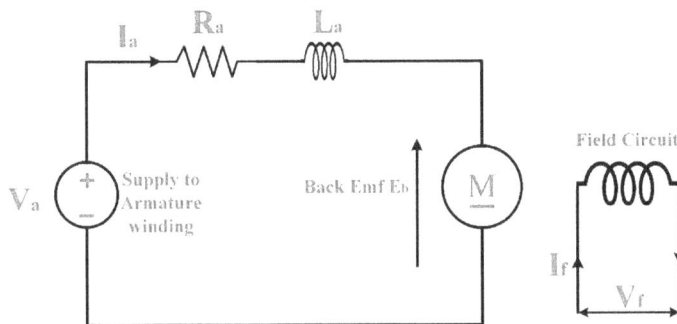

FIGURE 4.6 Equivalent circuit of separately excited DC motor.

the I_a. Assume V_a is the armature voltage in volts, the current in armature is amps, motor back emf is in (E_b) volts, inductance of the armature in H, and Ra is the resistance of the armature in ohms in the diagram.

The equation for armature is represented in the

$$V_a = E_b + I_a R_a + L_a \frac{dI_a}{dt} \tag{13}$$

This torsion formula should be written as

$$T_d = J \frac{dw}{dt} + B_\omega + T_L \tag{14}$$

It will provide B = 0 if there is no friction in the motor's rotor. The equivalent back emf and torque equations may be found by designating for

$$T_d = J \frac{dw}{dt} + T_L \tag{15}$$

The back emf of the motor will be calculated as follows:

$$E_b = K\Phi\omega \tag{16}$$

Also developed torque by the motor is

$$T_d = K\Phi I_a \tag{17}$$

The following equations are produced by the use of the over formula with the Laplace transform applied in (1).

$$I_a(s) = \frac{V_a - E_b}{R_a + L_a s} = \frac{V_a - K\Phi\omega}{R_a(1 + L_a / R_a s)} \tag{18}$$

$$\omega(s) = \frac{T_d - T_L}{Js} = \frac{K\Phi I_a - T_L}{J_s} \tag{19}$$

$T_a = L_a / R_a$. Figure 4.1 illustrates the corresponding DC motor model. The resulting transfer function will look like this after block reduction.

$$\frac{\omega(s)}{V_a(s)} = \frac{\dfrac{K\Phi / R_a}{Js(1 + sT_a)}}{1 + \dfrac{K^2\Phi^2}{Js(1 + sT_a)}} = \frac{1 / K\Phi}{sT_m(1 + sT_a) + 1} \tag{20}$$

$T_m = JR_a / (K\Phi)^2$ Equation (20) might be decreased even more (substituting KΦ or K$_m$):

$$\frac{\omega(s)}{V_a(s)} = \frac{1/K_m}{(1+sT_m)(1+sT_a)} \tag{21}$$

Here $K\Phi = K_m$; the T$_m$ with T$_e$ for the beyond system t.f, that is responsible for the system's reply, are T$_m$ and T$_a$, respectively.

$$w = \frac{(V_a - I_a R_a)}{K\Phi} \tag{22}$$

As can be seen from the preceding equation, the N of a DC motor is determined by the applied voltage, Ia, Ra, and field flux. As a result, Va control, Ra control and field flux control are the three methods for governing the N of a DC motor.

4.2.3 ELEVATOR MODELING

This elevator is an advanced transportation vehicle that is mostly utilized in high-rise buildings to transfer people and commodities. Because today's cities are thought to rise vertically, having effective and simple transit within a building is critical. In most cases, an elevator's hoisting function is performed by a three-phase induction motor (IM). However, much research has been conducted in order to replace the traditional motor with one that is more efficient, reliable and fast. Elevators have become increasingly prevalent in high-rise structures in recent years. The most crucial factor is to provide simple and safe transit within a facility. Because more elevators are utilized in cities, energy consumption is also a factor to consider. Elevators travel up and down a vertical shaft, transporting freight or passengers between floors of multi-story structures. We examined passenger, freight, residential, and vehicle elevators in the scope, which are used in commercial, residential and industrial applications. The gearless elevator lifting mechanism's electric drive comprises of a winch, a supplying power component and a control system. In an elevator, there are four torque-speed quadrants. When both the quantity of T and the angular N is both symbols, working zones termed forward and reverse modes, individually, according to a mechanical power of motor formula P = T. Also, forward braking arises after the N value is positive and the T value is negative, and reverse braking occurs when the N and T values remain negative and positive, respectively. The rotor N surpasses the N of the stator spinning magnetic field in braking zones, to the P stream way is inverted linked to motoring, traveling motor towards DC link, causing the motor to behave as a generator.

4.2.3.1 Gearless Elevator Arrangement

This 1:1 roping configuration reduces the force need through nearly a factor of dual. The 2:1 roping preparation reduces the force demand by roughly a factor of dual. The motor speed is a 2:1 attaching configuration is double of a 1:1 attaching system for a given elevator speed. The 2:1 configuration, on the other hand, doubles the rope speed and is employed for speeds up to 800 feet. A 1:1 rope configuration is employed for velocities above 800 feet per minute.

4.2.3.2 Gearless Elevator Energy Utilization

The energy usage of an elevator is determined by a variety of factors, including the kind of drive utilized, the elevator's capacity, speed, complete weight mass of the scheme, category of gearless motor employed, rope and so on. As a result, a robust analytical tool for evaluating energy use is of tremendous relevance. This chapter presents an analytical technique to determining average energy use [24].

4.2.3.3 Elevator Friction System for Dynamical Modeling

A friction motor, counter load, wire rope, and electromagnetic stop are all part of the direct-drive elevator's construction, as illustrated in Figure 4.7. For rotor of a PM friction motor, for traction sheave and stop pulley are fitted, and the electromagnetic brake is placed around the stop pulley. After crane is in break method, the electromagnetic brake is utilized as a safety precaution to keep the adhesion system stationary by retaining the brake crane. The brake will be released once the elevator begins to move. Because the eight of the passengers is unknown, the T_L of the system is unidentified. As a result, in order to prevent the elevator car from sliding, the traction device must provide the electromagnetic torque to equalize and unpredictable weight. Among the most cases, the brake drive issue entirely in 100–200 milliseconds, and the adhesion system must maintain balance throughout this time. As a result, riding comfort is guaranteed. This time, known as the zero-servo procedure, is divided into three parts. The following is the dynamic mathematical model of the system at per step. At this point, the brake is keeping the traction system stationary. The relative gravity change among the automobile and the CW is balanced by the friction caused by the brake. The mathematical equation can be written as follows, assuming the automobile and passengers are heavier.

$$(M_c - M_w)g - F_b = 0 \tag{23}$$

where Mc and Mw indicate the mass of the automobile with travelers and the counterweight and Fb denotes the friction caused by the electromagnetic brake. The working mechanism of the ropes lift is predicated on the transfer of gravity possible energy into electric energy by exploiting the differential in load between the car load and the counterweight. The elevators have a highly modulated power demand and their actual consumption is complicated to estimate. They have a high number of starts and stops, a high starting torque, and the need for full-power braking. In motor operation, the lift system acts as a load; in regenerative operation, it acts as an electrical energy source [25]. The load theory is used in rope lifts. The counterweight serves to balance the elevator system and minimize power consumption. The mass of the counter-weight is equivalent to the sum of an empty vehicle and 40–50 percent of the optimum lift capacity. Based on the travel direction and the weight difference between the car and the counterweight, the elevator process could be divided into four categories.

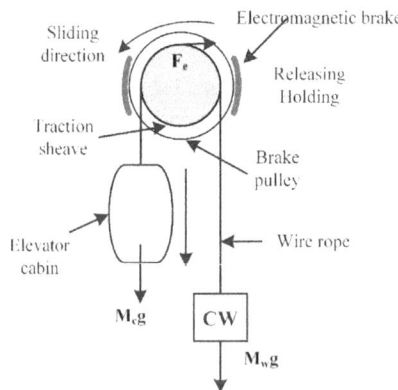

FIGURE 4.7 Structure of the elevator traction system.

The following is the dynamic mechanical equation for the vehicle side:

$$M_c g - (F_b + F_e) = k_1 \Delta x_1 + b_1 \Delta x_1 = M_c a \tag{24}$$

where k_1 and b_1 are the stiffness and damping coefficients from rope on the car part, x_1 remains rope's altered extension while sliding and a is the automobile's acceleration. The active force Fe corresponds to traction motor's electromagnetic torque Te. In the same way, the active mechanical equation for a counterweight part could be written as follows, for example.

$$(F_b + F_e) - M_w g = k_2 \Delta x_2 + b_2 \Delta x_2 = M_w a \tag{25}$$

where k_2 and b_2 are the rope's stiffness and damping coefficients on the counterbalance side, respectively, and x_2 is the rope's modified elongation while sliding. Fb takes around 100–200 ms to decline from its starting value to zero, resulting in non-linear changes in Fb and Fe [26].

4.2.4 CHOPPER CONTROL

A chopper is a high-speed semiconductor switch that can turn on or off. It fast connects the supply to the load and removes the load from the supply. A chopped load voltage is created in this manner from a steady DC supply of magnitude VS.

$$\text{Average voltage, } V_o = \frac{T_{on}}{T_{on} + T_{off}} * V_s, \ \delta = \frac{T_{on}}{T_{on} + T_{off}} \text{ and} \tag{26}$$

$$\text{Chopping frequency, } f = \frac{1}{T} = \frac{1}{T_{on} + T_{off}} \tag{27}$$

Thus, $V_o = \delta * V_s$. The average output voltage VO may be regulated by duty cycle by repeatedly opening and shutting the switch, as shown in equation (26).

4.2.5 TYPE E CHOPPER

A DC motor can activate in four different modes as exposed in Figure 4.8. The supply V is larger than the back emf when a DCM is working in I and III quadrants, indicating forward and reverse driving modes, respectively, although the current flow direction is different. The rate of the back emf generated by the motor must be more than the applied voltage after the motor works in the II and IV quadrants, that is, forward braking and reverse braking means of operation [27].

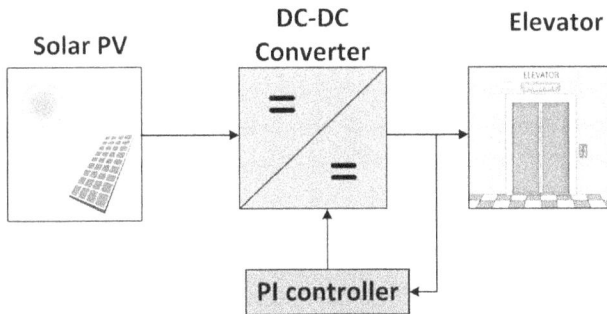

FIGURE 4.8 Solar PV with DC-DC converter for elevator.

4.2.6 FOUR QUADRANT OPERATION OF AN ELEVATOR

An elevator has two modes of operation, each with four working rounds of cabin relative direction. Motoring method: It indicates also that engine is exerting force and torque on the heavy side of the balancing system, and is the hoist. Generating method: Its engine completely not for create a positive torque in this condition, however the system moves with the load [28]. In traditional cranes, a relay monitors a current path while the elevator is in the producing mode, allowing to know after the brake resistor is coupled ckt. In this is beneficial since inspecting brake resistor is not an appropriate procedure, technically with economically. Furthermore, the way of delivering extra electricity to the mains power necessitates the plan and deployment of a holdup conversion scheme, putting significant expense on grid system authorities, with the resulting advantages being insignificant in comparison to the initial expenditure. As a result, this chapter presents an HESS for using a significant amount of regenerative energy in elevators.

Based on I, the relative load of the electric car with counterweight, and (II), the direction of the EC's movement, four modes are used in this chapter. Figure 4.9 depicts the elevator's four-quadrant functioning and Figure 4.8 shows the solar PV based with DC-DC converter for elevator application. It's important to note that the EC is full with passengers in quadrants I and IV, but empty in quadrants II and III. Forward rotation of the DC machine correlates to upward motion of the elevator cabin, although reverse rotation corresponds to downward motion of the elevator cabin.

The electric car weight is supposed to be greater over the counterweight in the first quadrant, implying that the electric car must go higher. As a result, the DC machine acts like motoring mode and generating mode shown by Figure 4.7. Also, in third quadrant, the DC machine acts as a motor (reverse) since the net electric car weight is lower than the counterweight and the electric car goes downward [29]. The electric car weight is expected to be less with the counterweight in the second quadrant, therefore electric car must travel higher. As a result, the DC machine, in conjunction with the worm gear, works for brake through torque and rotational speed with in reverse way. Worm

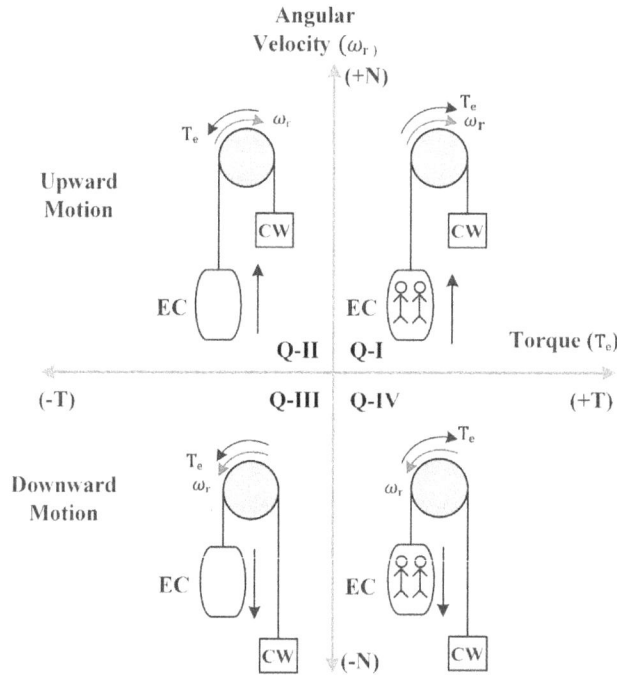

FIGURE 4.9   Elevator operation in four quadrants.

gear's major purpose in an elevator scheme is to act as a natural brake, preventing the electric car from moving owing for change for nets electric car load by counterweight with gravity. As a result, the existence of gear wheel prevents powered control movement in the reverse direction. In the II quadrant, the DC motor can serve as a generator if a proper gear procedure is used instead of gear wheel. Also, for DC motor act as a generator in the IV mode. Elevators are the greatest example of electric drive's four quadrant operations. In Figure 4.7 indicates for four quadrant actions of an electric drive. shows the regenerative operation of elevators in that quadrant 1 and 3 are electrical power consumption and quadrant 2 and 4 is the electrical power generation. It is in forward motoring mode when the driving torque and rotational direction are same. It is in forward breaking mode when the driving torque changes in the opposite direction. When the way of rotation and the driving T both shift and apply in the equal way, it is said to be in reverse motoring. This is reverse breaking mode if the direction of torque is reversed from the preceding mode. Due to the existence of gravitational pull, a weighted elevator travelling down needs to descend faster. At this point, the motor acts as a generator, producing electricity. This electricity will be used for other projects. This energy is turned to heat by conventional drives, which must subsequently be evacuated from the building by air conditioning systems [30]. A regenerative drive reduces an elevator's energy usage significantly, making it a should-have quality when choosing an eco-friendly elevator for a green building. There are several more advantages to using this approach. Depending on the building height and elevator speed, regenerative drive can reduce elevator energy consumption by 20–35 percent on average. In some circumstances, such as when there is heavy traffic and automobiles are fully loaded, this method can reduce energy use by up to 60 percent. A lift may be thought of as a structural system that consists of a stationary pulley on one side, a customer car on another, and a counterweight on the other. In most cases, the counterbalance is equivalent of the empty car's load plus the entire load value. This configuration can help to reduce the maximum load on the hoist, which is the elevator's driver's car. When we regard the elevator's forwards or backwards travel as the horizontal axis and the motor output or force as the vertical axis in this setup, we can split the elevator function in four situations, as illustrated in Figure 4.9 having four categories.

The regenerative approach is very efficient, as it removes the need to remove heat load from the building caused by driving or brake resistors. The regenerative drive generates energy that is both clean and safe.

4.2.7 Percentage of Regenerating versus Saved Power

They calculated the energy-saving rate of the test scenario as considered the lift completing a circular journey from one level to another level, we finished attaching the regeneration power to the device. Each round trip was made a number of times in the test design for accuracy, and afterwards the mean value was obtained; also, it was measured across different floors and the measurement of varied load situations under the contribution rate. The restoration of electricity power usage may be called to as this sort of numerator and the denominator. In regard with the preceding description, the energy saver rates could be recast as being changing the energy-saving rate for the regenerative energy ratio. Equation (29) indicates the regenerative ratio of the given system.

$$\text{Regenerative energy ratio} = \frac{\text{Regenerative energy feedback to electricity}}{\text{Power consumption}} \tag{28}$$

4.2.8 Controller Design

Since almost 60 years, PI controllers have been frequently employed in manufacturing operations. Although the control algorithm has changed from hydraulic to analog to digital controllers, the control method remains the same. The PI controller is a tried-and-true option for a wide range of industrial applications. Every sample time (T), the PI algorithm computes and sends a controller's output signal to the final switching device. I and T are the two ratios parameters for PI controllers. The PI controller will not increase the reaction time; therefore, the DC motor will continue to run at the same speed. Therefore, it has a detrimental influence on the system's total quickness of reaction as general reliability. This regulator is typically utilised in applications wherein system speed is not a concern. Because the PI controller is unable to estimate future system failures, it is unable to reduce the rise time and avoid vibrations. Any quantity of I, if used, ensures set point excess. In manufacturing, PI controllers are frequently employed, particularly where speed of reaction is not a concern. The fundamental purpose of a PI controller is to reduce the steady state error caused by a P controller. DC motor field terminals are energised individually in separately excited DC motors. Adjusting the armature voltage allows the motor to run at rated or below rated speed. Using bode analysis or other control system design techniques, create the PI controller for the present loop. The design of the speed controller is often the following phase. In most cases, the 0-db intercept of 1/Js (1+Tis) is far too short. A fundamental reason for this is that it has a very basic structure that is easy to understand and execute in reality, and it is the foundation for many advanced control systems, such as model predictive control. A high-speed application demands different PI improvements than a fixed-speed one. furthermore, to eliminate overshoots and oscillations, machines and equipment that operates at a wide range of speeds needs differing gains at the minimum and maximum end of the speed range. Tuning the PI constants for a large-scale variable speed technique is often costly and complex. Insufficient technique understanding, occasionally produced incorrect PI values for making the work much more difficult.

Figure 4.10 shows a block schematic of a closed-loop speed control system. An internal audit loop inside an outer speed loop was employed in this system. The internal audit loop maintains a safe level of motor current and torque. Regard a reference speed 1ω*m that results in a positive error *m. This N* is controlled by N controller and sent into current limiter, that is full level when the speed error is minor. Current was usual for internal current control loop by current limiter. The drive then increases, and the drive N reaches the appropriate speed, the motor torque equals the T_L. This lowers ref N, resulting in -N mistake. In braking mode, although the current limiter saturates, the drive develops de-accelerate. This drive is changed from braking to driving if current limiter develops desaturated [31]. This simulation results show that the suggested elevator system can operate in four quadrants.

The speed sensor produces a signal proportional to N_M, which are reduced to eliminate ac wave with checked to the ref N. This N controller is provided the N error signal. The error signal creates Ec, which regulates the duty cycle of the chopper ckt so as motor speed is equivalent to the ref N. The N controller is commonly a proportional integral controller, which uses entire action to stabilize drive, reduce steady state speed to near zero and filter noise. When the armature current is

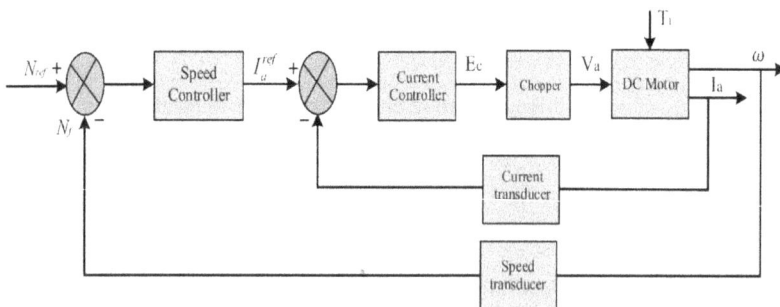

FIGURE 4.10   Speed control of a separately excited DC motor in a closed loop.

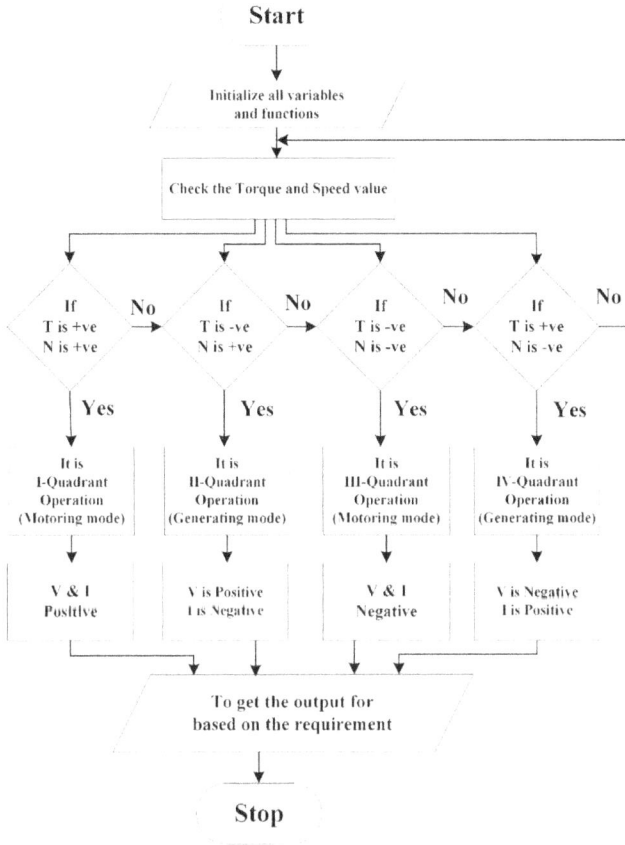

FIGURE 4.11   Four quadrant operation of chopper.

within legal limits, the drive has current limit control, which affects motor operation. The flowchart in Figure 4.11 indicates the four-quadrant operation of the chopper.

4.3   FOUR QUADRANT CHOPPER DESIGN

Most typical lifts are built in structures that cannot handle reversed power transfer and rely on diode converters. Due to this, the produced electrical energy is used by resistors, and therefore is great possibility for power reduction. The regenerative mode of operation on a permanent magnet motor in an elevator system is when the motor rotates without power or when the motor is braking. This generative mode is able to produce electricity for the elevators. The situation of regenerative mode in elevator can be possible when the elevator load is less than the counterweight or the elevator load is greater than counterweight. Both cases rely on the force of gravity pulling the motor rotation. Drive to the lift by consuming any electric power.

Based on if the T_M is less than or higher than the T_L, the motor accelerates or decelerates in multi-quadrant or four-quadrant operation.

(a)   Forward Motoring

The operating V is positive and larger than the motor's back emf in this mode of operation resulting in a positive I flow into the motor. The power is positive since both I and V are positive.

(b) Forward Braking

The motor goes forward in this mode, the induced emf remains positive. However, the provided V is abruptly lowered to a level below the back emf.

(c) Reverse Motoring

This is the motor's III quadrant activity, and both the V and I are negative. As a result, the P is positive, that is, it is provided from the supply to load.

(d) Reverse Braking

As energy is transferred back from the load to the supply, the chopper works as a step-up chopper. The inductance drives the current via D3 and D4 when the CH2 is switched off.

It should provide not just load torque but also an I_L component to overcome inertia after motor acceleration. The forward acceleration is caused by the motor's positive torque [32]. When the motor is rotating in a forward way, the motor N is positive. The resultant or dynamic T is negative while the motor is decelerating. This torque aids in the development of motor torque and helps to keep the motion going by extracting energy from the stored energy. As a result, if a motor creates deceleration, it is deemed negative torque. A motor can be operated to run in one of two modes: motor action or brake action. The four-quadrant operation of class E choppers is linked in Figure 4.12. The motor load is active in a class E chopper, and the motor's rotational way could be changed without retreating the sign of its excitement. Chopper-1 is made up of CH1, CH2, D4 and D3, whereas chopper-2 is made up of CH4, CH3, D1 and D2. As a result, by adjusting voltage that is less than or greater than the back emf, positive or negative torque may be generated. As a result, the four-quadrant process of the individually excited DC motor is inherent. Figure 4.12 depicts four quadrants functioning of DCM, with sign of this T indicated by a dot symbol on one of the motor terminals. If I flow into the dot, the machine creates a positive T. If I flow out of the dot, the T is also negative. These four quadrants are described in more detail later. Both positive and negative load currents and voltages are possible [33].

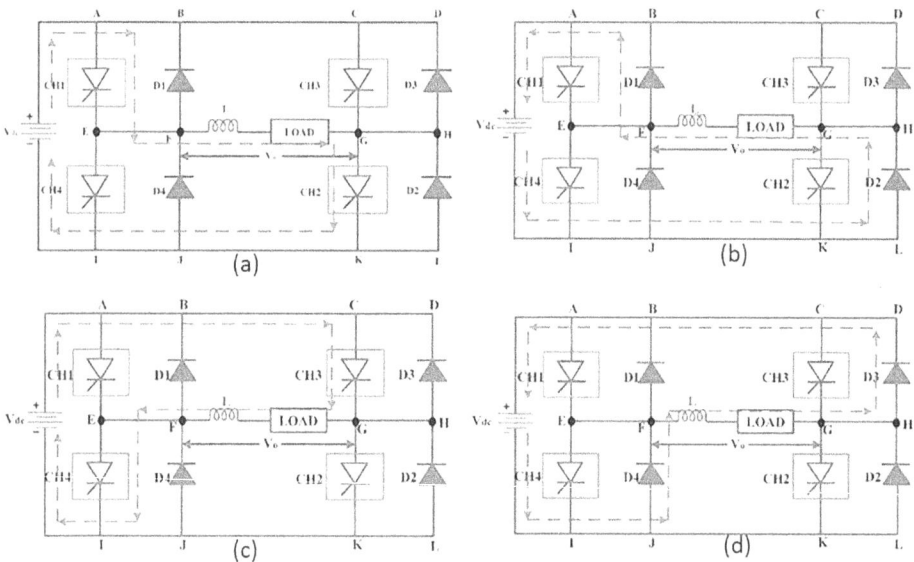

FIGURE 4.12 Four quadrant chopper: (a) forward motoring, (b) forward braking, (c) reverse motoring, (d) reverse braking.

4.3.1 Forward Motoring

Furthermore, in this quadrant, both N and T are positive. As a result, the motor spins in the way of forward motion. Forward driving or correcting will take place in this region. Switches CH1 and CH2 are both switched on at this location. The voltage across the motor $V_0 = V_S =$ supply voltage is now equal to the supply voltage, and the load current I_O is flowing. V_O and I_O are now both moving in the right direction. The electricity from the source is used by the load. The load current Io freewheels via CH2 and D4 when CH1 is shut off, as indicated in Figure 4.12(a), Inductance. When CH2 is turned on, L stores energy. As a result, both V_O and I_O were regulated to be positive, putting them in the I quadrant. Positive I_L is the flow of current from the source to the load [34].

4.3.2 Forward Braking

As a result, the current (which is caused by torque) will reverse direction. The energy flow is reversed due to the negative torque. Because the T_L and the T_M are in opposing directions, the mutual action reduces the motor's speed, and the back emf drops under the supplied voltage rate. Figure 4.12(b) shows the Inductance drives load current through D1 and D2 when CH4 is switched off. Supply is where this current flows. Because ($V_O = L.di/dt$) is greater than V_S. As a result, the source of energy is replenished. It is a second quadrant operation of the chopper because V_O is positive and I_O is negative. Current flows from load to supply when the load current is negative. Regenerative braking is the process of returning mechanical energy from a motor to the supply. Regenerative braking is seen in this quadrant action.

4.3.3 Reverse Motoring

Figure 4.12(c) shows that the motor begins to rotate in a counter-clockwise direction due to the supply's reverse polarity. The functioning of this quadrant is similar to that of the first, with the exception of the rotation orientation. The amount of the voltage applied to the motor determines the optimum reverse speed. CH3 and CH4 are switched on to reverse the motor's rotational orientation. The sign of the V_O changes, as does the current flow.

4.3.4 Reverse Braking

CH4 will be turned on, while CH1, CH2 and CH3 will be turned off. When the chopper CH4 is activated, positive current flows through CH4, D2 and L stores the power. When if switch 4 is absent, current is sent back to the supply via the diodes 2 and diodes 3 at Figure 4.12(d). Because V_L represents negative but the I_L is positive, the process is in the fourth quadrant. The energy is recycled back into the supply by this current flowing through the supply Vs. Note that ($Vo = L.di/dt$) is superior than the supply voltage Vs at this time. The V_L is negative, while the I_L is positive, resulting in the fourth quadrant's chopper operation. After the motor is in the II and IV quadrants of operation, power is given back to the source and the motor is in braking mode.

4.4 PROPOSED METHODOLOGY

The four-quadrant chopper model was developed in Simulink with ideal switches and diodes. A DC machine was connected as the load. The torque of the motor was measured as in closed-loop control. The pulse generation subsystem has the pulse generator giving switching pulses to the four switching elements as in order [35]. The armature voltage provided is 240 V and the field voltage is given to be 150 VDC as per the standard specification of the present model (5 HP 240 V, 1750 RPM 150 V). The switches are triggered with 50 percent duty cycles. At the output side, the speed output in rad/s is converted to its equivalent speed in RPM. Closed-loop control of torque of the motor is implemented [36]. Figure 4.13 shows the simulation model for a four-quadrant chopper fed DC motor for elevator application. It is connected to the output of the solar power-based dc to dc converter with the dc motor for linked to the elevator load profile for a day. At first, 1000 W/m^2 of irradiance is recorded at a temperature of 35°C and 54 voltages at maximum power point. The output current from the solar PV is then regularly increased and decreased in accordance with the variation of the day in a weather [37].

4.5 RESULT AND DISCUSSION

Figure 4.13 shows a Matlab/Simulink model of the solar-powered elevator drive control utilizing a four-quadrant chopper. The PV solar array, DC to DC power converter, elevator load, and control system are the key components of the model. At a constant ambient temperature of 25°C, the Sun PV array is exposed to solar irradiances of 1000 W/m^2 (0–1 second interval), 500 W/m^2 (1–2 second interval) and 100 W/m^2 (2–3 second interval). The following findings were achieved utilizing a four-quadrant chopper in the model to simulate the MPPT integrated solar-powered elevator motor control.

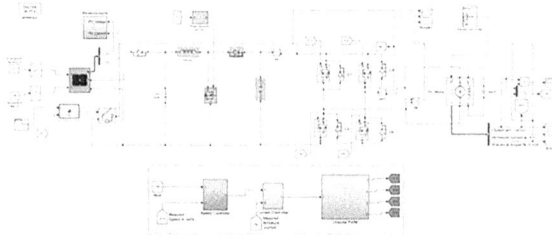

FIGURE 4.13 Simulation for four quadrant chopper provided DC motor for elevator application.

TABLE 4.1
Parameter of the Motor Drive System

Parameters	Expression	Range/unit
$V_a(t)$	Armature voltage	240 volt
$I_a(t)$	Current armature	Ampere
$E_b(t)$	Back emf	V/m/s
R_a	Resistance to armature	0.78 Ohm
L_a	Inductance of the armature	0.016 Henry
J	Inertia moment	0.05 kg m^2
B	Viscous friction	0.01 Nm/s
T_L	Load Torque	5 Nm
ω	Angular velocity	rad/s
θ	Angular position	rads
$E_f(t)$	Constant excitation voltage	150 volts
P	Motor power	5 hp and 3.73 kW
Mosfet	Ideal switch	4 switches
V_{chop}	Chopper supply voltage	280 volts
DCM	DC Motor	5 HP, 240 V, 1750 RPM, 150 DC Ext.
Power gui	Power Gui	50e-6
Rs	Repeating sequence	1 kHz

4.5.1 I Quadrant – Motoring Mode

Because the applied torque and speed are both positive in this first quadrant action, the motor receives a positive current. The power is positive since both I and V are positive. Furthermore, in this quadrant, both N and T are positive. The motor in Figure 4.13 is rotating forward. In other words, $v_o = v_s$. The direction of io indicates that the load current flows from source to load.

Figure 4.14 depicts the simulation's outcomes. At low irradiance, the grid provides electricity to the motor as the irradiance changes. At irradiance 1000, the first quadrant of the four-quadrant operation is shown.

4.5.2 II Quadrant – Forward Regenerating Mode

In this second quadrant operation, the applied T is negative and N is positive so a negative current stream into the motor. Meanwhile I is negative and V is positive, the power converts negative. So, in this mode power fed back to the load to source Figure 4.15 the motor rotates in forward regenerative direction. The fact that the current is sent back to the source indicates that the load is merely transmitting power to the source. For a more extensive examination and comprehension, please see Step-up chopper.

4.5.3 III Quadrant – Reverse Motoring Mode

Both the motor voltage and current are negative during this third quadrant functioning of the motor. As a result, the power is positive. Figure 4.16 shows the torque and speed is negative. So it acts as

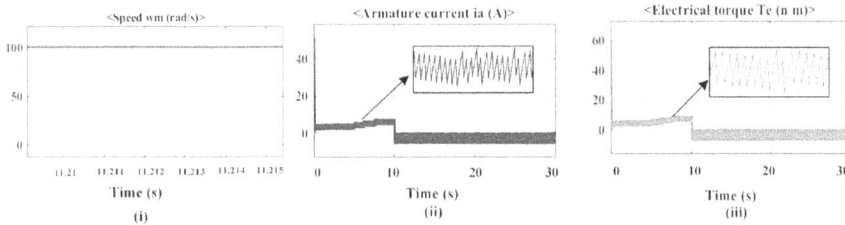

FIGURE 4.14 Four quadrant chopper provided DC motor – forward motoring.

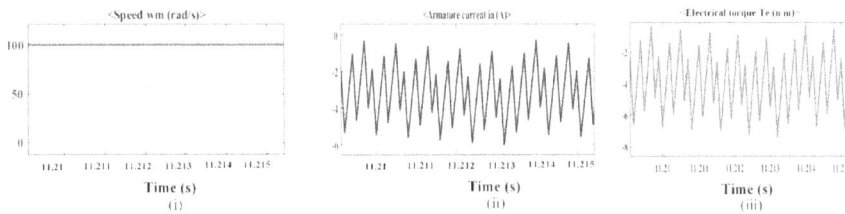

FIGURE 4.15 Four quadrant chopper provided DC motor – forward regenerative.

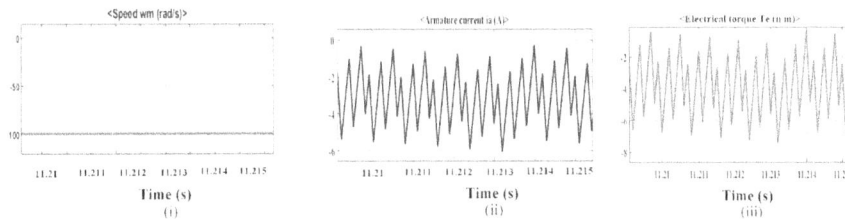

FIGURE 4.16 Four quadrant chopper fed DC motor – reverse motoring mode.

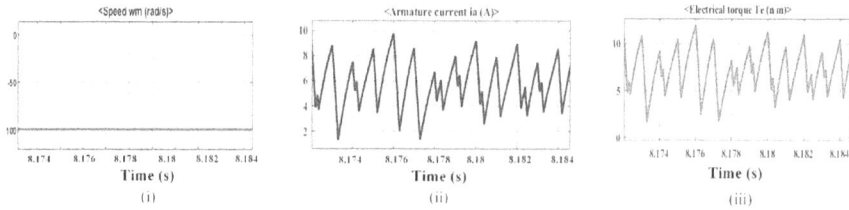

FIGURE 4.17   Four quadrant chopper fed DC motor – reverse regenerative mode.

the reverse motoring mode. However, note that the polarity of the load voltage v_o is the polarity represented in the circuit schematic. As a result, it's considered that v_o is negative. Let's take a look at the present state of the load current io. It can be observed that io is flowing in the reverse way as stated in the control circuit, indicating that it is negative.

4.5.4   IV Quadrant – Reverse Regenerating Mode

In this fourth quadrant process of the motor in V is negative and I is positive. Thus, the power is positive. Figure 4.17 shows the torque is positive and speed is negative. So it acts as the reversal regenerative form. The load current is always positive, even when the load voltage is negative. In the fourth quadrant, this results in chopper functioning. The load feeds power back to the source, and the chopper works as a step-up chopper. A four-quadrant helicopter, often known as a Class E chopper, is described in detail.

4.6   CONCLUSION

The need for vertical transportation is always expanding as a result of increased urbanization and innovations. It's more of a need than a luxury. As the world becomes more urbanized, the need for housing and commercial space will increase. This chapter includes a proposed elevator system that utilises a separately excited DC motor IV quadrant chopper with the help of a PI controller also without any loss of power. During the regenerative mode, the PI controller-based closed-loop strategy employed to regenerate electrical energy from kinetic energy and the produced voltage could be reverted to the supply mains, which will result in substantial saving of power. The preceding assumption is utilized in applications where the direction of rotation of the motor is often reversed, such as elevators and electric cars. Also in this chapter, the converter is created using calculated values and performed using MATLAB/Simulink. The PI-controller maintains the output voltage based on the estimated values.

REFERENCES

[1] N. Priyadarshi, S. Padmanaban, J. B. Holm-Nielsen, F. Blaabjerg, and M. S. Bhaskar. An experimental estimation of hybrid ANFIS-PSO-based MPPT for PV grid integration under fluctuating sun irradiance. *IEEE Systems Journal*, vol. 14, no. 1, pp. 1218–1229, 2020, doi: 10.1109/JSYST.2019.2949083.

[2] R. Nagarajan, S. Sathishkumar, K. Balasubramani, C. Boobalan, S. Naveen, and N. Sridhar. Chopper fed speed control of DC motor using PI controller. *Journal of Electrical and Electronic Engineering*, vol. 11, no. 3, pp. 65–69, 2016, doi: 10.9790/1676-1103016569.

[3] M. D. O. Rahul Baranwal, and O. Aftab. Four quadrant speed control of DC motor with the help of AT89S52 microcontroller. *International Journal of Advanced Research in Science, Engineering and Technology*, vol. 7, no. 1, pp. 111–122, 2015.

[4] S. Tiwari, and S. Rajendran. Four quadrant operation and control of three phase BLDC motor for electric vehicles. *2019 IEEE PES GTD Gd. Int. Conf. Expo. Asia, GTD Asia 2019*, pp. 577–582, 2019, doi: 10.1109/GTDAsia.2019.8715878.

[5] A. Boyko, and Y. Volyanskaya. Development of the gearless electric drive for the elevator lifting mechanism. *Eastern-European Journal of Enterprise Technologies*, vol. 4, no. 1–94, pp. 72–80, 2018, doi: 10.15587/1729-4061.2018.139726.

[6] M. Baranidharan, R. R. Singh, and R. Thirumalaivasan. Performance investigation of 3-level NPC inverter for solar PV application. pp. 1–6, 2022, doi: 10.1109/i-pact52855.2021.9696974.

[7] N. Priyadarshi, M. S. Bhaskar, S. Padmanaban, F. Blaabjerg, and F. Azam. New CUK-SEPIC converter based photovoltaic power system with hybrid GSA-PSO algorithm employing MPPT for water pumping applications. *IET Power Electron*, vol. 13, no. 13, pp. 2824–2830, 2020, doi: 10.1049/iet-pel.2019.1154.

[8] D. Das, N. Kumaresan, V. Nayanar, K. Navin Sam, and N. Ammasai Gounden. Development of BLDC motor-based elevator system suitable for DC microgrid. *IEEE/ASME Transactions on Mechatronics*, vol. 21, no. 3, pp. 1552–1560, 2016, doi: 10.1109/TMECH.2015.2506818.

[9] R. R. Kanna, M. Baranidharan, R. Raja Singh, and V. Indragandhi. Solar energy application in indian irrigation system. *IOP Conference Series: Materials Science and Engineering*, vol. 937, no. 1, 2020, doi: 10.1088/1757-899X/937/1/012016.

[10] P. Selvabharathi, S. Veerakumar, and V. Kamatchi Kannan. Simulation of DC-DC converter topology for solar pv system under varying climatic conditions with mppt controller. *IOP Conference Series: Materials Science and Engineering*, vol. 1084, no. 1, p. 012084, 2021, doi: 10.1088/1757-899x/1084/1/012084.

[11] V. S. Patil, S. Angadi, and A. B. Raju. Four quadrant close loop speed control of DC motor. *2016 International Conference on Circuits, Controls, Communications and Computing (I4C)*, pp. 1–6, 2017, doi: 10.1109/CIMCA.2016.8053305.

[12] K. Patel, S. Borole, K. Ramaneti, A. Hejib, and R. Raja Singh. Design and implementation of sun tracking solar panel and smart wiping mechanism using tinkercad. *IOP Conference Series: Materials Science and Engineering*, vol. 906, no. 1, 2020, doi: 10.1088/1757-899X/906/1/012030.

[13] D. Shah, J. Verma, R. R. Kanna, and R. R. Singh. Energy optimization for solar powered rural agro loads with elevated energy storage system. pp. 1–6, 2022, doi: 10.1109/npec52100.2021.9672481.

[14] S. Padmanaban, N. Priyadarshi, M. S. Bhaskar, J. B. Holm-Nielsen, E. Hossain, and F. Azam. A hybrid photovoltaic-fuel cell for grid integration with jaya-based maximum power point tracking: Experimental performance evaluation. *IEEE Access*, vol. 7, pp. 82978–82990, 2019, doi: 10.1109/ACCESS.2019.2924264.

[15] A. Rajasekhar, R. Kumar Jatoth, and A. Abraham. Design of intelligent PID/PIλDμ speed controller for chopper fed DC motor drive using opposition based artificial bee colony algorithm. *Engineering Applications of Artificial Intelligence*, vol. 29, pp. 13–32, 2014, doi: 10.1016/j.engappai.2013.12.009.

[16] A. K. Singh, and R. R. Singh. An overview of factors influencing solar power efficiency and strategies for enhancing. In *2021 Innovations in Power and Advanced Computing Technologies (i-PACT)*, 2021, pp. 1–6, doi: 10.1109/i-PACT52855.2021.9696845.

[17] A. Narendra, N. Venkataramana Naik, and N. Tiwary. PV fed separately excited DC motor with a closed loop speed control. *8th IEEE Power India Int. Conf. PIICON 2018*, 2018, pp. 10–13, doi: 10.1109/POWERI.2018.8704449.

[18] K. Pathak, S. H. Trivedi, and M. H. Ayalani. Operation and control of non-isolated interleaved bidirectional DC-DC converter integrated with solar PV system. *2019 IEEE International Conference on Innovations in Communication, Computing and Instrumentation (ICCI)*, pp. 92–95, 2019, doi: 10.1109/ICCI46240.2019.9404483.

[19] N. Priyadarshi, M. S. Bhaskar, P. Sanjeevikumar, F. Azam, and B. Khan. High-power DC-DC converter with proposed HSFNA MPPT for photovoltaic based ultra-fast charging system of electric vehicles. *IET Renewable Power Generation*, pp. 1–13, 2022, https://doi.org/10.1049/rpg2.12513.

[20] S. Padmanaban et al. A novel modified sine-cosine optimized MPPT algorithm for grid integrated PV system under real operating conditions. *IEEE Access*, vol. 7, pp. 10467–10477, 2019, doi: 10.1109/ACCESS.2018.2890533.

[21] A. Muraleedharan Pillai, R. Rajesh Kanna, and R. Raja Singh. Advancement of inventive solar power based frameworks for rural India. *IOP Conference Series: Materials Science and Engineering*, vol. 906, no. 1, 2020, doi: 10.1088/1757-899X/906/1/012001.

[22] M. S. Bhaskar, S. Padmanaban, N. Priyadarshi, F. Azam, and B. Khan. A novel hybrid grey wolf optimized fuzzy logic control based photovoltaic water pumping system. *IET Renewable Power Generation*, pp. 1–12, 2022, doi: 10.1049/rpg2.12638.

[23] S. S. Menon, R. R. Prasad, and R. R. Singh. Performance analysis of MPPT integrated solar charger for electric vehicle battery. pp. 1–6, 2022, doi: 10.1109/i-pact52855.2021.9696742.

[24] I. N. Jiya, A. M. S. Ali, H. Van Khang, N. Kishor, and R. Ciric. Novel multisource DC-DC converter for all-electric hybrid energy systems. *IEEE Transactions on Industrial Electronics*, vol. 0046, 2021, doi: 10.1109/TIE.2021.3131871.

[25] A. Mostaan, and M. Soltani. A family of four quadrant DC/DC converters with reduced number of components. *INTELEC, International Telecommunications Energy Conference*, vol. 2016, no. 1, 2016, doi: 10.1109/INTLEC.2015.7572307.

[26] Z. Li, and Y. Ruan. A novel energy saving control system for elevator based on supercapacitor bank using fuzzy logic. *Proceedings of ICEMS 2008: the 11th International Conference on Electrical Machines and Systems*, vol. 1, no. d, pp. 2717–2722, 2008.

[27] G. Wang, B. Wang, and C. Dianguo Xu. IET Electric Power Appl—2017—Wang—Weight-transducer-less control strategy based on active disturbance rejection theory.pdf." *IET Electric Power Applications* 2017, doi: 10.1049/iet-epa.2016.0474.

[28] R. Rahimi, S. Habibi, P. Shamsi, and M. Ferdowsi. A high step-up Z-source DC-DC converter for integration of photovoltaic panels into DC microgrid. *IEEE Applied Power Electronics Conference and Exposition (APEC)*, pp. 1416–1420, 2021, doi: 10.1109/APEC42165.2021.9487463.

[29] M. Kermani, E. Shirdare, S. Abbasi, G. Parise, and L. Martirano. Elevator regenerative energy applications with ultracapacitor and battery energy storage systems in complex buildings. *Energies*, vol. 14, no. 11, pp. 1–16, 2021, doi: 10.3390/en14113259.

[30] K. Bi et al. A model predictive controlled bi-directional four quadrant flying capacitor DC/DC converter applied in energy storage system. *IEEE Trans Power Electron*, vol. 8993, no. c, 2022, doi: 10.1109/TPEL.2022.3146510.

[31] S. Alotaibi, A. Darwish, X. Ma, and B. W. Williams. A new four-quadrant inverter based on dual-winding isolated CUK converters for railway and renewable energy applications. *IET Conference Publication*, vol. 2020, no. CP766, pp. 926–931, 2020, doi: 10.1049/icp.2021.1033.

[32] L. Cheng, and F. Wang. A new transformerless four quadrant DC-DC converter with wide conversion ratio. *Proceedings 2021 International Conference on Power Electronics, Computer Applications ICPECA 2021*, pp. 486–491, 2021, doi: 10.1109/ICPECA51329.2021.9362696.

[33] Z. Ouyang, Z. Zhang, M. A. E. Andersen, and O. C. Thomsen. Four quadrants integrated transformers for dual-input isolated DC-DC converters. *IEEE Trans Power Electron*, vol. 27, no. 6, pp. 2697–2702, 2012, doi: 10.1109/TPEL.2012.2186591.

[34] A. B. Kancherla, and D. R. Kishore. Design of solar-PV operated formal DC-DC converter fed PMBLDC motor drive for real-time applications. *Proceedings, 2020 IEEE International Symposium on Sustainable Energy, Signal Processing & Cyber Security*, 2020, doi: 10.1109/iSSSC50941.2020.9358813.

[35] S. Dey, and T. Bhattacharya. A transformerless DC-DC modular multilevel converter for hybrid interconnections in HVDC. *IEEE Transactions on Industrial Electronics*, vol. 68, no. 7, pp. 5527–5536, 2021, doi: 10.1109/TIE.2020.2994889.

[36] W. Gong, H. Wang, and L. Pan. The simulation and analysis of high-power-factor and energy feedback of elevator drive system on matrix rectifier. *2010 International Conference on Intelligent Computing, Automation and Applications ICICTA 2010*, vol. 3, pp. 955–958, 2010, doi: 10.1109/ICICTA.2010.439.

[37] D. Das, N. Kumaresan, V. Nayanar, K. Navin Sam, and N. Ammasai Gounden. Development of BLDC motor-based elevator system suitable for DC microgrid. *IEEE/ASME Trans Mechatronics*, vol. 21, no. 3, pp. 1552–1560, 2016, doi: 10.1109/TMECH.2015.2506818.

5 Non-Isolated Quadratic Bidirectional DC-DC Converters for Renewable Energy Systems

S. V. K. Naresh and Sankar Peddapati

CONTENTS

5.1 INTRODUCTION

The research on bidirectional dc-dc (BDC) converters has gained a good amount of attention due to its significant contribution in various applications like renewable energy generation with storage system, electrical vehicles (EV), and uninterruptable power supplies (UPS) [1]–[4], as shown in Figure 5.1. In renewable generation applications, as illustrated in Figure 5.1(a), an energy storage system is required to provide smooth power to the loads during transient and overload conditions to mitigate the intermittent nature of renewable sources. In general, the BDC converter transfers the power between two buses operating at different voltages and frequencies. For instance, in electric vehicles, the BDC converters provide the interface between the energy storage system (ESS) and the three-phase inverter fed motor as shown in Figure 5.1(b); thus, the energy imbalance can be solved by providing power during the starting or accelerating and absorbing the power during the

DOI: 10.1201/9781003323471-5

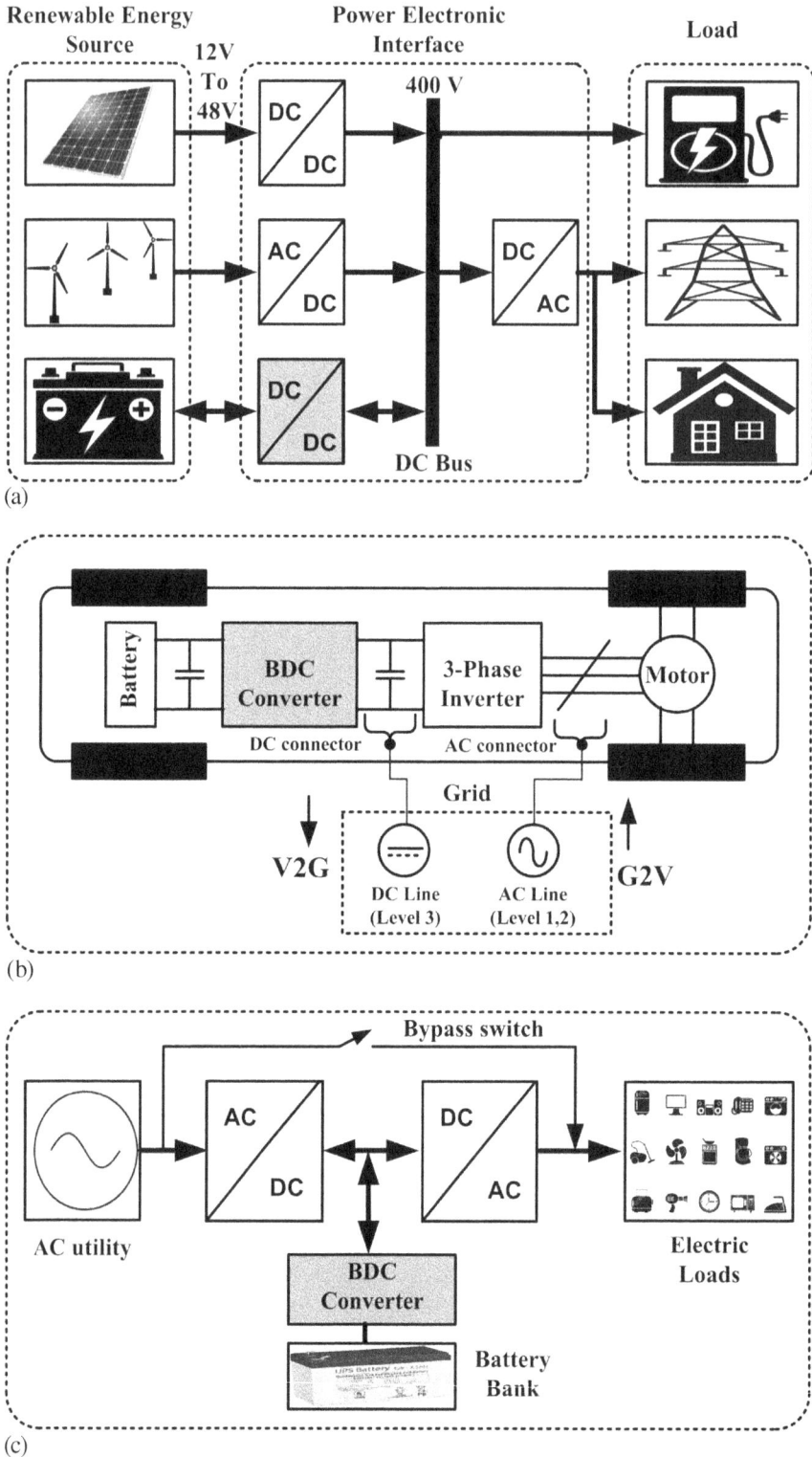

FIGURE 5.1 Applications of BDC converters: (a) renewable energy sources, (b) electric vehicles, and (c) uninterrupted power supplies.

regenerative braking period. Furthermore, the BDC converters also enable EVs to have the advanced V2G (vehicle-to-grid) and G2V (grid-to-vehicle) architecture capabilities to charge vehicle batteries at a low-price period and provide the power to the grid during high price or emergency periods.

Recently, the usage of uninterrupted power supplies (UPS) has increased due to the importance of data prevention in telecom applications, the protection of sensitive loads in health applications, and the requirement of backup supply in domestic and commercial applications, as shown in Figure 5.1(c). Here, the ac-dc and dc-ac converters enable the interface between the ac utility, load, and energy storage system. The BDC converter and associated Battery Management System (BMS) control are employed to charge the battery when the utility is available and to deliver the required load power during a power interruption. To sum up, in many applications, the BDC converter is used as an interfacing device between the power sources and ESS system to improve the system's performance, efficiency, and stability [1]–[4].

The conventional BDC converters are classified into isolated and non-isolated categories. However, non-isolated converters have the advantage of a simple and flexible structure, less weight due to the absence of a transformer, and lack of magnetic interference [1]. Due to less weight, cost, and size, the non-isolated BDC converters have received more attention among researchers. In literature various types of non-isolated BDC converters are reported, namely, buck-boost [5], [6], switched capacitor and switched inductor [7]–[9], interleaved [10], [11], coupled inductors [12], [13], soft-switching [14], [15], Quasi Z-source [16], [17], multilevel [18], and multi-input or/and multi-output converters [19], [20]. Each type of converter has its own merits and demerits; however, all these BDC converters suffer from the limited voltage conversion ratio (VCR).

In applications like distributed generation and electric vehicles, converters with high voltage conversion ratios and bidirectional capabilities are required. In literature, the converters with quadratic VCR have shown feasible solution to achieve high VCR due to their simple structure and control [21]–[25]. In this regard, recently, many works have been reported on the non-isolated BDC converters with quadratic VCR [26]–[35]. The cascaded boost converter with bidirectional switches is considered as the conventional quadratic bidirectional (CCQB-BDC) converter. In step-up mode, the CCQB-BDC converter produces quadratic boost gain, and it produces quadratic buck gain during the step-down mode of operation. However, higher electric stress on switches and low efficiency has been its major drawbacks. For a similar VCR in both modes, the QBDC converters are proposed [26]–[28] with reduced electric stress and input current ripple. Likewise, a new extendable QBDC is proposed [29] for electric vehicle charging applications. In [30], a new QBDC converter is proposed with coupled inductors to reduce the current ripple. To improve the efficiency, a soft-switching bidirectional converter with a quadratic VCR is proposed [31]; but it has the demerits of limited VCR and high switch count. Recently, a new BDC converter has been proposed [32] with quadratic buck-boost gain in both modes of operation with continuous current at both voltage ports. Furthermore, a high gain QBDC converter is proposed [33] with a high semiconductor utilization factor for EV charging applications. In [34], another high gain QBDC converter is proposed with an extended structure by using a switched-capacitor cell. In addition, a high gain QBDC converter is proposed [35] by adding an additional boost converter and switched capacitor to the converter in [27]. Despite a wide range of VCR, higher switch count and control complexity are the shortcomings of the converter.

In conclusion, limited research has been conducted on the QBDC converters compared to other types of BDC converters. However, this area has the potential to produce more topologies; thus, optimal topology can be selected for the given application. Therefore, this chapter aims to present a comprehensive review of the QBDC converters, which is not available in the existing literature.

5.2 QUADRATIC BIDIRECTIONAL (QBDC) CONVERTERS

The QBDC converters are constructed using bidirectional switches, inductors, and capacitors. The switch named S_1-S_n operates actively in step-up mode, and the switches named Q_1-Q_n operate in step-down mode. To simplify the analysis, it is assumed that converters are operating in steady-state, continuous conduction mode (CCM), all the components in the converter are ideal, and the size of the capacitor is adequate to maintain the constant voltage. The non-isolated quadratic bidirectional (QBDC) converters can be classified into three categories:

- Quadratic boost/buck BDC (QB-BDC) converters [26]–[31]
- Quadratic buck-boost BDC (QBB-BDC converters) [32]
- High quadratic gain BDC (HQG-BDC) converters [33]–[35]

5.3 QUADRATIC BOOST/BUCK BDC (QB-BDC) CONVERTERS

This section presents the details of six bidirectional converters, which give the quadratic voltage conversion ratio in both step-up (boost) and step-down (buck) mode. However, the conventional cascaded quadratic boost bidirectional converter is considered to demonstrate the operational and steady-state analysis on behalf of the converters in [26]–[31], as other converters follow the similar analysis and produce the same VCR.

5.3.1 CONVENTIONAL CASCADED QB-BDC CONVERTER

The conventional cascaded QB-BDC can be formed by cascading the two boost converters with bidirectional switches, as shown in Figure 5.2. It consists of four active bidirectional switches, two inductors, and three capacitors. In each direction of the power flow, two switches operate actively during the DT period, whereas body diodes of the other two switches contribute to the energy transfer during the (1-D)T period. Where D is the duty cycle of active switches, and T is the switching time. The operation of the CCQB converter is explained next.

5.3.1.1 Step-Up Mode Operation

Figure 5.3 shows the operating circuit of the CCQB-BDC converter in the step-up mode, and the key operating waveforms of the converter in CCM are shown in Figure 5.4.

During DT interval, the S_1-S_2 are ON while Q_1-Q_2 are OFF; thus, L_1 and L_2 are magnetized using V_L and C_1, respectively. The dashed line represents the current path direction in this interval, as shown in Figure 5.4, and the corresponding equations obtained in this interval are given here.

$$\begin{cases} V_{L1} = V_L \\ V_{L2} = V_{C1} \end{cases} \tag{1}$$

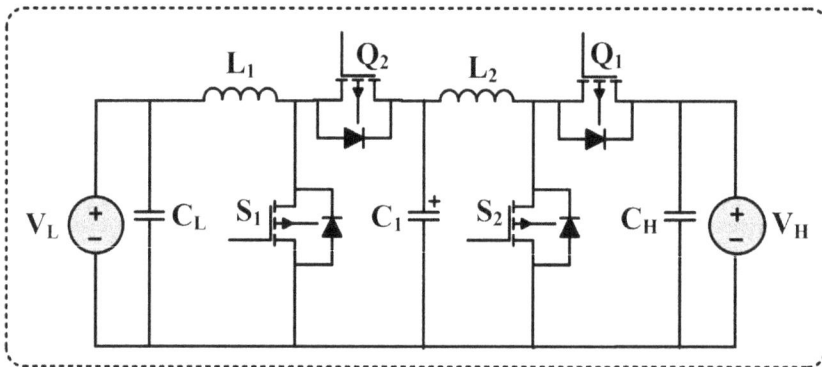

FIGURE 5.2 The conventional cascaded QB-BDC (CCQB-BDC) converter.

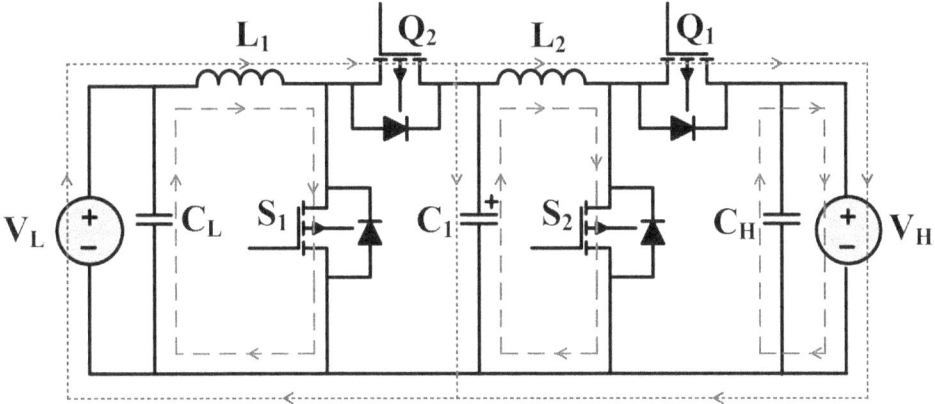

FIGURE 5.3   Step-up mode operation of CCQB-BDC converter.

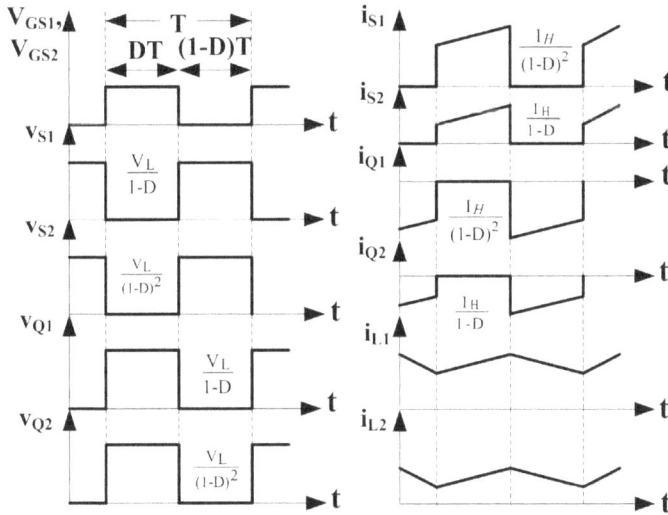

FIGURE 5.4   Key operating waveforms of the CCQB-BDC converter in the step-up mode.

During (1-D)T interval, the S_1-S_2 are OFF, and the anti-parallel diodes of Q_1-Q_2 will be ON. The inductor L_1 transfers energy to C_1 to charge, and inductor L_2 transfers energy to the load. The current flow direction is represented with a dotted line, as depicted in Figure 5.3. The inductor voltages in this interval are given here.

$$\begin{cases} V_{L1} = V_L - V_{C1} \\ V_{L2} = V_{C1} - V_H \end{cases} \tag{2}$$

The steady-state VCR of the CCQB-BDC converter can be obtained using the flux balance principle of the inductor voltages given in equations (1) and (2).

$$\begin{cases} V_{L1} = 0 = \left(V_L\right)DT + \left(V_L - V_{C1}\right)(1-D)T \\ V_{L2} = 0 = \left(V_{C1}\right)DT + \left(V_{C1} - V_H\right)(1-D)T \end{cases} \tag{3}$$

From equation (3), the steady-state capacitor's voltage and the step-up VCR (G_1) can be calculated and are given in equation (4).

$$\begin{cases} V_{C1} = \dfrac{1}{1-D} V_L \\[2mm] V_H = \dfrac{1}{1-D} V_{C1} \\[2mm] V_H = \dfrac{1}{(1-D)^2} V_L \\[2mm] G_1 = \dfrac{V_H}{V_L} = \dfrac{1}{(1-D)^2} \end{cases} \qquad (4)$$

5.3.1.2 Step-Down Mode Operation

The step-down operation circuit of the CCQB-BDC converter is shown in Figure 5.5, and the corresponding time-domain waveforms of the converter in the CCM are given in Figure 5.6.

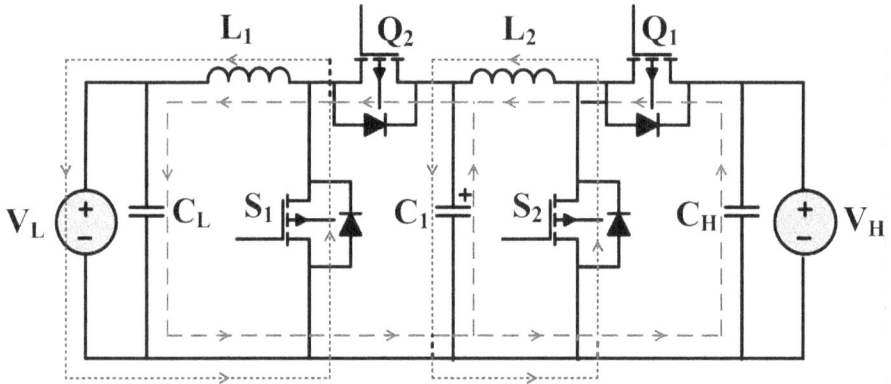

FIGURE 5.5   Step-down mode operation of CCQB-BDC converter.

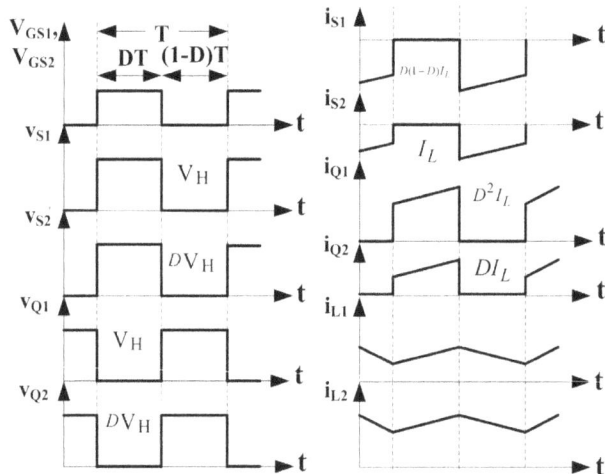

FIGURE 5.6   Typical operating waveforms of the CCQB-BDC converter in the step-down mode.

The active switches Q_1-Q_2 will be turned ON while S_1-S_2 are turned OFF during the $0 < t < DT$ interval. The high voltage source, V_H, magnetizes the inductors L_1, and the capacitor magnetizes the inductor L_2 and transfers energy to the capacitor C_L. The dashed line indicates the current flow in this interval, as shown in Figure 5.5; and the corresponding KVL equations are given here.

$$\begin{cases} V_{L1} = V_{C1} - V_L \\ V_{L2} = V_H - V_{C1} \end{cases} \tag{5}$$

In the $DT < t < T$ interval, the Q_1-Q_2 will be turned OFF, and the body diode of S_1-S_2 will be forward biased. In this interval, as shown in Figure 5.5, the inductor L_1 charges the capacitor C_1, and inductor L_2 transfer energy to the load. The current flow direction is represented with a dotted line. The following equations can be derived for this interval.

$$\begin{cases} V_{L1} = -V_L \\ V_{L2} = -V_{C1} \end{cases} \tag{6}$$

The steady-state step-down VCR of the CCQB-BDC converter can be obtained using the inductor voltages flux balance principle given in equations (5) and (6).

$$\begin{cases} V_{L1} = 0 = (V_{C1} - V_L)DT + (-V_L)(1-D)T \\ V_{L2} = 0 = (V_H - V_{C1})DT + (-V_{C1})(1-D)T \end{cases} \tag{7}$$

From equation (7), the steady-state capacitor's voltage and the step-down VCR (G_2) can be calculated and are given in equation (8).

5.3.1.3 Passive Components Design

$$\begin{cases} V_{C1} = DV_H \\ V_L = DV_{C1} \\ V_L = D^2 V_L \\ G_2 = \dfrac{V_L}{V_H} = D^2 \end{cases} \tag{8}$$

The design of circuit components is necessary to operate the converter in CCM over the range of output power. The maximum allowable current and voltage ripples can be calculated for known operating conditions. The minimum size of passive components required to maintain ripple standards can be estimated using the following expressions:

$$\begin{cases} L_1 = \dfrac{V_L D}{(\Delta i_{L1}) f_s} \\ L_2 = \dfrac{V_L}{(\Delta i_{L1}) f_s} \dfrac{D}{1-D} \end{cases} \text{ and } \begin{cases} C_1 = \dfrac{I_L}{(\Delta V_{C1}) f_s} \dfrac{D}{1-D} \\ C_{L/H} = \dfrac{I_L}{(\Delta V_{C1}) f_s} D(1-D)^2 \end{cases} \tag{9}$$

5.3.2   REMARKS ON QB-BDC CONVERTERS

Figure 5.7 shows the other bidirectional converters [26]–[30], which produce the same gain as the CCQB-BDC converter. These converters use four active switches, two inductors, and three capacitors. In each direction, two active switches and body diodes of the other two switches are responsible for the power transfer from one port to another. The following conclusions can be drawn from these converters.

- In [26] and [27], the converters are the modified cascaded quadratic converters. However, the converter in [27] produces less current ripple at low voltage port due to the current sharing between two inductors, which leads to power loss reduction. Therefore, the converter in [27] has better efficiency than the CCQB-BDC and [26] at a higher power operating condition.
- The converter in [28] uses two extra diodes to achieve a similar VCR as the CCQB-BDC converter. In this converter, to avoid switching losses, one active switch is turned ON continuously for the whole operation, and one active switch is operated with a duty cycle for the given operating mode. Thus, controlling the converter is complex as compared to other QB-BDC converters.

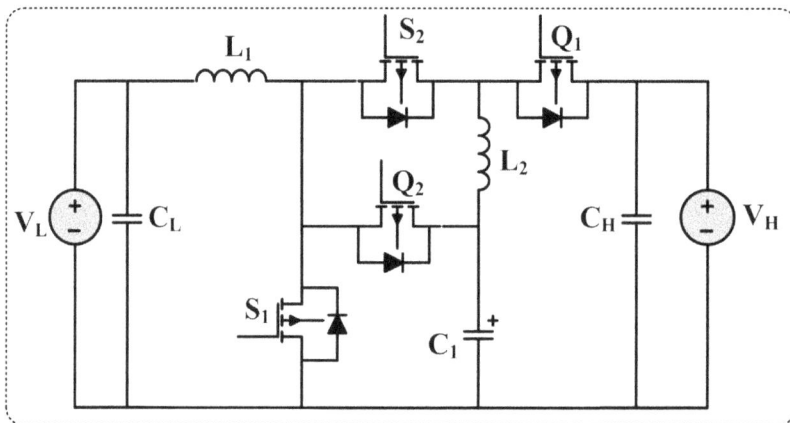

(a)

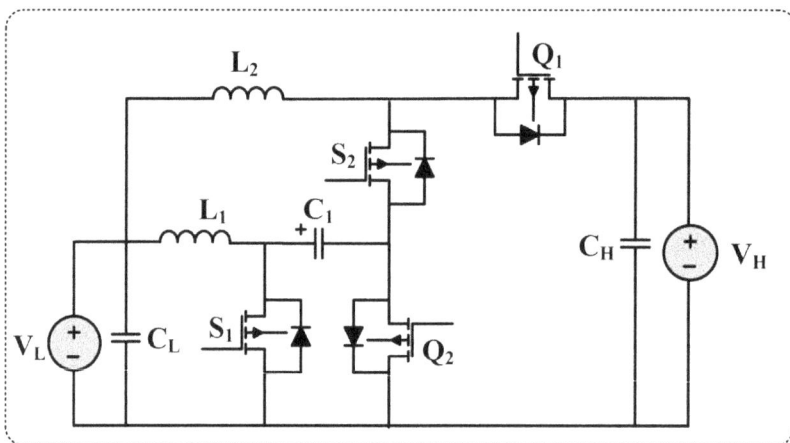

(b)

FIGURE 5.7   The BDC converters with quadratic boost/buck gain: (a) converter [26], (b) converter [27], (c) converter [28], (d) converter [29], (e) converter [30], (f) converter [31].

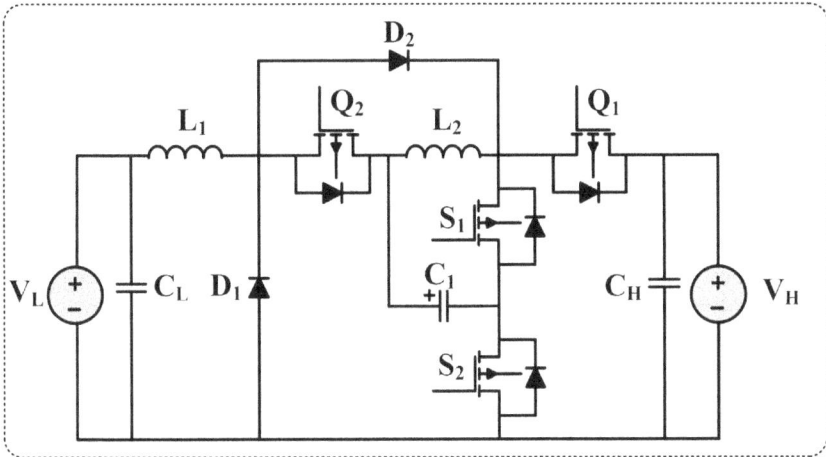

(c)

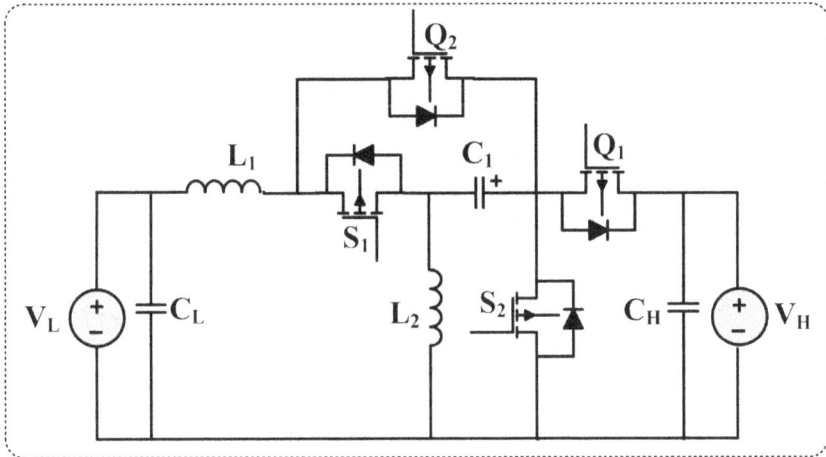

(d)

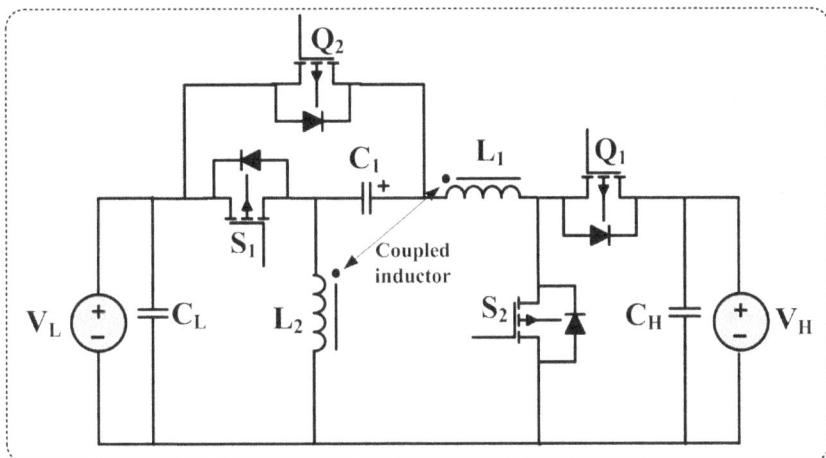

(e)

FIGURE 5.7   Continued

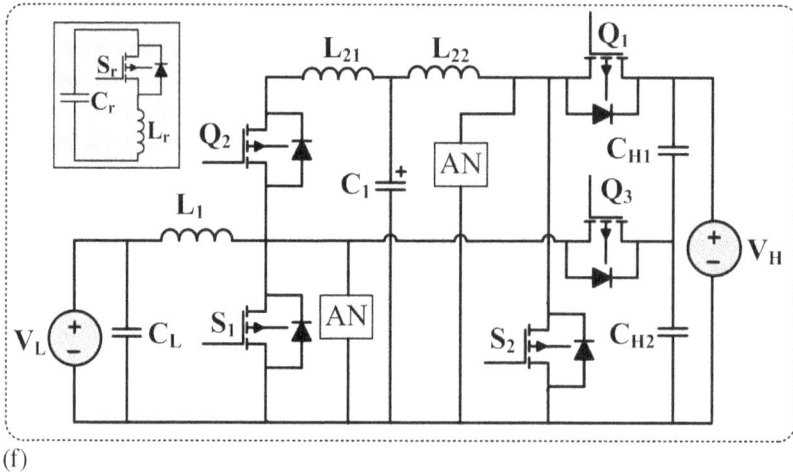

(f)

FIGURE 5.7 Continued

- In [29], the converter is proposed by introducing a switching cell to develop the extended quadratic BDC converter. The switching cell is highlighted in Figure 5.7(d), and the converter has the advantages of improved VCR, redundancy, and modularity. However, it has the drawbacks of a complex control model and is duty cycle sensitive.
- The converter in [30] is constructed with the help of coupled inductors, as shown in Figure 5.7(e). The coupled inductors in this converter help to reduce the current ripple significantly in both operating modes. In contrast, the designing of coupled inductor windings and magnetic interference has been its shortcomings.
- In [31], a bidirectional converter is proposed to obtain a high VCR with soft-switching capabilities in both directions of power flow, as given in Figure 5.7(f). The soft-switching is achieved by adding an auxiliary networks formed by a low power switch (S_r), capacitor (C_r), and inductor (L_r). Due to soft-switching and alleviating the reverse recovery losses, the converter can operate at a higher power rating with high efficiency but at the cost of high component count and limited VCR.

5.4 QUADRATIC BUCK-BOOST BDC (QBB-BDC) CONVERTER

The quadratic buck-boost bidirectional (QBB-BDC) converter [32] is shown in Figure 5.8, consisting of four active switches (S_1-S_2 and Q_1-Q_2), three inductors (L_1-L_3), and four capacitors (C_1-C_2 and C_L-C_H). This converter is formed by adding a voltage multiplier cell (VMC) to the Cuk-based circuit to achieve the continuous current at both ends, along with the quadratic VCR. This converter produces the quadratic buck-boost gain in both modes of operation; here, the V_L and V_H refer to the port L and H, respectively (not necessarily as LV and HV port), as the converter can give a buck-boost gain in both directions. It is worth mentioning that this representation applies to this converter only.

5.4.1 FORWARD OPERATION

In this mode, the energy will transfer from the V_L port to the V_H port through the QBB-BDC converter, as shown in Figure 5.9. Here, S_1-S_2 will be turned ON for DT interval; the L_1, L_2, and L_3 will be magnetized due to the positive voltage of V_L, C_1, and C_2, respectively. The dashed line represents the power flow path in this interval, as shown in Figure 5.9. In DT to T interval, the S_1-S_2 will be

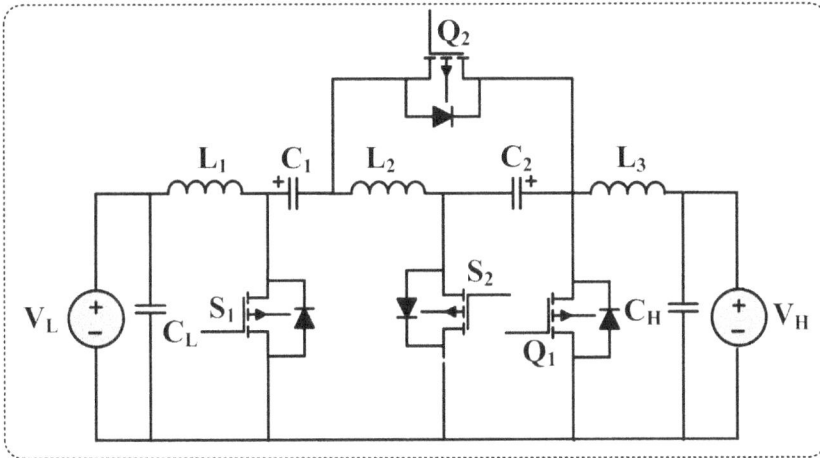

FIGURE 5.8 Quadratic buck-boost BDC (QBB-BDC) converter.

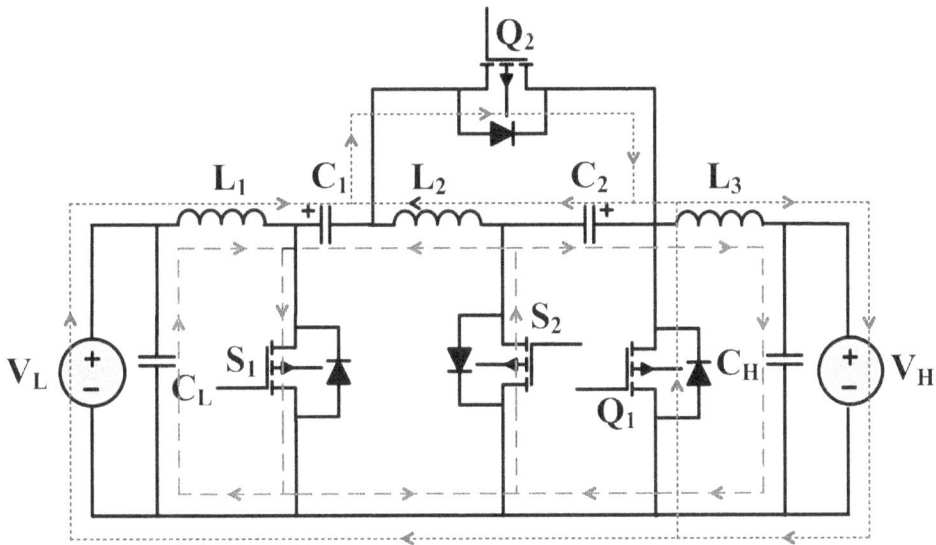

FIGURE 5.9 Forward operation of QBB-BDC converter.

turned OFF, and the body of the diodes of Q_1 and Q_2 will be in the conduction; and corresponding current direction is described with a dotted line, as shown in Figure 5.9.

The steady-state VCR of the QBB-BDC during the forward operation can be obtained by taking the flux balance of the inductors, as given here:

$$\begin{cases} V_{L1} = 0 = (V_L)DT + (V_L - V_{C1})(1-D)T \\ V_{L2} = 0 = (V_{C1})DT + (-V_{C2})(1-D)T \\ V_{L3} = 0 = (V_{C2} - V_H)DT + (-V_H)(1-D)T \end{cases} \tag{10}$$

From equation (10), the following steady-state expressions can be derived.

$$\begin{cases} V_{C1} = \dfrac{1}{1-D} V_L \\[2mm] V_{C2} = \dfrac{D}{(1-D)^2} V_L \\[2mm] G_1 = \dfrac{V_H}{V_L} = \left(\dfrac{D}{1-D} \right)^2 \end{cases} \tag{11}$$

5.4.2 BACKWARD OPERATION

The power will flow from V_H to the V_L port during backward operation, as shown in Figure 5.10. During DT interval, the switches Q_1-Q_2 will be turned ON, whereas the other two active switches are OFF; the high voltage source V_H, C_1, and C_2 magnetize the inductors L_3, L_1, and L_2, respectively. The Q_1-Q_2 will be turned OFF, and the body of the diodes of S_1-S_2 will be in the conduction during DT to T interval. The dashed, and dotted lines represent the power flow path in DT and (1-D)T intervals, respectively, as shown in Figure 5.10.

The steady-state VCR of the QBB-BDC during backward operation can be obtained by taking the flux balance of the inductors, as given here:

$$\begin{cases} V_{L1} = \left(V_{C1} - V_L \right) DT + \left(-V_L \right)(1-D)T \\ V_{L2} = \left(V_{C2} \right) DT + \left(-V_{C1} \right)(1-D)T \\ V_{L3} = \left(V_H \right) DT + \left(V_H - V_{C2} \right)(1-D)T \end{cases} \tag{12}$$

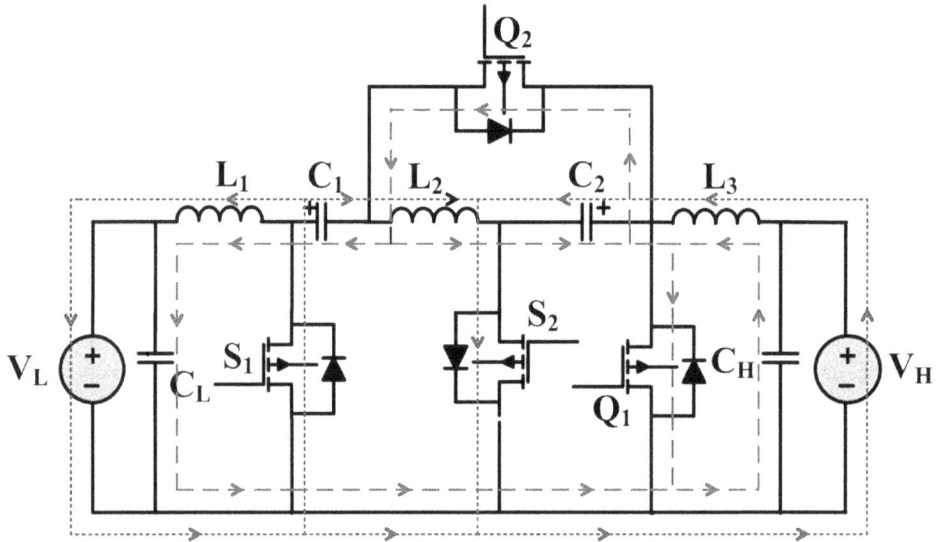

FIGURE 5.10 Backward operation of QBB-BDC converter.

From equation (12), the following steady-state expressions can be derived.

$$\begin{cases} V_{C1} = \dfrac{D}{(1-D)^2} V_H \\[3mm] V_{C2} = \dfrac{1}{1-D} V_H \\[3mm] G_2 = \dfrac{V_L}{V_H} = \left(\dfrac{D}{1-D}\right)^2 \end{cases} \tag{13}$$

5.4.3 Passive Components Design

The size of passive components depends on the operating power, switching frequency (f_s), and ripple specifications. The expressions of passive components of the QBB-BDC converter are given here:

$$\begin{cases} L_1 = \dfrac{D^2 V_H}{(1-D)(\Delta i_{L1}) f_s} \\[3mm] L_2 = \dfrac{D V_H}{(1-D)(\Delta i_{L2}) f_s} \\[3mm] L_3 = \dfrac{D V_H}{(\Delta i_{L3}) f_s} \end{cases} \text{ and } \begin{cases} C_1 = \dfrac{D I_L}{(\Delta v_{C1}) f_s} \\[3mm] C_2 = \dfrac{D^2 I_L}{(1-D)(\Delta v_{C2}) f_s} \\[3mm] C_{L/H} = \dfrac{D(I_{L1} - I_L)}{(\Delta v_{cL/H}) f_s} \end{cases} \tag{14}$$

Where D is the duty cycle, $\Delta i_{L1} - \Delta i_{L3}$ are the current ripple of the inductors L_1-L_3; and, $\Delta v_{L1}, \Delta v_{C2}$ and $\Delta v_{CL/H}$ are the voltage ripples of the capacitor C_1, C_2, and $C_{L/H}$, respectively. The maximum value of components is obtained by substituting the maximum duty cycle for a given power rating and switching frequency.

5.5 HIGH QUADRATIC GAIN BDC (HQG-BDC) CONVERTERS

The details of three high gain quadratic bidirectional converters are presented in this section. The converter in [34] is considered for operation demonstration, as it produces a high VCR with minimal circuit components, as shown in Figure 5.11.

5.5.1 High Quadratic Gain BDC Converter [34]

This converter is constructed by adding the switched capacitor network (Q_1, Q_4, C_{H2}, and C_2) to the quadratic converter (S_1, S_2, Q_2, Q_4, L_1, L_2, C_L, C_1, and C_{H1}), as depicted in Figure 5.11. The operation of this converter in both modes is explained in the following sections.

5.5.1.1 Step-Up Mode Operation

In the first state, the switches S_1-S_2 and the body diode of Q_1 will be conducted; and the switches Q_2-Q_3-Q_4 are OFF. In this state, the inductor L_1 and L_2 get magnetized, the capacitors C_2 and C_{H2} will get discharged, and capacitor C_{H1} will be charged. The line describes the current path in this state, as shown in Figure 5.12. The following KVL equations can be obtained in this state of operation.

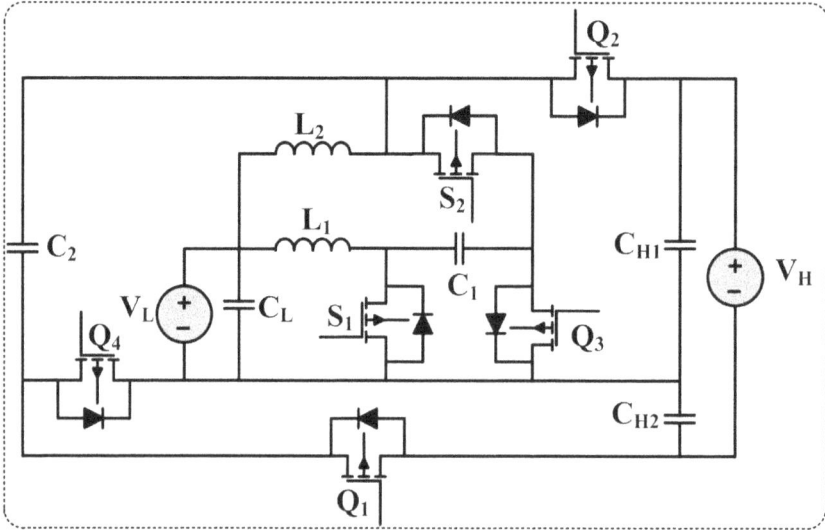

FIGURE 5.11 High quadratic gain BDC converter in [34].

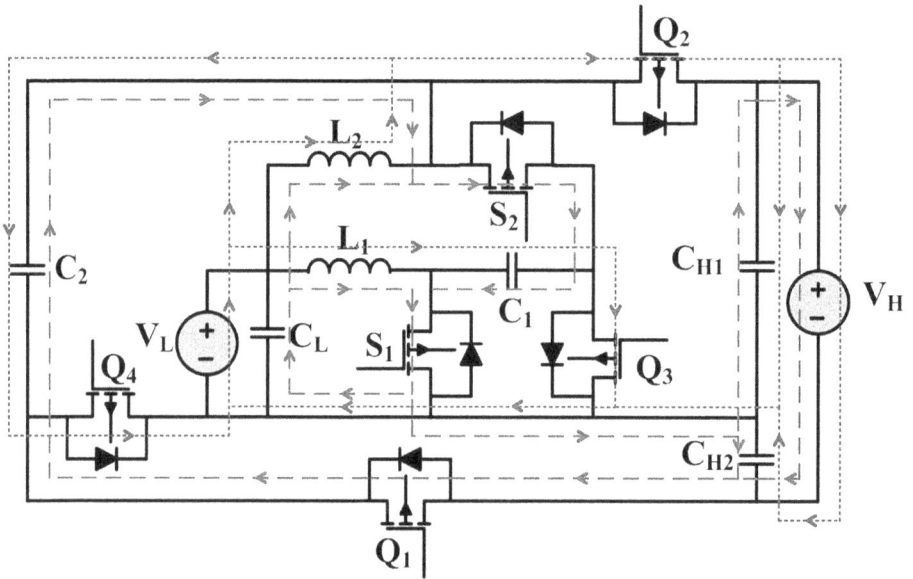

FIGURE 5.12 Step-up mode operation of high quadratic gain BDC converter in [34].

$$\begin{cases} V_{L1} = V_L \\ V_{L2} = V_L + V_{C1} \end{cases} \tag{15}$$

In the second state, the switches S_1-S_2 are OFF, and the body diodes of Q_2-Q_3-Q_4 will get forward biased. In this state, the inductors get demagnetized, the capacitors C_2 and C_{H2} will get charged, and the capacitor C_{H1} will be discharged. The current path in this state is represented with a dotted line in Figure 5.12. The KVL equations of inductor voltages can be written, as follows:

$$\begin{cases} V_{L1} = V_L - V_{C1}) \\ V_{L2} = V_L - V_{CH2} \end{cases} \tag{16}$$

The steady-state VCR and the capacitor's voltage can be calculated using the flux balance rule of the inductor's voltage.

$$\begin{cases} V_{C2} = V_{CH2} = \dfrac{1}{(1-D)^2} V_L \\[2ex] V_{C1} = \dfrac{1}{1-D} V_L \\[2ex] V_{CH1} = \dfrac{2-D}{(1-D)^2} V_L \\[2ex] V_H = V_{CH1} + V_{CH2} = \dfrac{3-D}{(1-D)^2} V_L \end{cases} \tag{17}$$

5.5.1.2　Step-Down Mode Operation

In the first state, the active switches S_1-S_2-Q_1 are OFF; and Q_2-Q_3-Q_4 are ON. In this state, the inductor L_1 and L_2 magnetized due to the voltage across capacitor C_1, and high voltage source, respectively. The capacitors C_1, C_2, and C_{H2} get discharged, and the capacitor C_{H1} gets charged. The dashed line describes the current path in this state, as shown in Figure 5.13.

In the second state, the Q_2, Q_3, and Q_4 are OFF; the switch Q_1 is turned ON along with body diodes of the S_1-S_2. In this state, the inductors and C_1 will transfer their energy to the other capacitor; thus, the capacitors C_1, C_2, and C_{H2} will be charged. The dotted line represents the current direction in this state, as shown in Figure 5.13.

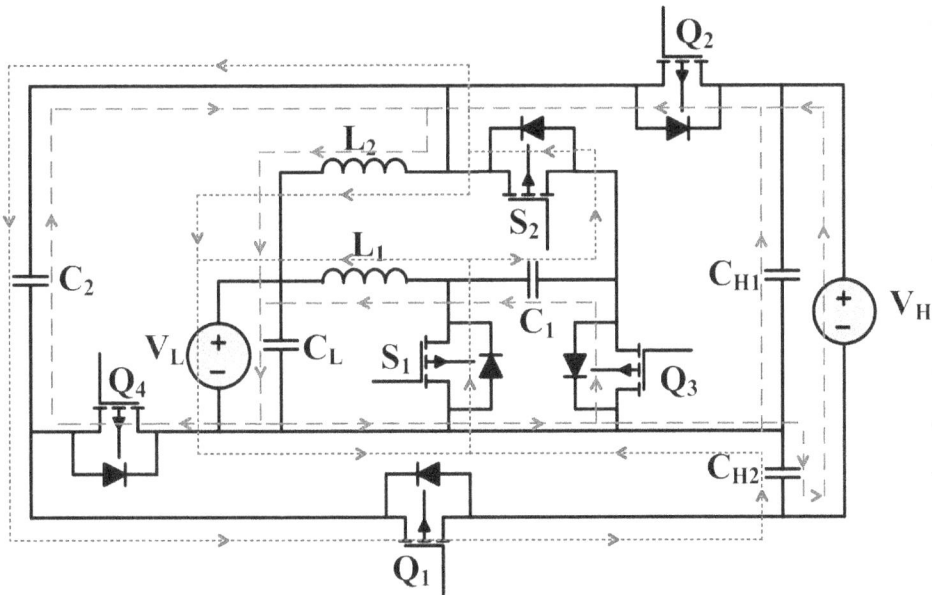

FIGURE 5.13　Step-down mode operation of high quadratic gain BDC converter in [34].

The two operating states can obtain the following steady-state voltage expressions during step-up operation.

$$
\begin{cases}
V_{C2} = V_{CH2} = \dfrac{1}{2+D} V_H \\[2mm]
V_{C1} = \dfrac{D}{2+D} V_H \\[2mm]
V_{CH1} = \dfrac{1+D}{2+D} V_H \\[2mm]
V_L = \dfrac{D^2}{2+D} V_H
\end{cases}
\tag{18}
$$

The steady-state VCR in equations (17) and (18) suggests that the converter in [34] produces the high VCR in both modes.

5.5.1.3 Design of Passive Components

The maximum allowable current and voltage ripples can be calculated for the known operating parameters. Afterward, the minimum size of passive components required to maintain ripple standards can be estimated for the given ripple specifications. The following expressions can be used to design the passive components of the converter.

$$
\begin{cases}
L_1 = \dfrac{DV_L}{f_s(\Delta I_{L1})} \\[3mm]
L_2 = \dfrac{DV_L}{f_s(\Delta I_{L1})}\left(\dfrac{2-D}{1-D}\right)
\end{cases}
\text{and }
\begin{cases}
C_L = \dfrac{DV_L}{8f_s^2(\Delta V_{C1})}\left(\dfrac{1}{L_1} + \dfrac{1}{L_2(1-D)}\right) \\[3mm]
C_1 = \dfrac{I_H}{f_s(\Delta V_{C1})}\left(\dfrac{1+D}{1-D}\right) \\[3mm]
C_2 = \dfrac{I_H}{f_s(\Delta V_{C2})} \\[3mm]
C_{H1} = \dfrac{DI_H}{f_s(\Delta V_{CH1})} \\[3mm]
C_{H2} = \dfrac{(1-D)I_H}{f_s(\Delta V_{CH2})}
\end{cases}
\tag{19}
$$

5.5.2 REMARKS ON HQG-BDC CONVERTERS

Figure 5.14 shows the other two high gain quadratic boost (HQG-BDC) converters. The steady-state VCR in both modes of operation of these converters can be derived from the analysis as mentioned earlier. The following conclusions can be derived from HQG-BDC converters.

- The high quadratic gain BDC converter in [33] is proposed for the V2G and G2V architecture of EVs, as shown in Figure 5.14(a). This converter has two inductors, four capacitors, and five active bidirectional switches; two switches (S_1-S_2) operate actively in the forward direction, and three switches (Q1-Q2-Q3) operate actively in the backward direction. This converter produces high VCR in both directions; further, it has low total voltage stress (TVS) and a high utilization factor.
- The converter in [35] presents three CQBDC converters on the LV side and a switched capacitor module on the HV side, as shown in Figure 5.14(b). This converter has eight

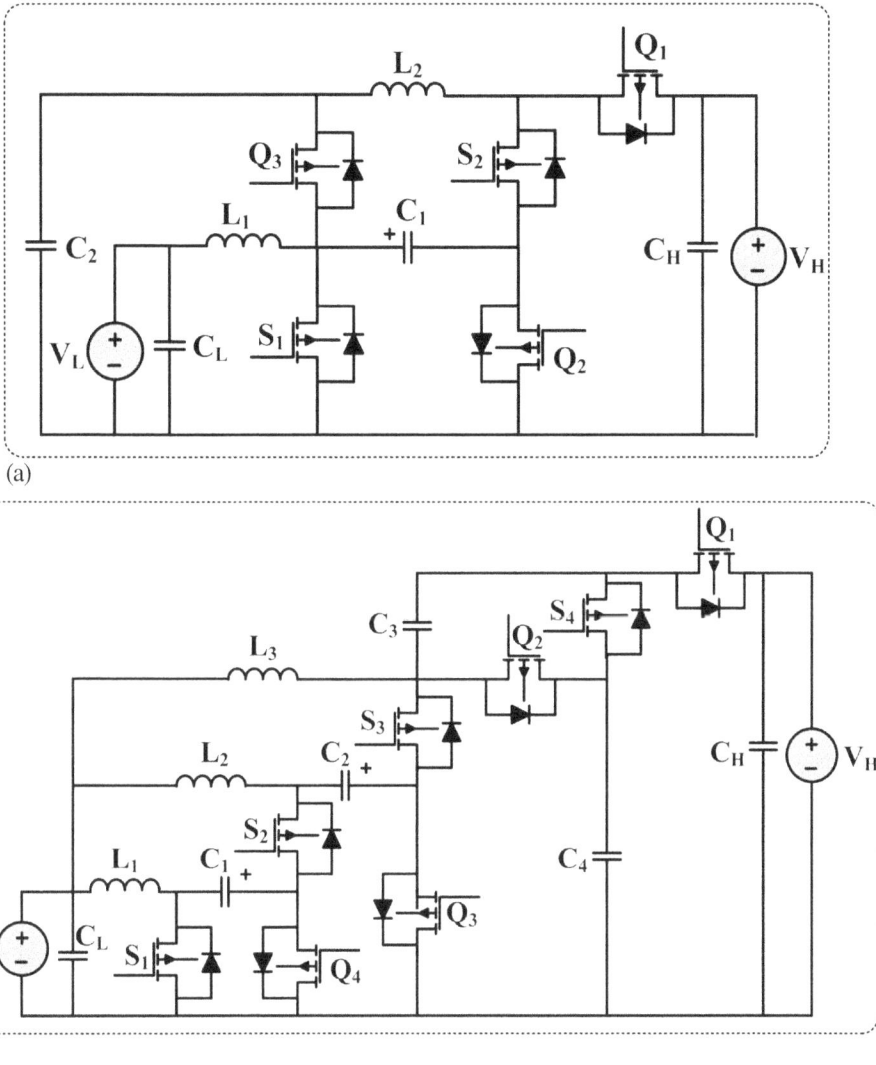

FIGURE 5.14 High quadratic gain BDC converters: (a) converter in [33], (b) converter in [35].

switches, three inductors, and six capacitors. Although it can produce high VCR, high component count, low efficiency, and complicated control circuit are its demerits.

5.6 COMPARATIVE ANALYSIS

This section of the chapter presents the comprehensive comparative analysis of the non-isolated quadratic bidirectional (QBDC) converters. The converters are compared with regard to a number of components, VCR, and efficiencies in both directions of power flow, and other features like soft-switching capability, continuous current at low and high voltage port, and availability of common ground tabulated in Table 5.1.

From Table 5.1, the CCQB-BDC and converters in [26]–[30] have a low component count; conversely, the converters in [31] and [35] have used a higher number of components to achieve the quadratic VCR, as compared to other converters. Understandably, the converter [31] requires more components to achieve the soft-switching capability in both directions. Likewise, the converter in

TABLE 5.1

Comparison of the Non-Isolated Quadratic Bidirectional DC-DC (QBDC) Converters

Topology	CCQB	[26]	[27]	[28]	[29]	[30]	[31]	[32]	[33]	[34]	[35]
Switches	4	4	4	4	4	4	7	4	5	6	8
Diodes	0	0	0	2	0	0	0	0	0	0	0
Inductors[*]	2	2	2	2	2	2	4	3	2	2	3
Capacitors[#]	3	3	3	3	3	3	7	4	4	5	6
Total	9	9	9	11	9	9	18	11	11	13	17
Step-up gain (G_1)	$\dfrac{1}{(1-D)^2}$	$\dfrac{1}{(1-D)^2}$	$\dfrac{1}{(1-D)^2}$	$\dfrac{1}{(1-D)^2}$	$\dfrac{1}{(1-D)^2}$	$\dfrac{1}{(1-D)^2}$	$1-\dfrac{D}{\frac{3}{(1-D)^2}}$	$\left(\dfrac{D}{1-D}\right)^2$	$\dfrac{1+D}{(1-D)^2}$	$\dfrac{3-D}{(1-D)^2}$	$\dfrac{D^2-3D+4}{(1-D)^3}$
Step-down gain (G_2)	D^2	D^2	D^2	D^2	D^2	D^2	$1.5D^2$	$\left(\dfrac{D}{1-D}\right)^2$	$\dfrac{D^2}{2-D}$	$\dfrac{D^2}{2+D}$	$\dfrac{D^3}{D^2+D+2}$
Output power (W)	100	100	160	200	500	250	200	500	500	2000	100
Switching frequency (kHz)	50	50	30	15	50	40	100	50	50	100	30
Step-up efficiency (%)	94.3	95.5	96.5	83.2	97.1	90	97	97.8	96.8	97.2	85.6
Step-down efficiency (%)	92.7	93.7	93.2	88.7	95.6	91	94.5	97.5	97.2	97.4	83.2
Soft-switching	*no*	*no*	*no*	*no*	*no*	*no*	*yes*	*no*	*no*	*no*	*no*
Continuous current at LV port	√	√	√	√	√	√	√	√	√	√	√
Continuous current at HV port	×	×	×	×	×	×	×	√	×	×	×
Common ground	√	√	√	√	√	√	√	√	√	×	√

Notes: *Included the coupled inductor, # included the capacitors at LV and HV ports

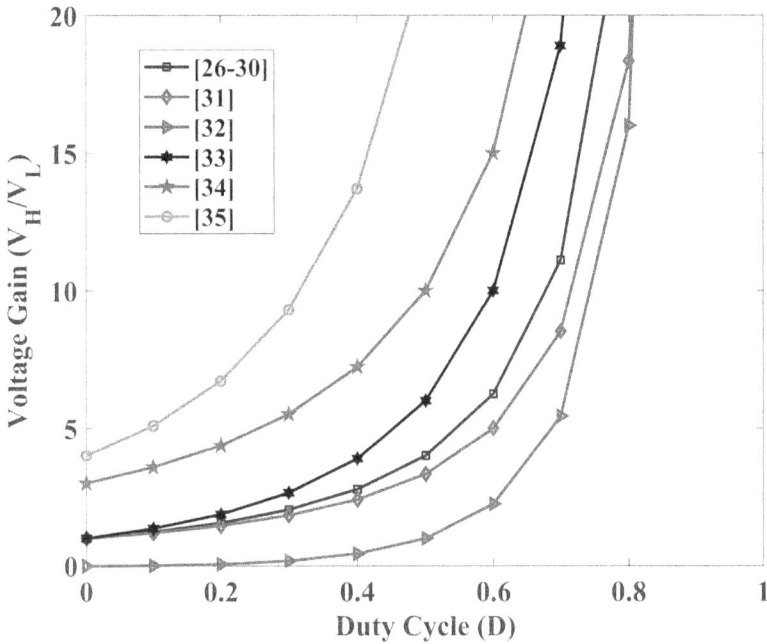

(a)

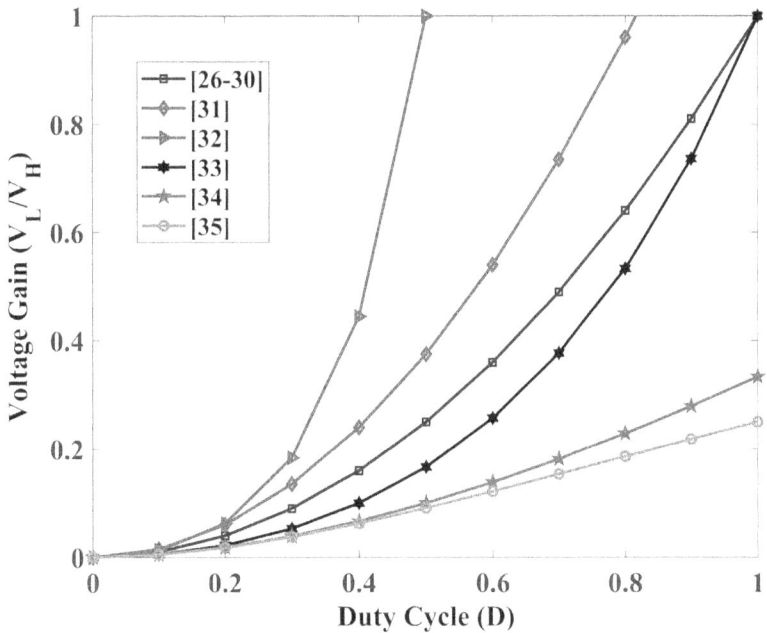

(b)

FIGURE 5.15 The VCR comparison of QB-BDC converters: (a) step-up mode, (b) step-down mode.

[35] achieves high VCR at the cost of a higher component count, low efficiency, and complex control. The VCR comparison of all the converters in step-up and step-down mode is plotted versus duty cycle and is shown in Figure 5.15. The converter in [35] produces a wide range of VCR in both directions at the cost of the component count. After that, the converter in [34] has a wide conversion ratio with minimal components. Despite the continuous current at both low and high voltage ports,

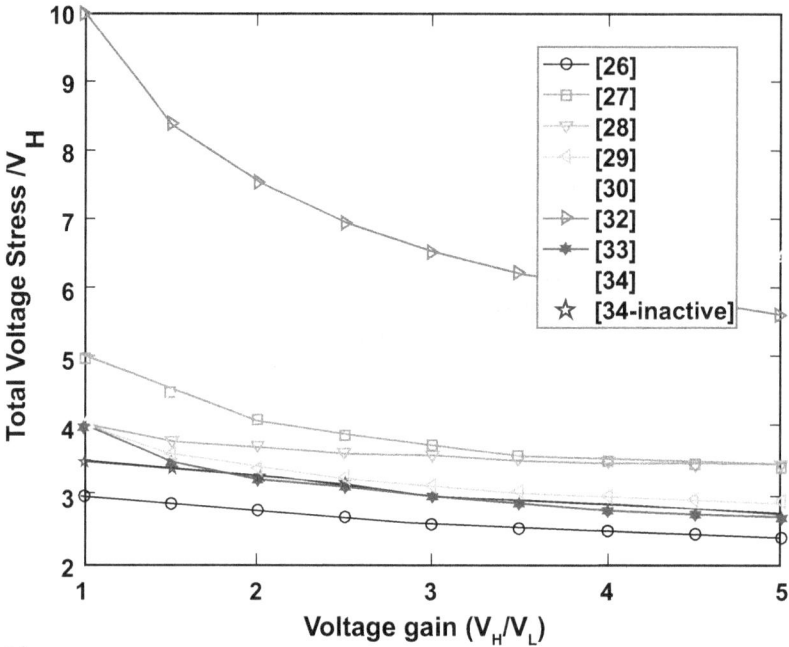

(a)

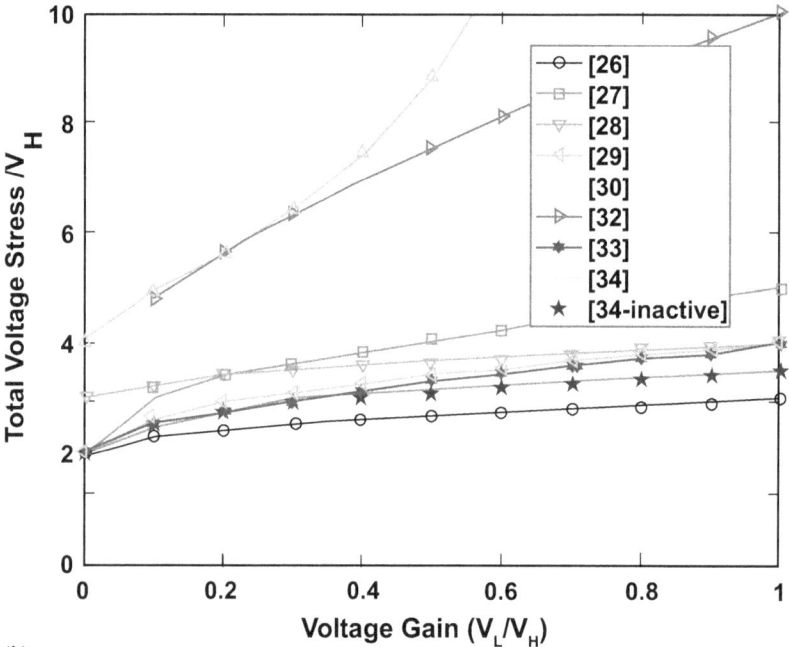

(b)

FIGURE 5.16   Comparison of TVS of switches: (a) step-up mode, (b) step-down mode.

the quadratic buck-boost (QBB-BDC) converter in [32] has low VCR in both modes, as compared to the other converters.

The total voltage stress (TVS) of the converter is vital performance indices to understand the efficiency and reliability of the converter, and it has to be minimal. Figure 5.16 shows the TVS on the active switches in both operating modes with respect to V_H or output port. From the figure, the converter in [32] has higher TVS in both modes of operation as compared to other converters. Understandably, the converter in [32] has less VCR in both modes, as it requires a high duty cycle to achieve the higher VCR. The converter in [26] exhibits good performance in terms of total voltage across active switches in both directions. Although the converter in [34] has a high VCR, lack of common ground and an inactive state in the VCR range from 1/3 to 3 have been its drawbacks. In a nutshell, the converter in [33] provides a wide range of VCR in both modes with less TVS. Further, the converter in [33] has the advantages of less voltage stress on capacitor for given gain range and an excellent semiconductor utilization factor compared to other quadratic bidirectional converters.

5.7 CONCLUSION

The dc-dc converters with bidirectional capabilities and a high VCR are required to enter the sustainable electric vehicles and their charging market. They will act as an interfacing device between the ESS and power sources. In this regard, this chapter presents the details of existing non-isolated quadratic bidirectional dc-dc (QBDC) converters. These converters are segregated into three categories, namely, quadratic boost/buck (QB-BDC), quadratic buck-boost (QBB-BDC), and high quadratic gain (HQG-BDC) converters. In addition, the operation, steady-state analysis, and passive components design are presented for one converter from each category. Furthermore, an attempt is made to present a comprehensive comparison of the QBDC converters. The comparison is made in terms of component count, the VCR in both directions of power flow, efficiencies, the TVS on the active switches, and other features like continuous current at both ports and availability of soft-switching capabilities and common ground. From the comparison analysis, the quadratic boost (QB-BDC) converters have used a minimal component to obtain the quadratic gain compared to other converters. In contrast, the HQG-BDC converters used a high component count to achieve the high VCR. Despite the continuous current at both ports, the quadratic buck-boost (QBB-BDC) converter has drawbacks of the limited VCR and higher TVS on semiconductor devices. To sum it up, the converter in [33] has better performance than other converters. It has the merits of a high VCR with minimal component count and low TVS on capacitor and semiconductor devices for a wide range of VCR. Moreover, the research on this area is limited and can be extended to develop more converter topologies to select the optimal topology for the given application.

REFERENCES

[1] S. A. Gorji, H. G. Sahebi, M. Ektesabi, and A. B. Rad. Topologies and control schemes of bidirectional DC–DC power converters: An overview. *IEEE Access*, vol. 7, pp. 117997–118019, 2019.

[2] S. V. K. Naresh, S. Peddapati, and M. L. Alghaythi. Non-isolated high gain quadratic boost converter based on inductor's asymmetric input voltage. *IEEE Access*, vol. 9, pp. 162108–162121, 2021.

[3] K. Tytelmaier, O. Husev, O. Veligorskyi, and R. Yershov. A review of non-isolated bidirectional dc-dc converters for energy storage systems. *IEEE International Young Scientists Forum on Applied Physics and Engineering (YSF)*, pp. 22–28, 2016.

[4] G. Lithesh, B. Krishna, and V. Karthikeyan. Review and comparative study of bi-directional DC-DC converters. *IEEE International Power and Renewable Energy Conference (IPRECON)*, pp. 1–6, 2021.

[5] C.-C. Lin, L.-S. Yang, and G. W. Wu. Study of a non-isolated bidirectional DC–DC converter. *IET Power Electron*, vol. 6, no. 1, pp. 30–37, 2013.

[6] P. Mounica, and S. Srinivasa Rao. Bipolar bidirectional DC-DC converter for bi-polar DC micro-grids with energy storage systems. *International Journal of Electronics*, vol. 108, no. 4, pp. 1–17, 2021.

[7] H. S. Chung, A. Ioinovici, and W.-L. Cheung, Generalized structure of bi-directional switched-capacitor DC/DC converters, *IEEE Transactions on Circuits and Systems I: Fundamental Theory and Applications*, vol. 50, no. 6, pp. 743–753, 2003.

[8] Y. Zhang, Y. Gao, L. Zhou, and M. Sumner. A switched-capacitor bidirectional DC–DC converter with wide VCR range for electric vehicles with hybrid energy sources. *IEEE Trans Power Electron*, vol. 33, no. 11, pp. 9459–9469, 2018.

[9] S. M. Fardahar, and M. Sabahi. New expandable switched capacitor/switched-inductor high-voltage conversion ratio bidirectional DC–DC converter. *IEEE Transactions on Power Electronics*, vol. 35, no. 3, pp. 2480–2487, 2020.

[10] Y. Wang, L. Xue, et al. "Interleaved high conversion ratio bidirectional DC DC converter for distributed energy storage systems circuit generation, analysis, and design. *IEEE Transactions on Power Electronics*, vol. 31, no. 8, pp. 5547–5561, 2016.

[11] Y. Zhang, Y. Gao, et al. Interleaved switched capacitor bidirectional DC DC converter with wide VCR range for energy storage systems. *IEEE Transactions on Power Electronics*, vol. 33, no. 5, pp. 3852–3869, 2018.

[12] L. Yang, and T. Liang. Analysis and implementation of a novel bidirectional DC–DC converter. *IEEE Transactions on Industrial Electronics*, vol. 59, no. 1, pp. 422–434, 2012.

[13] Y. Hsieh, J. Chen, L. Yang, C. Wu, and W. Liu. High-conversion ratio bidirectional DC–DC converter with coupled inductor. *IEEE Transactions on Industrial Electronics*, vol. 61, no. 1, pp. 210–222, Jan. 2014.

[14] S. Min-Sup, S. Young-Dong, and L. Kwang-Hyun. Non-isolated bidirectional soft-switching SEPIC/ZETA converter with reduced ripple currents. *Journal of Power Electronics*, vol. 14, no. 4, pp. 649–660, 2014.

[15] M. Vesali, M. Delshad, E. Adib, and M. R. Amini. A new non-isolated soft switched DC-DC bidirectional converter with high conversion ratio. *International Journal of Electronics*, vol. 107, no. 12, pp. 2006–2027, 2020.

[16] Y. Zhang, Q. Liu, Y. Gao, J. Li, and M. Sumner. Hybrid switched capacitor/switched-quasi-Z-source bidirectional DC–DC converter with a wide VCR range for hybrid energy sources EVs. *IEEE Transactions on Industrial Electronics*, vol. 66, no. 4, pp. 2680–2690, 2019.

[17] A. Kumar, X. Xiong, X. Pan, M. Reza, A. R. Beig, and K. A. Jaafari. A wide VCR bidirectional DC–DC converter based on quasi-Z-source and switched capacitor network. *IEEE Transactions on Circuits and Systems II: Express Briefs*, vol. 68, no. 4, pp. 1353–1357, 2021.

[18] X. Zhang, and T. C. Green. The modular multilevel converter for high step-up ratio DC–DC conversion. *IEEE Transactions on Industrial Electronics*, vol. 62, no. 8, pp. 4925–4936, 2015.

[19] F. Akar, Y. Tavlasoglu, E. Ugur, B. Vural, and I. Aksoy, "A bidirectional non isolated multi-input DC–DC converter for hybrid energy storage systems in electric vehicles," *IEEE Transactions on Vehicular Technology*, vol. 65, no. 10, pp. 7944–7955, 2016.

[20] K. Suresh et al. A multifunctional non-isolated dual input-dual output converter for electric vehicle applications. *IEEE Access*, vol. 9, pp. 64445–64460, 2021.

[21] S. V. K. Naresh, and S. Peddapati. New family of transformer-less quadratic buck-boost converters with wide conversion ratio. *International Transactions on Electrical Energy Systems*, vol. 31, no. 11, pp. 1–21, 2021.

[22] S. Peddapati, and S. Naresh. Quadratic boost converter for green energy applications. DC–DC converters for future renewable energy systems. *Energy Systems in Electrical Engineering*, vol. 1, pp. 173–202, 2022.

[23] S. Naresh, and S. Peddapati. Comparative analysis of quadratic buck-boost converters: Topology, electric stress, reliability. *IECON 2021–47th Annual Conference of the IEEE Industrial Electronics Society*, pp. 1–6, 2021.

[24] S. V. K. Naresh and S. Peddapati. Complementary switching enabled cascaded boost-buck-boost (BS-BB) and buck-boost-buck (BB-BU) converters. *International Journal of Circuit Theory and Applications*, vol. 49, no. 9, pp. 2736–2753, 2021.

[25] S. Naresh, and S. Peddapati. New continuous input buck-boost converter with quadratic voltage conversion ratio. *2021 National Power Electronics Conference (NPEC)*, 2021, pp. 1–6.

[26] H. Ardi, A. R. Ahrabi, and S. N. Ravadanegh. Non-isolated bidirectional DC–DC converter analysis and implementation. *IET Power Electronics*, vol. 7, no. 12, pp. 3033–3044, 2014.

[27] H. Ardi, A. Ajami, F. Kardan, and S. N. Avilagh. Analysis and implementation of a non-isolated bidirectional DC–DC converter with high VCR. *IEEE Transactions on Industrial Electronics*, vol. 63, no. 8, pp. 4878–4888, 2016.

[28] V. F. Pires, D. Foito, and A. Cordeiro. A DC–DC converter with quadratic gain and bidirectional capability for batteries/supercapacitors. *IEEE Transactions on Industry Applications*, vol. 54, no. 1, pp. 274–285, 2018.

[29] S. H. Hosseini, R. Ghazi, and H. Heydari-Doostabad. An extendable quadratic bidirectional dc-dc converter for V2G and G2V applications. *IEEE Transactions on Industrial Electronics*, vol. 68, no. 6, pp. 4859–4869, Jun. 2021.

[30] A. R. N. Akhormeh, K. Abbaszadeh, M. Moradzadeh, and A. Shahirinia. High-gain bidirectional quadratic DC–DC converter based on coupled inductor with current ripple reduction capability. *IEEE Transactions on Industrial Electronics*, vol. 68, no. 9, pp. 7826–7837, 2021.

[31] R. H. Ashique, and Z. Salam. A high-gain, high-efficiency nonisolated bidirectional DC–DC converter with sustained ZVS operation. *IEEE Transactions on Industrial Electronics*, vol. 65, no. 10, pp. 7829–7840, 2018.

[32] H. Heydari-Doostabad, S. H. Hosseini, R. Ghazi, and T. O'Donnell. Pseudo DC-link EV home charger with a high semiconductor device utilization factor. *IEEE Transactions on Industrial Electronics*, vol. 69, no. 3, pp. 2459–2469, 2022.

[33] H. Heydari-doostabad, and T. O'Donnell. A Wide-Range High-Voltage-Gain Bidirectional DC–DC Converter for V2G and G2V Hybrid EV Charger. *IEEE Transactions on Industrial Electronics*, vol. 69, no. 5, pp. 4718–4729, 2022.

[34] N. Elsayad, H. Moradisizkoohi, and O. A. Mohammed. A new hybrid structure of a bidirectional DC-DC converter with high conversion ratios for electric vehicles. *IEEE Transactions on Vehicular Technology*, vol. 69, no. 1, pp. 194–206, 2020.

[35] J. Mei, Q. Gao, and X. Cai. High gain bidirectional DC-DC converter with three boost converters and switched capacitor. *IEEE 1st International Power Electronics and Application Symposium (PEAS)*, 2021, pp. 1–6.

6 Modeling and MPPT Control of a PMSG-Based Wind Turbine System

Balaji Mendi, Monalisa Pattnaik,
and Gopalakrishna Srungavarapu

CONTENTS

Learning Objectives: After completing this module the readers will have better understanding on

- A small-scale standalone WTS
- Basic mathematical modeling of WT and PMSG
- Converter design for raising the voltage and extraction of peak power
- Understanding the characteristics of the system under duty cycle variation
- Simulation study of speed sensor based and speed sensorless MPPT control methods for maximum power extraction under different operating conditions

---

DOI: 10.1201/9781003323471-6

6.1   INTRODUCTION

In recent years, renewable energy sources (RES) are the best choice for electricity generation to meet the growing demand and they are naturally available and cause less environmental pollution. These are classified into different categories such as solar, wind, hydroelectric, ocean, geothermal, biomass, and hydrogen [1]. Among them, wind energy is prominently used, as it is sustainable and eco-friendly in nature.

In this chapter, variable speed wind energy system (VSWES) is focused more on electrical power production as they capture more energy from the available wind speed [2]. These systems can control the speed of rotation and also reduce the mechanical stress on the turbine during wind gusts. VSWES is mostly used in distribution systems and remote areas, where the grid network is highly uneconomical and impossible for rural people. The commonly used generators in the wind turbine system (WTS) are permanent magnet synchronous generators (PMSG) and different induction generators. But, the PMSG has a lot of advantages like high density of power, reduction in power loss, more reliability, no need for separate field excitation for rotor circuit, thereby saving energy. Also, it has no slip rings in the rotor, thus it requires less maintenance [3]–[5].

The PMSG based VSWES uses different power electronic circuit topologies which act as an intermediate stage between WT and grid. These circuits are useful to control the parameters of the system. In this chapter, the two-stage converter (ac-dc and dc-dc converter) is used. A boost circuit is designed for the system to maximize the output voltage [6]–[9]. To extract more power from WTS, different maximum power point tracking (MPPT) algorithms are used [10]–[24] which are briefly presented in the literature section. The maximum power point (MPP) is derived from the gradual increase of the duty cycle and the results are plotted in the MATLAB/Simulink software. The WTS performance is evaluated by the execution of perturb and observe (P&O) based MPPT methods under different scenarios. The simulation results are analyzed for speed sensor based and sensorles based P&O algorithms and the respective results are plotted and discussed.

In this chapter, the modeling of WT and PMSG are studied and detailed designing of a boost converter is also presented. The P&O based speed sensor based and sensorless MPPT techniques are explained using MATLAB/Simulink software, and the results are studied.

6.2   LITERATURE AND MOTIVATION

From the literature, the modeling of a wind turbine, modeling of PMSG, design of boost converter, and different MPPT algorithms are studied. This chapter examines the following research studies.

The WT modeling depends on the principle of aerodynamics, where the blades of the WT turns the wind energy into rotational energy. It causes the rotor to spin and generates electricity, which has been presented in [2]. The modeling of PMSG in the dq coordinate system is addressed in [3]. The comparison of wind generators used in wind turbines is necessary as wind energy is becoming more cost-competitive which is presented in [4]. The power electronic device is predominantly used in the VSWES as they are promising technologies for wind farms. The direct drive or less geared WT coupled with a permanent magnet generator is addressed in [5], thus, it reduces the cost.

The various power converter schemes for WTS are listed and explained in [7]. It makes the system suitable for high-frequency operation. Among them, buck and boost converters are efficient as it does not contain any isolation and the input energy is transferred directly to the output. The boost converter design is presented in [8] and states the main factors of the circuit which are inductor ripple current, output voltage ripple, power, and switching frequency. Based on these factors, the converter inductance, capacitance, and power rating of the switching devices can be calculated in the system.

The comparison of various MPPT control schemes in WTS are explained in [10]. MPPT algorithms based on TSR [11] and PSF require mechanical sensors for sensing the speed of the rotor and requires prevalent data about the parameters of WT. However, the method like hill-climbing search doesn't need any mechanical sensors which reduce the implementation cost [12]. The speed sensor-less P&O MPPT technique is addressed in [11]–[15]. This method uses the electrical sensors like voltage and current sensors to implement the algorithms. As there is no requirement for mechanical sensors, the robustness of the system can be increased.

6.3 SYSTEM DESCRIPTION

A standalone VSWES is depicted in Figure 6.1. The main units are WT coupled to PMSG, ac/dc (three-phase uncontrolled ac-dc converter), boost converter, and resistive load. The maximum power is extracted using the MPPT controller, which generates the desired duty cycle to operate at MPP.

6.4 MATHEMATICAL MODELING OF PMSG BASED VSWES

The PMSG based VSWES is growing its attention in the electrical generation sector, as it captures more power from the variable wind speed and reduces mechanical stress of the WT. Here, the mathematical model representing turbine parameters, and generator are modeled and analyzed briefly.

6.4.1 MATHEMATICAL MODELING OF WT

When air rushes through the WT, it creates two low-pressure and high-pressure zones surrounding the blades, and then it starts spinning the turbine blades which produces mechanical power. The turbine power is represented in terms of the wind speed, the diameter of the blades, power coefficient, etc. The modeling of the WT is based on the following mathematical equations.

$$P_m = \frac{1}{2} \rho A C_p V_\omega^3 \tag{1}$$

$$C_p = g_1 \left(\frac{g_2}{\lambda_i} - g_3 \beta - g_4 \right) e^{-\frac{g_5}{\lambda_i}} + g_6 \lambda \tag{2}$$

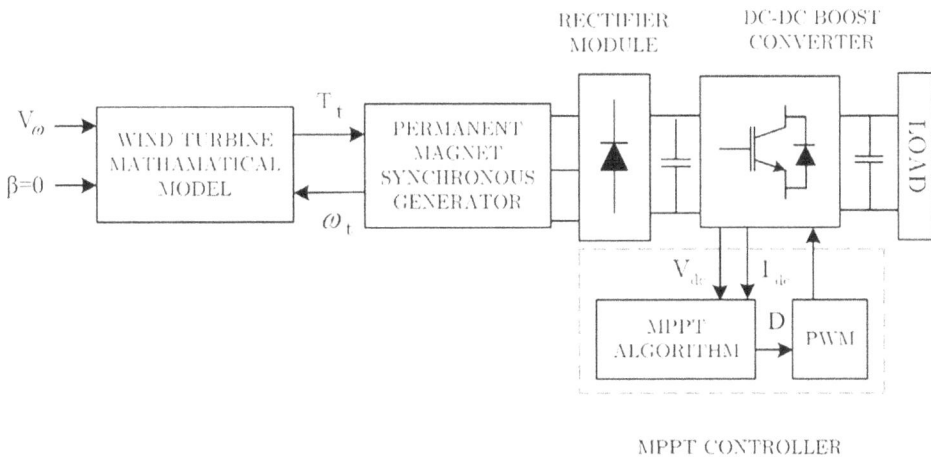

FIGURE 6.1 Configuration of PMSG based wind turbine system.

$$\frac{1}{\lambda_i} = \frac{1}{1+0.08\beta} - \frac{0.035}{\beta^3+1} \tag{3}$$

$$T_m = \frac{1}{2\lambda} \rho\pi R^3 C_p V_\omega^2 \tag{4}$$

Where P_m = aerodynamic mechancal power of the WT, ρ = density of air (kg/m^3), V_ω = velocity of air (m/s), $A = \pi R^2$ = swept area of the rotor (m^2), R = radius of the blades (m), $\lambda = \frac{\omega_t R}{V_\omega}$, ω_m = speed of the blade tip (rad/s), β = pitch angle, C_P = power coefficient, T_m = turbine rotor torque (N.m).

The torque derived from the turbine is presented in equation (4) which spins the generator for the available $V\omega$. Based on the preceding equations, the characteristics of the WT are plotted. The C_p vs λ for various β is shown in Figure 6.2, whereas, the P_m vs ω_g for various V_ω is plotted in Figure 6.3. It is observed that $C_{P\_max\,x} = 0.48$, $\lambda_{opt}(\beta = 0) = 8.1$. where C_{P_max} is the maximum value of the power coefficient and λ_{opt} is corresponding optimal value of the λ.

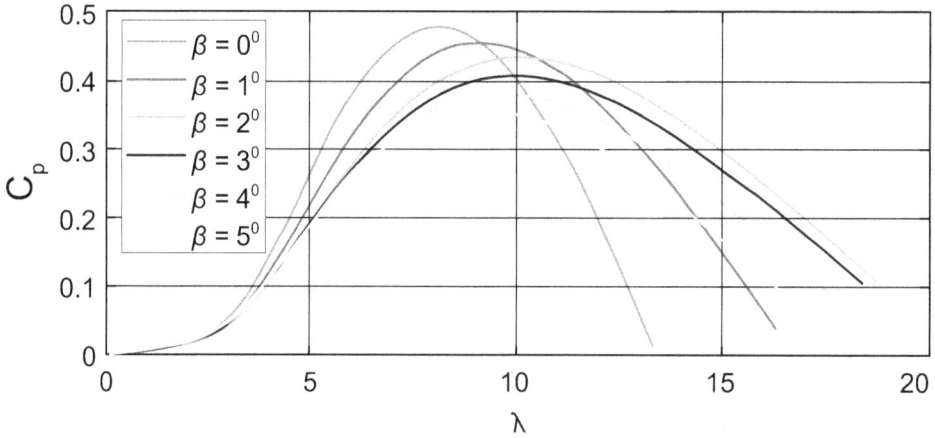

FIGURE 6.2 C_P vs λ for various β.

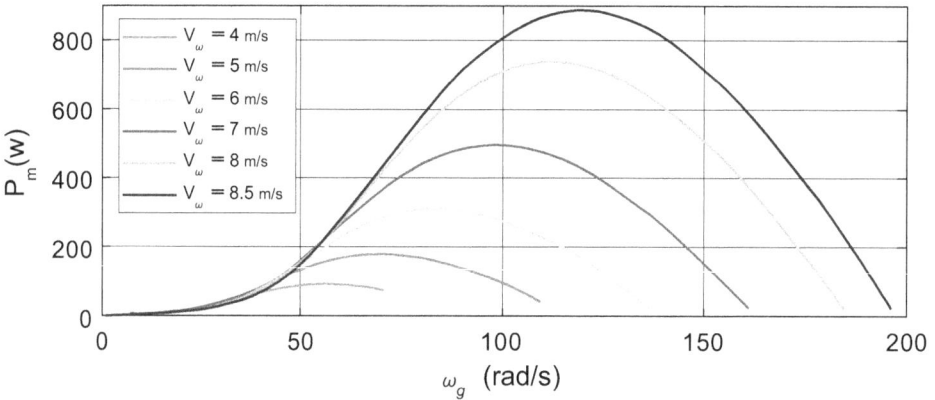

FIGURE 6.3 P_m vs ω_g for various V_ω.

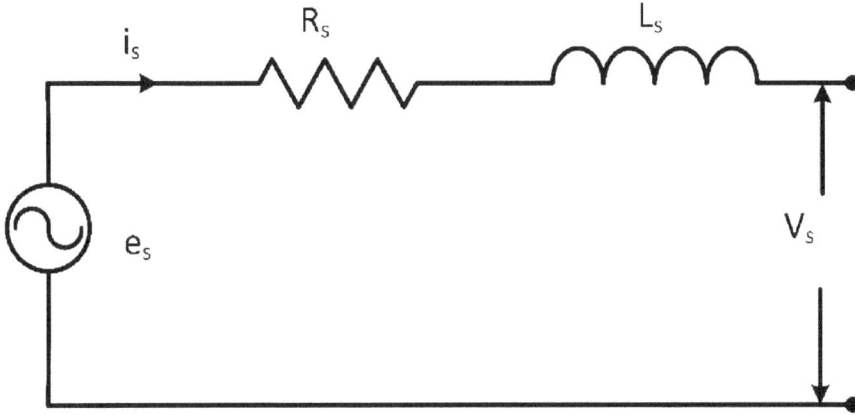

FIGURE 6.4 1-phase equivalent circuit of the PMSG.

6.5   MODELING OF PMSG

In PMSG, the generated voltage is directly related to the angular speed of the generator, Therefore, the output power can be easily extracted from the PMSG. The 1-phase equivalent circuit of the wind generator is depicted in Figure 6.4. It is a simple RLE model.

The generator voltage is a function of ω_g and magnetic flux of the rotor, $\varnothing$ is expressed as:

$$e_s = k\varnothing\omega_g = k\frac{\omega_e}{p} \tag{5}$$

Where $\omega_e = P\omega_g$,k is geneator voltage constant in (V/rpm), P is pole pairs of the rotor magnets, and ω_e is the electrical angular frequency of the generated armature voltage. From the equivalent circuit, the output voltage of the PMSG at steady-state is

$$V_s^2 = E_s^2 - \left(\omega_e L_s I_s\right)^2 \tag{6}$$

V_s is the per phase output voltage of the PMSG and I_s is the stator current. The total output power, P_g, electromagnetic torque, T_e of the PMSG are expressed in equations (7) and (8) respectively.

$$P_g = 3V_s I_s = 3\sqrt{E_s^2 I_s^2 - \left(\omega_e L_s\right)^2 I_s^4} \tag{7}$$

$$T_e = \frac{\sqrt[3]{E_s^2 I_s^2 - \left(\omega_e L_s\right)^2 I_s^4}}{\omega_e} \tag{8}$$

The general mechanical torque expression of the machine is

$$T_m = T_e + B\omega_m + J\frac{d\omega_m}{dt} \tag{9}$$

Where B is the viscous friction constant, and J is the machine inertia.

6.6   DESIGN OF THE DC-DC CONVERTER

The design of the boost converter is essential for enhancing the voltage and extracting the MPP from the wind power generation system. It depends on the factors like inductor ripple current, output voltage ripple, power, input voltage, output voltage, and switching frequency [7]. Moreover, depending on the output power requirement size of the capacitor, an inductor is decided. Nonetheless, the converter size is minimal for higher the operating frequency of the semiconductor device.

In this section, the converter is designed for WTS is 1000 W, and the switching frequency is 10 kHz. The output voltage rectifier is same as the no load voltage which is 200 V. At rated condition, the rectified voltage is 50 V and the converter current is 20 A as the machine operates at its full load. Therefore, the voltage range of the converter input voltages is 50–200 V. For, the lossless converter, the input power is same as output power which is 1000 V. The change in peak-to-peak ripple in inductor current is considered 20% of converter maximum input dc current, and the change in the ripple voltage across the capacitor is assumed to be 1.0% of the load voltage [9].

Assume ideal boost converter, $P_{in\ dc} = P_{o\ dc} = 1000$ W

Converter maximum input dc-link current, $I_{dcmax} = \dfrac{\text{Pin dc}}{V_{in\ min}} = 1000 / 50 = 20A$

Duty cycle, $D = 1- (V_{in\ min}/V_0) = 1 - (50/220) = 0.77$

Inductor peak to peak ripple, $\Delta I = (0.2)*(I_{dcmax}) = 4A$

Capacitor Voltage ripple, $\Delta V = (0.01) * (V_0) = 0.01*220 = 2.2$ V

Inductor $(L) = (V_{in\ min} *(V_0 - V_{in\ min})) / (f* \Delta I * V_0) = 50(220-50)/(10000*4*220) = 965\ \mu H \cong 1mH$

Capacitor $= (I_0 * D)/(V_0 * \Delta V) = 1200 \mu F$

6.7   CHARACTERISTICS OF THE SYSTEM UNDER DUTY VARIATION

Before apply of MPPT algorithm, MPP of the system is to be identified. This is achieved by gradually increasing the duty cycle (D) by keeping a constant V_ω. The variation of the dc-link voltage (V_{dc}), current (I_{dc}), and power (P_{dc}), with respect to duty cycle, are plotted in Figures 6.6, 6.7, and 6.8 respectively. All these results are simulated at a constant V_ω of 8.5 m/s and a resistive load (R_L) of 100 Ω. It is noticed from Figure 6.6 that as the duty cycle increases, the V_{dc} is decreased. But, the I_{dc}

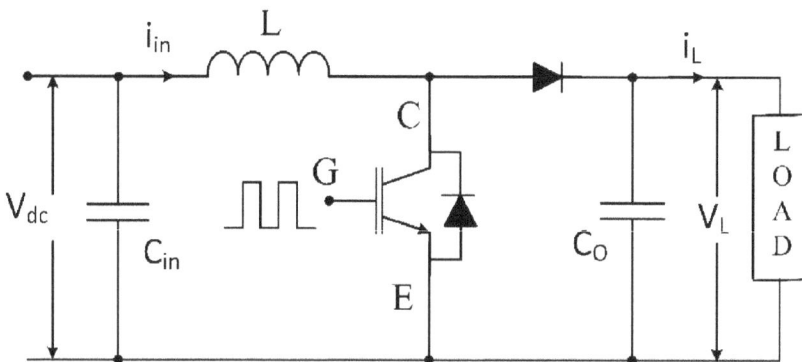

FIGURE 6.5   Dc-dc boost converter.

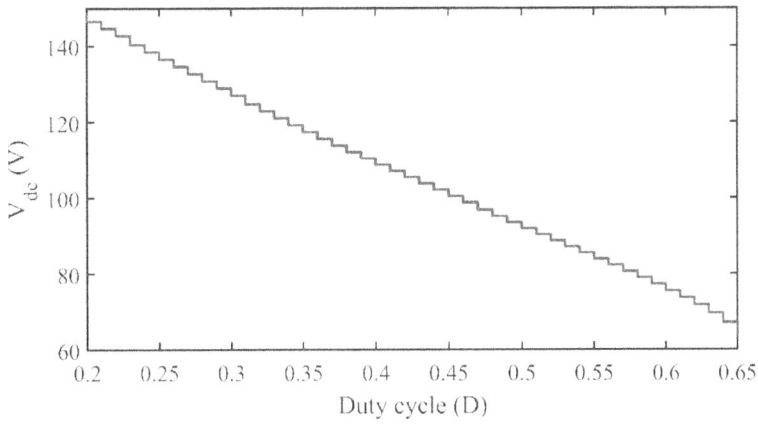

FIGURE 6.6 V_{dc} versus D.

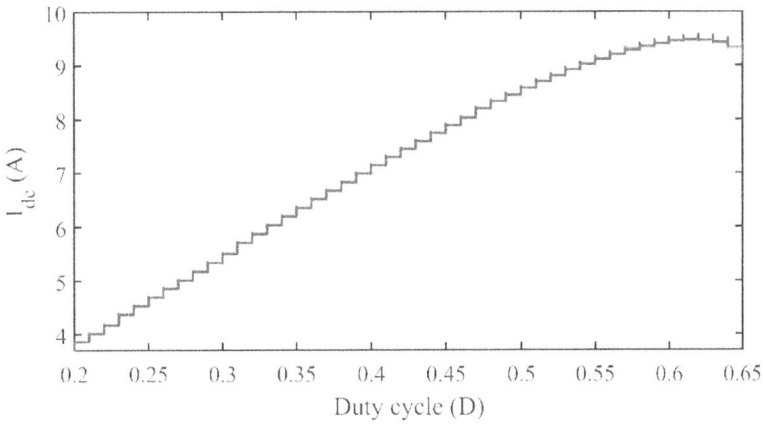

FIGURE 6.7 I_{dc} versus D.

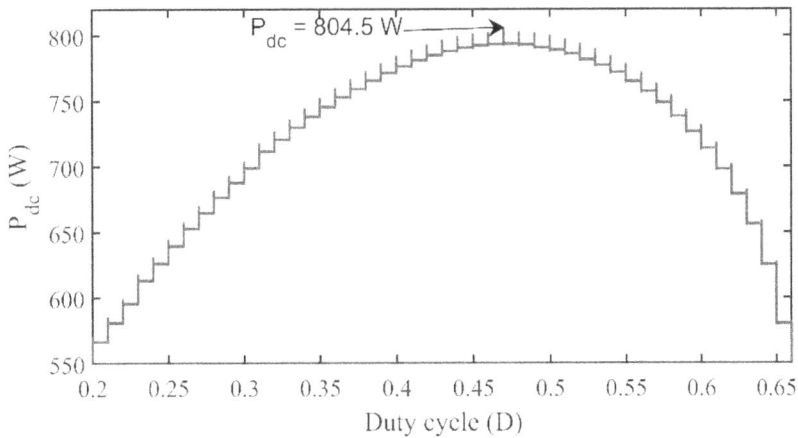

FIGURE 6.8 P_{dc} versus D.

increased, as it is shown in Figure 6.7. It is noticed from Figure 6.8, the maximum power obtained is 804.5 W at a duty cycle of 0.47.

6.8 MPPT CONTROLLER FOR WTS

The WTS considers the MPPT controller for the extraction of peak power from the available wind. The various MPPT schemes are studied in the literature section. Here, the speed sensor dependent and sensorless P&O MPPT methods are explained in detail. In general, the MPP of the system is derived from the characteristics of the P_{dc}–V_{dc} plot for traditional P&O method, and the P_{dc}–ω_m for speed sensor based P&O method.

6.8.1 SPEED SENSOR DEPENDENT P&O MPPT METHOD

This scheme requires mechanical speed sensor from the generator, voltage, and current sensors for the boost converter. The flow chart of this scheme is depicted in Figure 6.9. It is developed based on the the slopes of the P_{dc}–ω_m curve. Initially, set the perturbation time to 1 s, change in duty cycle to 1%, and initialize the P_{dc}, ω_m, and D. If the change in P_{dc}, ΔP_{dc} is +ve then algorithm

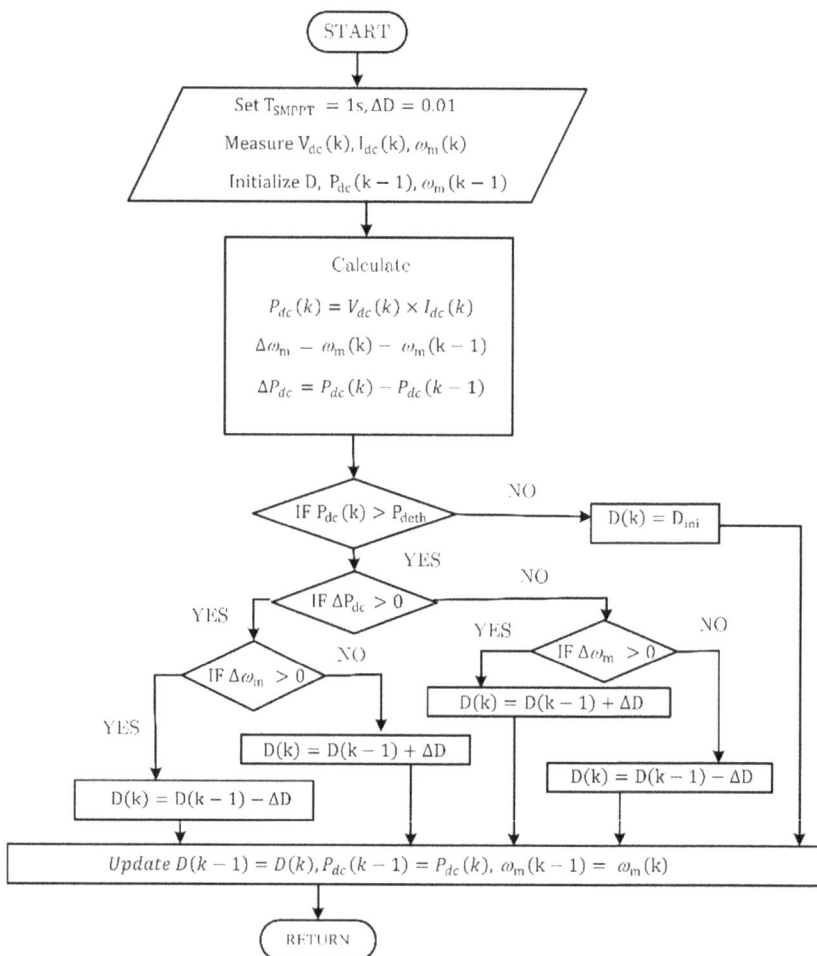

FIGURE 6.9 Flow chart of the speed sensor dependent P&O MPPT method

checks the sign of change in ω_m, $\Delta\omega_m$. If the $\Delta\omega_m$ is +ve then the D is decremented by ΔD in the next iteration; otherwise the D is incremented by ΔD. Similarly, for the ΔP_{dc} is -ve, and if the $\Delta\omega_m$ is +ve then the updated D is incremented by ΔD; otherwise it is decremented by ΔD in the next iteration.

6.8.2 Speed Sensorless P&O MPPT Method

The traditional P&O MPPT scheme is popular among speed sensorless methods. It doesn't require wind speed data, and any mechanical speed sensors, thus the scheme is more robust. This scheme senses the voltage and current from the input terminals of the boost converter and generates the D for the extraction of MPP. The flow sheet of the scheme is depicted in Figure 6.10. It is developed based on the slopes of the P_{dc}–V_{dc} curve. Initially, set the perturbation time to 1 s, change in duty cycle to 1%, and initialize the P_{dc}, V_{dc}, and D. If the ΔP_{dc} is +ve then algorithm checks the sign of change in V_{dc}, ΔV_{dc}. If the ΔV_{dc} is +ve then the D is decremented by ΔD in the next iteration; otherwise the D is incremented by ΔD. Similarly, for the ΔP_{dc} is -ve, and if the ΔV_{dc} is +ve then the updated D is incremented by ΔD; otherwise it is decremented by ΔD in the next iteration.

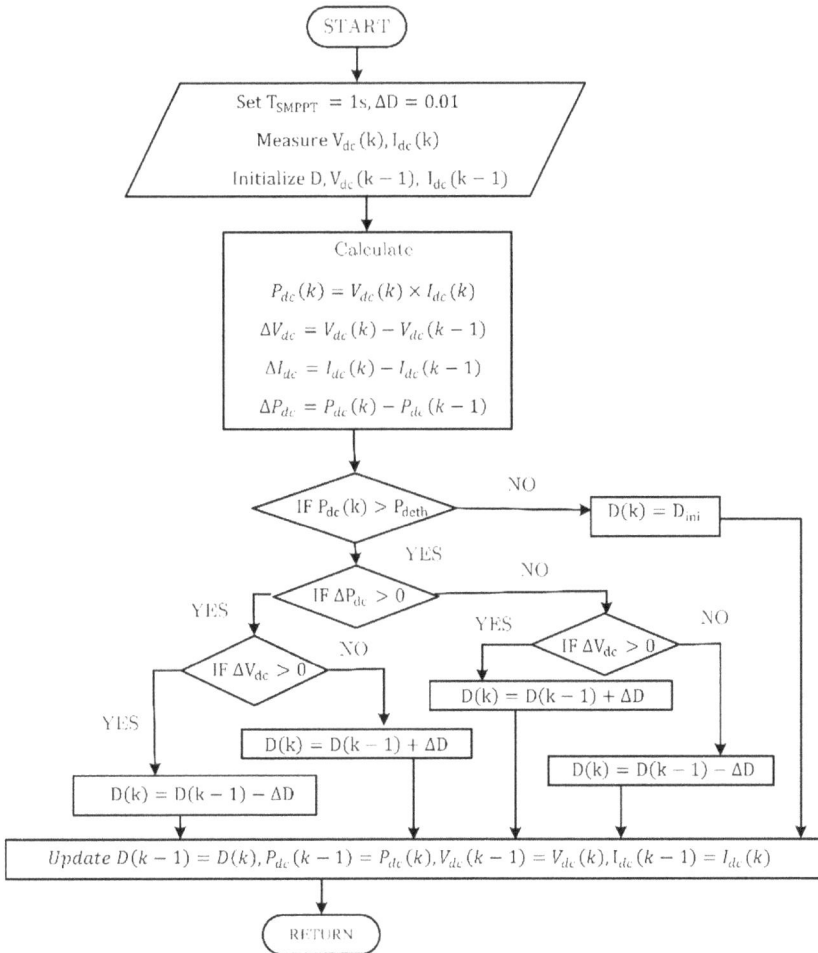

FIGURE 6.10 Flow sheet of the traditional P&O MPPT method.

6.9   SIMULATION RESULTS AND DISCUSSIONS

The speed sensor dependent and sensorless P&O MPPT algorithms are applied for PMSG based WTS in MATLAB Simulink software. These schemes are verified under different wind speed and load change conditions. The specifications of WT, PMSG, and boost converter are shown in Tables 6.1 and 6.2 respectively. A 100 Ω load resistor is connected across the boost converter output terminals, as shown in Figure 6.1.

6.9.1   SIMULATION RESULTS OF SPEED SENSOR BASED P&O MPPT ALGORITHM

6.9.1.1   Case-I: Wind Speed Change with Constant Load

The simulation results of the speed sensor dependent P&O MPPT method with constant load and step change in wind speed are depicted in Figure 6.11. Initially, the wind speed is set to 7.5 m/s which is shown in Figure 6.11(a), the respective P_{dc} (Figure 6.11(b)) is 547.2 W, V_{dc} (Figure 6.11(c)) is 90.31 V, I_{dc} (Figure 6.11(d)) is 6.119 A, V_{load} (Figure 6.11(e)) is 232.4 V and I_{load} (Figure 6.11(f)) is 2.33 A. The wind speed is increased to 8.5 m/s at time, t = 40 s and the corresponding results are P_{dc} = 789.4 W, V_{dc} = 97.25 V, I_{dc} = 8.092 A, V_{load} = 278.3 V, and I_{load} = 2.794 A.

TABLE 6.1

Specifications of a Wind Turbine and PMSG

Wind turbine

ρ = 1.225 kg/m^3, R = 1.25m, C_{P_max} = 0.48, $\lambda_{opt}(\beta=0)$ = 8.1, g_1 = 0.5175, g_2 = 116, g_3 = 0.4, g_4 = 5, g_5 = 21, g_6 = 0.0068

PMSG

3-phase, 1.5Kw, 110 V, 8.1A, R_s = 0.53 Ω, $L_d = L_q$ = 13.2 mH, Number of poles, P = 4, J (Kg. m^2) = 0.011295, B (Nm.s/rad) = 0.0078295.

TABLE 6.2

Boost Converter Parameters

L	C_{out}	ΔV	ΔI	f
1mH	1200 μF	2.2 V	4 A	10 kHz

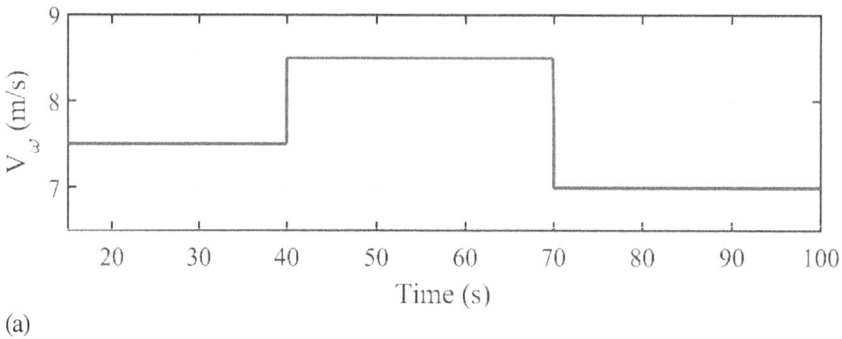

(a)

FIGURE 6.11   Simulation results of speed sensor based P&O MPPT algorithm with change in wind speed: (a)V_ω, (b) P_{dc}, (c) V_{dc}, (d) I_{dc}, (e) V_{load}, and (f) I_{load}.

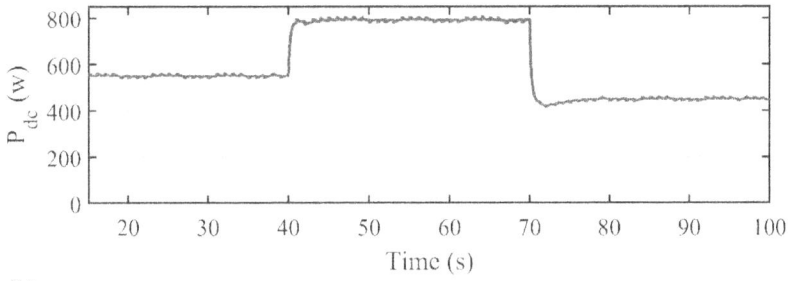

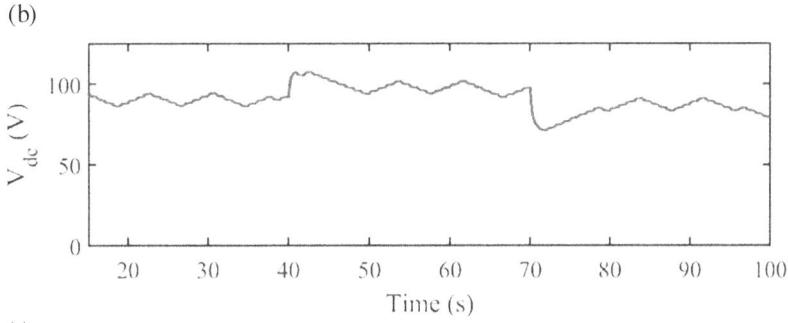

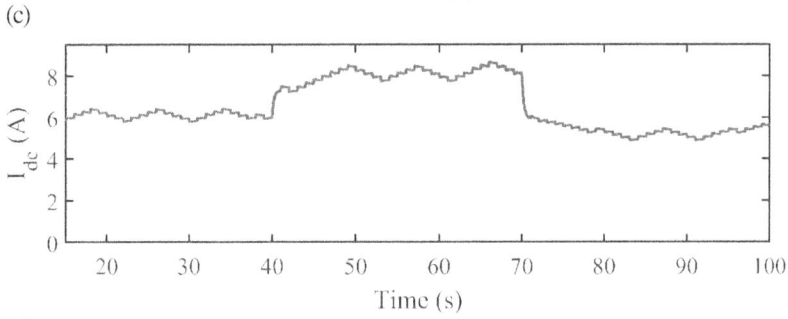

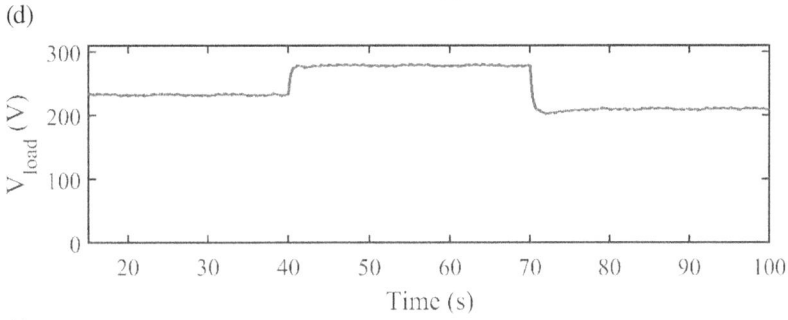

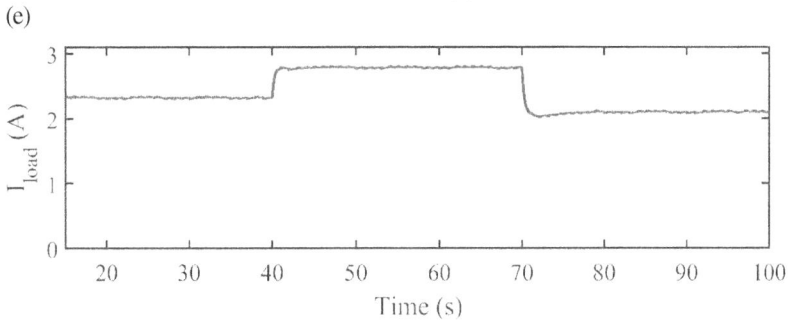

(b)

(c)

(d)

(e)

(f)

FIGURE 6.11   Continued

The wind speed is decreased to 7 m/s at time, t = 70 s and the corresponding results are P_{dc} = 445.1 W, V_{dc} = 85.29 V, I_{dc} = 5.109 A, V_{load} = 209.2 V, and I_{load} = 2.09 A. It is observed that the P_{dc} is settled to a steady state in 1 s during increased wind speed and 2 s during decreased wind speed conditions.

6.9.1.2   Case-II: Load Change with Constant Wind Speed

The simulation results of the speed sensor dependent P&O MPPT method with load change and constant wind speed of 8.5 m/s are depicted in Figure 6.12. Initially, the R_L of 100 Ω is applied to the WTS, the respective P_{dc} (Figure 6.12(a)) is 789.4 W, V_{dc} (Figure 6.12(b)) is 97.25 V, I_{dc} (Figure 6.12(c)) is 8.092 A, load voltage (Figure 6.12(d)) is 278.3 V and load current (Figure 6.12(e)) is 2.794 A. The R_L is decreased to 50 Ω at the time, t = 50 s. It is observed that the steady-state maximum P_{dc}, V_{dc}, and I_{dc} are settled to the same value. The amount of time it takes to attain a steady-state value of the P_{dc} for the load change is 7.94 s. The load voltage is decreased to 198.6 V and the load current

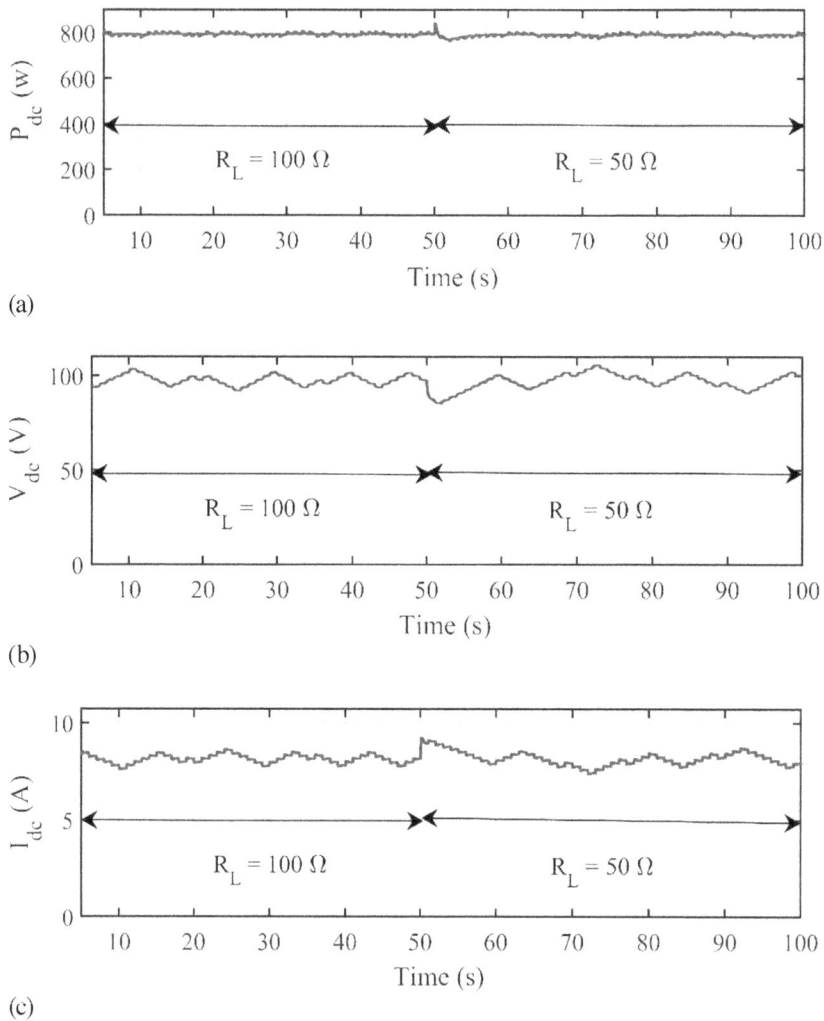

FIGURE 6.12   Simulation results of speed sensor dependent P&O MPPT scheme with load change: (a) P_{dc}, (b) V_{dc}, (c) I_{dc}, (d) V_{load}, and (e) I_{load}.

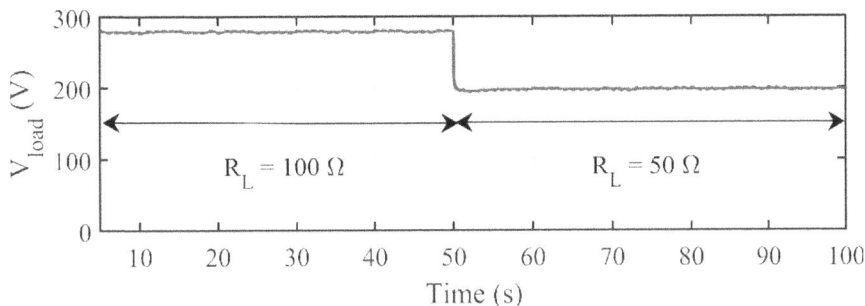

(d)

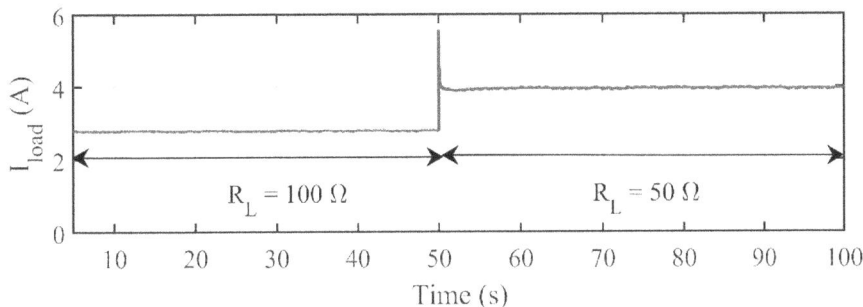

(e)

FIGURE 6.12 Continued

is increased to 3.959 A as the load is reduced by half. The speed sensor based P &O MPPT scheme provides the steady-state power oscillation is 33.1 W.

6.9.2 Simulation Results of Conventional P&O MPPT Algorithm

6.9.2.1 Case-I: Wind Speed Change with Constant Load

The simulation results of the traditional P&O MPPT method with constant load and step change in wind speed are depicted in Figure 6.13. In the beginning, the V_ω is set to 7.5 m/s which is depicted in Figure 6.13(a), the respective P_{dc} (Figure 6.13(b)) is 550.6 W, V_{dc} (Figure 6.13(c)) is 90.9 V, I_{dc} (Figure 6.13(d)) is 6.055 A, V_{load} (Figure 6.13(e)) is 232.1 V and I_{load} (Figure 6.13(f)) is 2.316 A. The wind speed is set to 8.5 m/s at time, t = 40 s and the corresponding results are P_{dc} = 785.5 W, V_{dc} = 95.79 V, I_{dc} = 8.23 A, V_{load} = 278 V, and I_{load} = 2.77 A. The V_ω is decreased to 7 m/s at time, t = 70 s and the corresponding results are P_{dc} = 452.5 W, V_{dc} = 85.12 V, I_{dc} = 5.36 A, V_{load} = 209.4 V, and I_{load} = 2.11 A. It is observed that the P_{dc} is settled to a steady state in 0.6 s during increased wind speed and 1 s during decreased wind speed conditions.

6.9.2.2 Case-II: Load Change with Constant Wind Speed

The simulation results of the traditional P&O MPPT method with load change and constant of 8.5 m/s are depicted in Figure 6.14. Initially, the R_L of 100 Ω is applied to the WTS, the respective P_{dc} (Figure 6.14(a)) is 785.5 W, V_{dc} (Figure 6.14(b)) is 95.79 V, I_{dc} (Figure 6.14(c)) is 8.23 A, V_{load} (Figure 6.14(d)) is 278 V and I_{load} (Figure 6.14(e)) is 2.77 A. The R_L is decreased to 50 Ω at the time, t = 50 s. It is observed that the steady-state maximum P_{dc}, V_{dc}, and, I_{dc} are settled to the same value. The amount of time it takes to attain a steady state value of the P_{dc} for the load change is 7.94 s. The

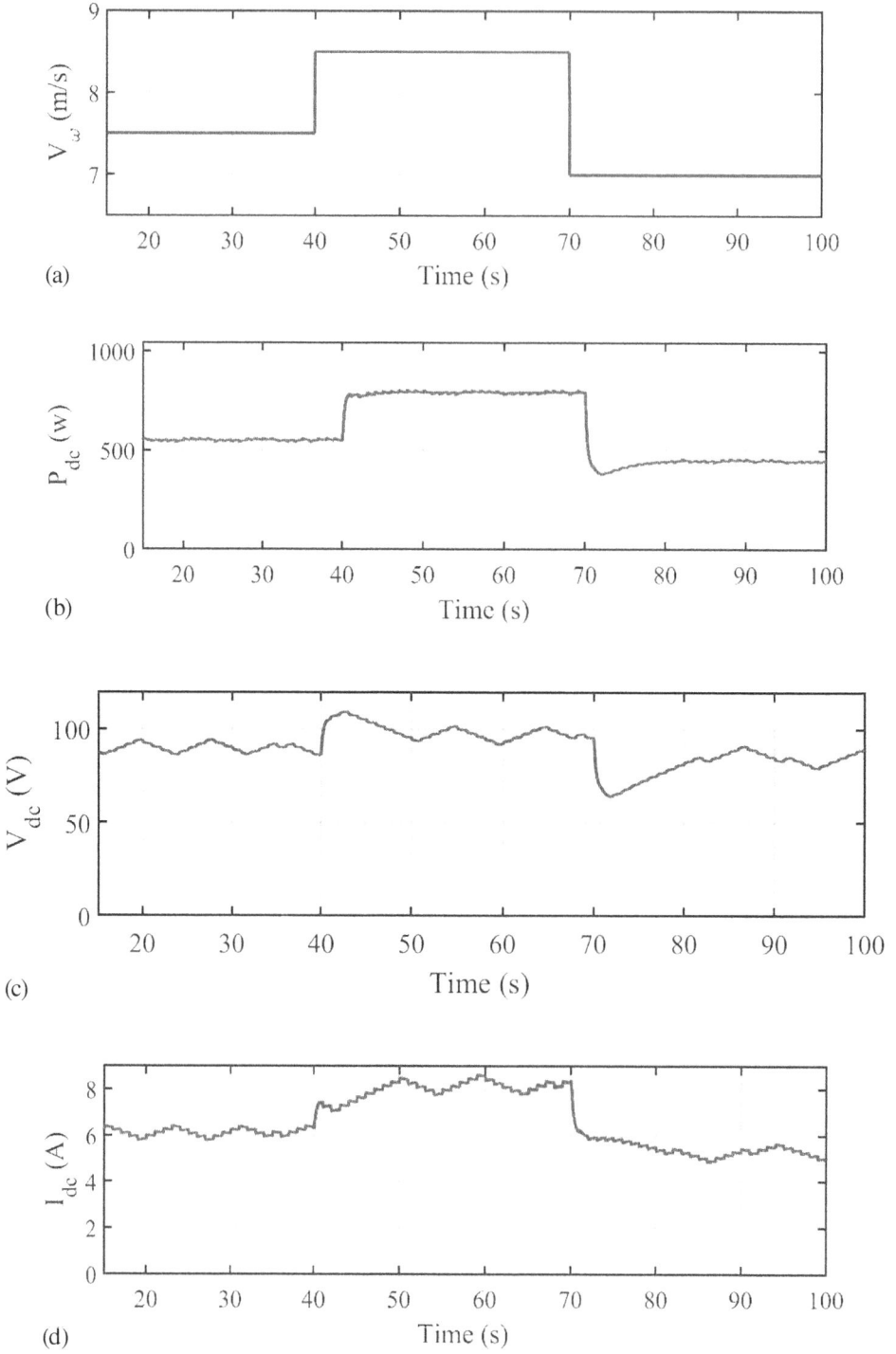

FIGURE 6.13 Simulation results of conventional P&O MPPT algorithm with change in wind speed: (a), V_ω, (b) P_{dc}, (c) V_{dc}, (d) I_{dc}, (e) V_{load}, and (f) I_{load}.

(e)

(f)

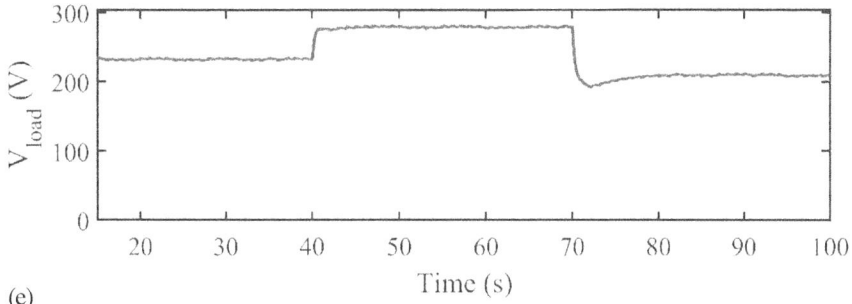

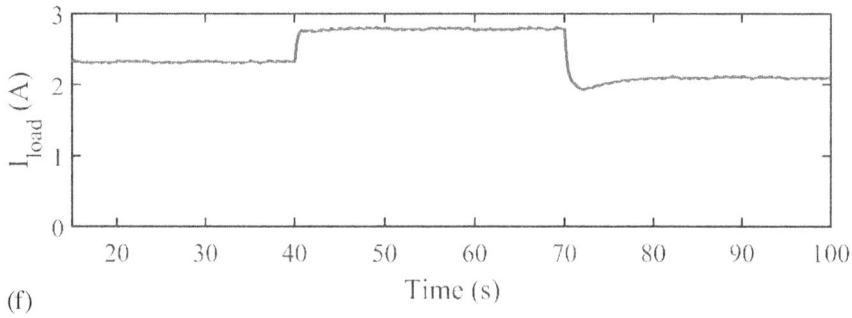

FIGURE 6.13 Continued

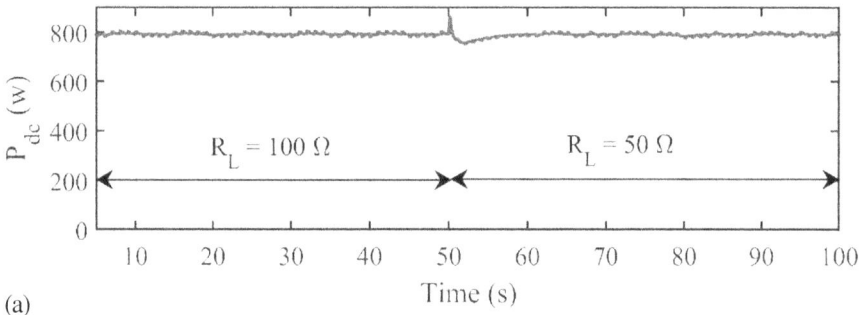

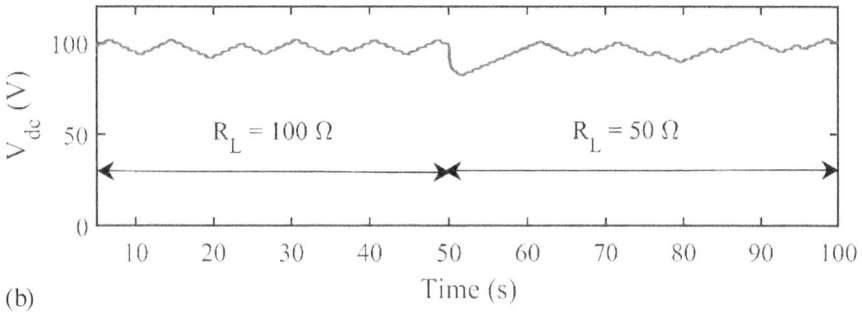

(a)

(b)

FIGURE 6.14 Simulation results of conventional P&O MPPT algorithm with load change: (a) P_{dc}, (b) V_{dc}, (c) I_{dc}, (d) V_{load}, and (e) I_{load}.

(c)

(d)

(e)

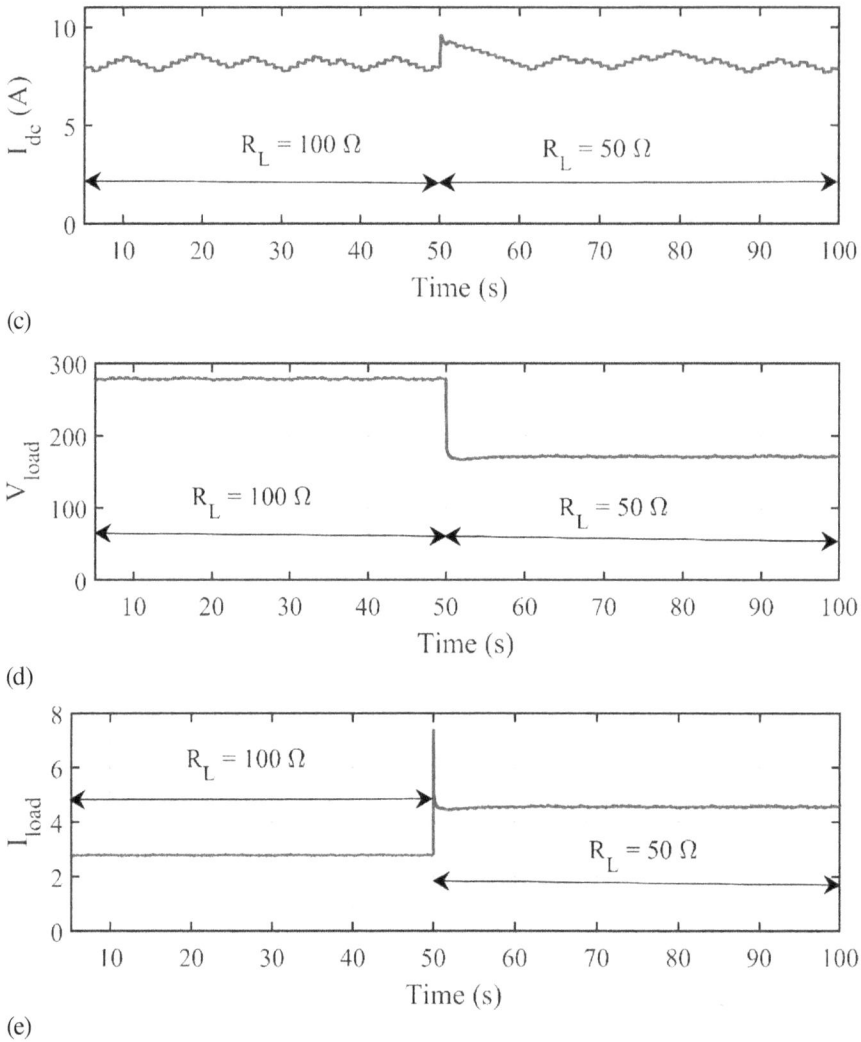

FIGURE 6.14 Continued

load voltage is decreased to 170.8 V and the I_{load} is increased to 4.6 A as the load is reduced by half. The traditional P &O MPPT scheme provides the steady-state power oscillation is 35.3 W.

6.10 CONCLUSION

In this book chapter, the modeling of the turbine and PMSG are explained in detail. Two-stage power electronic converters (ac-dc & dc-dc) are used for the WTS and a dc-dc boost converter is designed. The MPP is determined by the gradual variation of the duty cycle and the respective results are plotted. The speed sensor based and speed sensorless MPPT schemes are applied to the WTS and validated under wind speed and load change scenarios. The simulations results are examined considering transient and steady-state dynamic characteristics which confirm the satisfactory performance of the MPPT algorithms with standard WTS.

REFERENCES

[1] G. L. Johnson, *Wind Energy Systems*. Englewood Cliffs, NJ: Prentice-Hall Inc, 1985.

[2] A. Soriano, L. Wen Yu, and J. de Jesus Rubio. "Modeling and control of wind turbine." *Mathematical Problems in Engineering*, vol. 2013, 2013.

[3] W. Chia-Nan, L. Wen-Chang, and L. Xuan-Khoa. "Modelling of a PMSG wind turbine with autonomous control." *Mathematical Problems in Engineering*, vol. 2014, 2014.

[4] H. Li, and Z. Chen. "Overview of different wind generator systems and their comparisons." *IET Renewable Power Generation*, vol. 2, no. 2, pp. 123–138, 2008.

[5] H. Polinder, F. F. A. van der Pijl, G. de Vilder, and P. J. Tavner. Comparison of direct-drive and geared generator concepts for wind turbines. *IEEE Transactions on Energy Conversion*, vol. 21, no. 3, pp. 725–733, 2006.

[6] Y. Venkata, A. Dekka, M.J. Durán, S. Kouro, and B. Wu. "PMSG-based wind energy conversion systems: survey on power converters and controls." *IET Electric Power Applications*, vol. 11, no. 6, pp. 956–968, 2017.

[7] N. Priyadarshi, M. S. Bhaskar, P. Sanjeevikumar, F. Azam, and B. Khan. High-power DC-DC converter with proposed HSFNA MPPT for photovoltaic based ultra-fast charging system of electric vehicles. *IET Renewable Power Generation*, 2022.

[8] V. Adithya, D. Haribabu, and J.N. Sakamuri. "Modeling and control of DC/DC boost converter using K-factor control for MPPT of solar PV system." In *2015 International Conference on Energy Economics and Environment (ICEEE)*, pp. 1–6. IEEE, 2015.

[9] Instruments Texas. "Basic calculation of a boost converter's power stage." *Application Report*, vol. SLVA327B, 2009.

[10] T. Jogendra Singh, and M. Ouhrouche. "MPPT control methods in wind energy conversion systems." *Fundamental and Advanced Topics in Wind Power*, vol. 1, pp. 339–360, 2011.

[11] S. Tekeshwar Prasad, T. V. Dixit, and R. Kumar. "Simulation and analysis of perturb and observe MPPT algorithm for PV array using ĊUK converter." *Advance in Electronic and Electric Engineering*, vol. 4, no. 2, pp. 213–224, 2014.

[12] F. Nicola, G. Petrone, G. Spagnuolo, and M. Vitelli. "Optimization of perturb and observe maximum power point tracking method." *IEEE Transactions on Power Electronics*, vol. 20, no. 4, pp. 963–973, 2005.

[13] M. A. Elgendy, B. Zahawi, and D.J. Atkinson. "Assessment of perturb and observe MPPT algorithm implementation techniques for PV pumping applications." *IEEE Transactions on Sustainable Energy*, vol. 3, no. 1, pp. 21–33, 2011.

[14] K. Mekalathur, B. Hemanth, B. Saravanan, P. Sanjeevikumar, and F. Blaabjerg. "Review on control techniques and methodologies for maximum power extraction from wind energy systems." *IET Renewable Power Generation*, vol. 12, no. 14, pp. 1609–1622, 2018.

[15] S. Dezso, L. Mathe, T. Kerekes, S. Viorel Spataru, and R. Teodorescu. "On the perturb-and-observe and incremental conductance MPPT methods for PV systems." *IEEE Journal of Photovoltaics*, vol. 3, no. 3, pp. 1070–1078, 2013.

[16] N. Priyadarshi, S. Padmanaban, P. K. Maroti, and A. Sharma. An extensive practical investigation of FPSO-based MPPT for grid integrated PV system under variable operating conditions with anti-islanding protection. *IEEE Systems Journal*, vol. 13, no. 2, pp. 1861–1871, 2018.

[17] N. Priyadarshi, S. Padmanaban, J. B. Holm-Nielsen, F. Blaabjerg, and M. S. Bhaskar. An experimental estimation of hybrid ANFIS–PSO-based MPPT for PV grid integration under fluctuating sun irradiance. *IEEE Systems Journal*, vol. 14, no. 1, pp. 1218–1229, 2019.

[18] S. Padmanaban, N. Priyadarshi, J. B. Holm-Nielsen, M. S. Bhaskar, F. Azam, A. K. Sharma, and E. Hossain. A novel modified sine-cosine optimized MPPT algorithm for grid integrated PV system under real operating conditions. *IEEE Access*, vol. 7, pp. 10467–10477, 2019.

[19] S. Padmanaban, N. Priyadarshi, M. S. Bhaskar, J. B. Holm-Nielsen, E. Hossain, and F. Azam. A hybrid photovoltaic-fuel cell for grid integration with jaya-based maximum power point tracking: Experimental performance evaluation. *IEEE Access*, vol. 7, pp. 82978–82990, 2019.

[20] N. Priyadarshi, V. K. Ramachandaramurthy, S. Padmanaban, and F. Azam. An ant colony optimized MPPT for standalone hybrid PV-wind power system with single Cuk converter. *Energies*, vol. 12, no. 1, p. 167, 2019.

[21] N. Priyadarshi, S. Padmanaban, L. Mihet-Popa, F. Blaabjerg, and F. Azam. Maximum power point tracking for brushless DC motor-driven photovoltaic pumping systems using a hybrid ANFIS-FLOWER pollination optimization algorithm. *Energies*, vol. 11, no. 5, p. 1067, 2018.

[22] N. Priyadarshi, A. K. Sharma, and F. Azam. A hybrid firefly-asymmetrical fuzzy logic controller
based MPPT for PV-wind-fuel grid integration. *International Journal of Renewable Energy Research
(IJRER)*, vol. 7, no. 4, pp. 1546–1560, 2017.

[23] N. Priyadarshi, S. Padmanaban, M. S. Bhaskar, F. Azam, B. Khan, and M. G. Hussien. A novel hybrid
grey wolf optimized fuzzy logic control based photovoltaic water pumping system. *IET Renewable
Power Generation,* 2022.

[24] B. Mendi, M. Pattnaik, and G. Srungavarapu. A speed sensorless modified perturb and observe MPPT
scheme for stand-alone PMSG based wind turbine system. In *2022 IEEE IAS Global Conference on
Emerging Technologies (GlobConET)*, 2022, pp. 338–342.

7 DC/AC Power Converters for Hybrid Renewable Energy-Based Applications

T. Abhilash, Kuncham Sateesh Kumar,
Jammy Ramesh Rahul, A. Kirubakaran
and V. T. Somasekhar

CONTENTS

7.1 INTRODUCTION

Electrical energy has become a basic necessity for humankind. The conventional resources generating it are going to be depleted soon. Renewable Energy Sources (RES) are being promoted, in the interest of conservation of natural resources, taking the future load demand into consideration. All of the renewable energy sources require the support of power electronic systems to facilitate their optimal use and to achieve compatibility with the requirements of the load. The power electronic systems are mainly classified into four types based on the type of power conversion, namely: (1) DC-DC converters, (2) DC-AC converters, (3) AC-DC converters, and (4) AC-AC converters. This research deals with the DC-AC converters, which are also called inverters. Basically, a power

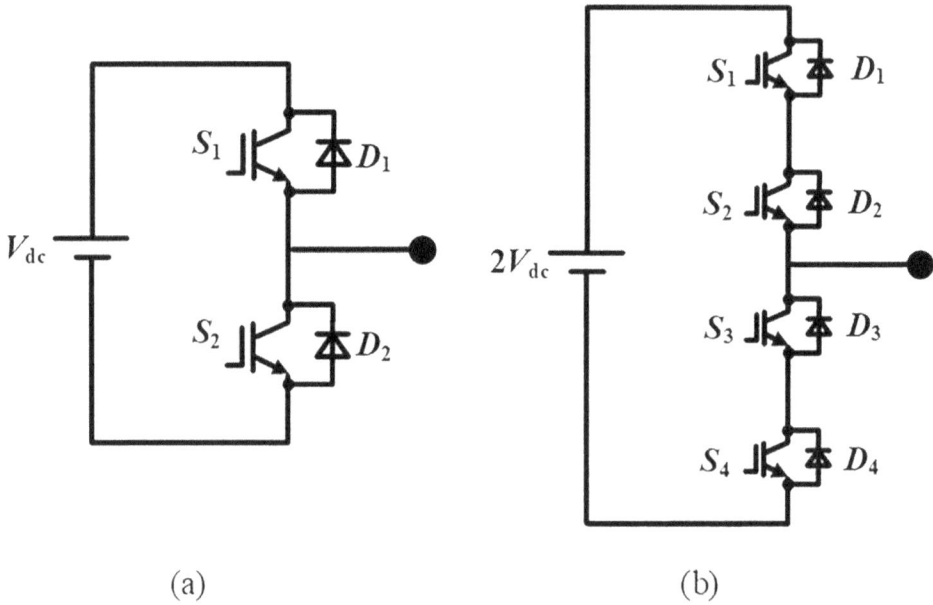

(a)   (b)

FIGURE 7.1 Two-level VSIs: (a) low and medium power and (b) high power.

inverter (or simply an inverter) is a circuit that converts DC power to AC power by means of power electronic devices.

The input and output voltages, frequency, performance, reliability, and the power handling capability of power electronic converters depend on the selection of the devices and the design of the circuitry. According to the type of the energy-storing element at the DC side, inverters are classified as (i) Current Source Inverters (CSI) and (ii) Voltage Source Inverters (VSIs). In this chapter, VSI based power converters have been considered as they have much higher employment and their remarkable development in the last two decades. The VSIs are further classified as (i) two-level inverters and (ii) multilevel inverters (MLIs).

Figure 7.1 shows one of the phase legs of a traditional 3-Φ two-level VSI. The former is appropriate for both low and medium power, while the latter is suitable for high power applications such as large rated machine drives, high-voltage direct current (HVDC) transmission, flexible ac transmission systems (FACTS), and renewable power sources. The series connection of switching devices enables the 2-L VSI to achieve a high power level. The addition of a few more switching devices along with either diodes or capacitors or DC power supplies enhances the quality of the output voltages, originating the multilevel VSI (ML-VSI) technology.

7.1.1 CLASSIFICATION OF MULTILEVEL INVERTERS

The market share of multilevel inverters is increasing rapidly than the two-level inverters in the area of medium-voltage and high power applications.

*   Better quality of output voltage and current
*   Reduction in filter size due to greater number of output levels
*   Ability to handle larger DC-link voltages with power devices of lower voltage ratings
*   Better electromagnetic compatibility

Figure 7.2 illustrates the classification of MLIs. MLIs are mainly classified into single DC and multiple DC source types. The diode clamped, flying capacitor, modular multilevel, and so on, fall

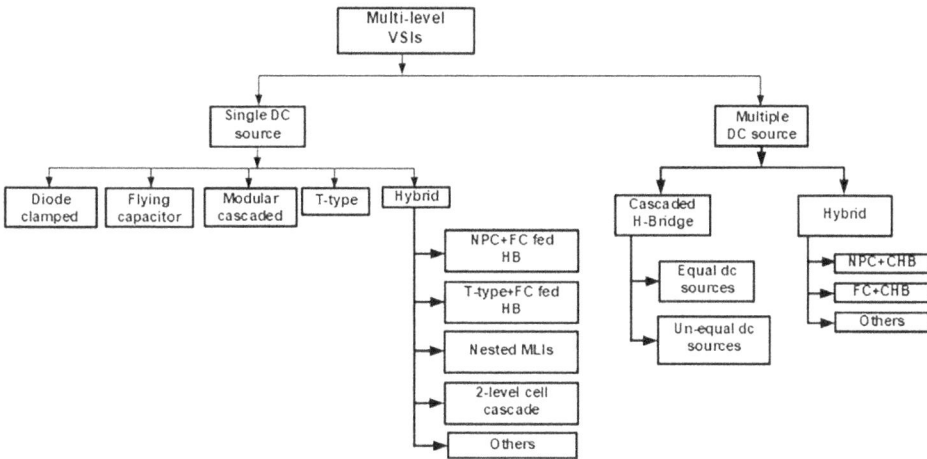

FIGURE 7.2 Classification of multilevel inverters.

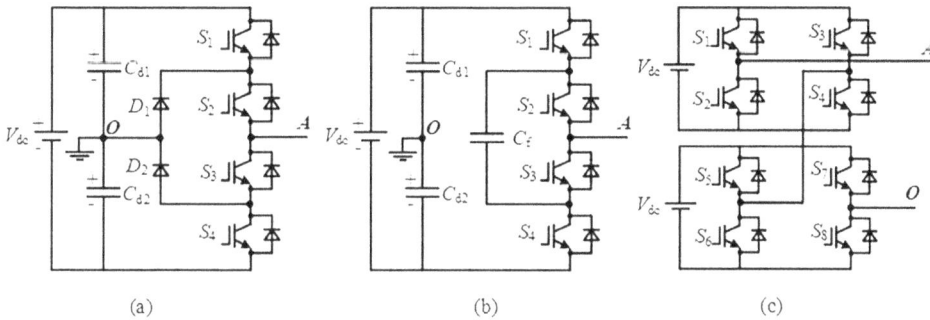

FIGURE 7.3 Classical MLI topologies: (a) NPC [1], (b) FC [1], and (c) CHB [2].

under the category of single DC source-based MLIs. On the other hand, the cascaded MLI structure and some hybrid MLIs use more than one DC power supply, as shown in Figure 7.2. Figure 7.3 illustrates the circuit diagrams of the traditional MLIs named as neutral point clamped (NPC), flying capacitor (FC), and cascaded H-bridge (CHB) inverters. These are typically used in motor drives, active filters, and distributed generation systems.

Recent research articles recommended the reduced switch count (RSC) MLI topologies for various applications, wherein the *Total Blocking Voltage* (TBV) of the semiconductor switches and the efficiency of the circuit are considered as Figure of merit. Most of the 1-Φ and 3-Φ topologies presented in the literature use classical structures such as NPC, FC, CHB, and their variants. Some of the topologies presented in the literature use isolated DC power supplies for the generation of multilevel voltage in the output. Using isolated DC power supplies results in an increased requirement of components and bulky transformers, which increase the cost. Thus, the present research focuses on the employment of a single DC power supply for a multilevel generation.

7.1.2 Hybrid Multilevel Inverters

A hybrid topology is an integration of two or more topologies in a single converter. The primary objective of a hybrid inverter structure is to achieve best trade-off in the assortment of power components in terms of operating frequency and voltage withstanding capability. Thus, the hybrid inverter structure is more economical and more efficient for the same number of level generations by employing fewer power components and DC sources as compared to classical MLI topologies.

Advantages of Hybrid MLIs:

1. Reduced number of DC sources
2. Low output switching frequency that results in low switching power losses
3. High conversion efficiency
4. Flexibility to enhance the ratings for different applications

Owing to the these merits, the investigation in this chapter is aimed to realize single-source based hybrid multilevel inverter (HMLI) topologies. The block diagrams and the basic working principles of the HMLI topologies are available in the literature. As a pre-requisite, the HMLI topologies reported in the past are briefly explained in the following subsections.

7.1.3   FC Fed H-Bridge Based HMLI

The topology of HMLI shown in Figure 7.4 uses a primary converter, which can be either an NPC, an FC, or of a T-type converter. It typically generates a three-level output per phase across the terminals A' and O that is denoted as $V_{A'O}$. An FC fed H-bridge circuit, which also generates a three-level output across the terminals A and A' (denoted as $V_{AA'}$) is cascaded to the primary converter, which results in pole-voltage (V_{AO}) waveform with seven levels.

$$V_{AO} = V_{AA'} + V_{A'O} \qquad (1)$$

The input DC source charges the DC-link capacitors. The flying capacitors are charged through the redundant switching states of the converter. A balancing scheme is employed to equalise the DC-link capacitor voltages ($V_{Cd1} = V_{Cd2} = 0.5V_{dc}$). The FC voltage is maintained at $0.25V_{dc}$ to generate the seven-level output. Table 7.1 lists the voltage levels generated by the primary converter, FC fed H-bridge, and the overall converter.

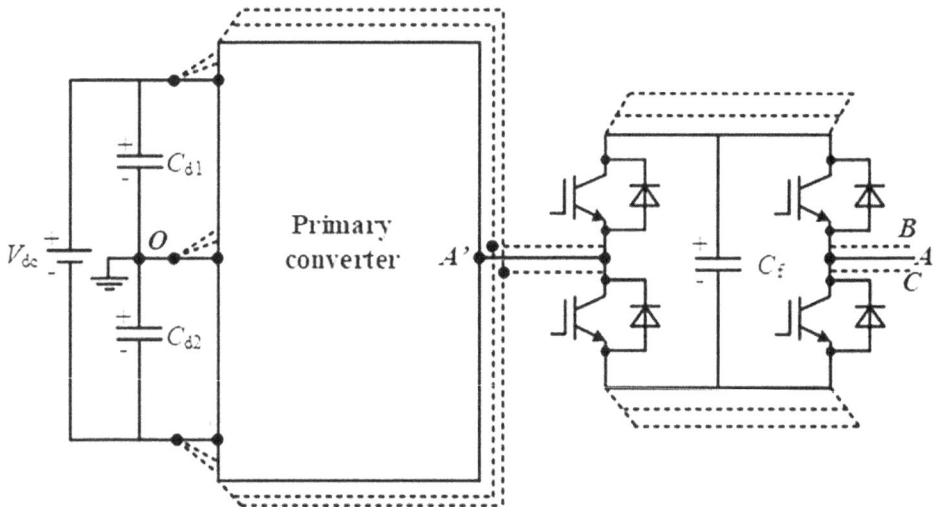

FIGURE 7.4   Block-diagram of the FC fed H-bridge based MLI.

TABLE 7.1
Seven-Level Output Voltage

$V_{A'O}$	$V_{AA'}$	V_{AO}
$0.5V_{dc}$	$0.25V_{dc}$	$0.75V_{dc}$
$0.5V_{dc}$	0	$0.5V_{dc}$
0	$0.25V_{dc}$	$0.25V_{dc}$
$0.5V_{dc}$	$-0.25V_{dc}$	
0	0	0
0	$-0.25V_{dc}$	$-0.25V_{dc}$
$-0.5V_{dc}$	$0.25V_{dc}$	
$-0.5V_{dc}$	0	$-0.5V_{dc}$
$-0.5V_{dc}$	$-0.25V_{dc}$	$-0.75V_{dc}$

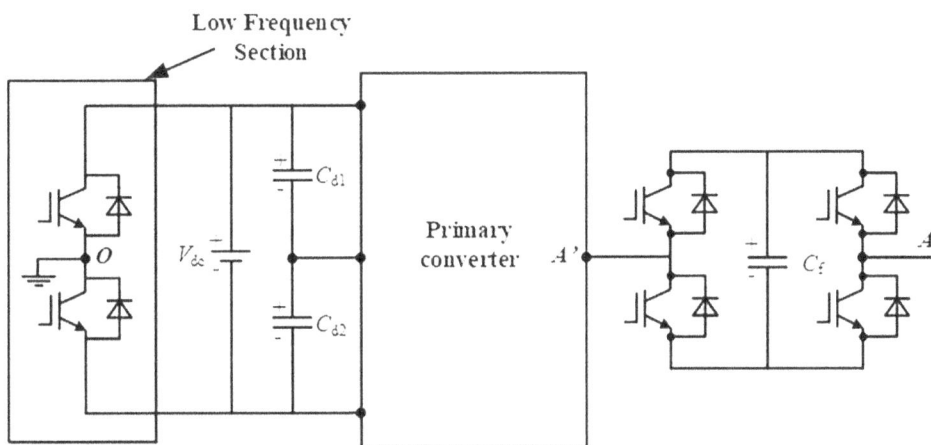

FIGURE 7.5   Block-diagram of the FC fed H-bridge and LFS2 Based MLI.

7.1.4   LFS² AND FC FED H-BRIDGE BASED HMLI

The inverter in Figure 7.5 uses a primary converter, which can be either an NPC, an FC, or a T-type converter. An additional section, consisting of two switching devices operates at the fundamental frequency is added to one phase of the converter in Figure 7.4. This section is denoted as low-frequency switching section (LFS²). Therefore, the primary converter, along with LFS², typically generates a five-level output across the terminals A' and O (denoted as $V_{A'O}$).

An FC fed H-bridge circuit, which generates a three-level output across the terminals A and A' that is denoted as $V_{AA'}$ is cascaded to the primary converter, which results in output-voltage (V_{AO}) waveform with nine levels. Equation (1) is applicable to this converter also, for output voltage level generation. Flying capacitor charges through the redundant switching states, while the charging cycle of DC-link capacitors. The FC voltage is maintained as $V_{dc}/4$ to generate the nine-level output. Table 7.2 lists the voltage levels generated by the primary converter, FC fed H-bridge, and the overall converter.

7.1.5   NESTED HMLIS

The inverter configuration shown in Figure 7.6 uses a nested converter which can be either an NPC, an FC, the combination of NPC and FC, or of a T-type converter. An additional section, consisting

TABLE 7.2

Nine-Level Output Voltage

$V_{A'O}$	$V_{AA'}$	V_{AO}
V_{dc}	0	V_{dc}
$0.5V_{dc}$	$0.25V_{dc}$	$0.75V_{dc}$
V_{dc}	$-0.25V_{dc}$	
$0.5V_{dc}$	0	$0.5V_{dc}$
0	$0.25V_{dc}$	$0.25V_{dc}$
$0.5V_{dc}$	$-0.25V_{dc}$	
0	0	0
0	$-0.25V_{dc}$	$-0.25V_{dc}$
$-0.5V_{dc}$	$0.25V_{dc}$	
$-0.5V_{dc}$	0	$-0.5V_{dc}$
$-0.5V_{dc}$	$-0.25V_{dc}$	$-0.75V_{dc}$
$-V_{dc}$	$0.25V_{dc}$	
$-V_{dc}$	0	$-V_{dc}$

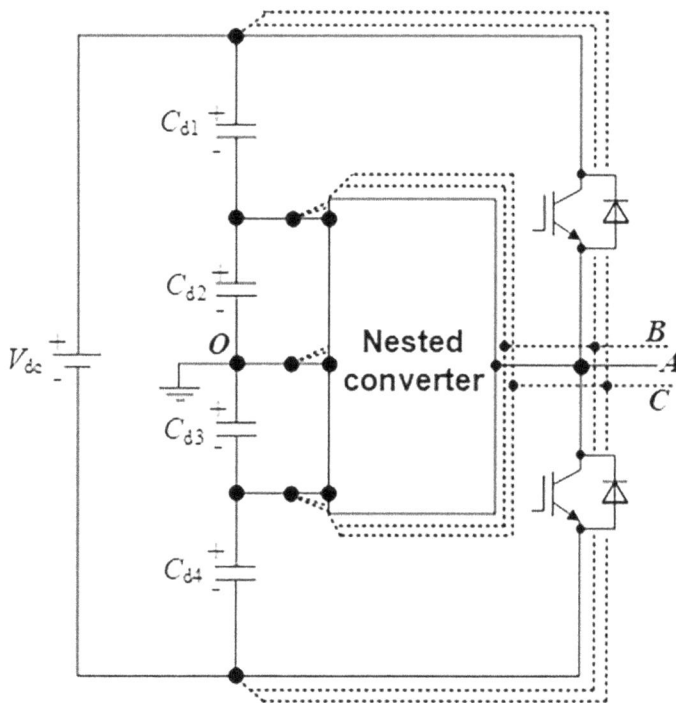

FIGURE 7.6   Block-diagram of the Nested MLI Structure.

of two switching devices operating at the fundamental frequency is cascaded to a nested structure. This section is denoted LFS^2. The switches in LFS^2 operate only during peak voltage levels ($\pm 0.5V_{dc}$). Therefore, the nested converter, along with LFS^2, typically generates a five-level output ranging from $-0.5V_{dc}$ to $0.5V_{dc}$ with a step size of $0.25V_{dc}$, across the terminals A and O (denoted as V_{AO}).

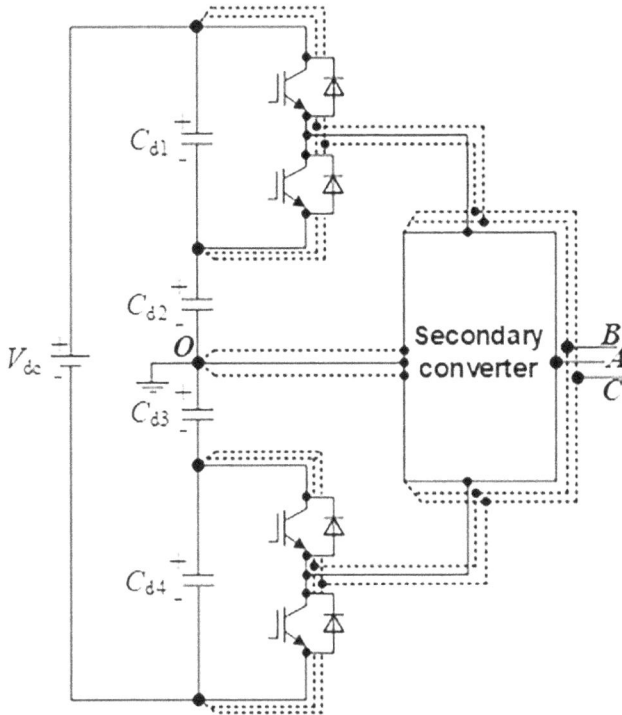

FIGURE 7.7   Block diagram of the half-bridge cell based MLI.

During the charge cycle of DC-link capacitors via input DC source; the voltage balancing of capacitors is achieved using an Auxiliary Voltage Balancing Circuit (AVBC). The details of AVBC (which are not shown in Figure 7.6) will be presented in later chapters.

7.1.6   Half-Bridge Cell Based HMLIs

The power circuit configuration shown in Figure 7.7 uses half-bridge cells clamped to the top and bottom DC-link capacitors. The outputs of the half-bridge cells are connected to a secondary converter to increase the number of levels in the output. Different types of circuits to act as secondary converters are explained in Chapter 2. The overall converter can generate a five-level output ranging from $-0.5V_{dc}$ to $0.5V_{dc}$ with a step size of $0.25V_{dc}$, across the terminals A and O (denoted as V_{AO}). During the charge cycle of DC-link capacitors via input DC source; the voltage balancing of capacitors is achieved using an AVBC.

7.2   LITERATURE

Multilevel inversion is emerging as a viable solution for medium-voltage/high power DC to AC power conversion and the advantages offered by multilevel inverters (MLIs) are described in [1]–[5]. MLIs have been used for many applications such as distributed generation [6], [7], micro-grid systems [8], power systems [9], [10], static reactive power compensation [11], and adjustable speed drives [12]. Some of the promising features of multilevel inverters are as follows:

1. They can generate output voltages and currents with low distortion and lower dv/dt stress across the switching devices.

2. They draw input current with lower distortion.
3. They can operate with lower switching frequencies.
4. They provide redundant switching states for capacitor voltage balancing.
5. They offer a high degree of flexibility in interfacing renewable energy sources.

7.2.1 VARIOUS TYPES OF MULTILEVEL INVERTERS

Multilevel technology started with the three-level inverter proposed in [13]. The most popular and classical MLI topologies are:

1. Neutral point clamped (NPC) or diode clamped converter (DCC) [13],
2. Flying capacitor clamped (FCC) converter [14]–[16], and
3. Cascaded H-bridge (CHB) converter [14], [15]–[19].

All these classical MLI topologies are acceptable to use for low voltage applications. The topologies such as NPC and FC require large number of diodes and capacitors respectively and it restricts their use in higher voltage applications. NPC and FC can be used for high power applications with the complex control schemes required for balancing the DC-link capacitor voltages. In CHB, the requirement of DC sources and active switches for each module which further increases the complexity of controller and cost.

These classical topologies have attracted the attention of both industry and academia. Still, no specific MLI seems to be completely advantageous as the choice of multilevel topology is principally determined by the end application and cost constraints. Because of its inherent characteristics, a given topology could be best solution for few applications and not suitable for some others. Thus, the precise topology is decided on a case-to-case basis. Hence, researchers continued (and still continue) to evolve new circuit configurations with an application-oriented approach. A detailed literature survey of MLIs is presented in this chapter.

In recent years, many types of MLI topologies under the classification of symmetrical (equal magnitude of DC sources), asymmetrical (unequal magnitude of DC sources), and hybrid (equal and unequal voltage sources) converters have been proposed in the literature to conquer the limitations of classical topologies [20]–[22]. Asymmetrical CHB based MLI topologies utilize less number of unequal voltage sources to realize the multilevel operation in comparison with the symmetrical CHB based MLI topologies. However, both CHB based topologies requires more number of DC sources compared to NPC and FC topologies. To overcome these limitations, a five-level MLI topology proposed in [23], which comprises of a three-level active NPC converter (ANPC^2) with additional FC and its balancing circuit. To refine the FC balancing in the previous topology [23], another interesting modulation scheme with a proportional controller is proposed for a 3-Φ five-level FC MLI in [24]. But, the sizing of FC and the tuning of controller plays a vital role in this topology to obtain the highest quality of output.

Contemporarily, MLI topologies in modular structures are getting popular in 3-Φ applications. Modular structure generates higher number of voltage levels by connecting each module in series. Further, each module consists of a DC source and the switches associated with individual module have lesser voltage rating than the output which enables the operation of modular MLIs in HVDC applications [25]. To improve the efficiency in comparison with CHB based topologies, a novel T-type three-level inverter and seven-level ANPC^2 for 3-Φ applications are illustrated in [26], [27]. But, the use of a DC source in each two-level module could be a drawback for modular MLIs when they are used for low- and medium-power applications.

To obtain high efficiency and lower number of DC sources, a novel T-type hybrid converter topology is presented in [28]. Reduction in the total blocking voltage (approximately half) is the advantage of this topology in comparison with the T-type inverter. Further it enhances the efficiency by reducing the switching losses. Unfortunately, it requires a greater number of active switches.

Another interesting asymmetric five-level MLI topology is reported in [29], which is formed by cascading the full-bridge and half-bridge submodules. FC used in this topology will be charged during positive half-cycle and discharged during negative half-cycle. But, the ripple content in both FC and output voltage is higher in this topology due to the non-availability of switching states.

In refs [30]–[32], a generalized CHB based MLI topology is proposed with reduced number of switches and of lower blocking voltage. The topology proposed in [33] employs cascaded cells. Each cell is made up of six active switches and two independent DC supplies. This configuration suffers from a high blocking voltage of the switches. Some of the other varieties of T-Type MLI topologies have been presented in [34], [35]. These topologies are suitable for both single-phase as well as three-phase applications and utilizes reduced number of switches compared to CHB based MLIs. Unfortunately, T-type structures need several isolated DC sources, restricting their applicability.

To solve the issues associated with classical and T-type MLI topologies, a modified T-type CHB based hybrid nine-level MLI is proposed in [36]. This topology effectively utilizes the total DC-link voltage for obtaining output levels with high rated switches in the CHB. Another interesting hybrid MLI topology is proposed in [37], to produce higher number of levels across the output with lesser number of switching devices. On the negative side, both the topologies demands two separate DC sources. The configuration presented in [38] comprises two cascaded branches of FC fed H-bridges, in which the number of output levels increases to enhance the quality of output waveform. But, the utilization of more number of components and conducting devices for a given voltage level restricts the use of full DC-link voltage.

Several hybrid topologies are reported in the literature in the interest of utilizing full DC-link voltage. A modified nine-level MLI configuration is presented in [39] to utilize total DC-link voltage. Also, the modulation scheme is proposed with more number of switching states for balancing of FC. An interesting nine-level dual boost MLI is proposed in [40], which utilizes the concept of switched capacitors and single DC source. Further it produces polarity without the use of H-bridge module. In [41], a novel packed U-cell based MLI is proposed with a single flying capacitor to generate the nine-level output. However, both the preceding topologies require high rated voltage switches, large capacitors, and low switching states to realize the FC voltage balancing. A single-phase nine-level switched capacitor topology is proposed in [42] to utilize the full DC-link voltage with the advantage of reduced switch count. It suffers from the disadvantage of a reduced number of switching states to balance the FC voltage. Thus, its applicability is restricted to fundamental frequency switching.

Another classification of various hybrid MLI topologies with multi-winding transformers is reported in [43] where the transformer is used to create isolated DC sources. But, it demands higher number of switches and complexity in control. The work reported in [44] introduces a four-level nested NPC inverter with lower number of components compared to a four-level DCC or a four-level FCC. Moreover, this topology is formed by joining all the mid points of the inner and outer legs to the output terminals of the phase legs. The same nested connection is further extended to T-type four-level inverter in [45] with lower number of switches which makes the inverter to use in high power and mid-voltage range applications. It demands voltage sensors in each phase for FC voltage balancing. A generalized nested MLI with an inner leg T-type structure is proposed in [46]. This topology produces a higher number of levels with reduced switches, and utilizes a single DC source. However, the sum of the blocking voltages of the switches required in this topology is higher.

Thus, in the latest literature, [47], [48] different MLI topologies have been proposed to reduce the switch count, voltage balancing issues, and the requirement of isolated DC sources as compared to the aforementioned classical MLI topologies. The single-phase topology presented in [47] can produce a seven-level output with lower number of components in comparison with the CHB MLI topology. A 3-Φ five-level topology is illustrated in [48] produces multilevel output with the reduced components but demands for a large number of isolated DC sources. In general, most of the MLI topologies, the primary issue lies in balancing the voltages of the DC-link capacitors and FCs. A novel PWM scheme is proposed in [49] to balance the DC-link capacitor voltages. However, this

PWM scheme is too cumbersome and it is difficult to implement. To balance the DC-link capacitors and to boost the input source voltage another interesting solution is proposed with front-end boost converter in [50]. However, the integration of a boost converter further increases the component count and size. The *five-level hybrid clamped* (5L-HC) converter proposed in [51], avoids the serial connected devices and ripple content of the FC voltages which makes this converter suitable for medium-voltage applications. The drawback of this MLI is that, it demands a higher number of capacitors and switches.

Stacked cell based five-level inverter is presented in [52]. Though this inverter utilizes a lower number of capacitors compared to [51], it demands an increased number of switches per phase and also requires two isolated DC sources. To overcome the preceding problems, capacitor fed H-bridge five-level inverter is proposed in [53]. It has the advantage of increased DC bus utilization. But, requires an additional sensor circuitry in each phase for balancing the FC voltages similar to [51], [52]. Thus, an unique solution is proposed in [54] with lower switch count using T-type structure. This configuration eliminates the FCs and DC-link capacitors for producing the multilevel output. However, the use of two unequal DC sources increases the implementation cost of the topology. Another interesting MLI topology is presented in [55], [56], named as five-level back-to-back E-type converter. This topology also realizes the multilevel output without using FCs and has the benefits of reduced power losses. Unfortunately, the auxiliary circuit used for balancing the DC-link capacitors increases the overall system size and complexity of controller.

In view of the preceding survey, it can be commented that the hybrid MLIs are developing enormously and are categorized into the four types for further analysis in the forthcoming sections.

7.3 VARIOUS TYPES OF FC-FED H-BRIDGE BASED HMLIS

Several topologies of HMLIs have been reported in the literature, which are derived from the classical topologies (NPC, FC, and CHB). Various seven-level HMLI topologies were proposed, which vary in terms of (i) devices and (ii) circuit combinations. Across the HMLIs, it is noted that the H-Bridge (HB) plays a pivotal role in these topologies [57], [58]. The voltage across the FC ensures identical voltage stress across all the four devices, which constitute the HB, making it a highly attractive solution to realize HMLI. In this category of HMLI topologies, an FC fed H-bridge (FCFHB) circuit is cascaded to a primary converter.

Figure 7.8(a) shows an HMLI, called NPC-HB, which is proposed in [59], consists of the cascade connection of a three-level NPC converter and an FC fed HB (FCFHB). It can produce a seven-level output in pole-voltage (V_{AO}). The limitation in this topology is that each phase leg requires clamping diodes of blocking voltage $V_{dc}/2$, which increases the switching losses of the overall system. The Diode Clamped Converter (DCC) also suffers from the drawback of unequal loss distribution among the switching devices. A combination of ANPC² and FCFHB is proposed in [23], as shown in Figure 7.8(b), to overcome the problems in NPC structure. This topology can produce a seven-level output and avoids the problem of unequal loss distribution. However, it demands more number of components for the generation of same number of levels in comparison with the NPC.

Figure 7.8(c) depicts the FC based primary converter cascaded to FCFHB [60]. This topology has the same switch count as DCC and eliminates the need for clamping diodes. However, it requires an additional flying capacitor. Consequently, its PWM algorithm becomes complex as there exists a necessity to balance the two flying capacitor voltages. A hybrid 3-Φ seven-level inverter was illustrated in [61], where a T-type converter (T²C) is joined with an H-bridge inverter. Thus, this circuit configuration is referred to as a *Hybrid 7-level Converter* (H7LC) and is shown in Figure 7.8(d). This topology avoids the use of clamping diodes and additional FCs by maintaining the same number of switching devices like DCC and FCC. However, in this topology, the voltage blocking capability of the switches $S1$ and $S2$ is V_{dc}, and it is same in each phase which increases overall switching losses of the topology.

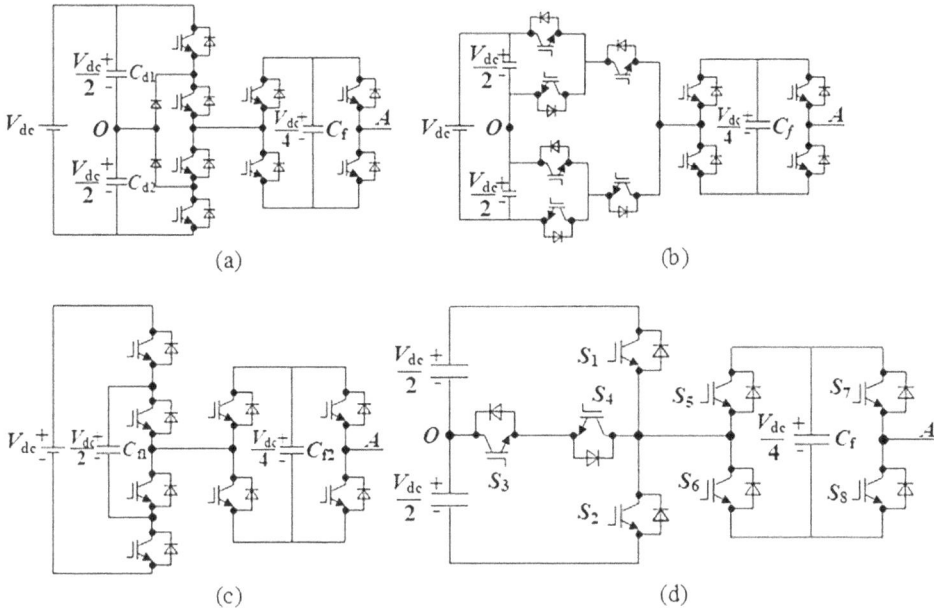

FIGURE 7.8   (a) Three-level DCC cascaded to FCFHB [59], (b) three-level ANPC[2] cascaded to FCFHB [23], (c) three-level FC cascaded to FCFHB [60], and (d) three-level T[2]C cascaded to FCFHB [61].

7.3.1   Various Types of LFS[2] and FC-Fed H-Bridge Based HMLIs

In this category of HMLI topologies, an FC fed H-bridge (FCFHB) circuit and a Low-Frequency Switching Section (LFS[2]) are cascaded to a primary converter at opposite ends. The FCFHB is connected at the output end, whereas the LFS[2] is connected at the source end. The topologies proposed in [62], [63] employ two low-frequency switching devices (LFSDs) across the DC-link. This facilitates the total DC voltage utilization to produce nine-level output voltage. The research reported in [62] uses an ANPC[2] structure to match the losses in the semiconductor switches at the cost of larger component count and equalize the power losses in the switching devices. In contrast, the work described in [63] comprises a T-type structure, which requires high voltage rated switches. Thus, the switching losses in both of the aforementioned topologies are higher.

In [64], double flying capacitor (DFC) topology is proposed to produce nine-level output using two Low-Frequency Devices (LFDs) by removing the mid-point of the DC-link. But, the use of more number of FCs, more number of active switches and capacitor balancing issues for the generation of multilevel output results in an increased control complexity and a reduced efficiency. Figure 7.9(a) shows the configuration of a nine-level DFCM converter. The limitation of this topology is the requirement of two flying capacitors. Thus, this circuit requires a complex capacitor voltage balancing algorithm and results in increased switching losses. Figure 7.9(b) shows a nine-level hybrid cascaded MLI [65] with reduced switch count compared to [64].

This topology avoids the use of FCs, but, it requires three independent DC sources. The topology proposed in [65] which are shown in Figure 7.9(c), uses an ANPC[2] to generate nine-level output. It employs a single DC source and a flying capacitor. As mentioned earlier, this topology suffers from the requirement of a higher switch count. The ANPC[2] is replaced with a T[2]C, as shown in Figure 7.9(d). It has the advantages of less switch count, single DC source and a flying capacitor to generate the nine-level output. However, it demands two high voltage blocking switches in the T-type structure.

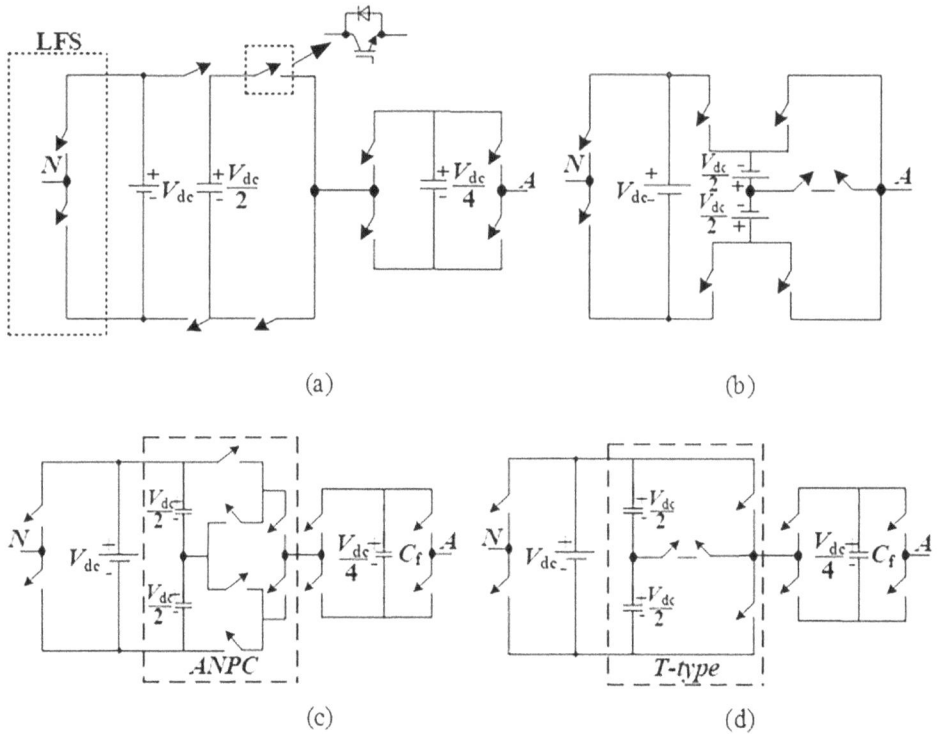

FIGURE 7.9 (a) Nine-level hybrid DFCM converter [64], (b) nine-level hybrid cascaded MLI [65], (c) nine-level DHANPC inverter [66], and (d) nine-level T²C cascaded MLI [65].

7.3.2 Various Types of Nested HMLIs

Figure 7.10(a) shows the phase leg of a 3-Φ nested NPC MLI [66]. It has the advantage of handling a higher DC voltage with the additional capacitor and switches in each leg. Similar to the DCC and FCC, this topology has the advantage of equal voltage rated switches. However, each phase of the NNPC requires three flying capacitors and two clamping diodes. It also demands more sensors and complex control algorithms to achieve FC voltage balancing. This configuration also demands two split dc sources of $0.5V_{dc}$ each. The generalized topology proposed in [46], paved way to the proposal of four-level, five-level, and six-level nested HMLIs [67], without any flying capacitors. Of these three topologies, the nested five-level inverter comprises of a T-type structure and it results in higher voltage rating of the switches, higher power losses, and increased cost of the system. The 3-Φ nested five-level topology, shown in Figure 7.10(b), employs a nested T-type structure and a low frequency switching section. The switches in the T²C operate to produce $-0.25V_{dc}$, 0, and $0.25V_{dc}$ levels. The switches S_1 and S_2 operate at fundamental frequency to clamp $\pm0.5V_{dc}$ levels, respectively across the pole-voltage (V_{AO}).

This configuration eliminates the need for clamping diodes and FCs. However, this topology demands high voltage rated switching devices and three independent dc sources. The number of dc sources can be reduced by employing single dc source with four dc-bus capacitors. The voltage across the capacitors is balanced using an optimized Auxiliary Voltage Balancing Circuit (AVBC) proposed in [68].

7.3.3 Various Types of Two-Level Cell Based HMLIs

A novel six-level hybrid inverter using a combination of two-level cells and three-level FC cells is presented in [69]. But, it requires two asymmetrical DC voltage sources which increases the cost of

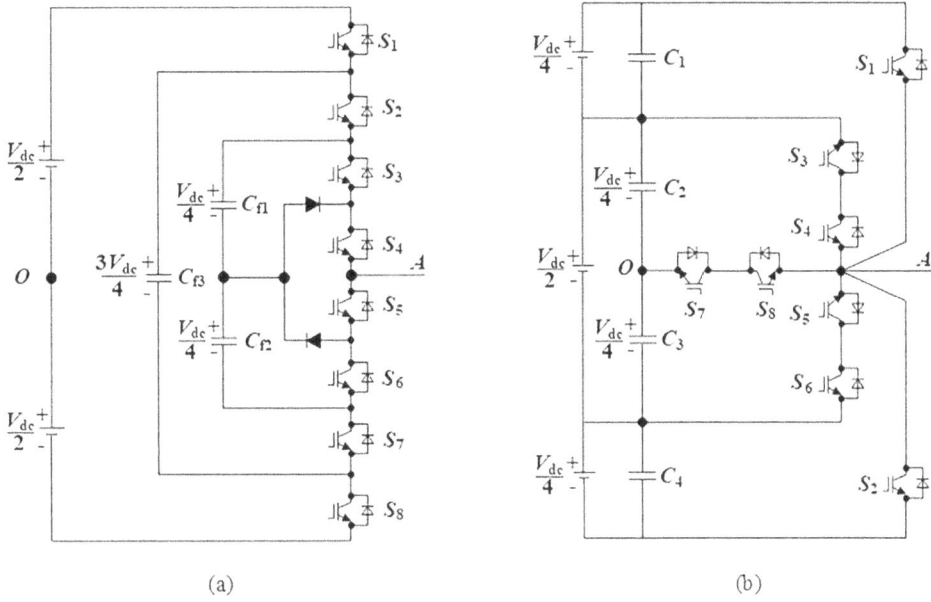

FIGURE 7.10   (a) Five-level NNPC [66], and (b) five-level nested T-type MLI [67].

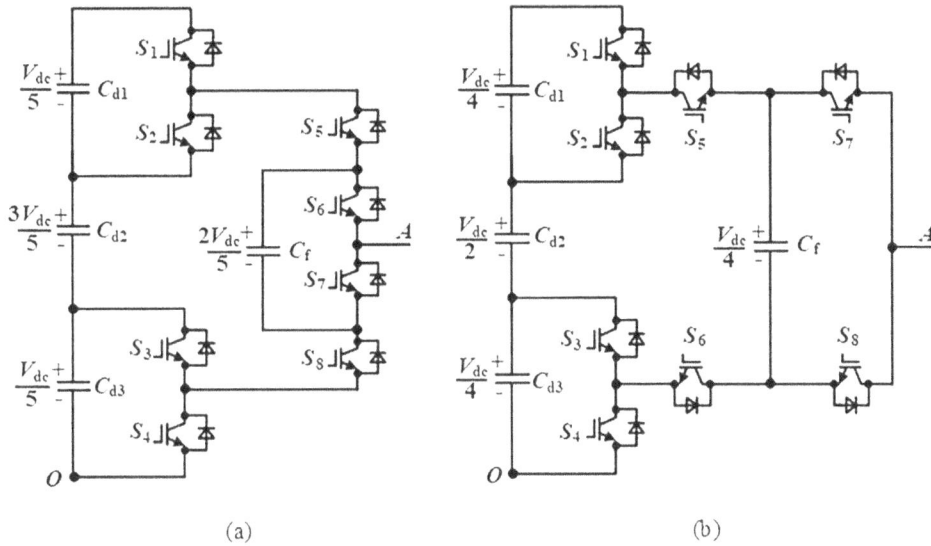

FIGURE 7.11   (a) Six-level hybrid Inverter [69], and (b) five-level HFC inverter [70].

overall system. Figure 7.11(a) shows one phase leg of a 3-ph six-level hybrid inverter based on two-level cells clamped to the top and bottom capacitors.

The secondary converter is of an FC structure. Figure 7.11(b) depicts a similar five-level hybrid flying capacitor (HFC) inverter [70]. In this converter, different output voltage levels are obtained due to the difference in voltage distribution across the DC-link and FCs. It is to be observed that both of the configurations require a flying capacitor in each phase, demanding the employment of an independent voltage sensor in each phase. In these inverter configurations, the number of DC sources can be reduced by employing a single dc source with four DC-bus capacitors. The voltage across the capacitors is balanced using an optimized Auxiliary Voltage Balancing Circuit (AVBC).

7.3.4 CONTROL METHODS USED IN MLIs

For multilevel inverters (MLIs), the control techniques play a prominent role in determining the quality of the output waveforms. Based on the switching frequencies employed by MLIs, the modulation schemes can be broadly categorized into two classes, namely, (i) low-frequency and, (ii) high-frequency modulation techniques. Low-frequency modulation methods mainly include staircase wave generation, selective harmonic elimination and lower switching losses [72]. However, this scheme has the drawbacks of higher THD and poor dynamic performance. High-frequency modulation methods majorly consist of Carrier Phase Shifted-Sinusoidal PWM (CPS-SPWM) [17], In Phase Disposition-Sinusoidal PWM (IPD-SPWM) [73], and Space Vector Modulation (SVM) [74] techniques. It is reported in [75], [76] that, the use of CPS-SPWM technique would achieve the natural output power balance between the cascaded cells (Ex. CHB inverter). However, the quality of the output level voltage is higher. In contrast, IPD-SPWM scheme reduces the harmonic content of the line voltage, but it fails to balance the output power naturally [77], [78]. With SVPWM, the switching sequences can be controlled and optimized to result in lower THD [79], [80]. However, implementation of SVPWM scheme is cumbersome, compared to the implementation of carrier based PWM schemes.

A hybrid multi-carrier PWM technique is proposed in [81], [82], which can achieve the dual objectives of balancing the output power and improving the spectral performance, compared to the CPS-SPWM technique, at lower modulation indices. However, the THD of the line-voltage at higher modulation indices remains unaltered, compared to the CPS-SPWM. The scheme reported in [83] mitigates the imbalance in the output power with a cyclic movement of the modulation signals in different carrier regions. However, this technique has restricted applicability, wherein the modulation index is less than 0.5. The research work described in [84] proposes a power balance method, which is based on one output period. This work shows that, it is possible to compensate for the power imbalance by controlling the power output of each cascaded cell, with the modification of the carrier amplitude in one output period. However, power balance would be a time-consuming process for a higher number of cascaded cells.

The topologies proposed in the next section of this chapter do not employ multiple-source based cascaded cells. Thus, the IPD-SPWM scheme could have been used to achieve better spectral performance. This technique is simple and does not require any optimization algorithm for the production of switching pulses [85]–[90].

7.4 RESULTS AND DISCUSSION

From the literature review on various topologies of hybrid MLIs, the following observations have been made:

1. Recent research is focused to synthesize hybrid topologies for power converters due to the requirement of less number of power devices and higher efficiency.
2. On the other hand, the traditional two-level VSIs were replaced by MLIs with the added advantages of lower THD and lower dv/dt stress.
3. Combining hybrid topologies with MLI technology could be an interesting proposition for medium-voltage applications in industries.
4. Several power conversion structures for Hybrid MLIs (HMLIs) are proposed since last decade to enhance their reliability and performance.

Hence, there is an adequate scope for further research in the area of hybrid MLI topologies. Recently, some of the HMLI topologies were proposed in the literature based on cascaded two two-level inverters, T-type inverters, and FC fed H-bridges with minimum component count AVBC suitable for medium voltage applications. It aims to provide a reduced number of the component count, size, cost, and voltage stress; suitability for all type of loads, improved efficiency, and better harmonic profile. To fulfil the afore-said objectives, different recently proposed HMLI topologies in both 3-Φ and 1-Φ applications have been explained with simulation results as follows.

7.4.1 A Seven-Level 3-Φ Hybrid VSI with Cascaded Three-Level Inverter and Flying Capacitor Fed H-Bridge

In this investigation, a 3-Φ hybrid cascaded MLI is introduced and it is shown in Figure 7.12 [91]. The topology is able to produce seven-level output voltage with lower number of active devices and also uses a common DC-link constructed by a H-bridge cell with FC at the output of the cascaded three-level inverter as shown in [22]. The voltage level of the FCs used in H-bridge is V_{dc} and the voltage across the DC-link capacitors are $0.5V_{dc}$. The employed modulation scheme produces the switching pulses to generate seven levels in the pole voltages (ranging from $-0.75V_{dc}$ to $+0.75V_{dc}$). Moreover, this topology requires a single DC source and also having enough switching states to balance the FC without using any additional source. But, the DC-link capacitor voltages are balanced by implementing separate control strategies. Simulation result of the three-phase pole voltage is shown in Figure 7.13.

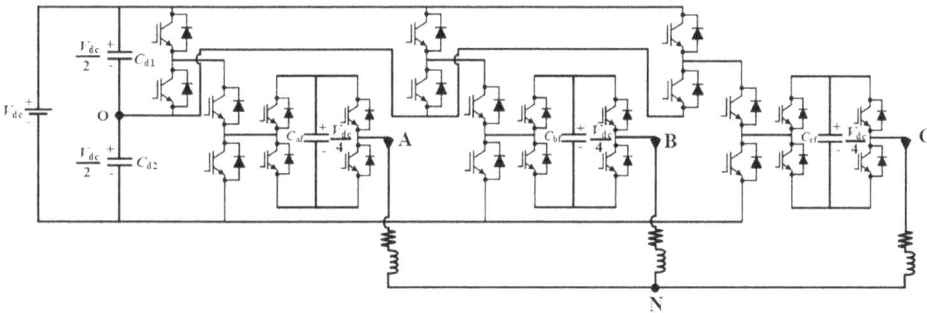

FIGURE 7.12   Topology [91].

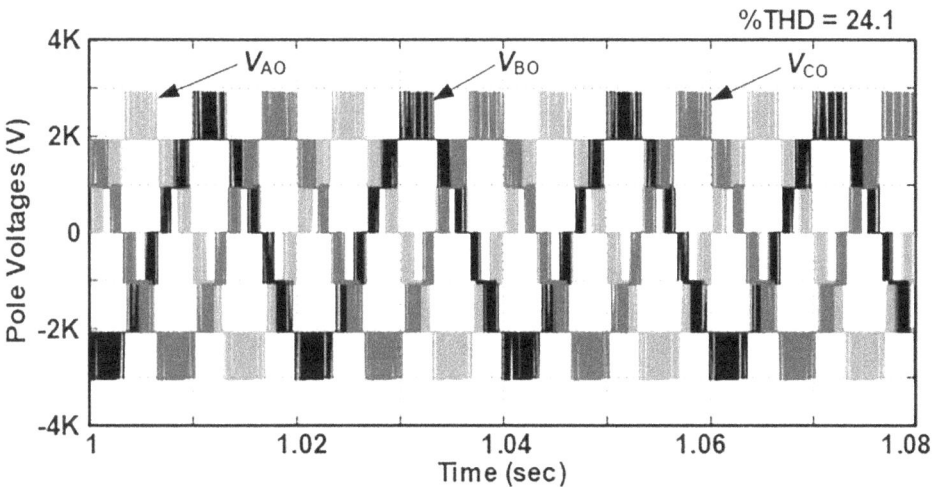

FIGURE 7.13   Simulation result of three-phase pole voltages of the topology [91].

The significant features of the proposed VSI are described as follows:

1. Self balancing FC voltage with additional switching states offered by the modulation scheme in each phase.
2. The dc-link capacitor voltages are well regulated by simple control technique.
3. Reduction of total blocking voltage of the switches compared to a T-type converter.
4. It uses a single dc source to convert dc power to three-phase ac power.

7.4.2 A NOVEL HYBRID FLYING CAPACITOR BASED 1-Φ NINE-LEVEL INVERTER

Figure 7.14 illustrates the 1-Φ 9-level hybrid MLI topology proposed in [92], which comprises of a total ten IGBTs for generating the nine-level output. This topology is constructed by using cascade connection of three different parts. Part 1 comprises low-frequency operating switches. Part 2 is a cascaded two two-level inverter (CT2-LI). Part 3 comprises of a traditional H-bridge with FC. Moreover, the output consists of nine levels which is ranging from $-V_{dc}$ to $+V_{dc}$ by triggering the IGBTs in appropriate manner. In addition, the FC used in this topology is balanced by the additional switching states offered by the modulation scheme during the voltage levels of and by sensing load current and FC voltage.

Figure 7.15 shows the inverter response of the output voltage (V_{AN}), load current (i_A), and FC voltage. It can be noticed that the FC voltage has remained constant without any ripple. Extensive

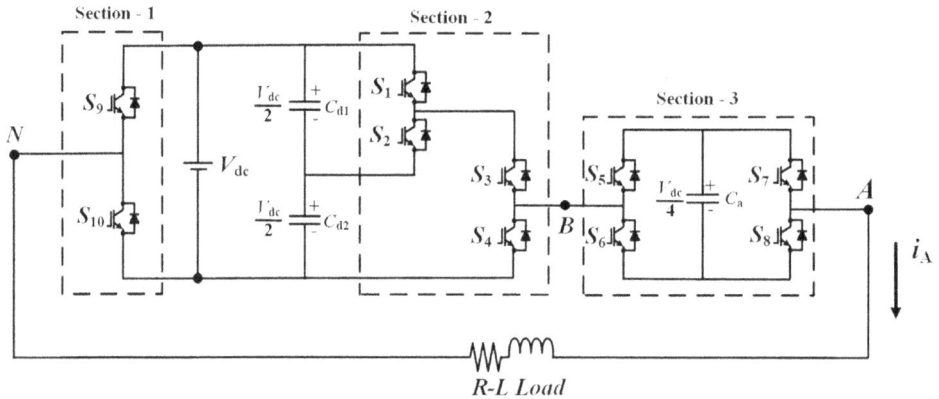

FIGURE 7.14 Nine-level hybrid inverter [92].

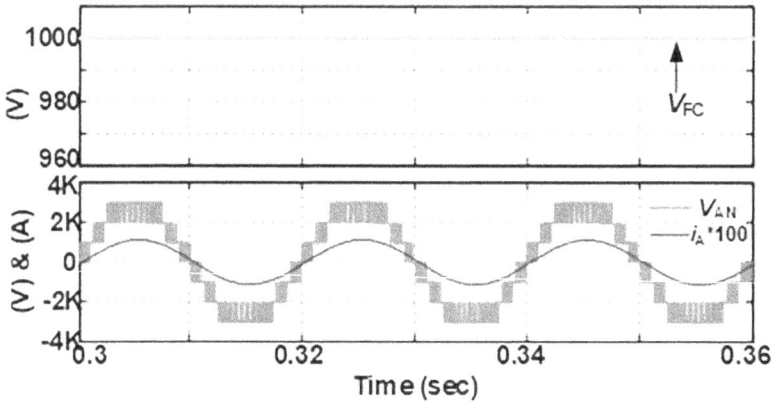

FIGURE 7.15 Simulation results of nine-level output voltage and the load current [92].

results based on simulation and experiment for step changes in load and the response of FC voltage and DC-link capacitor have been addressed in [92].

The significant features of the proposed topology are described as follows:

1. Availability of a rich number of redundant switching states in each output voltage level in which FC is involved.
2. Reduced voltage stress of the switches due to the presence of CT2L inverter compared to classical T-type inverter.
3. Inherent DC-link capacitor voltage balancing with a single dc source.
4. Reduced number of component count to produce the nine-level output compared to some of the other topologies presented in the literature.

7.4.3 A 3-Φ Five-Level VSI with Nested Two-Level Cells

This topology is derived by connecting three two-level cells (2LCs) in each phase, parallel to the common DC-link. Single DC source is used to feed all the series-connected DC-link capacitors. Switches S_{a1} and S_{a2} are connected across the capacitor C_2 in 2LC1. Two bidirectional switches are connected in 2LC2 and the switches S_{a7} and S_{a8} are connected in 2LC3. 2LC$_1$ and 2LC$_2$ form the internal leg and 2LC$_3$ is formed the external leg. The DC-link capacitor voltages are charged with a single DC source by using an AVBC [23]. Figure 7.17 illustrate the simulation results of line voltage and current waveforms. It can be noted that the voltage levels are uniform and the current is purely sinusoidal.

The significant features of the proposed topology are described as follows:

1. Reduced total blocking voltage of the switches due to the presence of CT2L inverter compared to the classical T-type inverter.
2. Simple and less component count based AVBC to balance the DC-link capacitor voltages.

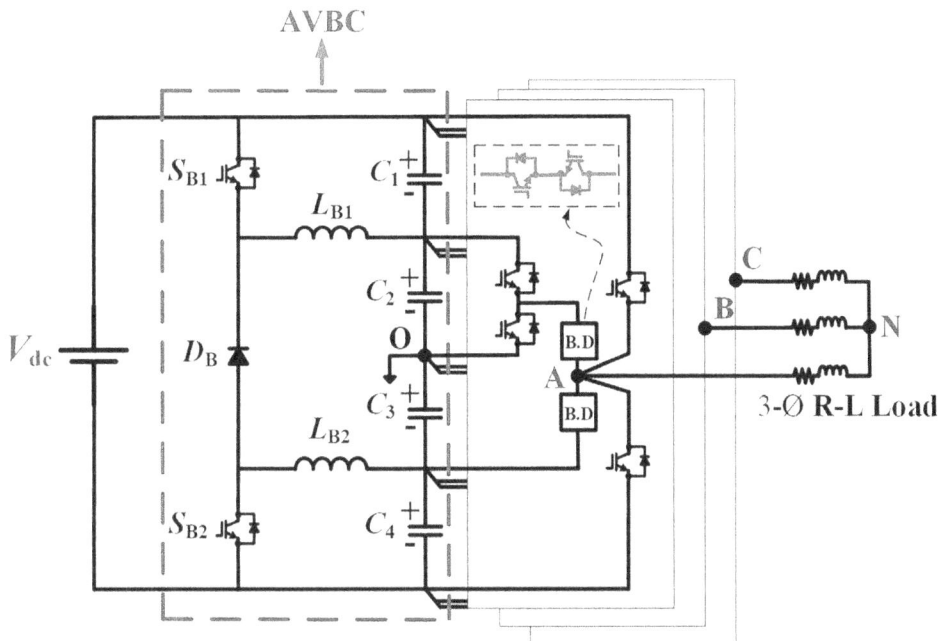

FIGURE 7.16 Nested three-phase five-level inverter [93].

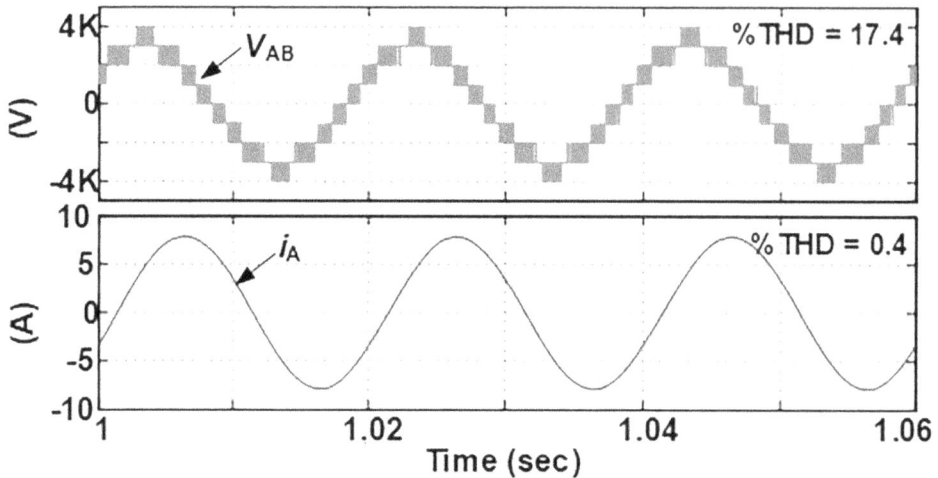

FIGURE 7.17 Simulation results of line voltage and current [93].

3. Reduced number of component count to produce the five-level output compared to some of the other topologies presented in the literature.
4. Less number of conducting devices ranging from one to three to generate 5-L output.

7.5 CONCLUSION

MLIs play a vital role in DC-AC conversion for various industrial applications. Hybrid MLI (HMLI) technology, which is still emerging, promises to play a pivotal role in the present scenario of power conversion. However, the development of suitable power circuit configurations with reduced device count, size, and higher efficiency is very important. To address these issues, contemporary research, focused on hybrid multilevel inverters, endeavours to realise the objectives of reduced voltage rating of the power semiconductor devices, improved reliability due to the requirement of single independent dc source, and improved efficiency compared to the conventional Voltage Source Inverters. Although they are suitable for all forms of power conversion, they are most popular for dc-ac power conversion.

Mindful of these objectives, this chapter discussed importance of HMLI topologies in dc-ac power conversion in various applications. Derivation of various HMLI topologies is explained with figures and analysis. Moreover, detailed literature study is presented on various HMLI topologies for dc-ac power conversion in medium and high voltage applications. Further, in this chapter the most recently proposed HMLI topologies and its significant merits has been discussed for the better understanding of the reader.

7.6 FUTURE RESEARCH DIRECTIONS

The proposed research can be extended further in the following areas:

- The proposed Cascaded three-level+FC fed H-Bridge and LFS2+Cascaded three-level+FC fed H-Bridge configurations can further be investigated with different control strategies such as Space Vector PWM, Model predictive control, etc.
- The proposed Nested five-level and Two Level Cells+T-NPC five-level configurations can further be studied for back-to-back converter systems to overcome the AVBC requirement.

- All the proposed topologies can further be studied using renewable energy sources for isolated loads as well as grid-connected applications.
- It would be interesting to investigate if these converters are amenable for the application of control techniques such as fuzzy logic, artificial neural networks, etc.

REFERENCES

[1] L. G. Franquelo, J. L. Rodriguez, J. Leon, S. Kouro, R. Portillo, and M. A. Prats. The age of multilevel converters arrives. *IEEE Industrial Electronics Magazine*, vol. 2, no. 2, pp. 28–39, 2008.
[2] M. Malinowski, K. Gopakumar, J. Rodriguez, and M. A. Pérez. A survey on cascaded multilevel inverters. *IEEE Transactions on Industrial Electronics*, vol. 57, no. 7, pp. 2197–2206, 2010.
[3] Y. Liu, and F. L. Luo. Multilevel inverter with the ability of self-voltage balancing. *IEE Proceedings on Electric Power Applications*, vol. 153, no. 1, pp. 105–115, 2006.
[4] J. Rodriguez, L. G. Franquelo, S. Kouro, J. I. Leon, R. C. Portillo, Ma. A. M. Prats, and M. A. Perez. Multilevel converters: an enabling technology for high-power applications. *Proc. IEEE*, vol. 97, no. 11, pp. 1786–1817, 2009.
[5] S. De, D. Banerjee, K. Siva Kumar, K. Gopakumar, R. Ramchand, and C. Patel. Multilevel inverters for low-power application. *IET Power Electronics*, vol. 4, no. 4, pp. 384–392, 2011.
[6] Z. Ye, P. K. Jain, and P. C. Sen. Circulating current minimization in high-frequency AC power distribution architecture with multiple inverter modules operated in parallel. *IEEE Transactions on Industrial Electronics*, vol. 54, no. 5, pp. 2673–2687, 2007.
[7] J. M. Guerrero, J. Matas, L. G. D. V. De Vicuna, M. Castilla, and J. Miret. Wireless-control strategy for parallel operation of distributed-generation inverters. *IEEE Trans. Ind. Electronics*, vol. 53, no. 5, pp. 1461–1470, 2006.
[8] X. Yu, H. H. Wang, and A. M. Khambadkone. Control of paralleled PEBBs to facilitate the efficient operation of microgrid. *IEEE International Symposium on Industrial Electronics*, pp. 2217–2222, 2010.
[9] E. Romero-Cadaval, M. I. Milanés-Montero, E. Gonzalez-Romera, and F. Barrero-González. Power injection system for grid-connected photovoltaic generation systems based on two collaborative voltage source inverters. *IEEE Transactions on Ind. Electronics*, vol. 56, no. 11, pp. 4389–4398, 2008.
[10] R. Turner, S. Walton, and R. Duke. Stability and bandwidth implications of digitally controlled grid-connected parallel inverters. *IEEE Transactions on Industrial Electronics*, vol. 57, no. 11, pp. 3685–3694, 2010.
[11] L. Liu, H. Li, Z. Wu, and Y. Zhou. A cascaded photovoltaic system integrating segmented energy storages with self-regulating power allocation control and wide range reactive power compensation. *IEEE Transactions on Power Electronics*, vol. 26, no. 12, pp. 3545–3559, 2011.
[12] K. Ramani, and A. Krishnan. New hybrid multilevel inverter fed induction motor drive-A diagnostic study. *International Review of Electrical Engineering*, vol. 5, no. 6, pp. 2562–2569, 2010.
[13] A. Nabae, I. Takahashi, and H. Akagi. A new neutral-point clamped PWM inverter. *IEEE Transactions on Industry Applications*, vol. IA-17, pp. 518–523, 1981.
[14] J. S. Lai, and F. Z. Peng. Multilevel converters – A new breed of power converters. *IEEE Transactions on Industry Applications*, vol. 32, pp. 509–517, 1996.
[15] T. A. Meynard, and H. Foch. Multi-level choppers for high voltage applications. *European Power Electronics and Drives Journal*, vol. 2, no. 1, pp. 45–50, 1992.
[16] C. Hochgraf, R. Lasseter, D. Divan, and T. A. Lipo. Comparison of multilevel inverters for static var compensation. in *Conference Record IEEE-IAS Annual Meeting*, pp. 921–928, 1994.
[17] P. Hammond. A new approach to enhance power quality for medium voltage ac drives. *IEEE Transactions on Industry Applications*, vol. 33, no. 1, pp. 202–208, 1997.
[18] E. Cengelci, S. U. Sulistijo, B. O. Woo, P. Enjeti, R. Teodorescu, and F. Blaabjerg. A new medium voltage PWM inverter topology for adjustable speed drives. *Conference Record IEEE-IAS Annual Meeting*, St. Louis, MO, 1998, pp. 1416–1423.
[19] R. H. Baker, and L. H. Bannister. Electric power converter. U.S. Patent 3–867–643, 1975.
[20] S. A. Gonzalez, M. I. Valla, and C. F. Christiansen. Five-level cascade asymmetric multilevel converter. *IET Power Electronics*, vol. 3, no. 1, pp. 120–128, 2010.
[21] A. Taghvaie, J. Adabi, and M. Rezanejad. A multilevel inverter structure based on a combination of switched-capacitors and DC sources. *IEEE Transactions on Industrial Informatics*, vol. 13, no. 5, pp. 2162–2171, 2017.

[22] S. Sabyasachi, V. B. Borghate, R. R. Karasani, S. K. Maddugari, and A. H. M. Suryawanshi. Hybrid control technique-based three-phase cascaded multilevel inverter topology. *IEEE Access*, vol. 5, pp. 26912–26921, 2017.

[23] S. R. Pulikanti, and V. G. Agelidis. Hybrid flying-capacitor-based active neutral-point clamped five-level converter operated with SHE-PWM. *IEEE Transactions on Industrial Electronics*, vol. 58, no. 10, pp. 4643–4653, 2011.

[24] A. M. Y. M. Ghias, J. Pou, M. Ciobotaru, and V. G. Agelidis. Voltage balancing method using phase-shifted PWM for the flying capacitor multilevel converter. *IEEE Transactions on Power Electronics*, vol. 29, no. 9, pp. 4521–4531, 2014.

[25] M. Hagiwara, and H. Akagi. Control and experiment of pulse width modulated modular multilevel converters. *IEEE Transactions on Power Electronics*, vol. 24, no. 7, pp. 1737–1746, 2009.

[26] M. Schweizer, and J. W. Kolar. Design and implementation of a highly efficient three-level T-type converter for low-voltage applications. *IEEE Transactions on Power Electronics*, vol. 28, no. 2, pp. 899–907, 2013.

[27] Ji. Yipeng, and G. Chen. A novel three-phase seven-level active clamped converter using H-bridge as a level doubling network. *Proceedings of the IEEE 8th International Power Electronics Motion Control Conference*, 2016, pp. 479–484.

[28] S. Xu, J. Zhang, X. Hu, and Y. Jiang. A novel hybrid five-level voltage-source converter based on T-type topology for high-efficiency applications. *IEEE Transactions on Industry Applied*, vol. 53, no. 5, pp. 4730–4743, 2017.

[29] S. K. Chattopadhyay, and C. Chakraborty. A new asymmetric multilevel inverter topology suitable for PV applications with varying irradiance. *IEEE Transactions on Sustainable Energy*, vol. 8, no. 4, pp. 1496–1506, 2017.

[30] R. J. Satputaley, and V. B. Borghate. Performance analysis of DVR using new reduced component multilevel inverter. *International Transactions on Electrical Energy Systems*, vol. 27, no. 4, pp. 1–11, 2017.

[31] N. K. Dewangan, S. Gupta, and K. K. Gupta. Fault-tolerant operation of some reduced device-count multilevel inverters with improved performance. *International Transactions on Electrical Energy Systems*, vol. 29, no. 2, pp. 1–15, 2018.

[32] A. Hota, S. Jain, and V. Agarwal. An optimized three-phase multilevel inverter topology with separate level and phase sequence generation part. *IEEE Transactions on Power Electronics*, vol. 32, no. 10, pp. 7414–7418, 2017.

[33] K. K. Gupta, and S. Jain. Topology for multilevel inverters to attain maximum number of levels from given DC sources. *IET Power Electronics*, vol. 5, no. 4, pp. 435–446, 2012.

[34] C. I. Odeh, D. and B. N. Nnadi. Single phase 9-level hybridized cascaded multilevel inverter. *IET Power Electronics*, vol. 6, no. 3, pp. 468–477, 2013.

[35] S. S. Lee, M. Sidorov, C. S. Lim, N. R. N. Idris, and Y. E. Heng. Hybrid cascaded multilevel inverter (hcmli) with improved symmetrical 4-level sub module. *IEEE Transactions on Power Electronics*, vol. 33, no. 2, pp. 932–935, 2018.

[36] S. P. Gautam, L. Kumar, and S. Gupta. Hybrid topology of symmetrical multilevel inverter using less number of devices. *IET Power Electronics*, vol. 8, no. 11, pp. 2125–2135, 2015.

[37] S. P. Gautam, L. Kumar, and S. Gupta. Single phase multilevel inverter topologies with self-voltage balancing capabilities. *IET Power Electronics*, vol. 11, no. 5, pp. 844–855, 2018.

[38] J. Li, S. Bhattacharya, and A. Q. Huang. A new nine-level active NPC converter for grid connection of large wind turbines for distributed generation. *IEEE Transactions on Power Electronics*, vol. 26, no. 3, pp. 961–972, 2011.

[39] K. Wang, Z. Zheng, D. Wei, B. Fan, and Y. Li. Topology and capacitor voltage balancing control of a symmetrical hybrid nine-level inverter for high-speed motor drives. *IEEE Transactions on Industry Applications*, vol. 53, no. 6, pp. 5563–5572, 2017.

[40] M. Saeedian, E. Pouresmaeil, E. Samadaei, E. M. G. Rodrigues, R. Godina, and M. Marzband. An innovative dual-boost nine-level inverter with low-voltage rating switches. *Energies*, vol. 12, no. 2, pp. 1–15, 2019.

[41] H. Vahedi, K. Al-Haddad, Y. Ounejjar, and K. Addoweesh. Crossover Switches Cell (CSC): A new multilevel inverter topology with maximum voltage levels and minimum DC sources. *Annual Conference of the IEEE Industrial Electronic Society*, 2013, pp. 54–59.

[42] N. Sandeep, and R. Udaykumar. A switched-capacitor-based multilevel inverter topology with reduced components. *IEEE Transactions on Power Electronics*, vol. 33, no. 7, pp. 5538–5542, 2018.

[43] R. Islam, A. M. Mahfur-Ur-Rahman, K. M. Muttaqi, and D. Sutanto. State of the art of the medium-voltage power conversion technologies for grid integration of solar photovoltaic power plants. *IEEE Transactions on Energy Conversion*, vol. 34, no. 1, pp. 372–384, 2019.

[44] M. Narimani, B. Wu, Z. Cheng, and N. R. Zargari. A New Nested Neutral-Point Clamped (NNPC) Converter for Medium-Voltage (MV) Power conversion. *IEEE Transactions on Power Electronics*, vol. 29, no. 12, pp. 6375–6382, 2014.

[45] A. Bahrami, and M. Narimani. A Sinusoidal Pulse width Modulation (SPWM) technique for capacitor voltage balancing of a nested T-type four-level inverter. *IEEE Transactions on Power Electronics*, vol. 34, no. 2, pp. 1008–1012, 2019.

[46] P. M. Bhagwat, and V. R. Stefanovic. Generalized structure of a multilevel PWM inverter. *IEEE Transactions on Industry Applications*, vol. IA-19, no. 6, pp. 1057–1069, 1983.

[47] E. Babaei, S. Laali, and S. Alilu. Cascaded multilevel inverter with series connection of novel H-bridge basic units. *IEEE Transactions on Industrial Electronics*, vol. 61, no. 12, pp. 6664–6671, 2014.

[48] A. Masaoud, H. W. Ping, S. Mekhilef, and A. Taallah. Novel configuration for multilevel DC-link three phase five-level inverter. *IET Power Electronics*, vol. 7, no. 12, pp. 3052–3061, 2014.

[49] Y. H. Lee, B. S. Suh, and D. S. Hyun. A novel PWM scheme for a three-level voltage source inverter with GTO thyristors. *IEEE Transactions on Industry Applications*, vol. 32, no. 2, pp. 260–268, 1996.

[50] K. A. Corzine, and S. K. Majeethia. Analysis of a novel four-level DC/DC boost converter. *IEEE Transactions on Industry Applications*, vol. 36, no. 5, pp. 1342–1350, 2000.

[51] Z. Wang, C. Gao, C. Chen, J. Xiong, and K. Zhang. Ripple analysis and capacitor voltage balancing of five-level hybrid clamped inverter (5L-HC) for medium-voltage applications. *IEEE Access*, vol. 7, pp. 86077–86089, 2019.

[52] A. Karthik, and U. Loganathan. A reduced component count five-level inverter topology for high reliability electric drives. *IEEE Transactions on Power Electronics*, vol. 35, no. 1, pp. 725–732, 2020.

[53] T. T. Davis, and A. Dey. Investigation on extending the DC bus utilization of a single-source five-level inverter with single capacitor-fed h-bridge per phase. *IEEE Transactions on Power Electronics*, vol. 34, no. 3, pp. 2914–2922, 2018.

[54] A. Hota, S. Jain, and V. Agarwal. An improved three-phase five-level inverter topology with reduced number of switching power devices. *IEEE Transactions on Industrial Electronics*, vol. 65, no. 4, pp. 3296–3305, 2017.

[55] M. Di Benedetto, L. Solero, F. Crescimbini, A. Lidozzi, and P. J. Grbović. 5-Level E-type back to back power converters- A new solution for extreme efficiency and power density. *13th IEEE Conference on Ph.D. Research in Microelectronics and Electronics*, 2017, pp. 341–344.

[56] M. Di Benedetto, A. Lidozzi, L. Solero, F. Crescimbini, and P. J. Grbović. Five-Level E-type inverter for grid-connected applications. *IEEE Transactions on Industry Applications*, vol. 54, no. 5, pp. 5536–5548, 2018.

[57] G. Adam, I. Abdelsalam, K. Ahmed, and B. Williams. Hybrid multilevel converter with cascaded H-bridge cells for HVDC applications: Operating principle and scalability. *IEEE Transactions on Power Electronics*, vol. 30, no. 1, pp. 65–77, 2015.

[58] P. Kumar, R. Kaarthik, K. Gopakumar, *et. al*. Seventeen-level inverter formed by cascading flying capacitor and floating capacitor H-bridges. *IEEE Transactions on Power Electronics*, vol. 30, no. 7, pp. 3471–3478, 2015.

[59] M. Veenstra, and A. Rufer. Control of a hybrid asymmetric multilevel inverter for competitive medium-voltage industrial drives. *IEEE Transactions on Industry Applications*, vol. 41, no. 2, pp. 655–664, 2005.

[60] P. Roshankumar, P. P. Rajeevan, K. Mathew, K. Gopakumar, J. I. Leon, and L. G. Franquelo. A five-level inverter topology with single-DC supply by cascading a flying capacitor inverter and an H-bridge. *IEEE Transactions on Power Electronics*, vol. 27, no. 8, pp. 3505–3512, 2012.

[61] H. Yu, B. Chen, W. Yao, and Z. Lu. Hybrid seven-level converter based on T-Type converter and H-bridge cascaded under SPWM and SVM. *IEEE Transactions on Power Electronics*, vol. 33, no. 1, pp. 689–702, 2018.

[62] N. Sandeep, and R. Udaykumar. Operation and control of an improved hybrid nine-level inverter. *IEEE Transactions on Industry Applications*, vol. 53, no. 6, pp. 5676–5686, 2017.

[63] N. Sandeep, and R. Udaykumar. Design and implementation of a sensor less multilevel inverter with reduced part count. *IEEE Transactions on Power Electronics*, vol. 32, no. 9, pp. 6677–6683, 2017.

[64] A. K. Sadigh, S. H. Hosseini, *et. al*. Double flying capacitor multi-cell converter based on modified phase-shifted pulse width modulation. *IEEE Transactions on Power Electronics*, vol. 25, no. 6, pp. 1517–1526, 2010.

[65] C. I. Odeh, E. S. Obe, and O. Ojo. Topology for cascaded multilevel inverter. *IET Power Electronics*, vol. 9, no. 5, pp. 921–929, 2016.

[66] M. Narimani, B. Wu, *et al*. A Novel five-level voltage source inverter with sinusoidal pulse-width modulator for medium-voltage applications. *IEEE Transactions on Power Electronics*, vol. 31, no. 3, pp. 1959–1967, 2016.

[67] E. C. D. Santos, J. H. G. Muniz, E. R. C. Da Silva, and C. B. Jacobina. Nested Multilevel Topologies. *IEEE Transactions on Power Electronics*, vol. 30, no. 8, pp. 4058–4068, 2015.

[68] R. Rojas, T. Ohnishi, and T. Suzuki. PWM control method for a four-level inverter. *IEE Proceedings on Electric Power Applications*, vol. 142, no. 6, pp. 390–396, 1995.

[69] Q. A. Le, and D. Lee. A novel six-level inverter topology for medium-voltage applications. *IEEE Transactions on Industrial Electronics*, vol. 63, no. 11, pp. 7195–7203, 2016.

[70] N. D. Dao, and D. C. Lee. Operation and control scheme of a five-level hybrid inverter for medium voltage motor drives. *IEEE Transactions on Power Electronics*, vol. 33, no. 12, pp. 10178–10187, 2018.

[71] M. Perez, S. Kouro, J. Rodriguez, and B. Wu, "Modified staircase modulation with low input current distortion for multicell converters," *Proceedings IEEE Electronics Spec. Conf. (PESC)*, 2008, pp. 1989–1994.

[72] M. S. A. Dahidah, and V. G. Agelidis, "Selective harmonic elimination PWM control for cascaded multilevel voltage source converters: A generalized formula," *IEEE Transactions on Industrial Electronics*, vol. 23, no. 4, pp. 1620–1630, 2008.

[73] A. Razi, A. S. A. Wahab, and S. A. A. Shukor, "Improved wave shape pattern performance using phase opposite disposition (POD) method for cascaded multilevel inverter," *Proc. IEEE 2nd Annu. Southern Electron. Conf. (SPEC)*, Auckland, 2016, pp. 1–6.

[74] W. Yao, H. Hu, and Z. Lu, "Comparisons of space-vector modulation and carrier-based modulation of multilevel inverter," *IEEE Transactions on Power Electronics*, vol. 23, no. 1, pp. 45–51, 2008.

[75] B. P. McGrath, and D. G. Holmes, "Multicarrier PWM strategies for multilevel inverters," *IEEE Transactions on Industrial Electronics*, vol. 49, no. 4, pp. 858–867, 2002.

[76] Y. Li, Y. Wang, and B. Q. Li, "Generalized theory of phase-shifted carrier PWM for cascaded H-bridge converters and modular multilevel converters," *IEEE Journal of Emerging and Selected Topics in Power Electronics*, vol. 4, no. 2, pp. 589–605, 2016.

[77] G. Carrara, S. Gardella, M. Marchesoni, R. Salutari, and G. Sciutto, "A new multilevel PWM method: A theoretical analysis," *IEEE Transactions on Power Electronics*, vol. 7, no. 3, pp. 497–505, 1992.

[78] J. Liu, Y. Sun, Y. Li, and C. Fu, "Theoretical harmonic analysis of cascaded H-bridge inverter under hybrid pulse width multilevel modulation," *IET Power Electron.*, vol. 9, no. 14, pp. 2714–2722, 2016.

[79] T. Boller, J. Holtz, and A. K. Rathore, "Neutral-point potential balancing using synchronous optimal pulsewidth modulation of multilevel inverters in medium-voltage high-power AC drives," *IEEE Transactions on Industry Applications*, vol. 50, no. 1, pp. 549–557, 2014.

[80] E. Babaei, S. Alilu, and S. Laali, "A new general topology for cascaded multilevel inverters with reduced number of components based on developed H-bridge," *IEEE Transactions on Industry Applications*, vol. 61, no. 8, pp. 3932–3939, 2014.

[81] I. Sarkar, and B. G. Fernandes, "Modified hybrid multi-carrier PWM technique for cascaded H-bridge multilevel inverter," *Proc. IEEE 40th Annu. Conf. Ind. Electron. Soc.*, 2014, pp. 4318–4324.

[82] V. G. Agelidis, and M. Calais, "Application specific harmonic performance evaluation of multicarrier PWM techniques," *Proc. IEEE Rec. 29th Annu. Power Electron. Spec. Conf. (PESC)*, vol. 1, pp. 172–178, 1998.

[83] L. M. Tolbert, F. Z. Peng, and T. G. Habetler, "Multilevel PWM methods at low modulation indices," *IEEE Transactions on Power Electronics*, vol. 15, no. 4, pp. 719–725, 2000.

[84] D. Sreenivasarao, P. Agarwal, and B. Das, "Performance evaluation of carrier rotation strategy in level-shifted pulse-width modulation technique," *IET Power Electron.*, vol. 7, no. 3, pp. 667–680, 2013.

[85] J. Liu, K. W. E. Cheng, and Y. Ye, "A cascaded multilevel inverter based on switched-capacitor for high-frequency AC power distribution system," *IEEE Transactions on Power Electronics*, vol. 29, no. 8, pp. 4219–4230, 2014.

[86] A. Kumar, and V. Verma, "Performance enhancement of single-phase grid connected PV system under partial shading using cascaded multilevel converter," *IEEE Transactions on Industry Applications*, vol. 54, no. 3, pp. 2665–2676, 2018.

[87] C. D. Fuentes, C. A. Rojas, H. Renaudineau, S. Kouro, M. A. Perez, and M. Thierry, "Experimental validation of a single DC bus cascaded H-bridge multilevel inverter for multistring photovoltaic systems," *IEEE Transactions on Industrial Electronics*, vol. 64, no. 2, pp. 930–934, 2017.

[88] V. Sonti, S. Jain, and S. Bhattacharya, "Analysis of the modulation strategy for the minimization of the leakage current in the PV grid-connected cascaded multilevel inverter," *IEEE Transactions on Power Electronics*, vol. 32, no. 2, pp. 1156–1169, 2017.

[89] N. Priyadarshi, S. Padmanaban, DM. Ionel, L. Mihet-Popa, and F. Azam "Hybrid PV-wind, micro-grid development using quasi-Z-source inverter modeling and control—experimental investigation. *Energies*, vol. 11, no. 9, p. 2277, 2018.

[90] N. Priyadarshi, F. Azam, AK. Sharma, Ch. Pradeep, and P. R. Thakura. An interleaved ZCS supplied switched power converter for fuel cell-based electric vehicle propulsion system. *Advances in Smart Grid Automation and Industry 4.0*, pp. 355–362. Springer, Singapore, 2021.

[91] T. Abhilash, A. Kirubakaran, and V. T. Somasekhar. A seven-level VSI with a front-end cascaded three-level inverter and flying capacitor fed H-bridge. *IEEE Transactions on Industry Applications*, vol. 55, no. 6, pp. 6073–6088, 2019.

[92] T. Abhilash, A. Kirubakaran, and V. T. Somasekhar "A new hybrid flying capacitor based single phase nine-level inverter. *International Transactions on Electrical Energy Systems*, e12139, 2019.

[93] T. Abhilash, A. Kirubakaran, and V. T. Somasekhar. A new structure of three-phase five-level inverter with nested two-level cells. *International Journal of Circuit Theory and Applications*, vol. 47, no. 9, pp. 1435–1445, 2019.

8 Single-Stage Grid-Connected Solar Inverter MPPT Based on Direct DC-Link Voltage Control Using Hiking-PSO Algorithm

Jia Shun Koh, Rodney H. G. Tan, Wei Hong Lim, and Nadia M. L. Tan

CONTENTS

8.1 INTRODUCTION

Conventional power generation technologies, which are mainly based on fossil fuels, are viewed to be unsustainable in the long-term future, as the Earth's limited proven fossil fuel reserves are being depleted. The increasing power demand has led to burning of more fossil fuels, increasing the carbon density in the atmosphere, accelerating the global warming and climate change problems. Concerning these imminent threats, power sectors are actively searching for alternative renewable energy sources in replacing fossil fuels. In recent years, there is an increasing effort in different countries on adaption of renewable energy to the electrical grid, such as solar photovoltaic, hydro, wind, biomass, geothermal, and tide power generations. At the end of 2020, total cumulative solar installed capacity has reached 767.2GW with approximately 33.1% in China, 12.5% in the USA, 9.4% in Japan, 7% in Germany, and 6.2% in India [1].

DOI: 10.1201/9781003323471-8

Solar PV systems are currently the most prominent renewable and clean energy source due to their long-life span, modular structure, non-moving parts (except those with solar trackers), and minimum maintenance. The design of PV systems can vary to suit different applications and locations. For example, most residential areas are receiving single-phase supply, therefore single-phase grid-connected PV systems are installed. Conversely, for large factories or PV farms, three-phase grid-connected PV systems are installed. The single-phase or three-phase PV system is further classified into single-stage or double-stage as presented in Figure 8.1.

In the double-stage PV system, the PV arrays are connected to the utility grid through an intermediate DC-DC converter and DC-AC inverter. The presence of a DC-DC converter allows broader operating voltage range and flexible number of panels in a PV string, as different PV string voltage can be stepped up to a value above grid voltage. However, the use of DC-DC converters incurs a higher cost and lowers the conversion efficiency of PV power injected into grid. In contrast, single-stage PV system removes the intermediate DC-DC conversion stage, thereby is more efficient and economical than the double-stage PV system [2]. In this book chapter, the single-phase single-stage PV system is considered.

In the 20th century, Blake and Hanson detected the hotspot failure on space PV modules and recommended the use of bypass diode to overcome the issue. During the time PV modules are partially blocked by adjacent building, tree, cloud or bird faeces, the shaded PV cells group will dissipate higher power from the non-shaded PV cells group as heat. In consequences, local overheating can bring irreversible damage to the shaded regions. Therefore, the connection of bypass diodes provides an alternative path in avoiding the shaded PV cells group.

Despite the advantage of bypass diode in solving the hot-spot problem, the use of bypass diode can create multiple-peaks PV curve, comprising of multiple local peaks and one global peak.

The recent study of [3] explained how the shading heaviness on different number of PV modules impacts the IV and PV curve in a comprehensive way. The working mechanism of the bypass diodes in two group of serially connected PV strings are presented in Figure 8.2.

The unshaded PV string exposed to solar irradiation of 900W/m^2 is capable generate current of 7.857A, whereas the shaded PV string received solar irradiation of 350W/m^2 manage to generate current of only 3.057A. Whenever the current generated by both the PV strings exceed 3.057A, the bypass diode parallel to the shaded PV strings will be activated. An I-V current curve with shape like a staircase and multiple-peaks P-V power curve will be generated. The lower peaks are called local maximum power point (LMPP) and the highest peak is called global maximum power point (GMPP). The voltage, V_{MPP}, and power, P_{MPP}, at maximum power point (MPP) are 332.4V and 2460.8W respectively

According to [4], Perturb and Observe (P&O) and incremental conductance (INC) algorithms are commonly implemented MPPT algorithm because they are simple in implementation and cost-effectiveness. However, the algorithm suffered from trade-off between steady-state oscillations and tracking speed. Besides, the algorithms are unable to solve the partial shading conditions and tend to trap at local peak. Recently, the utilization of artificial intelligent method such as artificial neural network (ANN) can solve the multiple-peaks issue but requires large computational power during

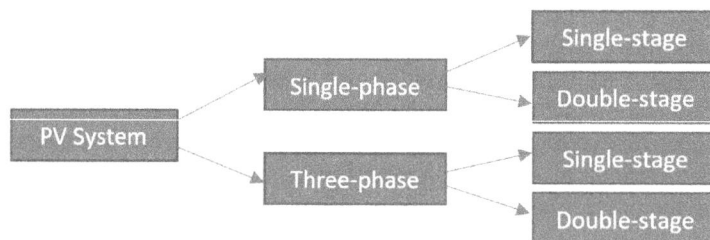

FIGURE 8.1   Different type of grid-connected PV system.

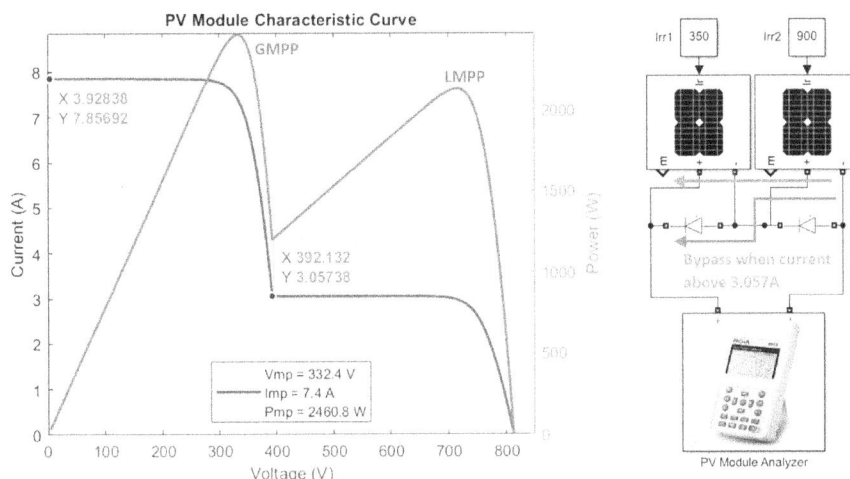

FIGURE 8.2 Analysis on PV string with shaded solar cell groups.

training process and suffers from overfitting issue. Particle Swarm Optimization (PSO), a nature inspired optimization algorithm is simple in implementation and proved to be effective in tracking the global peak of complex multiple-peaks PV curve. The algorithm has fast tracking speed and high tracking accuracy. In this book chapter, a proposed HPSO-based MPPT algorithm is implemented as the main MPPT algorithm of single-phase single-stage grid-connected PV system.

8.2 RELATED WORKS

Since the last decade, numerous studies related to grid-connected PV system have been published. These works are mainly to enhance the power efficiency of PV system as well as grid stability by proposing different converter topologies, MPPT control techniques, DC-bus voltage control techniques, inverter current control techniques and types of output filters connected to the utility grid.

In 2008, [5] combined the use of INC algorithm and varying power of output capacitor in tracking maximum power of voltage controlled PV inverters. The power fluctuation on the capacitor of the output LCL filter has been analysed at different operating points of PV curves and it is found to have lowest power fluctuation at the MPP. Therefore, the step size of INC algorithm is reduced according to the smaller power fluctuation when approaching MPP. Repetitive controller based on finite-impulse response digital filter together with proportional-integral (PI) controller is used to ensure grid current and grid voltage are in phase. Simulation on a 1.68kW PV model, and the proposed algorithm successfully achieved the maximum tracking efficiency of 99.48% and 2s tracking time. A new current control method for interfacing the grid with three-phase VSI through an output LCL filter is developed in [6]. In this study, a PI controller is employed in controlling the DC-link voltage to deliver maximum PV power to the grid. The ac side PI controllers are based on dq-frame and require the measurement of grid-side current, inverter-side current, and filter capacitor voltage to ensure that the VSI is operating at unity power factor. A 550V, 2.5kW PV generator is used in the simulation work, and the results obtained high stability with tracking error and time of 0.1% and 1.5s, respectively. However, the proposed control method requires many input parameters from several voltage and current sensors, reducing controller reliability.

The authors of [7] has designed a single-stage, three-phase grid-connected PV system with P&O as the MPPT algorithm. According to their study, the DC-link voltage control is integrated with power flow control, aiming to track the MPPT while regulating the active and reactive power of the grid based on the reference value given. Reverse blocking diodes are serially connected to the

200kW PV arrays for prohibiting reverse current flow at low irradiance. However, it tends to reduce the efficiency of the PV system due to voltage drop across it. LC filters and an isolating transformer are used to keep the harmonics as low as possible. The P&O algorithm successfully tracked the MPP within 0.1s and the tracking accuracy is 95% under the single peak condition. A single-phase, double-stage grid-connected PV systems is designed in [8]. The double-stage PV system consists of a DC-DC boost converter for MPPT tracking using P&O algorithm and a H-bridge inverter for transmitting the power from 24Vdc solar panels to 230V AC grid. The author proposed a switching strategy that combines the sinusoidal pulse width modulation (SPWM) with square wave signals of the inverter at 10kHz to minimize the switching losses. A low pass LC filter connected the point of common coupling (PCC) successfully lowers the total harmonic distortion (THD) below 8%, whereas the THD requirement of less than 5% is necessary according to IEEE Std-519.

In [9], a 1.5kW cost-efficient single-phase, double-stage grid-connected PV converter that eliminates the use of DC-link high voltage sensor has been developed. A boost converter is employed as the intermediate dc-dc converter and L filter is used at the AC output. Different from conventional PV inverters that utilized both grid current control loop with proportional-resonant (PR) controller and DC-link voltage control loop with PI controller, the suggested approach only employs the former control loop integrated with an ANN. Simulation and experimental studies reported that the suggested approach is effective in stabilising the dc-link voltage during transient conditions with varying irradiances, albeit a slower start-up time. The modified variable step-size INC algorithm has shown promising result of tracking efficiency as high as 93% under varying irradiance conditions. In [10], a backstepping control technique based on Lyapunov function together with INC algorithm are used in PV voltage regulation to track the MPP. This technique shows good tracking performance with near zero steady state error under varying climatic conditions. However, the main disadvantage is the requirement of large amount of control parameters.

Deadbeat or predictive current control loop is used in the single-phase, single-stage inverter topology to eliminate the DC-link double grid frequency voltage ripple [11]. The system includes grid angle detection techniques to operate at different power factor and INC as the main MPPT algorithm. The algorithm shows superior MPPT performance in simulation testing using a 7.08kW PV inverter tied to grid with a 5mH inductor filter. It is known that the use of a PI controller for inverter current regulation is dependent on the load parameter and the PI constants need to be readjusted in different load conditions. Therefore, the authors of [12] proposed the use of sliding mode controller (SMC) to achieve robust performance with varying load conditions. Fuzzy logic control optimized by genetic algorithm (FLC-GA) is implemented in the DC-DC boost converter of the double-stage, single-phase, grid coupled 5.4kW PV system. The FLC-GA managed to track all the peaks under different irradiances within 0.1s.

The recent study of [13] investigated a double-stage, single-phase PV system consisting of DC-DC Positive Output Super Lift Luo Converter (POSLLC) and a H-bridge inverter coupled to the grid using a RL filter. P&O MPPT algorithm combined with SMC is employed in regulating the PV voltage to its MPP. On the AC side, the SMC controller is used to ensure unity power factor. The proposed system is proven to have higher efficiency and greater stability than a conventional system that used boost converter and sign function controller (SFC). The use of a non-linear fractional-order back stepping controller (FOBSC) plus fractional-order PI controller (FOPI) is presented in [14]. The slow and intense decay characteristic of the proposed controller enables better control of DC-link voltage based on reference voltage provided by P&O algorithm. The proposed controller is verified in the 1.4kW PV system to have less THD, less output power oscillations and higher power factor compared to integral order (IO) controller under changing irradiation.

The need of designing a PV system in discrete-time domain instead of continuous-time domain in the case of small sampling frequency or small dc-link capacitor is discussed in [15]. In this study, a state-feedback current controller with discrete linear-quadratic regulator (DLQR) is proposed for current control loop of the three-phase grid-tied PV system with LCL filter. Simulation was carried out with DC-link capacitor with low capacitance of 60uF and sampling frequency of 4kHz.

The proposed controller presented satisfactory results with high robustness and stability in different DC-link voltage operating points. A new MPPT solution proposed in [16] divides MPPT into three stages where P&O, golden section search (GSS) and INC techniques are used progressively in each stage. The proposed technique is adapted in a single-phase PV system comprising of 1.8kW PV array, DC-DC boost chopper and inverter tied to the utility grid through a L filter. Simulation results verified the use of P&O with large step size initially to approach MPP, followed by GSS to narrow the search space, and INC method in last stage can successfully localize the MPP in 36ms and tracking efficiency of 99.95%.

As discussed in [17], DC-link voltage ripple can reduce the PV power extracted by the MPPT controller. The DC-link voltage ripple is due to the unavoidable double grid frequency ripple and is dependent on the dc-link voltage reference, output power of inverter and the size of dc-link capacitor used. An improved MPPT controller is recommended in the research work, where a compensating duty cycle is added to the reference duty cycle from ripple correlation control (RCC) and INC MPPT algorithm to eliminate DC-link voltage ripple. In [18], the author integrated PSO algorithm in the MPPT controller in single-stage grid-tied PV inverter to solve the mismatching problem due to shading effects. Furthermore, synchronous reference frame (SRF) algorithm and proportional-integral multi-resonant (PI-MR) controller are proposed in computing the reference current. The controllers allow maximum active power injection to the utility grid while mitigating the reactive power and harmonics current. The PSO algorithm used in simulation outperforms P&O algorithm with tracking accuracy of 99.93%. However, it takes approximately 150s to reach MPP and is not suitable for real time application.

The feasibility of Artificial Bee Colony combined with P&O (ABC-PO) algorithm has been examined in a double-stage PV system connected to the utility grid through LCL filters [19]. The ABC-PO algorithm is feasible under multiple-peaks PV curves, with fast tracking speed of 0.08s and 99.93%, respectively. However, the system is highly complex for practical implementation as compared to conventional algorithm. Moth Flame Optimization (MFO), a meta-heuristic algorithm that imitates the behaviour of moth converging towards the light source, is applied in [20] to solve extensive partial shading conditions of PV plants. The work also introduced a distributed MPPT control in a 1MW farm, where the MFO algorithm is applied in the DC-DC boost converter of every 100kW PV string. Simulation results showed that the MFO algorithm is proficient in tracking GMPP with tracking time around 0.05s and maximum tracking efficiency of 99.91%.

The disadvantages of DC-DC boost converter having a low voltage gain and overheating effect are highlighted in [21]. Therefore, the author proposed the use of SEPIC converter that provides less electrical stress and heat dissipation during the regulation of DC-link voltage. In their work, Grey wolf Optimization (GWO) combined with PI controllers successfully extract maximum power of PV curve with tracking efficiency of 98.5%. A firefly asymmetrical fuzzy logic controller (FAFLC) is proposed in [22] for MPPT control of grid connected power system using CUK converter. The proposed algorithm is based on natural-inspired firefly algorithm (FA) for tuning the membership function of FLC which is essential to track the MPP. In terms of inverter control approach, a space vector PWM (SVPWM) based on d-q rotating synchronized frame is suggested to minimize the THD. Experimental validation through dSPACE recorded the improved tracking accuracy of 97%.

In [23], an adaptive neuro-fuzzy inference system PSO (ANFIS-PSO) based MPPT algorithm in double-stage PV system is proposed. A fourth order non-linear Zeta converter is utilized for its flexibility to operate in buck or boost mode as well as continuous or discontinuous mode. A space vector modulation hysteresis current controller (SVMHCC) is recommended for inverter current control. The robustness of the proposed system is validated through MATLAB interfaced dSPACE and recorded best tracking time of 0.3s and accuracy of 98.35%. Similar works with fuzzy PSO (FPSO) as MPPT algorithm and modified ripple-factor compensation or fuzzy based SVPWM current control method are presented in [24], [25]. The proposed system successfully alleviates the ripple effect of dc bus voltage and provides better tracking efficiency.

In [26], the deficiency of CUK converter having slower response in stepping up and down voltage is discussed. Therefore, an integrated cuk converter with added two diodes and capacitors is proposed and able to deliver sharp transient. Jaya-based MPPT algorithm is employed in a 200W hybrid PV-fuel cell ultra-capacitor grid integrated system and managed to achieve fast tracking within 2s. In [27], a modified sine-cosine optimized (MSCO) based MPPT algorithm and adaptive fuzzy sliding mode controller (AFSMC) as inverter current controller is implemented in a single phase inverter system. In contrast to classical SCO, MSCO is redesigned to have exponential and linear decreasing parameters resulting in better tracking performance. In this research work, MSCO is proved to perform more superior to PSO and P&O algorithm with average tracking efficiency of 98.4%. In [28], an adaptive TS-fuzzy based RBF neural network (ATSFRBFNN) MPPT algorithm is implemented in boost converter connected to utility grid through a three-phase inverter system. Besides, an improved damping circulating current limiting inverter (IADCCLIC) is proposed to improve the power quality by mitigating the harmonics and stabilizing the load.

Table 8.1 summarizes the difference in the topologies of grid-connected PV system and control approaches proposed by different researchers mentioned in the literature review. Based on the summary, most researchers favoured the use of double-stage PV system utilizing DC-DC boost converter. However, the use of DC-DC converter will result in higher conversion losses and costing of the overall PV system. Moreover, the MPPT algorithms are only tested under single peak PV curve in most literatures, where they are not proven to be effective in fast varying partial shading conditions that require real-time capability in tracking the global peak. Besides, HPSO has not been implemented together with single-stage solar inverter system in any of the literatures. Therefore, the main contribution in this book chapter includes a proposed HPSO-based MPPT algorithm is implemented in single-phase single-stage PV system, with PI controller in DC-link voltage control and PR controller in inverter current control. The overall system implementation methodology and simulation analysis under varying shading conditions are discussed in the following sections.

8.3   METHODS

The proposed single-stage grid-connected solar PV inverter system using HPSO-based MPPT algorithm is modelled in MATLAB and Simulink platform as presented in Figure 8.3. The PV inverter system includes a 5.5kW PV string with two bypass diodes, a DC-link capacitor, a DC-AC voltage source inverter coupled to the utility grid with an output LCL filter, a HPSO-based MPPT algorithm block, DC-link voltage and grid current feedback control. Solar measurement solar measurement scope is used to monitor and track the PV dc voltage, PV output power and DC-link voltage reference. Moreover, a grid measurement scope is used to observe grid-side voltage and current. The injection of active power into the grid and the total current harmonic distortion are also monitored.

In the simulation, different sets of irradiances in W/m^2 are used to reproduce partial shading conditions. A PV characteristic curve with single peak will be generated if PV strings are receiving same irradiances. On the other hand, when mismatch irradiances acted on PV strings with bypass diodes, the activation of bypass diode connected to the shaded string with lower irradiance will result in multiple-peaks PV characteristic curve. The HPSO-based MPPT algorithm initialized its particles uniformly across the solution space of DC-link voltage reference to search for optimal solution that can lead to the maximum power. Particularly, the HPSO-based MPPT algorithm receives the inputs of PV string voltage and current before evaluating the fitness value of individual particle in terms of PV power tracked. The positions of all HPSO particles are updated and their fitness values are computed every iteration until the global peak is identified. In the meantime, the DC-link voltage reference encoded into the position of each HPSO particle is fed into the control loops. In the control loops comprising of DC-link voltage control and grid current control, the DC-link voltage is adjusted according to the reference voltage from HPSO-based MPPT algorithm and the grid current is ensured to be in phase with grid voltage. An output of SPWM switching to the voltage source inverter converts solar DC power into AC power delivers to the utility grid.

TABLE 8.1

List of Publications with Different Proposed Technique of Grid-Connected PV System

Reference	Type of PV system	DC-DC converter	Filter	MPPT control	Voltage control	Current control
[5]	Single-phase, single-stage	N/A	LCL	INC based on Output capacitor power ripple	Repetitive Controller	PI Controller
[6]	Three-phase, double-stage	–	LCL	–	PI Controller	DQ-frame with PI controller
[7]	Three-phase, single-stage	N/A	LC	–		DC-link voltage + Power flow control
[8]	Single-phase, double-stage	Boost	LC	P&O	PI Controller	–
[9]	Single-phase, double-stage	Boost	L	Variable step-size INC	Removed	PR Controller + ANN
[10]	Single-phase, double-stage	Boost	L	INC	PI Controller	Backstepping control based on Lyapunov function
[11]	Single-phase, single-stage	N/A	L	INC	PI Controller	Dead-beat
[12]	Single-phase, double-stage	Boost	RL	FLC-GA	PI Controller	SMC controller
[13]	Single-phase, double-stage	POSLLC	RL	P&O	SMC controller	SMC controller
[14]	Single-phase, double-stage	Boost	L	P&O	FOPI controller	FOBSC controller
[15]	Three-phase	–	LCL	–	PI Controller	Discrete State-Feedback
[16]	Single-phase, double-stage	Boost	L	P&O + GSS + INC	PI Controller	Predictive control
[17]	Single-phase, double-stage	Boost	L	RCC + INC	PI Controller	–
[18]	Single-phase, single-stage	N/A	L	PSO	PI Controller	SRF and PI-MR
[19]	Single-phase, double-stage	Boost	LCL	ABC-PO	PID Controller	PI Controller
[20]	Three-phase, double-stage	Boost	–	MFO	–	–
[21]	Single-phase, double-stage	SEPIC	L	GWO	PI Controller	PI Controller
[22]	Double-stage	CUK	LC	FAFLC	–	SVPWM + DQ-frame
[23]	Three-phase, double-stage	ZETA	–	ANFIS-PSO	–	SVMHCC
[24]	Three-phase, double-stage	ZETA	L	FPSO	–	Modified SVPWM
[25]	Three-phase, double-stage	ZETA	L	FPSO	–	Fuzzy SVPWM
[26]	Three-phase, double-stage	Integrated CUK	–	JAYA	PI Controller	PI Controller
[27]	Single-phase, double-stage	ZETA	LC	MSCO	–	AFSMC
[28]	Three-phase, double-stage	Boost	L	ATSFRBFNN	IADCCLJC	AFSMC

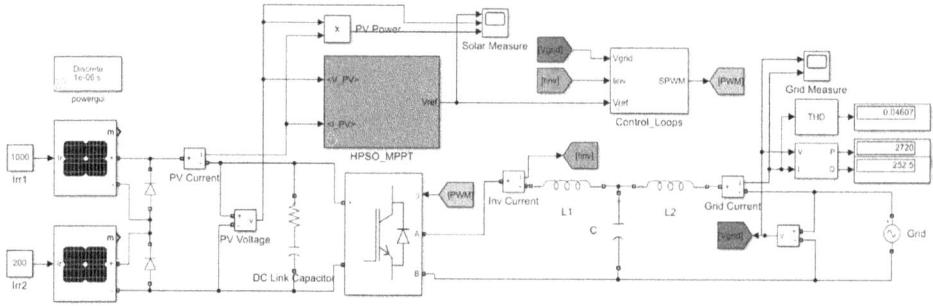

FIGURE 8.3 Single-stage grid-connected solar inverter system using HPSO-based MPPT algorithm.

The PV modules considered in current study is implemented based on commercial JA Solar 250W 60 cells panel. Referring to datasheet, the short circuit current and open circuit voltage of PV modules are obtained as 8.65A and 37.85V, respectively. Other models parameters known as diode ideality factor, series resistance and shunt resistance are set as 1.0251, 0.38, and 1372.42Ω, respectively. Twenty-two of these PV modules are serially connected to produce PV string with the maximum voltage and power of 832.7V and 5.5kW, respectively. The PV model is also divided into two sets of PV strings with each consists of 11 modules and a bypass diode to simulate the scenarios of multiple peaks. A DC-link capacitor of 1000μF is parallelly connected to smoothen the output of PV strings. A single-phase full-bridge VSI consisting of four insulated gate bipolar transistors (IGBTs) is employed to convert DC power from PV strings to AC power. The VSI is switched using SPWM technique at the switching frequency of 30kHz. An LCL filter is designed using the following equations:

$$I_o = \frac{P_o}{V_o} \tag{1}$$

$$\Delta I_{pp} = 0.2 \times I_o \times \sqrt{2} \tag{2}$$

$$L_1 = \frac{V_{dc}}{4 \times f_{sw} \times \Delta I_{pp}} \tag{3}$$

$$L_2 = \frac{0.1 \times V_o^2}{S \times 2\pi f} - L_1 \tag{4}$$

$$C = \frac{0.05 \times S}{V_o^2 \times 2\pi f} \tag{5}$$

where V_o, I_o, P_o and S are the inverter output voltage, current, active power and apparent power respectively. The inverter ripple current, ΔI_{pp} is limited at 20%, V_{dc} is the dc-link voltage, f_{sw} is the inverter switching frequency, f is the grid frequency. With utility grid voltage of 230V_{rms} at frequency of 50Hz and 5kVA inverter with switching frequency of 30kHz, the LCL filter is designed with L_1, L_2 and C of 1mH, 2.4 mH, and 15.04μF respectively. The LCL filter is mainly to reduce the harmonics injection into the AC side due to inverter switching. The grid-connected PV system parameters can be summarized in Table 8.2.

The details of DC-link voltage control loop, grid current control loop and HPSO-based MPPT algorithm are elaborated in the following subsections.

TABLE 8.2

PV System Parameter

Parameter	Maximum PV strings voltage	Maximum PV strings current	Maximum PV strings power	DC-link capaci-tor	Inductors of LCL filter	Capacitor of LCL filter	Utility grid voltage	Switching frequency
Value	832.7V	8.65A	5.5kW	1000uF	$L_1 = 1\text{mH}$ $L_2 = 2.4\text{mH}$	$C = 15.04\text{uF}$	230V_{rms}	30kHz

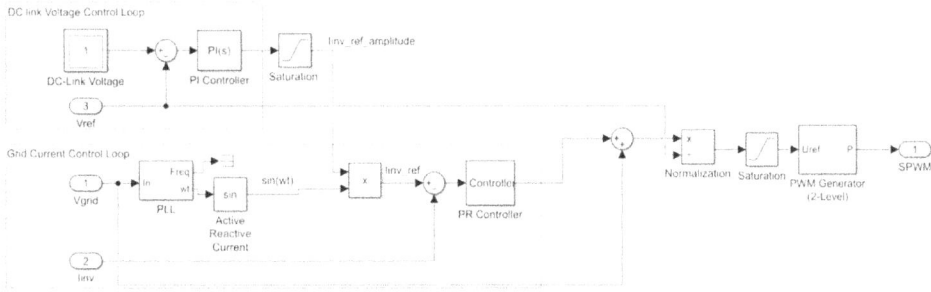

FIGURE 8.4 DC-link voltage and grid current control loops.

8.3.1 DC-Link Voltage and Grid Current Control Loops

Solar PV system interfaced to grid is achieved using two control loops, the DC-link voltage control loop and grid current control loop as illustrated in Figure 8.4. The DC-link voltage control loop is dedicated for the regulation of DC-link voltage by altering the amplitude of inverter current reference, $I_{inv\_ref\_amplitude}$. Deviation between the measured DC-link voltage and voltage reference, V_{ref} provided by the HPSO-based MPPT algorithm is fed as the input of PI controller. This control loop allows direct control of DC-link voltage in tracking the V_{MPP} of the PV curve that can provide maximum power and minimize the steady-state error of DC-link voltage. The PI controller gains, G_{PI} can be expressed as:

$$G_{PI}(s) = P + \frac{I}{s} \tag{6}$$

where P is the proportional gain and I is the integral gain. A saturation block is used to limit the $I_{inv\_ref\_amplitude}$ from -80A to 80A for preventing the large oscillation of DC-link voltage during the regulation process.

For grid current control loop, a phase-locked loop (PLL) is applied to obtain the sinusoidal unit vector of grid voltage represented as $sin(wt)$. The output current of the DC-link voltage control loop, $I_{inv\_ref\_amplitude}$ is then multiplied by $sin(wt)$ to enable the synchronization with the grid voltage. With the presence PR controller, the output current of inverter, I_{inv} is forced to match the sinusoidal inverter current reference, I_{inv_ref}. The PR controller gain, G_{PR} can be represented as:

$$G_{PR}(s) = k_p + k_r + \frac{s}{s^2 + w^2} \tag{7}$$

where k_p is the proportional part gain, k_r is the resonance part gain and ω is the resonance frequency of the controller. The resonance frequency is set equivalent to the grid frequency at 50Hz.

Lastly, the output of the grid current control loop is added to the grid voltage and divided by V_{ref} to obtain the normalized value within the range of -1 to 1. A desired SPWM switching signal is output from the SPWM generator to the inverter, thus ensuring maximum power is extracted from the PV string.

8.3.2 Proposed MPPT Algorithm

8.3.2.1 Overview of Conventional PSO

PSO is a nature-inspired intelligent optimization algorithm proposed in [29] to simulate the social behaviors of birds flock in searching for food sources. As a population-based stochastic algorithm, PSO employs a group of particles to perform searching in the solution space with varying position, velocity and fitness values. Every particle is considered to be potential solution to resolve a given optimization problem. During the k-th iteration of the searching process, the velocity and position of each i-th particle can be adjusted according to the personal best position, $P_{best,i}^k$ and the global best position, G_{best}^k found by itself and the population, respectively, where $k = 1, \ldots, K_{max}$ and K_{max} refers to maximum iteration numbers. Particularly, the new velocity, v_i^k and position, x_i^k of each i-th particle can be updated in every $(k + 1)$-th iteration as follows:

$$v_i^{k+1} = wv_i^k + c_1\, r_1 \left(P_{best,i}^k - x_i^k \right) + c_2 r_2 \left(G_{best}^k - x_i^k \right) \tag{8}$$

$$x_i^{k+1} = x_i^k + v_i^{k+1} \tag{9}$$

where c_1 and c_2 are the acceleration coefficient; w is the inertia weight used in sustaining the inertial motion of particles on its current search trajectory; r_1 and r_2 are both uniformly distributed random values in the interval of 0 to 1. In this study, w, c_1 and c_2 are set to 0.5, 1 and 1.2, respectively. For each i-th particle, the fitness value of its new position x_i^{k+1} is evaluated as $f\left(x_i^{k+1}\right)$ and compared with its personal best fitness denoted as $f\left(P_{best,i}^k\right)$. As shown in Eq. (10), the new position x_i^{k+1} can be used to update the personal best position $P_{best,i}^k$ of i-th particle if the former position has better fitness than the latter one. Otherwise, x_i^{k+1} is discarded if it has worse fitness than $P_{best,i}^k$. Similar criterion is used to update the current global position in every $(k + 1)$-th iteration as shown in Eq. (11), where $f\left(G_{best}^k\right)$ refers to the global best fitness of population.

$$P_{best,i}^{k+1} = \begin{cases} x_i^{k+1}, & \text{if } f\left(x_i^{k+1}\right) \text{ is better than } f\left(P_{best,i}^k\right) \\ P_{best,i}^k, & \text{otherwise} \end{cases} \tag{10}$$

$$G_{best}^{k+1} = \begin{cases} P_{best,i}^{k+1}, & \text{if } f\left(P_{best,i}^{k+1}\right) \text{ is better than } f\left(G_{best}^k\right) \\ G_{best}^k, & \text{otherwise} \end{cases} \tag{11}$$

8.3.2.2 Conventional PSO-Based MPPT Algorithm and Its Drawbacks

Since its inception, PSO has been implemented to overcome various real-world optimization problems such as MPPT tracking due to its desirable features such as simplicity of implementation, fast convergence rate and good exploration search capability. For MPPT problem using direct DC-link voltage control of grid-tied PV system, DC-link voltage reference is considered as the decision variable to be optimized, hence encoded into the position of each particle. Referring to DC-link voltage reference determined from the position of each particle, the measured average PV string

TABLE 8.3

The Relationship between PSO and Power Electronics Parameters

PSO parameter	Power electronic parameter
Current Positions (x_i^k)	DC-link voltage reference
Current Velocity (v_i^k)	Change in DC-link voltage reference
Fitness value ($f\left(x_i^{k+1}\right)$)	PV power generated
Personal best position $\left(P_{best\ i}\right)$	Best DC-link voltage reference found by individual particle
Global best position $\left(G_{best}\right)$	Best DC-link voltage reference found by the best particle

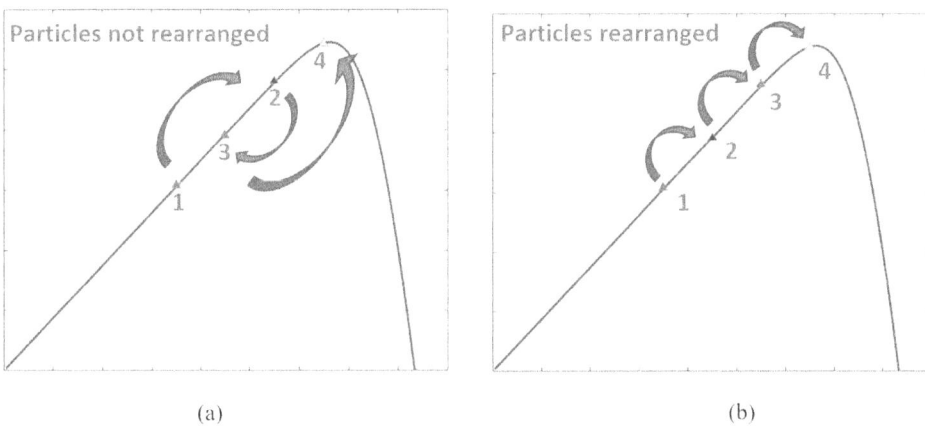

FIGURE 8.5 Comparison of DC-link voltage reference encoded into the particles of (a) conventional PSO and (b) proposed HPSO-based MPPT algorithm along with the PV power tracked.

voltage and current are used in the computation of corresponding PV power as fitness value of each particle. Given that the DC-link capacitor is parallelly connected to PV string, the direct control of DC-link voltage is considered equivalent to the control of PV string voltage to its GMPP. For current study, the solution space is limited to between voltage values of 280V to 720V to cover the potential regions of GMPP. The sampling time is set as 0.01s.

Table 8.3 summarizes the relationship between each PSO parameter and the equivalent power electronics parameter when using PSO or its variants to tackle MPPT problem based on direct DC-link voltage control of grid-tied PV system.

Despite of the popularity of using conventional PSO to solve MPPT tracking problem, the goal of achieving proper trade-off between short tracking time and high tracking accuracy remain a challenging issue due to inherent drawbacks of conventional PSO-based MPPT method. Given the stochastic nature of PSO, there is high tendency for each particle's position encoded with different values of DC-link voltage reference without referring to its population index. If these encoded DC-link voltage reference values fluctuate with the population indices of particles, the power tracked by different particles tend to be scattered around the PV curve as illustrated in Figure 8.5(a). The presence of these undesirable voltage oscillations can jeopardize the accuracy of sampled PV signals and compromise the MPPT performance of solar inverter system.

8.3.2.3 Proposed HPSO-Based MPPT Algorithm

In this study, a modified PSO variant known as hiking PSO (HPSO) is designed as a more robust MPPT algorithm to address the undesirable effects brought by large oscillation of DC-link voltage reference encoded in each particle. The search mechanism of HPSO is inspired by the concept of mountain hiking as presented in Figure 8.5(b), where the value of DC-link voltage reference encoded into the position of each particle can increase or decrease gradually with the population index during the search process. In contrary to conventional PSO, the modification introduced in HPSO-based MPPT algorithm offers a more systematic approach to track GMPP by avoiding the occurrence of large voltage oscillation in solar inverter system that might lead to the inaccurate sampling of PV power.

The search mechanism of proposed HPSO-based MPPT algorithm is presented in Figure 8.6 and explained as follows. At the beginning of search process, the initial velocity of each i-th HPSO particle is set as zero. Meanwhile, the position of each i-th particle is uniformly initialized across the solution space within a search range of $\left[V_{ref}^{min}, V_{ref}^{max} \right]$ as follow:

$$x_i = V_{ref}^{min} + \left(V_{ref}^{max} - V_{ref}^{min} \right) \left(\frac{i-1}{PS-1} \right), \quad i = 1, 2, \dots PS \tag{12}$$

where PS is population size of HPSO; $V_{ref}^{min} = 350V$ and $V_{ref}^{max} = 650V$ refer to the maximum and minimum values of DC-link voltage reference during initialization. In order to speed up the transient of particle to its desired voltage reference and ensure correct sampling of fitness values, the initial parameters of PI controller, that is, P and I are set as 4 and 30 respectively. Meanwhile, the initial parameters of PR controller are set as $k_p = 1.5$ and $k_r = 1$, respectively.

During the iterative search process, all HPSO particles are sorted based on the DC-link voltage reference encoded in their respective positions before performing fitness evaluation. When the

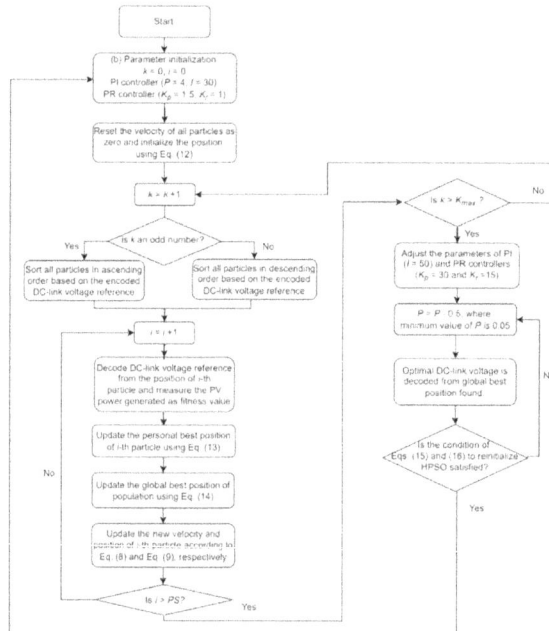

FIGURE 8.6 Flowchart of HPSO-based MPPT algorithm.

iteration number of k is an odd value, these particles are arranged in ascending order with their DC-link voltage reference values. Otherwise, the particles are arranged in descending order by referring to their DC-link voltage reference values for the even number of k-th iteration. The proposed sorting process incorporated into HPSO ensures the gradual increasing or decreasing of DC-link voltage reference values along with the population indices of particles, therefore is expected to alleviate detrimental effect as such inaccurate sampling of PV power due to the large voltage fluctuation. After completing the sorting process, the fitness value $f(x_i)$ of each i-th HPSO particle is evaluated as PV power generated based on the DC-link voltage reference encoded. In current study, MPPT is considered as a maximization problem because it aims to search for an optimal DC-link voltage reference that can generate the largest PV power. Hence, the i-th particle with larger value of $f(x_i)$ is more desirable. If the current position of i-th particle has larger $f(x_i)$ value than its personal best fitness $f(P_{best,i})$, the latter solution is replaced by the former one according to Eq. (*13*). Similarly, the global best solution can be updated by any i-th particle if its $f(P_{best,i})$ is larger than $f(G_{best})$ as shown in Eq. (*14*).

$$P_{best,i} = \begin{cases} x_i, & if \ f(x_i) > f(P_{best,i}) \\ P_{best,i,} & otherwise \end{cases} \tag{13}$$

$$G_{best} = \begin{cases} P_{best,i}, & if \ f(P_{best,i}) > f(G_{best}) \\ G_{best}, & otherwise \end{cases} \tag{14}$$

Referring to the updated $P_{best,\ i}$ and G_{best}, the new velocity v_i and new position x_i of each i-th HPSO particle can be updated using Eqs. (8) and (9), respectively.

The aforementioned procedures are repeated to enable convergence of HPSO particles towards the global peak power. In this study, the maximum iteration number of $K_{max} = 5$ is defined as convergence criterion of HPSO. When current iteration number k exceeds K_{max}, the proposed HPSO-based MPPT algorithm is terminated, followed by the decoding of optimal DC-link voltage reference from The parameters of both PI and PR controller, P is reduced by 0.5 in every time step to 0.05, whereas I, k_p and k_r are set as 50, 30 and 15, respectively. The adjustment of these parameters enables the output grid current with sine wave to have low THD and the power factor close to unity. It is notable that the reinitialization process of HPSO is performed to track new global MPP due to changes of environmental factors such as irradiances. Let Pow_{new} and Pow_{old} be the new PV power and the PV power at global maximum point of the last operating point, respectively. The changes of power ΔPow is then calculated as:

$$\Delta Pow = \frac{(Pow_{new} - Pow_{old})}{Pow_{old}} \times 100\% \tag{15}$$

As shown in Eq. (16), the changes of shading patterns or insolation status can be detected if ΔPow exceeds the predefined threshold value of $\varepsilon = 5\%$. The HPSO population is reinitialized to search for the new global peak when the condition of Eq. (16) is satisfied.

$$\Delta Pow > \varepsilon \tag{16}$$

The parameters of the proposed HPSO-based MPPT algorithm, PI and PR controllers can be summarized in Table 8.4.

TABLE 8.4

Control Loops Parameter

Parameter	PS	K_{max}	w	c_1	c_2	$V_{min\,ref}$ (V)	$V_{max\,ref}$ (V)	Sampling time (s)	ε (%)	P	I	k_p	k_r
										In = Initial value			
										F = Final value			
Value	4	5	0.5	1	1.2	350	650	0.01	5	In = 4 / F = 0.05	In = 30 / F = 50	In = 1.5 / F = 30	In = 1 / F = 15

8.4   RESULTS AND DISCUSSION

From methodology, it is understood that software simulation of single-stage grid-connected solar inverter system based on direct DC-link voltage control using HPSO-based MPPT algorithm can be realized in MATLAB Simulink. The parameter and component settings for software implementation will follow the settings mentioned in the methodology section.

8.4.1   PREPARATION OF DIFFERENT TEST

In an operating PV system, the GMPP of multiple-peaks PV curve due to partial shading condition can be either located on the low-voltage side (LVS) or high-voltage side (HVS) depending on the shading heaviness on the PV strings. Hence, a robust MPPT algorithm is vital in tracking the GMPP under dynamic shading patterns.

In order to prove that the proposed HPSO-based MPPT algorithm can work in the single-stage grid-connected solar inverter under different partial shading conditions, different set of irradiance data in W/m² are used to generate PV curve with different global peak positions. Three shading conditions, PSC1, PSC2 and PSC3 with their respective PV curves as shown in Figure 8.7 are tested. Figure 8.7(a) presents PV curve with single peak (V_{MPP} = 664.76V, P_{MPP} = 5455.33W). Furthermore, Figure 8.7(b) and Figure 8.7(c) show PV curves with double peaks, where the global peaks can be found at LVS (V_{MPP} = 334.37V, P_{MPP} = 2717.38W) and HVS (V_{MPP} = 705.12V, P_{MPP} = 3575.93W) respectively.

Figure 8.8(a) to (e) demonstrate how the HPSO consisting of four particles managed to gradually track down the global peak in five iterations. The particles are introduced uniformly in the search space as shown in Figure 8.8(a). It can be noticed that the updated positions of particles are being rearranged after each iteration. Every particle is progressively shifting towards the location of GMPP as presented in Figure 8.8(b) to (e). After the fifth iteration when the convergence criterion is achieved, the particle with global best position or best voltage reference is used to track GMPP consistently until reinitialization is triggered.

8.4.2   PERFORMANCE COMPARISON OF TRACKING USING CONVENTIONAL PSO ALGORITHM AND PROPOSED HPSO-BASED MPPT ALGORITHM

This section aims to prove the effectiveness of proposed HPSO-based MPPT algorithm in reducing DC-link voltage fluctuation during the tracking process. Conventional PSO is applied on PSC1 as shown in Figure 8.9 which is tracking under single peak condition. The solar voltage, solar power, DC-link voltage reference, grid current and THD are recorded. The starting DC-link voltage reference is set at 500V until 0.05s as it is considered the middle of the search space, and the searching process starts after that.

It can be observed that throughout the five iterations, when the positions of the particles are being updated with new velocities after each iteration, the particles are not being rearranged according to the DC-link voltage reference encoded into their position. During the transition from iteration to the next iteration in the exploration process, there will be sudden large voltage change from high to

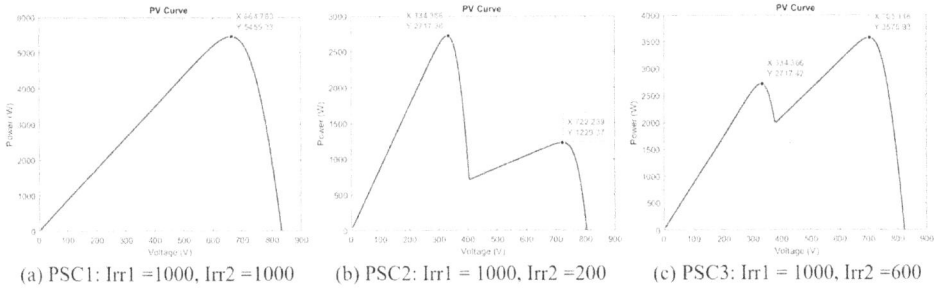

(a) PSC1: Irr1 =1000, Irr2 =1000   (b) PSC2: Irr1 = 1000, Irr2 =200   (c) PSC3: Irr1 = 1000, Irr2 =600

FIGURE 8.7 Different test cases with single and double peak PV curve.

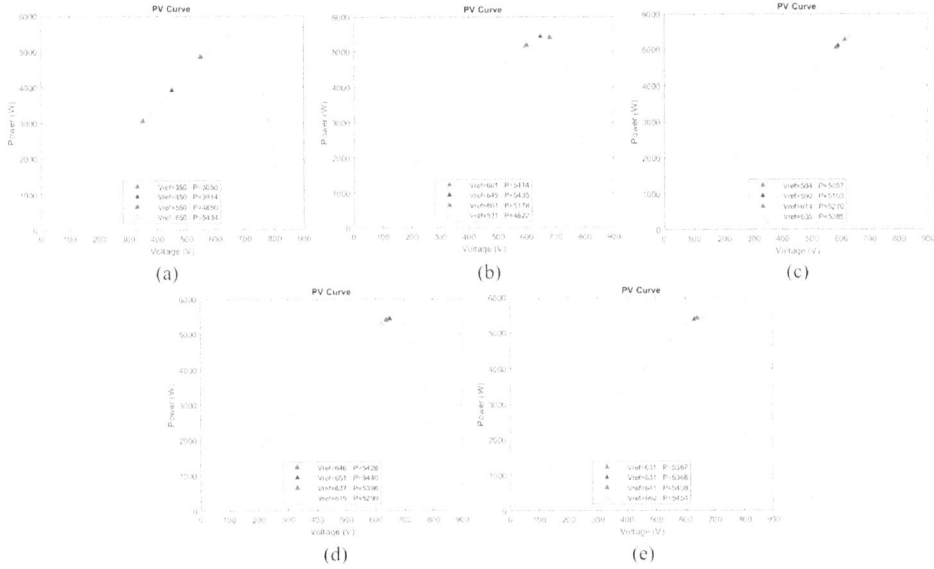

FIGURE 8.8 HPSO-based MPPT tracking of GMPP in five iterations.

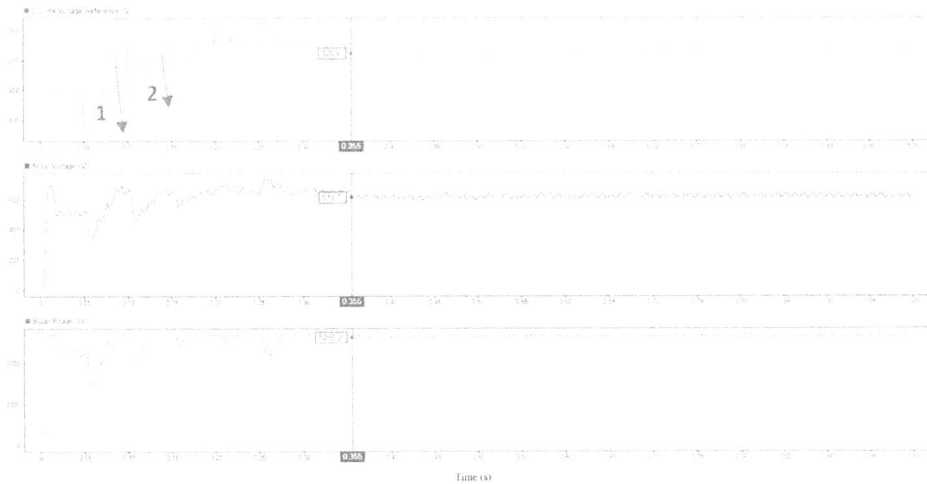

FIGURE 8.9 Tracking process under PSC1 using conventional PSO.

low voltage or vice versa. In consequences to the larger voltage fluctuation, the solar output power or fitness values are inaccurately sampled leading to false convergence at 626V and 5298.2W, where the exact V_{MPP} should be at 664.76V. As presented in Figure 8.10, the recorded grid current, I_{grid} is 33.8A with THD of 2.24%, meanwhile the active power injected into the grid, P_{grid} is 5291W, and the inverter conversion efficiency, I_{eff} can be calculated as 99.86%.

In comparison to tracking using conventional PSO-based MPPT algorithm, Figure 8.11 presents the tracking process under same PSC1 using the proposed HPSO-based MPPT algorithm. The tracking pattern reveals the uniqueness of the proposed algorithm, where the particles track the GMPP in a way similar to climbing up and down the mountains.

The proposed HPSO-based MPPT algorithm allows steady change of solar voltage by rearranging their positions in alternate ascending and descending order each iteration before analysing their fitness value. The solar voltage operating points are being updated in smaller steps and it prevents large overshoots and DC-link voltage oscillation problems during the tracking of global peak, thus providing more accurate sampling of fitness values and higher tracking accuracy. The convergence stops after the fifth iteration at 0.36s and the recorded V_{MPP} and P_{MPP} are at 654V and 5438.2W, respectively. The recorded I_{grid} is 34.1A with THD of 1.93% as illustrated in Figure 8.12. The recorded P_{grid} is 5426W, and I_{eff} is calculated as 99.78%.

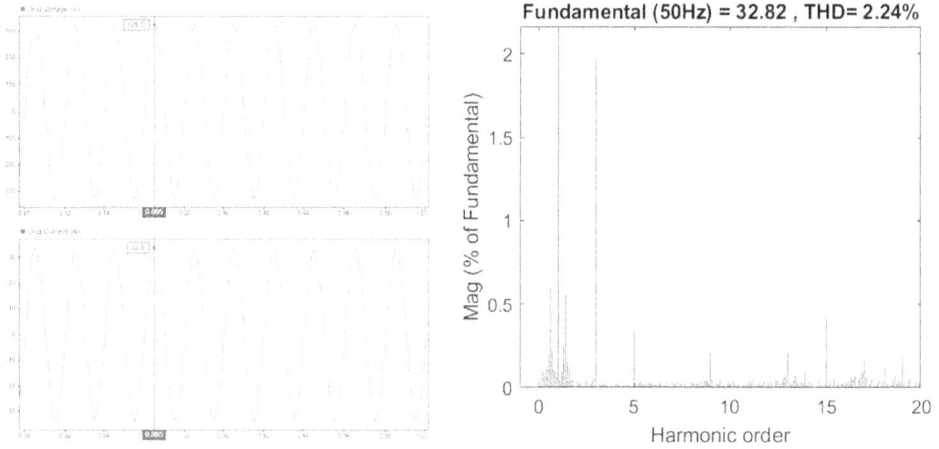

FIGURE 8.10   Grid measurement under PSC1 using conventional PSO.

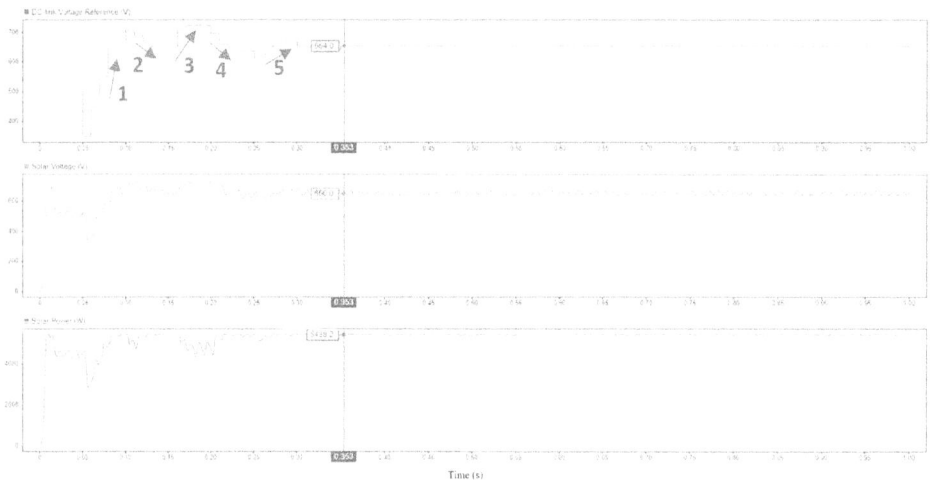

FIGURE 8.11   Tracking process under PSC1 using proposed HPSO-based MPPT algorithm.

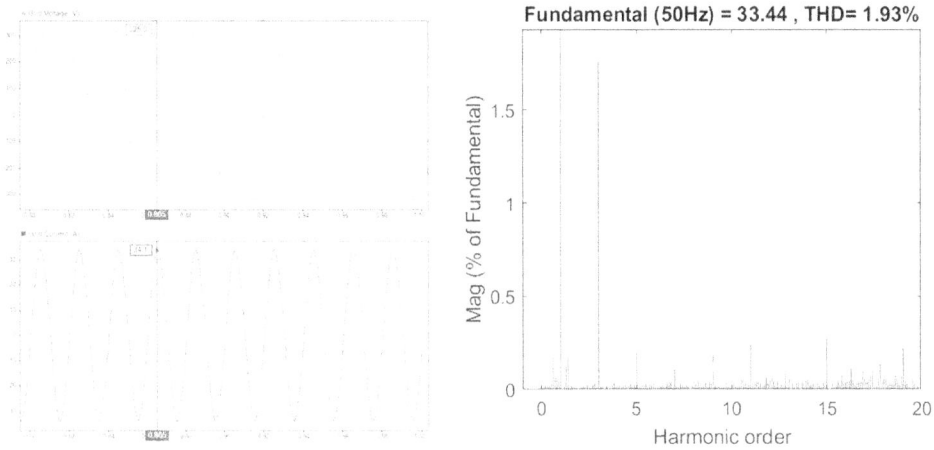

FIGURE 8.12   Grid measurement under PSC1 using proposed HPSO-based MPPT algorithm.

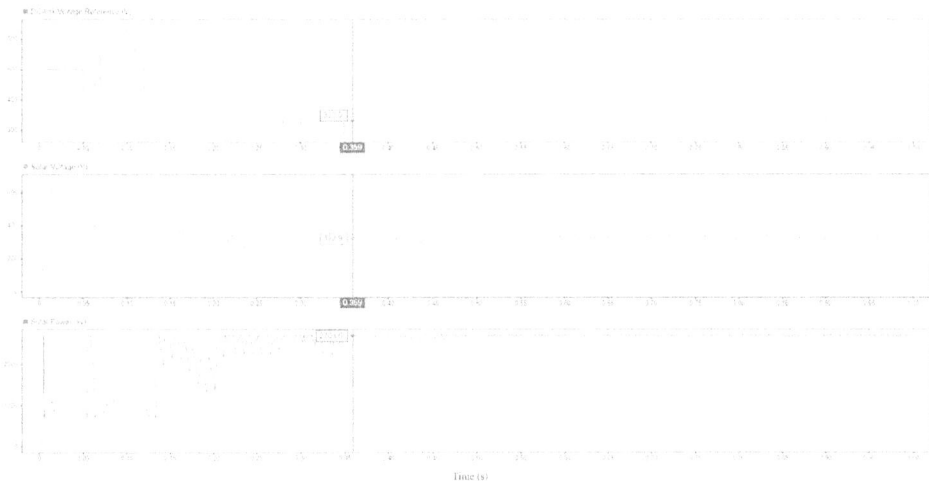

FIGURE 8.13   Tracking process under PSC2 using proposed HPSO-based MPPT algorithm.

In addition, the proposed HPSO-based MPPT algorithm has been verified working on multiple-peaks PV curves of PSC2 and PSC3 as presented in Figure 8.7. Referring to Figure 8.13 and Figure 8.14, HPSO-based MPPT algorithm is applied under PSC2 with double peaks where the global peak is located at the LVS of 334.37V. It can be seen that the solar power oscillations during the tracking process is larger compared to single peak condition, due to the exploration of global peak between the two MPP position. The result presents the tracked GMPP at V_{MPP} of 330V, P_{MPP} of 2708.6W, I_{grid} of 17.2A with THD of 4.60%. The recorded P_{grid} is 2700W, and I_{eff} is calculated as 99.68%.

Similarly in Figure 8.15 and Figure 8.16, the tracking process on PSC3 successfully reached global peak at V_{MPP} equals 693V, P_{MPP} equals 3568.2W, I_{grid} equals 21.9A and THD equals 1.85%. The recorded P_{grid} is 3554W, and the I_{eff} is 99.60%.

From the preceding simulations, it is ascertained that under different solar shading conditions resulting in distinct GMPP locations, HPSO-based MPPT algorithm managed to track the global peak in single-stage grid-connected solar inverter by controlling the DC-link voltage directly. The recorded THD in all three PSCs are within 5% limit complying to the IEEE Std-519 and it is notable that the dominant third order harmonics are due to the switching of inverter in single-phase system.

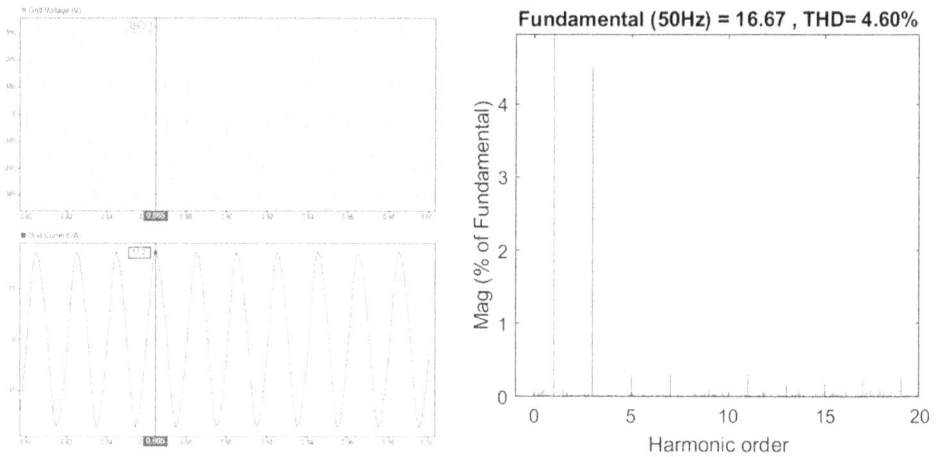

FIGURE 8.14   Grid measurement under PSC2 using proposed HPSO-based MPPT algorithm.

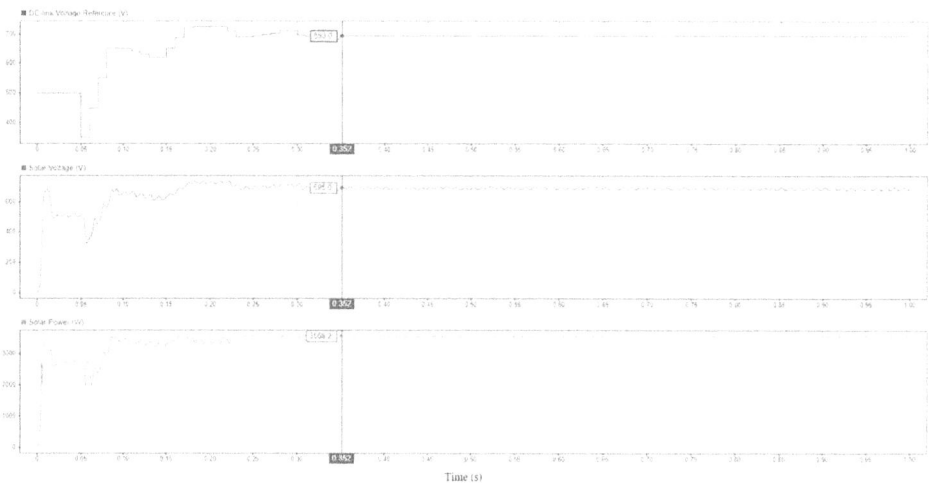

FIGURE 8.15   Tracking process under PSC3 using proposed HPSO-based MPPT algorithm.

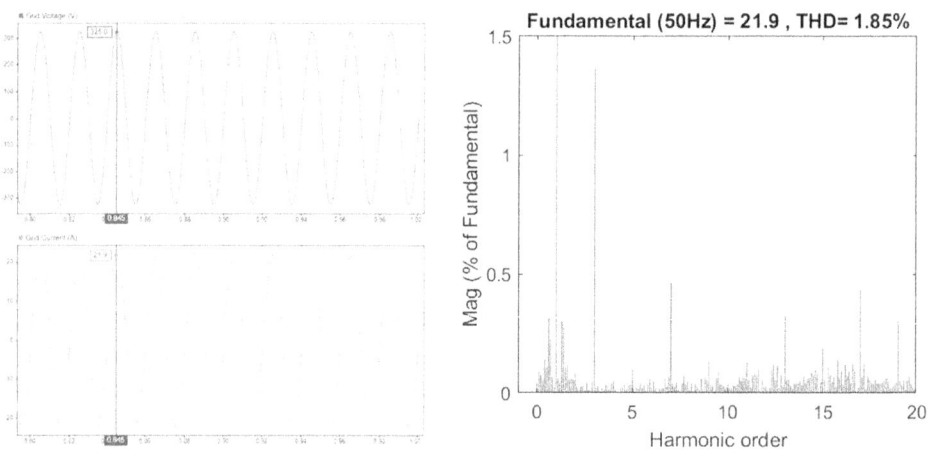

FIGURE 8.16   Grid measurement under PSC3 using proposed HPSO-based MPPT algorithm.

8.4.3 Performance Evaluation of Tracking Using Conventional PSO Algorithm and Proposed HPSO-Based MPPT Algorithm in 30 Consecutive Attempts

As mentioned in the earlier studies, most conventional MPPT algorithms tend to be trapped at the LMPP of multiple-peaks PV curves. Therefore, it is essential to evaluate the robustness of the proposed HPSO-based MPPT algorithm in tracking the GMPP of multiple-peaks PV curve based on the performance metrics known as tracking time, tracking accuracy and success rate of tracking. Three different sets of PV curves introduced in Section 8.4.1 (i.e., PSC1, PSC2 and PSC3) are used to thoroughly test the robustness of proposed MPPT algorithm. Particularly, the proposed HPSO-based MPPT algorithm is simulated for 30 consecutive times to perform tracking under the individual test cases of PSC1, PSC2 and PSC3 in order to reduce the random discrepancy brought by stochastic characteristics of PSO. The simulation results obtained by proposed HPSO-based MPPT algorithm are then compared with those of conventional PSO-based MPPT algorithm as shown in Figure 8.17 and Table 8.5.

The tracking accuracies of GMPP produced by conventional PSO-based MPPT algorithm and HPSO-based MPPT algorithm when solving each test case of PSC1, PSC2 and PSC3 for 30 consecutive times can be compared qualitatively using the boxplots illustrated in Figure 8.17. The

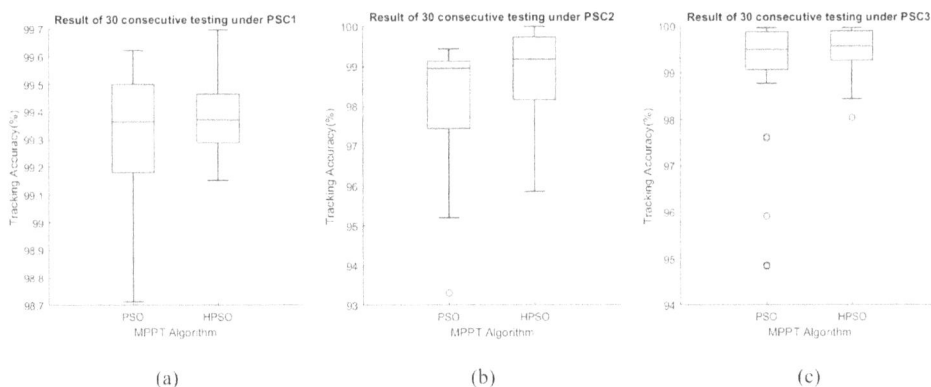

FIGURE 8.17 Boxplot of tracking accuracy produced by conventional PSO-based MPPT algorithm and HPSO-based MPPT algorithm when solving the test cases of (a) PSC1, (b) PSC2, and (c) PSC3 for 30 consecutive times.

TABLE 8.5
Quantitative Analysis Results Produced by Conventional PSO-Based MPPT and Proposed HPSO-Based MPPT under Different Test Cases for 30 Consecutive Simulation Runs

Irradiation (W/m²) [Irr1 Irr2]	Exact GMPP, P_E (W)	Tracking time (5 iterations)	30 consecutive simulation testing			
			HPSO		Conventional PSO	
			Average GMPP found, P_F (W)	Mean tracking accuracy $\frac{P_F}{P_E} \times 100\%$	Average GMPP found, P_F (W)	Mean tracking accuracy $\frac{P_F}{P_E} \times 100\%$
PSC 1 [1000 1000]	5455.33	0.36s	5421.26	99.38 ± 0.13%	5418.48	99.32 ± 0.23%
PSC 2 [1000 200]	2717.38	0.36s	2684.05	98.77 ± 1.20%	2665.35	98.09 ± 1.61%
PSC 3 [1000 600]	3575.93	0.36s	3556.67	99.46 ± 0.49 %	3535.55	98.87 ± 1.01 %

proposed HPSO-based MPPT algorithm is observed to outperform the conventional PSO-based MPPT algorithm in the test case of PSC1 because Figure 8.17(a) shows that the former algorithm is able to track the global peak with tracking accuracy higher than 99.1% for all 30 attempts. For the test case of PSC2 where global peak is located at the LVS of PV curve, Figure 8.17(b) reveals the superiority of proposed HPSO-based MPPT algorithm as evidenced by its capability to achieve 98% of tracking accuracy in locating the global peak of PV curve for 75% of its attempts. Figure 8.17(c) also demonstrates the competitive tracking accuracy of proposed HPSO-based MPPT algorithm in locating the global peak of PV curve for being able to achieve 98.5% of tracking accuracy for majority of its attempts. On the other hand, Figure 8.17 reveals that conventional PSO-based MPPT algorithm in general has more inferior tracking accuracy than that of proposed HPSO-based MPPT algorithm. In contrary to the latter approach, the DC-link voltage references encoded into the particles of former approach did not gradually change with their respective population indices, therefore has higher tendency to suffer with voltage oscillation issue during the fitness evaluation process. This undesirable scenario can lead to the inaccurate sampling of fitness value (i.e., PV power tracked) and global peak is then wrongly recorded due to the presence of large voltage fluctuation.

The tracking performances of proposed HSO-based MPPT algorithm and conventional PSO-based MPPT algorithm under different test cases of PSC1, PSC2 and PSC3 are also compared quantitatively in Table 8.5. Accordingly, the proposed HPSO-based MPPT algorithm can consistently outperform the conventional PSO-based MPPT algorithm in tracking the global peaks of all PSC1, PSC2 and PSC3 with different shading patterns as indicated by the high mean tracking accuracy and low standard deviation value of former approach. It is noteworthy that the proposed HPSO-based MPPT algorithm is also reported to achieve 100% success rate of tracking the GMPPs under different partial shading conditions for all 30 consecutives of simulation runs, implying the competitive tracking performance of proposed approach.

8.4.4 Performance Evaluation on Transition Tracking

Moving clouds or changes of solar irradiation acted on PV strings in different time of the day can lead to position change of global peak. Therefore, it is important to incorporate the reinitialization capability into HPSO-based MPPT algorithm to track the new global peak appears during sudden changes of irradiances.

In the proposed HPSO-based MPPT algorithm, the reinitialization process will be triggered when the PV output power changes more than 5% due to dynamic changes in the solar irradiances. The PV curves in previous section will be used for transition tracking test, where the transition will be from PSC2 to PSC1 and to PSC3 at one second per transition.

The transition tracking performance of HPSO-based MPPT algorithm is illustrated in Figure 8.18. During the interval of 0s to 1s, it is recorded the proposed algorithm managed to track the GMPP of PSC2 at 2673.92W. When the irradiances change at 1s, the sudden change in power will trigger the reinitialization process of HPSO-based MPPT algorithm and the new global peak of PSC1 at 5410.19W is successfully tracked. Similarly in the third transition, the algorithm managed to locate the global peak of PSC3 at 3541.92W.

In summary, the proposed HPSO-based MPPT algorithm, components and the parameter settings used in direct controlling of the DC-link voltage is proved to perform effectively in single-stage grid-connected solar inverter. The HPSO-based MPPT successfully tracked the GMPP under different PSCs in all the 30 consecutive attempts. In addition, the success in tracking under real-time transition of different PSCs further proven the feasibility of the algorithm in the real time application.

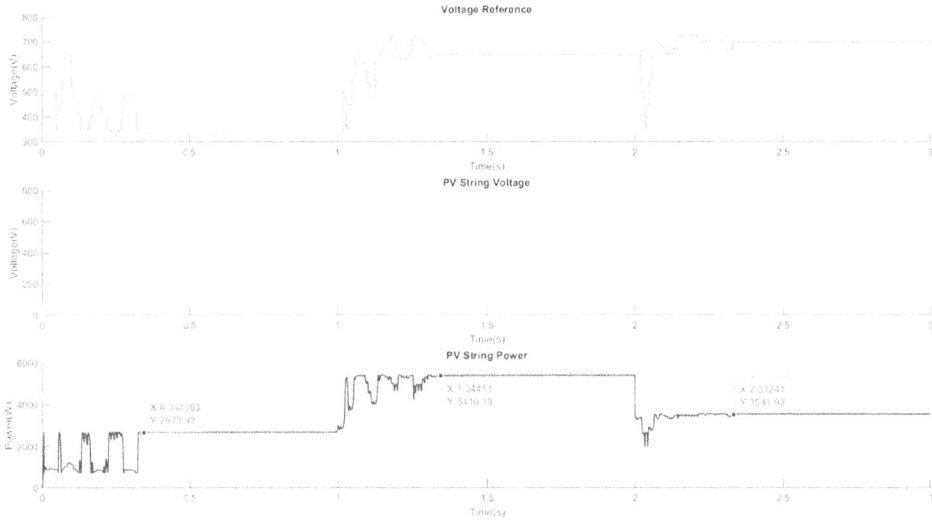

FIGURE 8.18   Transition tracking using HPSO-based MPPT algorithm.

8.5   CONCLUSION

A single-stage single-phase grid-connected solar inverter system based on direct DC-link voltage control is developed with MATLAB/Simulink. The design of the proposed HPSO-based MPPT algorithm, DC link voltage control loop with PI controller and grid current control loop with PR controller are presented in current study. The tracking robustness of HPSO-based MPPT algorithm is evaluated based on multiple-peaks PV characteristic curve produced by different PSC in 30 consecutive testing. Simulation results reported that the GMPP is achieved with tracking time of 0.36s and the best average tracking accuracy of 99.46%, signifying its practicability in real time PV system. The recorded THD of the grid current is also within the 5% limit as stated in IEEE Std-519. Lastly, proposed HPSO-based MPPT algorithm is also proved to be robust in dynamically changing partial shading conditions, reinitialization is triggered to locate the new location of GMPP each time weather condition changes. This book chapter contributes a complete idea on how direct DC-link voltage control using proposed HPSO-based MPPT algorithm can be realised in single-phase single-stage PV system for practical tracking application. In future research, the proposed grid-connected PV system may be incorporated with power control loop to monitor the injection of active and reactive power to the utility grid. Besides, fault ride-through capability such as voltage stabilizing during voltage dip condition and disconnection from the grid during critical fault conditions are not considered in this book chapter. In order to make the system more practical according to current renewable energy trend, the proposed PV system might need to be added with dynamic load, and battery energy storage system, while ensuring its MPPT stability in extracting maximum power under complex shading conditions.

ACKNOWLEDGEMENT

The authors would like to acknowledge the REIG-FETBE-2021/045 grant support from UCSI University research management unit CERVIE so that the research and publication of this work are made possible.

REFERENCES

[1] "Trends in PV applications 2021. *IEA-PVPS.* https://iea-pvps.org/trends_reports/trends-in-pv-applications-2021/ (accessed Feb. 22, 2022).

[2] Y. Yang, and F. Blaabjerg. Overview of single-phase grid-connected photovoltaic systems. *Electric Power Components and Systems*, vol. 43, no. 12, pp. 1352–1363, Jul. 2015, doi: 10.1080/15325008.2015.1031296.

[3] J. C. Teo, R. H. G. Tan, V. H. Mok, V. K. Ramachandaramurthy, and C. Tan. Impact of partial shading on the P-V characteristics and the maximum power of a photovoltaic string. *Energies*, vol. 11, no. 7, Art. no. 7, Jul. 2018, doi: 10.3390/en11071860.

[4] B. Pakkiraiah, and G. D. Sukumar. Research survey on various MPPT performance issues to improve the solar PV system efficiency. *Journal of Solar Energy,* vol. 2016, pp. 1–20, 2016, doi: 10.1155/2016/8012432.

[5] T. Kerekes, R. Teodorescu, M. Liserre, R. Mastromauro, and A. Dell'Aquila. MPPT algorithm for voltage controlled PV inverters. in *2008 11th International Conference on Optimization of Electrical and Electronic Equipment*, 2008, pp. 427–432, doi: 10.1109/OPTIM.2008.4602444.

[6] N. Hamrouni, M. Jraidi, and A. Chérif. New method of current control for LCL-interfaced grid-connected three phase voltage source inverter. *Renewable Energy Renouvelables*, vol. 13, 2010.

[7] G. Mehta, and S. P. Singh. Design of single-stage three-phase grid-connected photovoltaic system with MPPT and reactive power compensation control. *International Journal of Power and Energy Conversion*, vol. 5, pp. 211–227, 2014, doi: 10.1504/IJPEC.2014.063199.

[8] A. H. Mollah, Prof. G. K. Panda, and Prof. P. KSaha. Single phase grid-connected inverter for photovoltaic system with maximum power point tracking. *International Journal of Advanced Research in Electrical, Electronics and Instrumentation Engineering*, vol. 4, no. 2, pp. 648–655, 2015, doi: 10.15662/ijareeie.2015.0402021.

[9] N. E. Zakzouk, A. K. Abdelsalam, A. A. Helal, and B. W. Williams. PV single-phase grid-connected converter: DC-link voltage sensorless prospective. *IEEE Journal of Emerging and Selected Topics in Power Electronics*, vol. 5, no. 1, pp. 526–546, 2017, doi: 10.1109/JESTPE.2016.2637000.

[10] C. Aouadi, A. Abouloifa, I. Lachkar, M. Aourir, Y. Boussairi, and A. Hamdoun. Nonlinear controller design and stability analysis for single-phase grid-connected photovoltaic systems. *International Review of Automatic Control (IREACO)*, vol. 10, p. 306, Jul. 2017, doi: 10.15866/ireaco.v10i4.12322.

[11] O. M. Arafa, A. A. Mansour, K. S. Sakkoury, Y. A. Atia, and M. M. Salem. Realization of single-phase single-stage grid-connected PV system. *Journal of Electrical Systems and Information Technology*, vol. 4, no. 1, pp. 1–9, 2017, doi: 10.1016/j.jesit.2016.08.004.

[12] B. Abdelhalim, B. Abdelhak, B. Noureddine, A. Thameur, L. Abdelkader, and Z. Layachi. Optimization of the fuzzy MPPT controller by GA for the single-phase grid-connected photovoltaic system controlled by sliding mode. *AIP Conf. Proc.*, vol. 2190, no. 1, p. 020003, 2019, doi: 10.1063/1.5138489.

[13] N. Abjadi, S. A. A. Fallahzadeh, A. Kargar, and F. Blaabjerg. Applying Sliding-Mode Control to a Double-Stage Single-Phase Grid-Connected PV System. *J. Renew. Energy Environ.*, no. Online First, Sep. 2020, doi: 10.30501/jree.2020.233358.1114.

[14] S. Fahad, N. Ullah, A. J. Mahdi, A. Ibeas, and A. Goudarzi. An advanced two-stage grid connected PV system: A fractional-order controller. *International Journal of Renewable Energy Research (IJRER)*, vol. 9, no. 1, Art. no. 1, 2019.

[15] O. Kalmbach, C. Dirscherl, and C. M. Hackl. Discrete-Time DC-Link Voltage and Current Control of a Grid-Connected Inverter with LCL-Filter and Very Small DC-Link Capacitance. *Energies*, vol. 13, no. 21, Art. no. 21, Jan. 2020, doi: 10.3390/en13215613.

[16] S. Xu, R. Shao, B. Cao, and L. Chang. Single-phase grid-connected PV system with golden section search-based MPPT algorithm. *Chinese Journal of Electrical Engineering*, vol. 7, no. 4, pp. 25–36, Dec. 2021, doi: 10.23919/CJEE.2021.000035.

[17] M. M. Salem, Y. Atia, and O. Arafa. Improved MPPT control for single-phase grid-connected PV system. *International Journal of Power and Energy Conversion*, vol. 12, p. 1, Jan. 2021, doi: 10.1504/IJPEC.2021.113037.

[18] F. M. Oliveira, S. A. Oliveira da Silva, F. R. Durand, L. P. Sampaio, V. D. Bacon, and L. B. G. Campanhol. Grid-tied photovoltaic system based on PSO MPPT technique with active power line conditioning. *IET Power Electron.*, vol. 9, no. 6, pp. 1180–1191, 2016, doi: 10.1049/iet-pel.2015.0655.

[19] D. Pilakkat, and S. Kanthalakshmi. Single phase PV system operating under Partially Shaded Conditions with ABC-PO as MPPT algorithm for grid connected applications. *Energy Reports*, vol. 6, pp. 1910–1921, 2020, doi: 10.1016/j.egyr.2020.07.019.

[20] N. Aouchiche, M. S. Aitcheikh, M. Becherif, and M. A. Ebrahim. AI-based global MPPT for partial shaded grid connected PV plant via MFO approach. *Sol. Energy*, vol. 171, pp. 593–603, Sep. 2018, doi: 10.1016/j.solener.2018.06.109.

[21] S. Ravikumar, and D. Venkatanarayanan. Gwo based controlling of sepic converter in PV fed grid connected single phase system. *Microprocess Microsystem*, p. 103312, 2020, doi: 10.1016/j.micpro.2020.103312.

[22] N. Priyadarshi, A. K. Sharma, and F. Azam. A hybrid firefly-asymmetrical fuzzy logic controller based MPPT for PV-wind-fuel grid integration. *International Journal of Renewable Energy Research IJRER*, vol. 7, no. 4, Art. no. 4, Dec. 2017.

[23] N. Priyadarshi, P. Sanjeevikumar, J. Holm-Nielsen, F. Blaabjerg, and F. Azam. An experimental estimation of hybrid ANFIS–PSO employed MPPT for PV Grid integration under fluctuating sun irradiance. *IEEE System Journal*, 2019, doi: 10.1109/JSYST.2019.2949083.

[24] N. Priyadarshi, S. Padmanaban, P. Kiran Maroti, and A. Sharma. An extensive practical investigation of FPSO-based MPPT for grid integrated PV system under variable operating conditions with anti-islanding protection. *IEEE System Journal*, vol. 13, no. 2, pp. 1861–1871, 2019, doi: 10.1109/JSYST.2018.2817584.

[25] N. Priyadarshi, S. Padmanaban, M. Sagar Bhaskar, F. Blaabjerg, and A. Sharma. Fuzzy SVPWM-based inverter control realisation of grid integrated photovoltaic-wind system with fuzzy particle swarm optimisation maximum power point tracking algorithm for a grid-connected PV/wind power generation system: hardware implementation. *IET Electric Power Applications*, vol. 12, no. 7, pp. 962–971, 2018, doi: 10.1049/iet-epa.2017.0804.

[26] S. Padmanaban, N. Priyadarshi, M. S. Bhaskar, J. B. Holm-Nielsen, E. Hossain, and F. Azam. A hybrid photovoltaic-fuel cell for grid integration with jaya-based maximum power point tracking: Experimental performance evaluation. *IEEE Access*, vol. 7, pp. 82978–82990, 2019, doi: 10.1109/ACCESS.2019.2924264.

[27] S. Padmanaban et al. A novel modified sine-cosine optimized MPPT algorithm for grid integrated PV system under real operating conditions. *IEEE Access*, vol. 7, pp. 10467–10477, 2019, doi: 10.1109/ACCESS.2018.2890533.

[28] N. Priyadarshi, P. Sanjeevikumar, M. Bhaskar, F. Azam, I. B. M. Taha, and M. G. Hussien. An adaptive TS-fuzzy model based RBF neural network learning for grid integrated photovoltaic applications. *IET Renewable Power Generation*, vol. 16, no. 14, pp. 3149–3160, 2022, doi: 10.1049/rpg2.12505.

[29] R. Eberhart, and J. Kennedy. A new optimizer using particle swarm theory. *MHS'95. Proceedings of the Sixth International Symposium on Micro Machine and Human Science*, 1995, pp. 39–43, doi: 10.1109/MHS.1995.494215.

9 A Review on Cascaded H-Bridge and Modular Multilevel Converter: Topologies, Modulation Technique and Comparative Analysis

A. Sivapriya and N. Kalaiarasi

CONTENTS

DOI: 10.1201/9781003323471-9

9.1 INTRODUCTION

Global warming is a key issue for the universe in the current scenario, necessitating the search for alternatives to minimize carbon emissions. Renewable energy sources (RES) have consistently proven to be the most reliable option. Solar, wind, biomass, biogas, fuel cells, and hydro energy have all seen considerable growth in recent years in industrial and non-industrial purposes. Since these sources produce non-sinusoidal output voltages and currents, an interfacing circuit is necessary to decrease output distortion [1]. Generally, the voltage source converter can be utilized as an interface unit for power conversion which employs semiconductor devices such as IGBTs and diodes to produce sinusoidal voltage with minimal distortion. Many high-power applications use DC to AC power converters, including renewable energy, FACTS, high-voltage DC transmission systems, motor drives, and grid-connected or standalone photovoltaic (PV) applications [2]. The multilevel converter (MLC) is predominantly used in several applications because of its ability to build the staircase output waveform by combining various direct current (DC) sources. Additionally, it offers numerous unique attributes, including high-quality output power having very minimal distortion, excellent harmonic performance, modularity, and fault tolerance. The popularly used MLCs are neutral point clamped (NPC), flying capacitor (FC), cascaded h-bridge (CHB), modular multilevel converter (MMC) as shown in Figure 9.1. The NPC converter generates the output voltage across the load terminal using a sequence of diodes, capacitors, and semiconductor switches. The clamping diodes are used to distribute the blocking voltage. The configuration of the FC converter is identical to that of the NPC converter, except that it employs a series of capacitors rather than clamping diodes [3].

The primary issues of these topologies are the increased component count required for high voltage applications. This is due to the fact that these topologies have a lesser number of levels as well as a semiconductor increment, which makes them less likely to be used in high-power applications [4]. However, NPC and FC converter topologies are suitable for high power drives, but also confront various issues when more levels are added, such as neutral point voltage balancing, increased switching losses, and capacitor voltage balancing [5]. Among these topologies, CHB and MMC topology uses many low voltage sources as input to produce a high-quality sine waveform. Many low-voltage isolated dc inputs make it more appropriate for solar PV applications [6]. Besides, the CHB multilevel converter has garnered the most attention owing to its unique design, which offers for expansion to higher voltage levels with few additional components. Nonetheless, a phase-shifting transformer at the

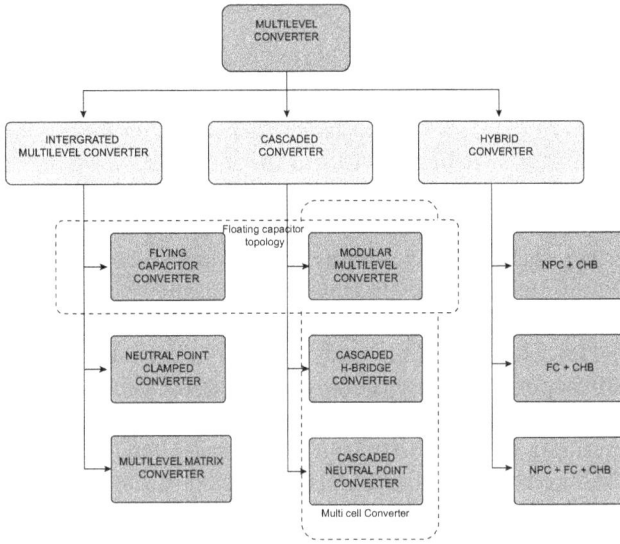

FIGURE 9.1 Multilevel converter topology.

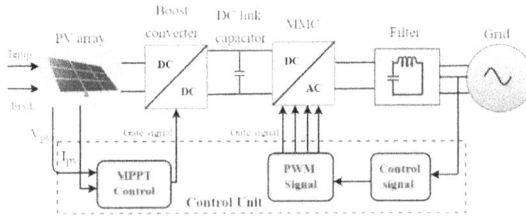

FIGURE 9.2 General structure of PV integrated grid system.

input end is required to generate independent dc sources, which is a main issue of the CHB converter. The modular multilevel converter (MMC) was proposed by Marquardt and Lesnicar in 2001 as a derivative of the CHB. The MMC generates a high voltage across the load side by connecting a large number of low-voltage modules in series. MMC retains all of the features of CHB without requiring an isolated DC source [7]. As a result, this topology is ideal for medium-voltage applications since it removes the necessity of coupling transformers. High-power HVDC, battery energy storage units (BESS), medium-power drives, STATCOM, grid-connected applications, and unified power flow controllers generally employ the modular multilevel converter [8]. Also due to the recent advancements in technology, multilevel converters are a good solution for high-power applications.

Large-scale grid integrated PV system demands a higher power conversion ratio at a low cost which allows solar technologies to the commercial energy market. A PV system includes PV panels, bidirectional converters, and a control unit for regulating the power extracted from the PV cells as depicted in Figure 9.2. Therefore, numerous studies have been performed to improve PV reliability, efficiency, and competitiveness. Converter topologies are thoroughly examined, with an emphasis on hardware complexity, the advanced approach of MPPT, and optimal topology for RES application [9].

Yet, another excellent, in-depth examination of the power unbalance issues in MMC and CHB converters used in PV, STATCOM, and BESS systems are discussed in [10]. Hence, the MLC-based power grid requires increased voltage and power ratings in its power equipment. Because of its scalability, MLC-based modular systems are now the most promising commercial alternatives in this area. Furthermore, the use of modular architectures reduces the need for a transformer and enhanced power quality [11]. The pulse width modulation (PWM) approach is the most basic switching strategy for MLC since it generates gate pulses primarily by comparing the sinusoidal waveform to a higher frequency carrier signal.

The PWM approach has many benefits, including ease of implementation, improved performance [12], and minimal power loss in switching devices. However, switching and conduction losses are the constraints of MLC, which are largely dependent on the modulation methods used to operate the semiconductor switching devices in the multilevel converter. Despite its benefits, MLC has several limitations by means of voltage balancing of the capacitor, circulating current reduction, power balancing, harmonics, and fault tolerance [13]. These concerns have a significant impact on the converter's stability and reliability. Thus, more research is being conducted in the domain of multilevel converter topologies to maximize their reliability by proper design, appropriate modulation and control techniques, improved power quality, and the addition of fault tolerance to the converter [14], [15]. Henceforth, the study's primary contribution is to concentrate on the various CHB and MMC topologies and the contemporary issues associated with power imbalances between phases and submodules, circulating current control, and voltage balancing. Consequently, it is focused on delivering a detailed overview of appropriate modulation and control strategies for MLC systems. Additionally, this manuscript describes in detail the fault diagnosis and fault tolerance techniques for various kinds of SM failures, and AC and DC side faults. The following sections comprise the manuscript: Section 9.2 discusses the topology of multilevel converters, their problems, and solutions for photovoltaic applications. Section 9.3 illustrates a modulation strategy for a multilevel converter. Section 9.4 exhibits several approaches for fault detection and mitigation in multilevel converters. Section 9.5 elaborates on fault-tolerant methods for CHB and MMC converters. Section 9.6 discusses the overall challenges and possible future directions, while Section 9.7 draws conclusions.

9.2   LITERATURE SURVEY ON MULTILEVEL CONVERTER TOPOLOGIES

Numerous industrial applications can benefit significantly from the low-cost multilevel converter architecture. The number of power supply units and semiconductor switches required vary according to the number of levels and their structure. The cascaded H-bridge converter requires several isolated DC sources and switches for power conversion, making it more suitable for photovoltaic applications.

The most promising way for producing large-scale PV integrated systems is to employ cascaded or modular multilevel converters [16] which exhibit high efficiency, modularity, independent MPPT control, and fault-tolerant characteristics. This section addresses the various topologies of CHB and MMC, as well as the problems and solutions associated with their use in solar applications.

9.2.1   CASCADED H-BRIDGE CONVERTER TOPOLOGY

The CHB topology has a set of h-bridge cells with distinct DC sources. Each h-bridge has four IGBT switches and a DC voltage source. This architecture has fewer components than the NPC and FC topologies because it doesn't use clamping diodes or clamping capacitors and doesn't have voltage balancing issues [17]. To obtain the desired output voltage, all the h-bridge cells are connected in series with separate DC sources which can also be solar cells, batteries, or a fuel cell, as displayed in Figure 9.3(a). The output voltage is the summation of all the voltages obtained in each h-bridge cell which is calculated using equation 1 and equation 2 gives the number of levels in the output voltage.

$$\text{Overall output voltage, } V_{output} = \Sigma V_{dc} = V_1 + V_2 + V_3 \tag{1}$$

$$\text{Output voltage levels, } N = 2S + 1 \tag{2}$$

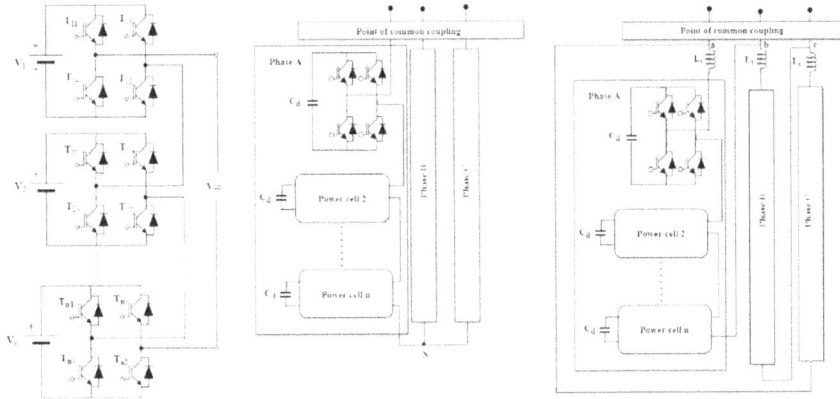

FIGURE 9.3 (a) Generalized CHB converter structure, (b) three-phase CHB star configuration, (c) three-phase CHB delta configuration.

Each h-bridge is configured to build three voltage levels: +Vdc, 0Vdc, -Vdc, and three such h-bridges are connected in series to produce 7 levels in output. Four switches (T1-T4) are connected in various ways to get the appropriate output. Turning on switches T1 and T2 generates a positive output voltage, whereas connecting switches T3 and T4 generate a negative voltage. Zero voltage may be accomplished by turning on the switches T1 and T3 or T2 and T4. By summing the output voltages of each H-bridge, the output voltage is obtained. Because the output terminals of each H-bridge are connected in series, the DC sources must be separated from one another. The CHBMLI enables the use of variable magnitude DC sources and produces a greater number of output voltage levels [18].

The CHB has numerous benefits for photovoltaic systems: it has several dc-links for connecting PV strings, each with its MPPT, and it quickly reaches medium voltage. Nonetheless, because each H-bridge cell has its photovoltaic system with its power source, an inherent power imbalance exists between the cells [19].

For the three-phase system, the star-to-delta CHB converter (Y_CHB) and the delta-to-star CHB converter (D_CHB) configuration are shown in Figure 9.3(b) and (c). The primary goal of a grid-integrated photovoltaic system is to enhance the amount of energy fed into the grid by monitoring the panel's maximum power point (MPPT).

Due to the requirement for several separate dc sources, the CHB architecture is particularly suited for solar PV applications. For low voltage applications, the single-phase CHB converter is proposed in [20], which eliminates the requirement of dc side sensors leads to cost reduction and higher efficiency of the solar system. In [21], a CHB for a PV system with BESS is addressed that provides constant power despite of changing irradiance. From the charging phase to the discharging phase, the suggested converter showed a seamless transition. In [22], the CHB converter for a large-scale PV system is investigated and the suggested converter was capable of delivering high-quality currents with minimal variation. Its robust architecture allowed it to function in high voltage circumstances with enhanced power quality. In [23], the author demonstrated an upgraded CMLI through a robust structure for reducing current leakage in transformer less photovoltaic systems; moreover, conduction and turn-on losses were minimized, allowing for operation at high frequency domain.

In [24], the CHB converter for PV systems is presented, which utilizes the H6-type power cell perhaps than the normal h-bridge type. The proposed system improves power quality and output voltage under partially shaded situations, hence increasing the efficiency of the system and energy fed to the grid. Using three different connections to optimize the performance and power quality of the cascaded converter, the PV-fed cascaded inverter arrangement illustrated in [25] reduces the number of circuit components required, allowing for a reduced cost. A total of twelve switches are used in a binary connection to generate fifteen levels, and 27 levels are obtained using a trinary configuration.

9.2.2 Modular Multilevel Converter Topology

Typically, MMC circuits include a dc side that is powered by a dc-link capacitor on the dc bus. The MMC is composed of cascaded submodules (SM) or cells linked through an arm inductor that creates two arms for each phase leg of the converter. Three-phase MMCs have three legs and two arms, referred to as upper and lower arms, for each leg. Each arm has a series inductor and a series of submodules, as seen in Figure 9.4. The SM is of two types: half-bridge (HB) cell or full-bridge (FB) cell. The voltage level of the MMC is governed by the number of cells in each arm and the utilization of PWM on both arms. It is widely applied in the fields of HVDC transmission, motor drives, renewable energy systems, and FACT devices which offer more flexibility than cascaded converters [26].

9.2.2.1 Submodule Configuration of MMC

Each submodule has capacitors and a certain number of semiconductor switches. The following subsections explore many essential configurations of modular multilevel converters with SM.

9.2.2.1.1 Asymmetric Half-Bridge SM (AHB-SM)

HB and FB are the two fundamental forms of SM, the half-bridge topology is most widely used which consists of two switches and a capacitor. As portrayed in Figure 9.5(a), the two switches (S1 and S2) work complementarily in asymmetrical half-bridge submodules (AHB-SM). AHB-SM provide a positive voltage when S1 is ON and S2 is OFF. Alternatively, we may acquire a zero-voltage level (0).

Thus, AHB-SM support bypass operation but just provide unipolar voltage outputs [27]. Notably, MMCs are unable to operate during dc-side breakdowns owing to their unipolar voltage outputs. However, as compared to H-bridge submodules, AHB-SM saves half the switches. Additionally, each asymmetrical half-bridge submodule transports the load current through a single switch, inevitably resulting in reduced conduction losses.

9.2.2.1.2 Symmetric Half-Bridge SM (SHB-SM)

Two switches and two energy storage devices make up the symmetrical half-bridge submodule as seen in Figure 9.5(b). When S1 is activated and S2 is deactivated, SHB-SM generate a positive

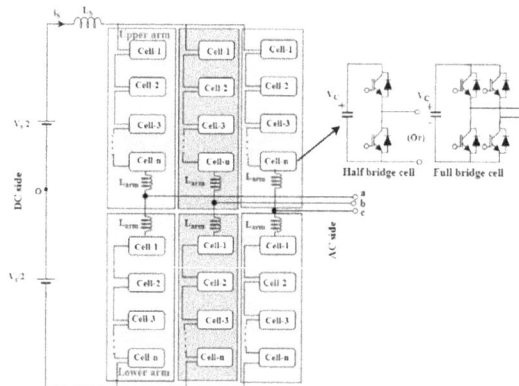

FIGURE 9.4 Schematic representation of modular multilevel converter.

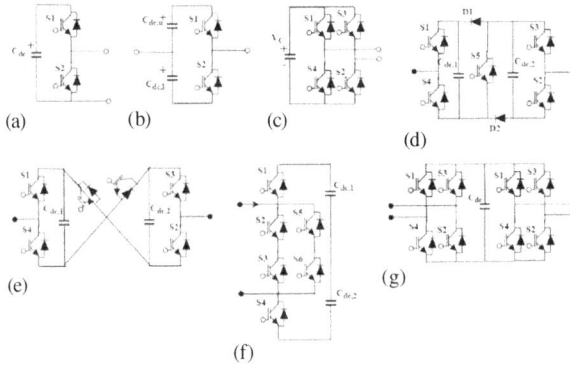

FIGURE 9.5   Various configuration of MMC-submodule: (a) asymmetrical half-bridge SM, (b) symmetrical half-bridge SM, (c) full bridge SM, (d) clamped double SM, (e) cross connected SM, (f) modified ANPC SM, and (g) double half-bridge SM with parallel connectivity.

voltage through their top capacitors. Alternatively, SHB-SM output their lower capacitor voltages inversely when S1 is OFF and S2 is ON. As a consequence, SHB-SM allow bipolar operation while eliminating the need for a bypass state. Additionally, with SHB-SM and converters, the balancing of top and bottom capacitor voltages poses a unique difficulty [28].

9.2.2.1.3   Full Bridge SM (FBSM)
As seen in Figure 9.5(c), the FBSM [29], which consists of four switches and a capacitor, can establish bipolar voltage levels when the switches S1 to S4 are operated properly. Thus, the FBSM-based MMC is compatible with both dc and ac systems. Additionally, the FBSM is capable of managing dc-link short circuit fault current (I_f) using reverse voltage generation through the capacitor voltage. The FBSM-based MMC can be completely detached by turning off all switches, but the HBSM-based MMC cannot be completely disconnected. The limitation of FBSM is that it has double the number of semiconductor devices as HBSM, making it more expensive and inefficient [30].

9.2.2.1.4   Clamped Double SM (CDSM)
The configurations include the addition of switches and diodes to two half-bridge submodules. In Figure 9.5(d), switch S5 is permanently ON, and the submodule functions as two HBSMs linked in series, although both capacitors display either a parallel or series connection for CDSM when the IGBT switches are blocked. Thus, the SM may be used as an FBSM by connecting both capacitors in parallel at separate voltages to eliminate paralleling concerns; the diodes can be replaced by switches.

CDSM is capable of generating reverse voltage and hence capable of handling dc-link current. CDSM has a loss profile similar to that of HBSM and FBSM when the same voltage values are utilized [31].

9.2.2.1.5   Cross Connected SM
In Figure 9.5(e), a cross-linked SM comprised of two HBSMs coupled back-to-back by additional switches and their anti-parallel diodes. Semiconductor losses are the same as those of the clamp-double SM.[32]. The clamped double SM and cross-connected SM are bipolar SM which also provides the general fault blocking capabilities and controllable negative voltages like FBSM.

9.2.2.1.6   Modified-ANPC SM
To increase the efficiency of the cells, multilevel structures such as NPC or FC, as illustrated in Figure 9.5(f), or a twin module may be used to replace the ordinary SM. A modified active-NPC (ANPC) SM has lower losses compared to HBSM [33].

9.2.2.1.7   Double Half-Bridge SM (DHB-SM)

Figure 9.5(g) depicts a DHB-SM with parallel connection, which includes four switch strands, single storage battery, and two sets of output terminals.

One of these SMs communicates with the preceding or next SM via a sets of output terminals, rather than a single output terminal as in traditional MMCs and CBCs. The pairs of module connectivity permits parallelization of many SM while maintaining 50% switch usage maintaining the modules' blocking-voltage constraints [34]. The merits and demerits of CHB and MMC topology are elucidated in Table 9.1.

9.2.3   Challenges Involved in Integrating CHB

Typically, the CHB converter is configured so that each phase leg gets the same amount of power. However, this presumption is no longer valid for large photovoltaic deployments.

Even though the quantity of photovoltaic modules coupled to each module is the same and their attributes are identical, the effective power provided to each module will vary. This unbalanced power distribution arises as an outcome of the photovoltaic string's rely on climatic conditions, partial shading caused by nearby objects, unequal temperature difference, fabrication variances, dirt, or non-homogeneous module deterioration [35]. Without countermeasures, the three-phase currents would become unbalanced.

The following are the primary issues and concerns associated with the deployment of cascaded h-bridge converters:

TABLE 9.1
Merits and Demerits of CHB and MMC Topology

Features	Cascaded H-bridge converter	Modular multilevel converter
Advantages	• Direct integration of PV with CHB • Connecting each cell to a low-voltage PV string • Easy extension, more suitable for large scale PV system • Modifications to the MPPT algorithm on an individual basis • Enhanced power quality with fault redundancy • Transformer less direct grid integration • High reliability	• Low voltage semiconductor switches • Generates good quality output waveform by increasing number of voltage steps • Scalability, transformer less operation, low operating cost • Effectively uses both fundamental and high switching frequency • Modular in construction • Suitable for low and high-power rating applications • Arm inductance present limits the fault current.
Drawbacks	• Utilization of isolated dc source • The power imbalance between phases • Requires a greater number of gate driver circuits • Switches must tolerate blocking voltage equivalent to the input voltage • Loss of modularity, high implementation cost with different voltage rating (asymmetric configuration)	• Circulating current causes second harmonics in the phase legs • Requirement of passive filter to limit circulating current • Balancing capacitor voltage at higher-order levels is problematic • Need output filter and interface transformer • Incapable of managing and riding through faults • Current circulation, high voltage ripple through submodule capacitor
References	[20]–[22], [35]–[37]	[3], [27]–[29], [38]

- The CHB converter requires many isolated dc sources to feed the h-bridge.
- The lack of a common dc link in CHB results in difficulties sharing energy across phases, which causes unbalanced voltage among h-bridge modules.
- Due to uneven irradiance, unequal power generation results in unbalanced grid currents.

9.2.3.1 Power Balancing in CHB

In a three-phase system, distorted PV power generates an imbalance in the functioning of the CHB converter, resulting in an unbalanced grid current.

The power balancing among the three-phase cascaded converter can be done in two ways: (a) active power balancing across each phase leg, which is employed when each phase leg supplies a variable amount of power; (b) individual phase leg active power balancing is used when the output of each power cell in a phase varies.

Active power balancing is a technique that makes use of fundamental frequency zero sequence injection (ZSI) to balance the grid current even when the photovoltaic power is imbalanced [36]. However, this approach results in a rise in the output voltage of the converter, which may easily surpass the limit imposed by the dc side when the power imbalance grows. In [37], the Y_CHB was implemented using Weighted Min-Max (WMM) zero-sequence voltage injection. The zero-sequence voltage is obtained by comparing weighted positive-sequence voltage references. However, WMM ZSI is insufficiently accurate because it generates an inaccurate fundamental-frequency zero-sequence component. As a result, the dynamic performance of the system is degraded. With the use of third-harmonic injection in a three-phase converter, dc voltage usage can be greatly improved (THI). Third harmonic injection of 1/6 of the fundamental-frequency component was reported in [38], [39] to attain high ac output voltage without overmodulation. Several more approaches, such as double THI, square wave ZSI, and optimum ZSI, have been suggested in [40].

9.2.4 Challenges Involved in Integrating MMC

The key problems and concerns involved in modelling the multilevel converter are as follows:

- Even while the MMC offers some unique advantages, it also gets suffered from circulating current inside the converter, which intensifies the conduction losses of the switches.
- Also, due to the submodule capacitor, the voltage ripples are detected due to variations of circulating current.
- Voltage stress across the submodule rises due to the capacitor's excessive voltage ripple. It also demands complicated control requirements due to a greater number of components.
- The modulation system and submodule selection strategy have a significant effect on efficiency.
- The additional issues are associated with concurrent output current management, creating the proper reference current for the input current required by the load, keeping dc voltages in accordance with the reference value, and balancing capacitor voltages among submodules.
- Proper designing of inductor and capacitor to maintain the short-circuiting of dc bus and dc to ac isolation.

9.2.4.1 Submodule Capacitor Voltage Balancing

The capacitor for the submodule should be chosen according to the capacitor's allowed voltage ripple. Owing to the existence of circulating currents in the arms, the floating capacitor charges, and discharges. As a result of this, arm currents generate voltage ripples. Voltage ripple generated in this manner is inversely proportional to capacitance [41]. While it is impossible to eliminate the voltage ripple, its magnitude can be reduced. Another critical design aspect is capacitor size, which is normally determined by the allowed dc component in the SM capacitor. The energy stored in an arm and the SM capacitor values are given by equations 3 and 4.

$$\Delta W_{arm} = \frac{P_d}{3\omega} m \left(1 - \frac{1}{m^2}\right)^{3/2} \tag{3}$$

$$C_{sm} = \frac{\Delta W_{arm}}{2N\varepsilon V_c^2} \tag{4}$$

where ΔW_{arm} denotes the change in energy stored in one phase arm, P_d is the converter's power rating, and ω is the alternating current frequency in radians per second.

As a result, numerous studies have been undertaken in recent years to address the disadvantages of MMC are available in the literature [3], [21], [42], [43]. The article [42], which includes a modulation approach for balancing the submodule capacitor voltages and obtaining suitable concurrent switching of the module, provides an upgrade on the intrinsic construction, operation, and control scheme for the MMC. A new estimating unit with a smaller number of sensors was proposed in [43] to stabilize the SM voltage of the MMC. Using this approach, there is no longer a requirement to directly monitor capacitor voltage. Hence it can be used to construct a significant number of SMs at minimal initial investments. By integrating with FC, the topology suggested in [3] to balance the capacitor voltage in the MMC between the upper arm and lower arm, the FC is tapped. Even at rated torque settings, voltage ripples in the submodule capacitor are decreased owing to the FC's ability to manage the AC flowing through it. Another article on voltage balancing of capacitor employing a low voltage rating clamping diode has been accomplished in [38].

9.2.4.2 Circulating Current Control

The submodules are linked in line with the inductor on the arm (L_{arm}). Because the circulating current is created while operating, a L_{arm} is necessary to isolate the upper and lower arms [44]. Apart from that, the L_{arm} filters off output current ripples. For a dc side short circuit, the L_{arm} must be employed to restrict the fault current. Typically, arm inductors have a per-unit value (L_{pu}) of between 0.1 and 0.2. Equation 5 is used to compute the L_{arm}.

$$L_{arm} = L_{pu} L_{base} = L_{pu} \frac{V_{base}}{\omega I_{base}} \tag{5}$$

where L_{arm} is the inductance of the arm in SI units and V_{base} and I_{base} are the voltage and current base values. This offers an acceptable compromise between the size of the inductor, fault current limiting, and inductive smoothing at the AC output terminals. This inductance may also contribute to the filtering effect of a few SMs. The circulating current is frequently anticipated to be reduced since it does not promote the AC, but instead alters the arm current and worsens the arm's power losses. Hence, many strategies for eliminating circulating currents have been developed. Raising the inductance of Larm is the simplest and most easy method to reduce the circulating current [45]. However, this strategy can only decrease circulating currents, not remove them. Additionally, it raises the quantity and overall cost of MMCs. A design approach for Larm and capacitor of SM may be utilized to inhibit the circulating current relying on the estimate of the AC constituents of the arm current [46], which enhances the inductor design to some point. Between the bridge inductors, an LC doubled switching frequency resonant filter may be connected to suppress the second order harmonic circulating current [47]. The control effect, on the other hand, is highly dependent on the characteristics of LC resonant filters. To completely remove the circulating current, voltage correction in MMC arms is often utilized. The voltage compensation is created using an open-loop technique based on the arm energy estimation. However, the control effect is contingent upon correct measurement of the ac load current and dc-link voltage, thus emphasizing the sensor's dependability and sensitivity. Based on the look-up table approach, circulating current injection control is utilized to optimize the capacitor voltage fluctuation [48].

9.3 PULSE WIDTH MODULATION AND CONTROL TECHNIQUES

9.3.1 Pulse Width Modulation

PWM methods are critical for modelling multilevel converters. The PWM approach directs the converter in such a way that the system's harmonics are reduced without reducing the converter's output power. The converter receives the gating signal through pulse width modulation, which employs a single carrier and reference signal [49]. The gate pulse is produced by comparing the reference signal to the carrier signal. When the carrier signal is less than the reference signal, the PWM output is high, whereas when the carrier signal is more than the reference signal, the PWM output is low. In a half-bridge MMC circuit, the PWM signal regulates the top switch, while its inverse controls the bottom switch.

It has multiple of these half-bridge circuits that must all be independently regulated. As a result, this issue may be resolved by employing multicarrier PWM approaches which are most often utilized for CHB and MMC topology.

Each half-bridge circuit (or submodule) in the converter requires its carrier waveform. The various forms of PWM techniques available for the multilevel converter are illustrated in Figure 9.6. This section gives an insight into numerous modulation methods that are extensively used to switch multilevel converters.

9.3.1.1 Sinusoidal PWM Technique

Switching techniques for power converters include the sinusoidal PWM approach (SPWM). The gating signal is generated at each sinusoidal-triangular waveform intersection. The MMC architecture has been modified to include a CHB module and extra chopper unit in each arm, which varies from earlier topologies [50].

By using a control approach for the MMC, PD, SPWM, and a control technique on the H-bridge, it has been shown that the output voltage may be increased twofold when compared to the standard method. In [51], investigated the effectiveness of a modified MMC that used an FC converter as its source of power. It is effective to evaluate THD at various modulation indices applying SPWM levels like PD, POD, and APOD.

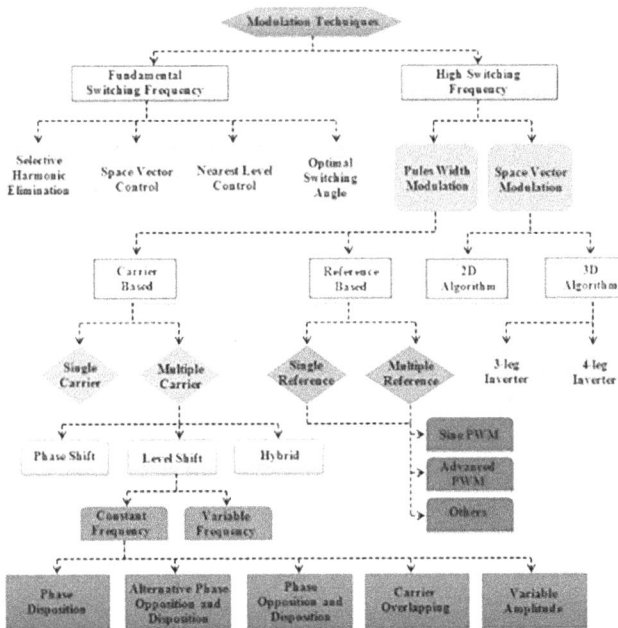

FIGURE 9.6   PWM technique for multilevel converter.

9.3.1.2 Phase Shifted (PS) PWM Technique

One of the most extensively used PWM techniques in power electronics is the phase-shift carrier (PSC). It is simple to implement, due to digital signal controllers (DSCs) that have their registers for phase angle adjustment and synchronization with other carrier waves. The carrier waves used in the PSC PWM approach have the same magnitude, frequency [52]. As depicted in Figure 9.7(a), the carrier waves have a phase angle of $2\pi/n$ between them. Incorporating the PSPWM approach into MMC will allow each submodule to contribute in a more consistent way to synthesize the output waveform. In other words, the operational powers and outputs of each submodule will be comparable. When many buses are connected, the voltage ripple on each bus will be comparable results in modularity [53].

However, as compared to the VF approach, the PSPWM methodology is not ideal for topologies such as NPC because it has a lesser sensitivity and efficacy to DC bus voltage ripples when compared to the VF method.

9.3.1.3 Level Shifted PWM (LS-PWM) Technique

Multicarrier PWM techniques are extensively utilized for MMC because they are very simple to implement in low voltage modules. Multicarrier PWM approaches are categorized as level shifted (LSPWM) and phase-shifted (PS PWM). Level Shifted Pulse Width Modulation (LS-PWM) is classified into three types: Phase Disposition PWM (PDPWM), Phase Opposition PWM (PODPWM), and Alternate Phase Opposition PWM (APODPWM).

The upper and lower triangle waves in the PDPWM approach are in phase about the zero references as represented in Figure 9.7(b). The triangle waves in the PODPWM technique will have the same frequency but has various amplitudes. The phase difference between the top and bottom triangle waves is 180 degrees as shown in Figure 9.7(c). The triangle and carrier frequencies are identical in APOD PWM; however, the carriers have a phase difference of 180 degrees between them as shown in Figure 9.7(d). Initially, level-shift PWM approaches were suggested for regulating MLCs in series with other IGBT devices, like NPC or FCMCs. When these approaches were used in cascaded topologies, they resulted in power imbalances between the various submodules. Unbalanced operation results in harmonic content being injected into the power grid and circulates current between submodules [54].

Numerous ways have been offered to address these issues, some of which are discussed here. To avoid power imbalance issues in MMC as discussed earlier, the balanced PD, POD, APOD

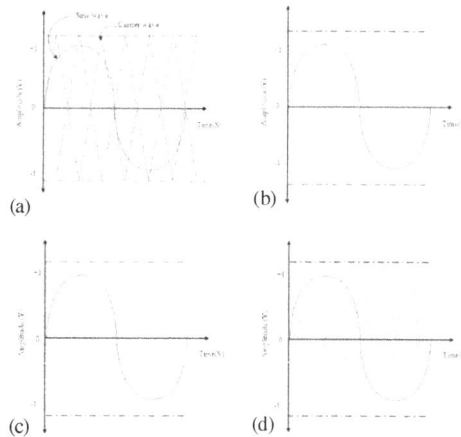

FIGURE 9.7 Various multicarrier based PWM strategy: (a) phase shifted carrier (PSC), (b) phase disposition (PD), (c) phase opposition disposition (POD), and (d) alternate phase opposition disposition (APOD).

technique proposed in [55]. This method of recombination may be used with any level shifting approach. Thus, in a short time, the contribution of each module to the synthesized waveform may be balanced. Take note that the recombination must occur across all levels, not just adjacent levels. Otherwise, the imbalance will persist as the number of layers increases and the recombination is done only on adjacent modules [56]. The enhanced PWM approach presented in [57] is intended to maximize the utilization and power quality of a grid-tied PV multilevel converter by reducing its power consumption. The suggested approach achieves a THD of 4.64% of converter output current, which complies with IEEE requirements. In addition, relative to the previous method, it reduces the converter losses significantly.

9.3.1.4 Space Vector Modulation (SVM) Method

SVM is portrayed as having major benefits in aspects of reliability, simplicity in operation, minimum harmonic current. As a result of these advantages, SVM is better suited for use in high-voltage applications, as switching redundancy and complexity with an increased number of levels [58]. The SVM has been used to regulate the output voltages of the inverter in any switching state to match the required values. The three instructions switch S1, S2, S3 in a three-phase converter provide for eight different switching state combinations. There are two feasible configurations, both of which result in zero magnitudes (zero vectors). An "active vector" is a non-null configuration. The duty cycles of the active vectors are chosen to ensure that the average output voltage vector is identical to the target voltage vector during switching periods. For example, if we assume that the required voltage is located in Division 1, then the nearby vectors are (0,0,1) and (0,1,1) as interpreted in Figure 9.8. Multilevel converters can be implemented using several different SVM algorithms. In [59], a novel optimum PWM-based SVM was developed employing sequential optimal PWM to decrease switching frequency, limiting harmonic instability, and eradication of zero sequence current.

Two- and three-stage induction motor drives supplied by NPC-MC are used to test the new approach. As a result of this suggested strategy, zero sequence components were eliminated and the switching frequency was limited to 200Hz. In addition, a new SVM technique for 5-level NPC was developed in [60] to minimize the drive system's common-mode voltage while also lowering switching losses. For a constant switching frequency, [61] presented a technique based on SVM to reduce the transition waves in the output currents. An SVM employed to a CHB MC modulates the selection of switching states for DC link voltage balancing [62]. SVM has been proposed as a new method for controlling CHB-MC [63], the prime objective of which is to prevent the converter's unnecessary high operating frequency during the converter's high, mid, and low voltage divisions.

Using redundant switching vectors, [64] presented a double SVM modulation approach of an MMC in combination with FC to regulate the capacitor voltage. This approach can be effectively scaled up with larger SMs without requiring any substantial changes. According to [65], an MMC SVM control technique might have been improved. The MMC's capacitor voltage and circulating current may be precisely controlled by exploiting this method's variable cycles and flexibility to their maximum potential. Then, [66] proposed a generalized SVM for any MMC. Compared to conventional modulation techniques, this unique extended approach provides the greatest output voltage, a better comparable switching frequency with less harmonic distortion.

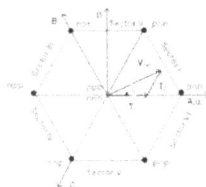

FIGURE 9.8 Space vector PWM technique.

9.3.1.5 Selective Harmonic Elimination (SHE) Technique

The modulation approach of SHE relies on voltage waveform using the Fourier transformation. Also, this method determines the switching interval to remove certain ripples from the output voltage while keeping the operating frequency constant [67]. The harmonics that must be removed in an n-level converter can be represented as ((n-3)/2). By using the equation 6 and 7 the output voltage and odd harmonic components are calculated.

$$\text{Output voltage, } V_{(\omega t)} = \sum_{n=1,3,}^{\infty} V_h \sin(n\omega t) \tag{6}$$

$$\text{Odd harmonic component, } V_h = \frac{4V_{dc}}{h\pi \sum_{i=1}^{m} \cos(h\alpha_i)} \tag{7}$$

SHEPWM primarily serves to remove some unwanted ripples from the AC output voltage of a CHB MLC, improve the input current quality, and mitigation of losses. Furthermore, in [68] proposed a unique approach to eradicate low harmonics (5th and 7th) by using a symmetric 7-level SHEPWM technique for controlling a 17-level CHB. The suggestion is considered to possess a simpler equation and can be utilized to a substantial range of switching angles. In [69], a study was conducted of controlling the MMC employing SHE with two alternative PWM schemes to run the converter at a constant switching frequency while preserving high output waveforms.

9.3.1.6 Nearest Level Modulation (NLM) Technique

Due to the higher number of voltage levels generated by the multilevel converter, the nearest-level modulation (NLM) enables straightforward implementation and computation of gating signals. Combining NLM and PWM, which modulates just one submodule per arm, can decrease the reduced current distortion caused by NLM without raising modulation complexity. Additionally, the NLM generates imbalanced capacitor voltages, which may be balanced again by rotating the modulated submodule periodically [70]. Moreover, a sequence of logical functions may be used to induce the insertion of submodules, so enhancing the capacitor voltage balance.

9.3.2 Control Techniques

Compared to several other converters, the controlling of MMC is highly complex and requires numerous control objectives as illustrated in Figure 9.9. MMC direct modulation and closed-loop control are discussed in [71]. Due to a large number of submodules in the control system, it might be a concern. In [71], shows that in these instances, open-loop control decreases the usage of communication and processing resources. Submodule failure may lead to a range of voltage and current irregularities, including order harmonics.

The MMC comprises two typically different currents, the load current as well as the circulating current. Whereas a defective submodule influences the MMC's energy balance, this issue is often addressed by the circulating current regulation. The circulating current control was employed to

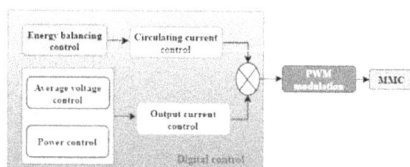

FIGURE 9.9 Digital control technique.

balance as well as divide the amount of energy evenly across the two arms and to decrease the circulating current's harmonics under asymmetric operation. In [72], proposes a CCS control capable of suppressing the circulating current's multiple frequency components under various submodule failure modes. Additionally, this strategy may assist in rapidly decreasing the instantaneous fault current generated at the instant the problematic submodule is bypassed.

9.4 TYPES OF FAULTS AND DETECTION METHODS

The chance of failure increases with the converter operating levels and passive components (primarily capacitors) in a system. The use of a significant number of power switches in circuit topologies reduces the converter reliability. The system must be resilient and fault-tolerant. Hence, the system is meant to operate continuously. In this approach, in the event of a component or submodule failure, another operational state would be employed to ensure that the system's function could continue flawlessly [15]. This section describes the types of an electrical fault in multilevel converters, fault detection (FD) techniques, and fault-tolerant (FT) methods.

9.4.1 TYPES OF ELECTRICAL FAULT

Submodules are more prone to failure in MMC architecture. Thermal fatigue is responsible for roughly 25% of SM failures, whereas semiconductor faults account for 33% of SM losses.

Open circuit (OC) and short circuit (SC) faults are two types of semiconductor failures. OC causes 18% of semiconductor defects, whereas SC causes 15%. The CHB faults are categorized into semiconductor fault, module failure or dc source failure, overvoltage, under-voltage, overcurrent, and gate drive failure [73]. MLC faults are classified and shown in Figure 9.10.

9.4.1.1 Open Circuit (OC) Fault

In MLC, destructions like bond wire lift-off caused by excessive temperature rise-mechanical strain and substantial junction temperature variations may result in OC failures. as shown in Figure 9.11(a). An OC fault can also be caused by a faulty gate driver, a broken electrical connection, or other factors. As a result, the voltages on the capacitors in the malfunctioning SM may rise significantly, which might harm the converter systems since there is no discharge circuit connection.

Furthermore, the lack of access to certain output voltage levels caused by an OC switching fault can produce severe distortions in output voltages and currents, resulting in further load stability degradations [74]. Thus, it is essential to detect and diagnose OC failures in multilevel converter so that a fault-isolation approach should be initiated in time.

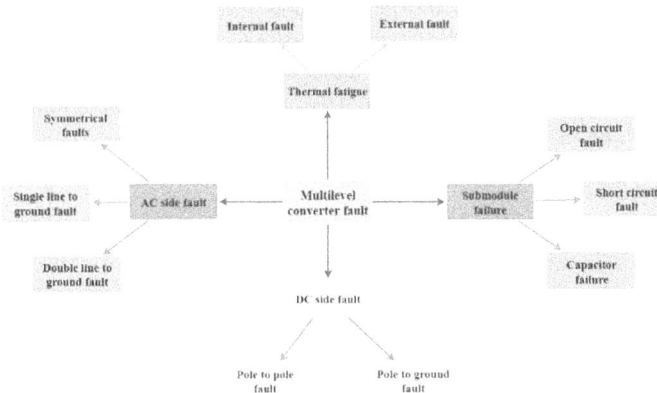

FIGURE 9.10 Faults on the multilevel converter.

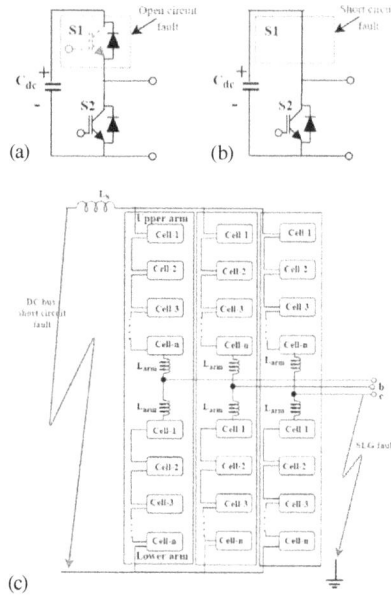

FIGURE 9.11   (a) OC fault, (b) SC fault in a submodule, (c) DC and AC side fault.

9.4.1.2   Short Circuit (SC) Fault

Multilevel converter systems are also damaged by the switching of SC faults. Due to parasitic turn-on, gate driver fault, inrush current, voltage unbalance, overheat, and other reasons, SC failures can arise in MMC systems [75] as shown in Figure 9.11(b). The SC resistance period of semiconductor devices is often < 10 μs, which necessitates the inclusion of additional devices in gate drivers to detect short-circuit switching failures.

9.4.1.3   Fault on DC Side

Half-bridge converters are not susceptible to dc short circuits or pole-to-pole faults. The fault current flowing between the converter phases and the dc bus is mostly derived by the discharging of SM capacitors when an SM fault occurs.

   Semiconductor switches are normally switched off when a fault is detected, allowing an ac source to supply fault current through anti-parallel diodes in half-bridge cells, resulting in a blocking mode. As a result of the restricted current capability of semiconductor devices or other power components, the converter will most likely be damaged if operating in an uncontrolled manner and must resist significant fault current from the ac network. It will be critical to analyse and find a dc SC fault and its post fault operation will be extremely difficult [76].

9.4.1.4   Fault on AC Side

In general, AC side faults are characterized as symmetrical or asymmetrical. Line to line (LL) and line to ground (LG) are asymmetrical faults. There are three different line-to-ground fault scenarios, depending on the location of the problem: a fault on the transformer-based system, or a fault in a transformer less system. In the case of a transformer wall bushing insulation failure, an LG fault might occur among the converter and the secondary of transformer itself. Immediately after an SLG failure occurs, ac voltage drops, and the MMCs lose power. The dc-bus voltage might also be subjected to a twice-line frequency oscillation. There is a high possibility that the MMC's power components will be subjected to additional stress due to the twice-line frequency oscillations, as well as the protective mechanism [77]. The dc side and ac side faults on the MMC structure are portrayed in Figure 9.11(c).

9.4.2 Fault Detection Methods

The fault diagnosis comprises two stages: fault detection (FD) and fault location. The fault diagnosis process should possess the following features: increasing the accuracy, low complexity, rapid diagnosis, inexpensive, non-invasive diagnosis, and maximum resilience. These thresholds will be used to conduct a study of the various fault diagnostic processes for the significant electrical faults in the CHB and MMC structures. Several fault detection methods are depicted in Figure 9.12. Hardware-based FD approaches require a higher number of sensors, increasing the system's cost and complexity. In large-scale systems, software or analytical-based FD approaches are much more effective [78]. It is further categorized into methodologies that are model-based, data-driven, and signal-processing-based. The fault indicator used in MMC topology is SM capacitor voltage, circulating current and SM output voltage likewise for CHB topology as module converter output voltage, current, and THD.

9.4.2.1 Open Circuit Fault Detection

The open-circuit fault detection technique is mainly based on non-invasive software method owing to the notion that OC faults can be identified for a longer period than SC faults, and most soft-computing FD methods are often less expensive to deploy.

The occurrences of OC switching faults may be identified using the model-based diagnostic approaches [79] by evaluating measured and estimated values of fault variables obtained from the derived computational models. Signal processing-based approaches [80] process the monitored variables in the time or spectral ranges and then compare those to predefined cut-off values. According to [81], a conventional two-step technique that relies on sliding-mode observers is presented. The uncertainty among the calculated arm currents with the predicted rates in a typical MMC arrangement is governed in the FD mode.

An isolation mode is initiated when the difference crosses a particular threshold for a specified amount of time. The defective switch will be identified if the calculations match the predicted values premised on a specific fault. The FD method based on circulating current [82] and SM capacitor voltage [83] is proposed to identify and locate the OC fault within the faulty leg. In [84], an OC fault detection approach that depends on SM voltages was presented in which the load voltages of a particular SM are examined with the SM gating signals. Because measuring the output voltages of SM is typically unnecessary for basic MMC control.

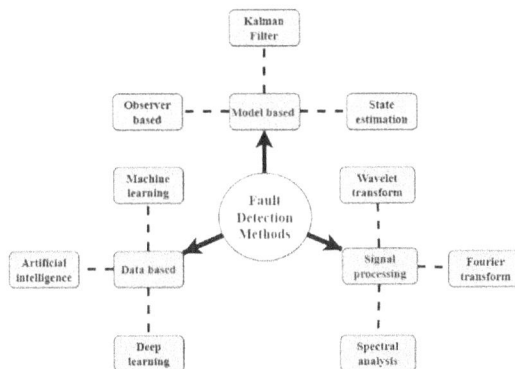

FIGURE 9.12 Fault detection methods in multilevel converter.

For CHB multilevel converter, a model-based fault diagnostic approach was devised in [85] by acquiring an observation of the phase current and computing a residual. As a result, it is an accurate approach for defect identification but adds to the system's mathematical complexity. The parameter estimating approach [86] defines a threshold value; if the real-time parameters surpass this value, a fault has emerged; otherwise, the system is assessed in fault-free mode. The converter's accuracy, on the other hand, is fully dependent on the precision of the model parameters used to develop the technique for parameter estimation. In [87], a method for diagnosing open circuit faults in a generalized CHB converter was proposed. This system gives bridge voltage data, as well as half-cycle mean voltage (HCMV) information, which is used to locate OC faults. Using a fault identifier, the issue may be diagnosed within one cycle of the output voltage, significantly reducing the amount of processing required.

9.4.2.2 Short Circuit Fault Detection

When an SC failure occurs, it is important to utilize a bypass mechanism to remove the failed SM from the circuitry. The fastest identification rate (i.e., a few µs) is the most critical requirement for the diagnosis of SC faults since a prolonged duration of switching SC faults would cause severe thermal strain on power equipment as a result of the overcurrent. For this reason, rather than relying on online FD methods for quick protection, short-circuit detection is often implemented in hardware circuits such as gate driver hardware circuits [88].

A recent study assessed gate voltage and current changes in SM to determine the presence of an SC failure [89]–[91]. These protection techniques have a high rate of FD and are thus preferred to be included in a gate control circuit. Nonetheless, these methods necessitate complex detection circuitry and are susceptible to electromagnetic interference and diode reverse-recovery current, making them unfeasible for practical applications. Several approaches for fault identification in CHB are suggested in [92], some of them are based on artificial intelligence (AI). Using spectrum analysis of the phase output voltage, the author of [93] suggested a method for identifying faults. According to [94], the SC fault detection in the CHB converter needs just one additional sensor per phase.

However, it is more difficult to identify a faulty cell in a multilevel converter because it is important to identify the defective module in the faulty phase also so that the appropriate fault-tolerant converter and controller reconfiguration should be used. An optimization approach such as a genetic algorithm may be used to train the neural network (NN) and obtain a minimal classification error in [95] to identify OC and SC faults with a diagnostic accuracy of 150ms.

9.4.2.3 DC Bus Short Circuit Fault Detection

A DC bus SC fault diagnosis approach is provided in [96], which provides a 1ms response to detecting the fault. Considering the propagation period of the signal over 200km, the mechanism assures that the protection is engaged after 2ms. Simple and affordable solutions for transient DC bus failures have been investigated and suggested [97].

In [98], the DC is monitored individually to increase the reliability of FD, although this takes a longer period and the precision of the fault classification is a challenge. Yet another way of diagnostic technique relies on the derivative of direct current voltage described in [99] for rapidly detecting and localizing direct current problems in grid systems. However, the arm inductor's effect was neglected, and it was considered that perhaps the converter output power remained constant instantly after a DC failure.

9.4.2.4 AC Side Fault Detection

A line-to-ground fault can be diagnosed by monitoring the fluctuations in frequency ripples in a dc voltage. Symmetric portions of the arm voltage and current of MMC may also detect a line to ground fault because of their unbalanced properties after an SLG fault occurs.

An SLG fault is detected by monitoring the dc-link overvoltage owing to the frequency instability of a dual-line and the sudden SM voltage rise [100]. The comparison of various fault detection

TABLE 9.2
Comparison of Various Fault Detection Methods for CHB and MMC Topology

Reference	Topology	Parameters used	FD method	Fault type	FD time	Multiple FD ability
[82]	HB-MMC	Arm current, capacitor voltage	Model based	OC	60ms	N/A
[83]	HB-MMC	Arm current, capacitor voltage	Model-based	OC	235ms	√
[85]	CHB	Phase current	Model-based	OC	100µs	√
[87]	CHB	HCMV	Model based	OC	1fund.cycle	N/A
[89]	HB-MMC	Capacitor voltage	Signal processing method	OC & SC	6ms	√
[93]	CHB	Output voltage	Model-based	SC	Few milliseconds	N/A
[94]	FB-MMC	Output voltage	Signal processing method	SC	50ms	N/A
[95]	HB-MMC	Output voltage	Neural network	OC and SC	150ms	N/A
[99]	HB-MMC	DC current	Model-based	Pole to pole	10ms	N/A

methods with emphasis on their diagnostic time is elucidated in Table 9.2. Due to the asymmetry of the SLG fault, [101] uses "negative-sequence components" and the doubling frequency variation in the active power as a diagnostic approach for SLG failures.

9.5   FAULT-TOLERANT METHODS

Fault-tolerant control of a multilevel converter is the capacity to continue its operation without substantial loss of performance despite the failure of a finite number of its composing sub-units. The fault-tolerant method for OC and SC fault is almost used the same methods are categorized into (a) redundancy in switching states, (b) hardware redundancy.

9.5.1   Fault-Tolerant Method for CHB Converter

The fault-tolerant control for CHB converter is based on hardware redundancy is presented in [102] to tolerate OC failure utilizing extra relay in each circuit. This makes the circuit complicated and more costly. Even though, in the reconfiguration process, it was noticed that the healthier switches had greater blocking voltages, resulting in increased switching stress. Alternatively, an auxiliary module may be used to compensate for the limits of the relay to create a stable output voltage As demonstrated in [103], in the event of system failure or partial shading, an ANN-based CHBMLI fault-tolerant system can distribute power to the load. OC and SC abnormalities (in MLI and PV systems) may be tolerated by the suggested method (partial shading). Two switches control the functioning of an auxiliary module in this technique. With both [103] and [104], the topology recommended increases weight, price, and inverter loss. An entirely new level-shifted carrier PWM technique has been developed in [105] that exploits the uneven (faulty) voltage level to supply emergency loads. Both CHB converter modes of operation are supported, as well as fault tolerance.

Reduced power loss and improved inverter performance may be achieved by using fewer conductive switches at all voltage levels. Aside from that, the proposed architecture enables the production of voltage levels that are missing from the output. This study presents a fault-tolerant method based on Fundamental Phase Shift Compensation PWM (FPSC-PWM) in [106]. This approach involves modifying both the phase voltages of the converter and the angles by varying the phase angles of the carrier pulses. This approach, on the other hand, cannot be used to MLCs that are regulated using the Space Vector Modulation (SVM) technique [107].

9.5.2   FAULT-TOLERANT METHOD FOR MMC CONVERTER

9.5.2.1   Using Redundant Switching States

In [84], a fault-tolerant control technique called "seamlessly ride-through" was presented in which the PWM inputs are updated online to compensate for voltage perturbations produced by OC failures while the defective SM is being localized. Once the defective SM has been located, it is segregated by shutting the bypass switch, and the associated fault indicator is removed. Following that, the fault-tolerant mode of operation will be enabled. In [108], a new fault-tolerant technique was devised by altering capacitor reference voltages (V_{Cref}) and PWM signals in the defective phase of MMC while the other phase was operating properly. This innovative technology may result in malfunctioning leg capacitors being overcharged. However, because the ratio of SM is often large, this overvoltage may be ignored.

9.5.2.2   Using Redundant Hardware

In [109], redundant SM are introduced to MMC to offer redundancy in the event of an SM failure. Therefore, it should be noticed that by shutting the bypass switches, these additional SMs are placed in a "cold reserve state". When SM systems fail, it is necessary to remove the affected SMs.

In [110], proposes a revolutionary spare method for emergencies in which the quantity of broken SM cells outweighs the spare units. Because structural redundancy may result in an unequal distribution of power among their modules, a mitigation method is implemented to maintain energy balance while preventing undesirable switching stress. The faulty switch bypassing mechanism for HB and FB submodule is represented in Figure 9.13(a) and (b).

9.5.2.3   DC Side Fault-Tolerant Methods

Since the SC failure on the dc bus creates a large amount of fault current, it is required to isolate the dc bus through a circuit breaker (CB). Nevertheless, the absence of zero intersection point in the DC and the sluggish protection response complicates the use of traditional mechanical breakers, despite the advantage of very low on-state conduction losses [111]. Hence, the literature proposes two kinds of solutions: hybrid dc breaker and solid-state circuit breaker (SSCB). Other methods for isolating and tolerating dc short-circuit faults in MMCs include the use of FBSM rather than HBSM [98]. The FBSMs may be designed to inject capacitor voltages with the reverse voltage, which could be utilized to restrict or completely inhibit the I_f. Moreover, this necessitates the increased use of semiconductor switches which increases the cost and complexity. The hybrid MMC [112] can also be used to suppress the inrush current which blocks dc fault current when all semiconductor switches are off. In [113], describes a CDSM structure that consists of two cascaded HBSM, a switch, and two diodes. This technique has higher accuracy of resolving dc faults, however, the accompanying device losses and overall costs will be raised.

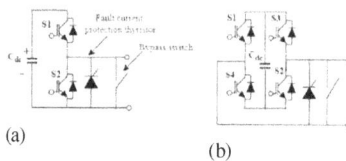

(a)   (b)

FIGURE 9.13   Bypass structure realization for (a) HBSM and (b) FBSM.

9.5.2.4   AC Side Fault-Tolerant Methods

Unwanted ac voltage sag results in a single line to ground fault, lowering the instantaneous power. The fault-tolerant technique, when combined with negative-sequence and zero-sequence current regulation proven in [114] to significantly improve the MMC's entire fault mitigation capabilities without incurring extra costs. In [115], proposed a feedback/feed-forward control technique for improving the voltage regulation of MMCs in uneven grid circumstances, thus improving the fault isolation possibility despite asymmetrical failures like a line to ground faults. In [116], voltage ripples in FBMMC during line to ground fault situations are suppressed by the current injection approach.

9.6   SUMMARY AND FUTURE DIRECTIONS

The evolution of power converters and other related technologies has led to an increase in the utilization of renewable energy systems in the power grid. Even though it's intermittent, this has raised several issues about the quality of electricity, protection, energy storage, and system dependability besides these issues.

Also, for this reason, numerous protocols have been created for grid-interfacing RES to guarantee that electricity quality on the grid remains stable. From the examined literature, the following areas in this respect demand more research.

- The fluctuating nature of the electricity generated by renewable energy continues to be a significant challenge with the grid-connected system. As the significance of renewable energy to the worldwide energy system is forecast to grow in the coming decades, this issue of power variations in grid integrated systems must be addressed.
- Compared with other multilevel converters, the CHB and MMC are most suitable for solar PV applications since it has direct interconnection with PV, CHB converter is highly recommended for PV integrated system.
- Even though the CHB and FBMMC are capable of bidirectional power flow, the raised switch count increases the system cost and failure rate. As a corollary, it is essential to examine reduced switch converter topologies without sacrificing performance.
- Additionally, power balancing between each phase remains a challenge for large-scale PV integration with CHB, owing to PV's intermittent nature. Thus, further study is needed to determine to maintain a balanced power distribution across each phase under both normal and abnormal conditions.
- The integration of the MMC system into the existing power system would greatly increase system reliability and performance, facilitate the integration of renewable energy, and flexibility of power transmission. However, still, significant technological complexities are remaining in the MMC system modelling, control, and protection.
- In MMC, the elimination of circulating current is amongst the most difficult, yet several techniques have been offered in the studies, none of which give a complete solution. Indeed, greater attention should be given to digital control systems to achieve better responses and more reliable performance.
- Another significant problem for MMC is balancing the SM capacitor voltage, which can be regulated with the proper design and control method.
- Additional research should be conducted on the effectiveness analysis of modulation approaches for MC to alleviate the various issues including power quality enhancement, harmonic mitigation, and voltage instability in grid integrated RES.
- According to the relatively high switch count in the MC, the fault-tolerant control structure should be equipped with SM topologies.
- Online-based fault detection and monitoring system has received more attention among researchers to furnish the complete solution for fault diagnosis and fault tolerance.

- Although, still there is a need for more research in the fault detection methods to locate the multiple switch failures among submodules.
- Overall, there is still a possible research gap in determining the accurate location of the fault, as well as its reliable fault-tolerant techniques for the multilevel converter, and the arena will see significant advancements in the upcoming years.

9.7 CONCLUSION

Multilevel converters (MLC) have gained more attention in medium-voltage and high-power applications due to their recent achievements in the industrial and commercial sectors. This chapter performs a comprehensive review of the emerging MLC topologies like CHB and MMC on various aspects. This chapter has reviewed the present state of the art of various submodules of MMC, CHB topologies, modulation, and control techniques for solar photovoltaic (PV) applications. This chapter also scrutinized the main issues and challenges encountered by the CHB and MMC topologies while integrating with large-scale PV systems. Also, the deep analysis of various electrical faults of MLC, and their fault detection and fault-tolerant schemes were thoroughly examined in the contemporary for the future advancements of high-reliability multilevel converters. The fault-tolerant MLC for photovoltaic integrated systems with power balancing technologies might be examined as a possible future scope.

REFERENCES

[1] N. Priyadarshi, S. Padmanaban, P. Kiran Maroti, and A. Sharma. An Extensive Practical Investigation of FPSO-Based MPPT for Grid Integrated PV System under Variable Operating Conditions with Anti-Islanding Protection. *IEEE Syst. J.*, vol. 13, no. 2, pp. 1861–1871, 2019, doi: 10.1109/JSYST.2018. 2817584.

[2] A. Salem, H. Van Khang, K. G. Robbersmyr, M. Norambuena, and J. Rodriguez. Voltage Source Multilevel Inverters with Reduced Device Count: Topological Review and Novel Comparative Factors. *IEEE Trans. Power Electron.*, vol. 36, no. 3, pp. 2720–2747, 2021, doi: 10.1109/TPEL.2020.3011908.

[3] S. Du, B. Wu, N. R. Zargari, and Z. Cheng. A Flying-Capacitor Modular Multilevel Converter for Medium-Voltage Motor Drive. *IEEE Trans. Power Electron.*, vol. 32, no. 3, pp. 2081–2089, 2017, doi: 10.1109/TPEL.2016.2565510.

[4] J. H. Lee, and K. B. Lee. A Fault Detection Method and a Tolerance Control in a Single-Phase Cascaded H-bridge Multilevel Inverter. *IFAC-PapersOnLine*, vol. 50, no. 1, pp. 7819–7823, 2017, doi: 10.1016/j. ifacol.2017.08.1058.

[5] N. Priyadarshi, S. Padmanaban, M. S. Bhaskar, F. Blaabjerg, and A. Sharma. Fuzzy SVPWM-based inverter control realisation of grid integrated photovoltaic wind system with fuzzy particle swarm optimization maximum power point tracking algorithm for a grid connected PV/wind power generation system: Harware implementation. *IET Electr. Power Appl. Res.*, vol. 12, no. 7, pp. 962–971, 2018, doi: 10.1049/ iet-epa.2017.0804.

[6] F. Azam, A. Biradar, N. Priyadarshi, S. Kumari, and S. Tangade. A Review of Blockchain Based Approach for Secured Communication in Internet of Vehicle (IoV) Scenario. *Int. Conf. Smart Technol. Comput. Electr. Electron.*, no. ii, pp. 1–6, 2022, doi: 10.1109/icstcee54422.2021.9708555.

[7] T. S. Basu, S. Maiti, and C. Chakraborty. A hybrid modular multilevel converter for medium voltage variable speed motor drives. *2016 IEEE 7th Power India Int. Conf. PIICON 2016*, vol. 32, no. 6, pp. 4619–4630, 2017, doi: 10.1109/POWERI.2016.8077444.

[8] N. Priyadarshi, S. Padmanaban, L. Mihet-Popa, F. Blaabjerg, and F. Azam. Maximum power point tracking for brushless DC motor-driven photovoltaic pumping systems using a hybrid ANFIS-FLOWER pollination optimization algorithm. *Energies*, vol. 11, no. 5, 2018, doi: 10.3390/en11051067.

[9] N. Priyadarshi, V. K. Ramachandaramurthy, S. Padmanaban, and F. Azam. An ant colony optimized mppt for standalone hybrid pv-wind power system with single cuk converter. *Energies*, vol. 12, no. 1, 2019, doi: 10.3390/en12010167.

[10] I. Marzo, A. Sanchez-Ruiz, J. A. Barrena, G. Abad, and I. Muguruza. Power balancing in cascaded H-Bridge and modular multilevel converters under unbalanced operation: A review. *IEEE Access*, vol. 9, pp. 110525–110543, 2021, doi: 10.1109/ACCESS.2021.3103337.

[11] S. Padmanaban, N. Priyadarshi, M. S. Bhaskar, J. B. Holm-Nielsen, E. Hossain, and F. Azam. A hybrid photovoltaic-fuel cell for grid integration with jaya-based maximum power point tracking: Experimental performance evaluation. *IEEE Access*, vol. 7, pp. 82978–82990, 2019, doi: 10.1109/ ACCESS.2019.2924264.

[12] N. S. Hasan, N. Rosmin, D. A. A. Osman, and A. H. Musta'amal@Jamal. Reviews on multilevel converter and modulation techniques. *Renew. Sustain. Energy Rev.*, vol. 80, pp. 163–174, 2017, doi: 10.1016/j. rser.2017.05.163.

[13] A. Nami, J. Liang, F. Dijkhuizen, G. D. Demetriades. Modular multilevel converters for HVDC applications: Review on converter cells and functionalities. *IET Power Electronics*, vol. 12, no. 2, pp. 170–183, 2019, doi: 10.1109/TPEL.2014.2327641.

[14] F. Azam, A. Biradar, N. Priyadarshi, S. Kumari, D. Almakhles, and S. Tangade. A framework for secured dissemination of messages in Internet of Vehicle (IoV) using blockchain approach. pp. 1–6, 2022, doi: 10.1109/icmnwc52512.2021.9688397.

[15] P. Tu, S. Yang, and P. Wang. Reliability and cost-based redundancy design for modular multilevel converter. *IEEE Trans. Ind. Electron.*, vol. 66, no. 3, pp. 2333–2342, 2019, doi: 10.1109/TIE.2018.2793263.

[16] P. Rajan, and S. Jeevananthan. A new partially isolated hybrid output of multiport multilevel converter for photovoltaic based power supplies. *J. Energy Storage*, vol. 45, p. 103436, 2022, doi: 10.1016/j. est.2021.103436.

[17] K. Wang, Z. Zheng, Y. Li, K. Liu, and J. Shang. Neutral-point potential balancing of a five-level active neutral-point-clamped inverter. *IEEE Trans. Ind. Electron.*, vol. 60, no. 5, pp. 1907–1918, 2013, doi: 10.1109/TIE.2012.2227898.

[18] E. Babaei, M. F. Kangarlu, M. Sabahi, and M. R. A. Pahlavani. Cascaded multilevel inverter using sub-multilevel cells. *Electr. Power Syst. Res.*, vol. 96, pp. 101–110, 2013, doi: 10.1016/j.epsr.2012.10.010.

[19] Y. Yu, G. Konstantinou, B. Hredzak, and V. G. Agelidis. Operation of cascaded H-bridge multilevel converters for large-scale photovoltaic power plants under bridge failures. *IEEE Trans. Ind. Electron.*, vol. 62, no. 11, pp. 7228–7236, 2015, doi: 10.1109/TIE.2015.2434995.

[20] G. Farivar, B. Hredzak, and V. G. Agelidis. A DC-side sensorless cascaded H-bridge multilevel converter-based photovoltaic system. *IEEE Trans. Ind. Electron.*, vol. 63, no. 7, pp. 4233–4241, 2016, doi: 10.1109/TIE.2016.2544243.

[21] A. Lashab, D. Sera, J. Martins, and J. M. Guerrero. Multilevel DC-link converter-based photovoltaic system with integrated energy storage. *Proc. 2018 5th Int. Symp. Environ. Energies Appl. EFEA 2018*, 2019, doi: 10.1109/EFEA.2018.8617110.

[22] W. Zhao, H. Choi, G. Konstantinou, M. Ciobotaru, and V. G. Agelidis. Cascaded H-bridge multilevel converter for large-scale PV grid-integration with isolated DC-DC stage. *Proc. — 2012 3rd IEEE Int. Symp. Power Electron. Distrib. Gener. Syst. PEDG 2012*, pp. 849–856, 2012, doi: 10.1109/PEDG.2012.6254100.

[23] S. Jain, and V. Sonti. A highly efficient and reliable inverter configuration based cascaded multilevel inverter for PV systems. *IEEE Trans. Ind. Electron.*, vol. 64, no. 4, pp. 2865–2875, 2017, doi: 10.1109/ TIE.2016.2633537.

[24] A. Lashab et al. Cascaded multilevel PV inverter with improved harmonic performance during power imbalance between power cells. *IEEE Trans. Ind. Appl.*, vol. 56, no. 3, pp. 2788–2798, 2020, doi: 10.1109/TIA.2020.2978164.

[25] A. Alexander, and M. Thathan. Modelling and analysis of modular multilevel converter for solar photovoltaic applications to improve power quality. *IET Renew. Power Gener.*, vol. 9, no. 1, pp. 78–88, 2015, doi: 10.1049/iet-rpg.2013.0365.

[26] A. Lesnicar, and R. Marquardt. An innovative modular multilevel converter topology suitable for a wide power range. *IEEE Trans. Electron Devices*, vol. 33, no. 10, pp. 1511–1517, 2003, doi: 10.1109/T-ED.1986.22701.

[27] S. Kouro et al. Recent advances and industrial applications of multilevel converters. *IEEE Trans. Ind. Electron.*, vol. 57, no. 8, pp. 2553–2580, 2010, doi: 10.1109/TIE.2010.2049719.

[28] A. Yadav, S. N. Singh, and S. P. Das. Modular multi-level converter topologies: Present status and key challenges. *2017 4th IEEE Uttar Pradesh Sect. Int. Conf. Electr. Comput. Electron. UPCON 2017*, vol. 2018-Janua, pp. 280–288, 2017, doi: 10.1109/UPCON.2017.8251061.

[29] M. Kurtoğlu, F. Eroğlu, A. O. Arslan, and A. M. Vural. Modular multilevel converters: A study on topology, control and applications. *2018 5th Int. Conf. Electr. Electron. Eng. ICEEE 2018*, vol. 2, no. c, pp. 79–84, 2018, doi: 10.1109/ICEEE2.2018.8391305.

[30] E. Solas, G. Abad, J. A. Barrena, S. Aurtenetxea, A. Cárcar, and L. Zaj. Modular multilevel converter with different submodule concepts—Part I : Capacitor voltage balancing method. *IEEE Transactions on Industrial Electronics,* vol. 60, no. 10, pp. 4525–4535, 2013, doi: 10.1109/TIE.2012.2210378.

[31] H. Nademi, A. Das, R. Burgos, and L. E. Norum. A new circuit performance of modular multilevel inverter suitable for photovoltaic conversion plants. *IEEE J. Emerg. Sel. Top. Power Electron.*, vol. 4, no. 2, pp. 393–404, 2016, doi: 10.1109/JESTPE.2015.2509599.

[32] G. Konstantinou, J. Zhang, S. Ceballos, J. Pou, and V. G. Agelidis. Comparison and evaluation of sub-module configurations in modular multilevel converters. *Proc. Int. Conf. Power Electron. Drive Syst.*, vol. 2015, pp. 958–963, 2015, doi: 10.1109/PEDS.2015.7203440.

[33] A. Dekka, B. Wu, R. L. Fuentes, M. Perez, and N. R. Zargari. Evolution of topologies, modeling, control schemes, and applications of modular multilevel converters. *IEEE J. Emerg. Sel. Top. Power Electron.*, vol. 5, no. 4, pp. 1631–1656, 2017, doi: 10.1109/JESTPE.2017.2742938.

[34] J. Fang, F. Blaabjerg, S. Liu, and S. Goetz. A review of multilevel converters with parallel connectivity. *IEEE Trans. Power Electron.*, vol. 36, no. 11, pp. 12468–12489, 2021, doi: 10.1109/TPEL.2021.3075211.

[35] Y. Yu, G. Konstantinou, B. Hredzak, and V. G. Agelidis. Power balance of cascaded H-Bridge multilevel converters for large-scale photovoltaic integration. *IEEE Trans. Power Electron.*, vol. 31, no. 1, pp. 292–303, 2016, doi: 10.1109/TPEL.2015.2406315.

[36] Y. Yu et al. Delta-Connected cascaded H-bridge multilevel converters for large-scale photovoltaic grid integration. *IEEE Transactions on Industrial Electronics*, vol. 64, no. 11, pp. 8877–8886, 2017, doi: 10.1109/TIE.2016.2645885.

[37] B. Xiao, L. Hang, J. Mei, C. Riley, L. M. Tolbert, and B. Ozpineci. Modular cascaded H-Bridge multi-level PV inverter with distributed MPPT for grid-connected applications. *IEEE Trans. Ind. Appl.*, vol. 51, no. 2, pp. 1722–1731, 2015, doi: 10.1109/TIA.2014.2354396.

[38] C. Gao, X. Liu, J. Liu, Y. Guo, and Z. Chen. Multilevel converter with capacitor voltage actively balanced using reduced number of voltage sensors for high power applications. *IET Power Electron.*, vol. 9, no. 7, pp. 1462–1473, 2016, doi: 10.1049/iet-pel.2015.0073.

[39] Y. Yu, G. Konstantinou, B. Hredzak, and V. G. Agelidis. On extending the energy balancing limit of multilevel cascaded H-bridge converters for large-scale photovoltaic farms. *2013 Australas. Univ. Power Eng. Conf. AUPEC 2013*, 2013, doi: 10.1109/aupec.2013.6725385.

[40] Z. Ye et al. A novel DC-power control method for cascaded H-Bridge multilevel inverter. *IEEE Trans. Ind. Electron.*, vol. 64, no. 9, pp. 6874–6884, 2017, doi: 10.1109/TIE.2017.2686798.

[41] M. M. Harin, V. Vanitha, and M. Jayakumar. Comparison of PWM techniques for a three level modular multilevel inverter. *Energy Procedia*, vol. 117, pp. 666–673, 2017, doi: 10.1016/j.egypro.2017.05.180.

[42] J. M. Kharade, and A. R. Thorat. Simulation of an alternate arm modular multilevel converter with over-lap angle control for capacitor voltage balancing. *2015 Int. Conf. Ind. Instrum. Control. ICIC 2015*, no. Icic, pp. 502–506, 2015, doi: 10.1109/IIC.2015.7150794.

[43] M. Abdelsalam, M. Marei, S. Tennakoon, and A. Griffiths. Capacitor voltage balancing strategy based on sub-module capacitor voltage estimation for modular multilevel converters. *CSEE J. Power Energy Syst.*, vol. 2, no. 1, pp. 65–73, 2016, doi: 10.17775/cseejpes.2016.00010.

[44] P. H. Riley, O. Dordevic, K. Pullen, L. DeLilo, and M. De Giorgio. A qualitative assessment of a modified multilevel converter topology M2LeC for lightweight low-cost electric propulsion. *Engineering*, vol. 12, no. 7, pp. 496–515, 2020, doi: 10.4236/eng.2020.127035.

[45] Y. Li, and F. Wang. Arm inductance selection principle for modular multilevel converters with circulating current suppressing control. *Conf. Proc.—IEEE Appl. Power Electron. Conf. Expo.—APEC*, pp. 1321–1325, 2013, doi: 10.1109/APEC.2013.6520470.

[46] K. Ilves, A. Antonopoulos, S. Norrga, and H. P. Nee. Steady-state analysis of interaction between harmonic components of arm and line quantities of modular multilevel converters. *IEEE Trans. Power Electron.*, vol. 27, no. 1, pp. 57–68, 2012, doi: 10.1109/TPEL.2011.2159809.

[47] B. Bahrani, S. Debnath, and M. Saeedifard. Circulating current suppression of the modular multilevel converter in a double-frequency rotating reference frame. *IEEE Trans. Power Electron.*, vol. 31, no. 1, pp. 783–792, 2016, doi: 10.1109/TPEL.2015.2405062.

[48] J. Wang, X. Han, H. Ma, and Z. Bai. Analysis and injection control of circulating current for modular multilevel converters. *IEEE Trans. Ind. Electron.*, vol. 66, no. 3, pp. 2280–2290, 2019, doi: 10.1109/TIE.2018.2808901.

[49] K. B, Senbakaraj, Periyasamy, and Poongkabilan. THD reduction in multi level inverters based on multi-carrier pulse width modulation technique. *Int. J. Eng. Adv. Technol.*, vol. 9, no. 4, pp. 1970–1977, 2020, doi: 10.35940/ijeat.d8994.049420.

[50] Z. Shu, X. He, Z. Wang, D. Qiu, and Y. Jing. Voltage balancing approaches for diode-clamped multi-level converters using auxiliary capacitor-based circuits. *IEEE Trans. Power Electron.*, vol. 28, no. 5, pp. 2111–2124, 2013, doi: 10.1109/TPEL.2012.2215966.

[51] M. K. Sheikh, and P. M. Meshram. Performance analysis of modular multilevel converter based on modi-fied flying-capacitor multicell converter using SPWM. *2015 Int. Conf. Ind. Instrum. Control. ICIC 2015*, no. Icic, pp. 1555–1560, 2015, doi: 10.1109/IIC.2015.7150997.

[52] E. J. Lee, S. M. Kim, and K. B. Lee. Modified phase-shifted PWM scheme for reliability improvement in cascaded H-bridge multilevel inverters. *IEEE Access*, vol. 8, pp. 78130–78139, 2020, doi: 10.1109/ACCESS.2020.2989694.

[53] A. António-Ferreira, C. Collados-Rodríguez, and O. Gomis-Bellmunt. Modulation techniques applied to medium voltage modular multilevel converters for renewable energy integration: A review. *Electr. Power Syst. Res.*, vol. 155, pp. 21–39, 2018, doi: 10.1016/j.epsr.2017.08.015.

[54] K. Deepa, P. Savitha, and B. Vinodhini. Harmonic analysis of a modified cascaded multilevel inverter. *IEEE, 1st International Conference on Electrical Energy Systems*, pp. 92–97, 2011.

[55] P. Sochor, and H. Akagi. Theoretical and experimental comparison between Phase-shifted PWM and lev-el-shifted PWM in a modular multilevel SDBC inverter for utility-scale photovoltaic applications. *IEEE Trans. Ind. Appl.*, vol. 53, no. 5, pp. 4695–4707, 2017, doi: 10.1109/TIA.2017.2704539.

[56] S. Haq, S. P. Biswas, M. K. Hosain, and M. R. I. Sheikh. Performance analysis of switching techniques in modular multilevel converter fed induction motor. *2019 4th Int. Conf. Electr. Inf. Commun. Technol. EICT 2019*, pp. 20–22, 2019, doi: 10.1109/EICT48899.2019.9068802.

[57] S. Haq et al. An advanced PWM technique for MMC inverter based grid-connected photovoltaic systems. *IEEE Trans. Appl. Supercond.*, vol. 31, no. 8, pp. 8–12, 2021, doi: 10.1109/TASC.2021.3094439.

[58] V. Jayakumar, B. Chokkalingam, and J. L. Munda. A comprehensive review on space vector modulation techniques for neutral point clamped multi-level inverters. *IEEE Access*, vol. 9, pp. 112104–112144, 2021, doi: 10.1109/ACCESS.2021.3100346.

[59] A. Edpuganti, and A. K. Rathore. New optimal pulsewidth modulation for single DC-link dual-in-verter fed open-end stator winding induction motor drive. *IEEE Trans. Power Electron.*, vol. 30, no. 8, pp. 4386–4393, 2015, doi: 10.1109/TPEL.2014.2353415.

[60] C. Tan, D. Xiao, and J. E. Fletcher. An improved space vector modulation strategy for three-level five-phase neutral-point-clamped inverters. *2015 17th Eur. Conf. Power Electron. Appl. EPE-ECCE Eur. 2015*, 2015, doi: 10.1109/EPE.2015.7309106.

[61] S. Choi, and M. Saeedifard. Capacitor voltage balancing of flying capacitor multilevel converters by space vector PWM. *IEEE Trans. Power Deliv.*, vol. 27, no. 3, pp. 1154–1161, 2012, doi: 10.1109/TPWRD.2012.2191802.

[62] A. K. Morya, A. Shukla, and S. Doolla. Control of grid connected cascaded H-bridge multilevel converter during grid voltage unbalance for photovoltaic application. *IECON Proc. (Industrial Electron. Conf.*, pp. 7990–7995, 2013, doi: 10.1109/IECON.2013.6700468.

[63] S. Mekhilef, and M. N. Abdul Kadir. Novel vector control method for three-stage hybrid cascaded multilevel inverter. *IEEE Trans. Ind. Electron.*, vol. 58, no. 4, pp. 1339–1349, 2011, doi: 10.1109/TIE.2010.2049716.

[64] A. Dekka, B. Wu, N. R. Zargari, and R. L. Fuentes. A space-vector PWM-based voltage-balancing approach with reduced current sensors for modular multilevel converter. *IEEE Trans. Ind. Electron.*, vol. 63, no. 5, pp. 2734–2745, 2016, doi: 10.1109/TIE.2016.2514346.

[65] Y. Deng, Y. Wang, K. H. Teo, and R. G. Harley. Space vector modulation method for modular mul-tilevel converters. *IECON Proc. (Industrial Electron. Conf.*, pp. 4715–4721, 2014, doi: 10.1109/IECON.2014.7049213.

[66] Y. Deng, M. Saeedifard, and R. G. Harley. An optimized control strategy for the modular multilevel con-verter based on space vector modulation. *Conf. Proc.—IEEE Appl. Power Electron. Conf. Expo.—APEC*, vol. 2015, pp. 1564–1569, 2015, doi: 10.1109/APEC.2015.7104555.

[67] M. Fakhry, A. Massoud, and S. Ahmed. Quasi seven-level operation of multilevel converters with selec-tive harmonic elimination. *Proc. Int. Conf. Microelectron. ICM*, vol. 2015, pp. 216–219, 2014, doi: 10.1109/ICM.2014.7071845.

[68] G. Konstantinou, M. Ciobotaru, and V. Agelidis. Selective harmonic elimination pulse-width modulation of modular multilevel converters. *IET Power Electron.*, vol. 6, no. 1, pp. 96–107, 2013, doi: 10.1049/iet-pel.2012.0228.

[69] S. R. Pulikanti, G. Konstantinou, and V. G. Agelidis. DC-link voltage ripple compensation for multilevel active-neutral-point- clamped converters operated with SHE-PWM. *IEEE Trans. Power Deliv.*, vol. 27, no. 4, pp. 2176–2184, 2012, doi: 10.1109/TPWRD.2012.2209207.

[70] S. M. Kim, M. G. Jeong, J. Kim, and K. B. Lee. Hybrid Modulation scheme for switching loss reduction in a modular multilevel high-voltage direct current converter. *IEEE Transactions on Power Electronics*, vol. 34, no. 4, pp. 3178–3191, 2018, doi: 10.1109/TPEL.2018.2848620.

[71] D. Siemaszko, A. Antonopoulos, K. Ilves, M. Vasiladiotis, L. Ängquist, and H. P. Nee. Evaluation of control and modulation methods for modular multilevel converters. *2010 Int. Power Electron. Conf.—ECCE Asia -, IPEC 2010*, pp. 746–753, 2010, doi: 10.1109/IPEC.2010.5544609.

[72] W. Wu, X. Wu, J. Yin, L. Jing, S. Wang, and J. Li. Characteristic analysis and fault-tolerant control of circulating current for modular multilevel converters under sub-module faults. *Energies*, vol. 10, no. 11, 2017, doi: 10.3390/en10111827.

[73] N. K. Dewangan, T. K. Tailor, R. Agrawal, P. Bhatnagar, and K. K. Gupta. A multilevel inverter structure with open circuit fault-tolerant capability. *Electr. Eng.*, no. 0123456789, 2021, doi: 10.1007/s00202-020-01149-6.

[74] A. Anand, A. Vinayak B, R. Nithin, and J. Raj. A generalized switch fault diagnosis for cascaded h-bridge multilevel inverters using mean voltage prediction. *IEEE Transactions on Industry Applications*, vol. 56, no. 2, pp. 1563–1574, 2020, doi: 10.1109/TIA.2019.2959540.

[75] H. Wang. A Short-Circuit Fault-Tolerant Strategy for Three- Phase Four-Wire Flying Capacitor Three-Level Inverters. *IEEE 10th Int. Symp. Power Electron. Distrib. Gener. Syst.*, no. 51777189, pp. 781–786, 2019, doi: 10.1109/PEDG.2019.8807759.

[76] H. O. A. Ahmed, Y. Yu, Q. Wang, M. Darwish, and A. K. Nandi. Intelligent Fault diagnosis framework for modular multilevel converters in HVDC transmission. *Sensors*, vol. 22, no. 1, 2022, doi: 10.3390/s22010362.

[77] S. Cui, H. J. Lee, J. J. Jung, Y. Lee, and S. K. Sul. A comprehensive AC-side single-line-to-ground fault ride through strategy of an MMC-based HVDC system. *IEEE J. Emerg. Sel. Top. Power Electron.*, vol. 6, no. 3, pp. 1021–1031, 2018, doi: 10.1109/JESTPE.2018.2797934.

[78] Z. Lu, Z. Chen, Y. Gong, J. Cao, and H. Wang. Sub-module fault analysis and fault-tolerant control strategy for modular multilevel converter. *IET Conf. Publ.*, vol. 2016, no. CP696, 2016, doi: 10.1049/cp.2016.0482.

[79] S. Shao, P. W. Wheeler, J. C. Clare, and A. J. Watson. Fault detection for modular multilevel converters based on sliding mode observer. *IEEE Trans. Power Electron.*, vol. 28, no. 11, pp. 4867–4872, 2013, doi: 10.1109/TPEL.2013.2242093.

[80] Q. Yang, J. Qin, and M. Saeedifard. Analysis, detection, and location of open-switch submodule failures in a modular multilevel converter. *IEEE Trans. Power Deliv.*, vol. 31, no. 1, pp. 155–164, 2016, doi: 10.1109/TPWRD.2015.2477476.

[81] Y. Zhang, X. Cheng, Z. Liu, Y. Zheng, and C. Cheng. Sliding mode observer based robust fault reconstruction for modular multilevel converter with actuator and sensor fault. *IFAC-PapersOnLine*, vol. 53, no. 2, pp. 13365–13370, 2020, doi: 10.1016/j.ifacol.2020.12.172.

[82] F. Deng, Z. Chen, S. Member, and M. R. Khan. Fault Detection and Localization Method for Modular Multilevel Converters. *IEEE Transactions on Power Electronics*, vol. 8993, no. c, pp. 1–11, 2014, doi: 10.1109/TPEL.2014.2348194.

[83] B. Li, S. Shi, B. Wang, G. Wang, W. Wang, and D. Xu. Fault diagnosis and tolerant control of single IGBT open-circuit failure in modular multilevel converters. *IEEE Trans. Power Electron.*, vol. 31, no. 4, pp. 3165–3176, 2016, doi: 10.1109/TPEL.2015.2454534.

[84] S. Haghnazari, M. Khodabandeh, and M. R. Zolghadri. Fast fault detection method for modular multilevel converter semiconductor power switches. pp. 165–174, 2015, doi: 10.1049/iet-pel.2015.0392.

[85] M. Salehifar, R. S. Arashloo, M. Moreno-Eguilaz, V. Sala, and L. Romeral. Observer-based open transistor fault diagnosis and fault-tolerant control of five-phase permanent magnet motor drive for application in electric vehicles. *IET Power Electron.*, vol. 8, no. 1, pp. 76–87, 2015, doi: 10.1049/iet-pel.2013.0949.

[86] E. M. Cimpoeşu, B. D. Ciubotaru, and D. Ştefănoiu. Fault detection and identification using parameter estimation techniques. *UPB Sci. Bull. Ser. C Electr. Eng. Comput. Sci.*, vol. 76, no. 2, pp. 3–14, 2014.

[87] M. Shahbazi, M. R. Zolghadri, M. Khodabandeh, and S. Ouni. Fast detection of open-switch fault in cascaded H-bridge multilevel converter. *Sci. Iran.*, vol. 25, no. 3D, pp. 1561–1570, 2018, doi: 10.24200/sci.2017.4371.

[88] S. Tang et al. Detection and identification of power switch failures using discrete fourier transform for DC-DC flying capacitor buck converters. *IEEE J. Emerg. Sel. Top. Power Electron.*, vol. 9, no. 4, pp. 4062–4071, 2021, doi: 10.1109/JESTPE.2020.3012201.

[89] M. A. Rodríguez-Blanco et al. A failure-detection strategy for IGBT based on gate-voltage behavior applied to a motor drive system. *IEEE Trans. Ind. Electron.*, vol. 58, no. 5, pp. 1625–1633, 2011, doi: 10.1109/TIE.2010.2098355.

[90] M. A. Rodríguez, A. Claudio, D. Theilliol, and L. G. Vela. A new fault detection technique for IGBT based on gate voltage monitoring. *PESC Rec.—IEEE Annu. Power Electron. Spec. Conf.*, pp. 1001–1005, 2007, doi: 10.1109/PESC.2007.4342127.

[91] Z. Wang, X. Shi, L. M. Tolbert, F. Wang, and B. J. Blalock. A di/dt feedback-based active gate driver for smart switching and fast overcurrent protection of IGBT modules. *IEEE Trans. Power Electron.*, vol. 29, no. 7, pp. 3720–3732, 2014, doi: 10.1109/TPEL.2013.2278794.

[92] S. Khomfoi, and L. M. Tolbert. Fault diagnostic system for a multilevel inverter using a neural network. *IEEE Trans. Power Electron.*, vol. 22, no. 3, pp. 1062–1069, 2007, doi: 10.1109/TPEL.2007.897128.

[93] P. Lezana, R. Aguilera, and J. Rodríguez. Fault detection on multicell converter based on output voltage frequency analysis. *IEEE Trans. Ind. Electron.*, vol. 56, no. 6, pp. 2275–2283, 2009, doi: 10.1109/TIE.2009.2013845.

[94] M. Shahbazi, M. Reza, P. Poure, and S. Saadate. Fast short circuit power switch fault detection in cascaded H-bridge multilevel converter. *IEEE*, pp. 1–5, 2013.

[95] S. Khomfoi, and L. M. Tolbert. Fault diagnosis and reconfiguration for multilevel inverter drive using AI-based techniques. *IEEE Trans. Ind. Electron.*, vol. 54, no. 6, pp. 2954–2968, 2007, doi: 10.1109/TIE.2007.906994.

[96] C. Li, A. M. Gole, and C. Zhao. A fast DC fault detection method using DC reactor voltages in HVdc grids. *IEEE Trans. Power Deliv.*, vol. 33, no. 5, pp. 2254–2264, 2018, doi: 10.1109/TPWRD.2018.2825779.

[97] X. Li, Q. Song, W. Liu et al. Protection of nonpermanent faults on DC overhead. *IEEE Transactions on Power Delivery*, vol. 28, no. 1, pp. 483–490, 2013. doi: 10.1109/TPWRD.2012.2226249.

[98] E. Kontos, R. T. Pinto, and P. Bauer. Providing dc fault ride-through capability to H-bridge MMC-based HVDC networks. *9th Int. Conf. Power Electron.—ECCE Asia "Green World with Power Electron. ICPE 2015-ECCE Asia*, pp. 1542–1551, 2015, doi: 10.1109/ICPE.2015.7167983.

[99] G. P. Adam, and I. E. Davidson. Robust and generic control of full-bridge modular multilevel converter high-voltage DC transmission systems. *IEEE Trans. Power Deliv.*, vol. 30, no. 6, pp. 2468–2476, 2015, doi: 10.1109/TPWRD.2015.2394387.

[100] H. Aji, M. Ndreko, M. Popov, and M. A. M. M. Van Der Meijden. Investigation on different negative sequence current control options for MMC-HVDC during single line to ground AC faults. *IEEE PES Innov. Smart Grid Technol. Conf. Eur.*, 2016, doi: 10.1109/ISGTEurope.2016.7856182.

[101] X. Shi, Z. Wang, B. Liu, Y. Liu, L. M. Tolbert, and F. Wang. Characteristic investigation and control of a modular multilevel converter-based HVDC system under single-line-to-ground fault conditions. *IEEE Trans. Power Electron.*, vol. 30, no. 1, pp. 408–421, 2015, doi: 10.1109/TPEL.2014.2323360.

[102] M. M. Haji-Esmaeili, M. Naseri, H. Khoun-Jahan, and M. Abapour. Fault-tolerant structure for cascaded h-bridge multilevel inverter and reliability evaluation. *IET Power Electron.*, vol. 10, no. 1, pp. 59–70, 2017, doi: 10.1049/iet-pel.2015.1025.

[103] A. A. Stonier, and B. Lehman. An Intelligent-Based Fault-Tolerant System for Solar-Fed Cascaded Multilevel Inverters. *IEEE Trans. Energy Convers.*, vol. 33, no. 3, pp. 1047–1057, 2018, doi: 10.1109/TEC.2017.2786299.

[104] H. Salimian, and H. Iman-Eini. Fault-tolerant operation of three-phase cascaded H-bridge converters using an auxiliary module. *IEEE Trans. Ind. Electron.*, vol. 64, no. 2, pp. 1018–1027, 2017, doi: 10.1109/TIE.2016.2613983.

[105] V. S. Prasadarao K, S. Peddapati, and S. Naresh. A new fault-tolerant MLI—Investigating its skipped level performance. *IEEE Trans. Ind. Electron.*, vol. 0046, no. c, 2021, doi: 10.1109/TIE.2021.3062259.

[106] M. Aleenejad, H. Mahmoudi, P. Moamaei, and R. Ahmadi. A new fault-tolerant strategy based on a modified selective harmonic technique for three-phase multilevel converters with a single faulty cell. *IEEE Trans. Power Electron.*, vol. 31, no. 4, pp. 3141–3150, 2016, doi: 10.1109/TPEL.2015.2444661.

[107] H. Mahmoudi, M. Aleenejad, and R. Ahmadi. A fault tolerance switching strategy based on modified space vector modulation method for cascaded multilevel converter. *2017 IEEE Power Energy Conf. Illinois, PECI 2017*, vol. 2, no. 2, pp. 3–8, 2017, doi: 10.1109/PECI.2017.7935760.

[108] A. H. E. A. Sallam, A. Ragi, R. Hamdy, M. Z. Moustafa. New measurement technique for modular multilevel converter with IGBT open-circuit failure detection and tolerance control for three-level submodule. In *5th International Conference on Renewable Energy Research and Applications*, vol. 5, pp. 1–6, 2016.

[109] R. Picas, J. Zaragoza, J. Pou, and S. Ceballos. Reliable modular multilevel converter fault detection with redundant voltage sensor. *IEEE Trans. Power Electron.*, vol. 32, no. 1, pp. 39–51, 2017, doi: 10.1109/TPEL.2016.2526684.

[110] K. Li, Z. Zhao, L. Yuan, S. Lu, and Y. Jiang. Fault detection and tolerant control of open-circuit failure in MMC with full-bridge sub-modules. *ECCE 2016 — IEEE Energy Convers. Congr. Expo. Proc.*, 2016, doi: 10.1109/ECCE.2016.7855109.

[111] J. Häfner, and B. Jacobson. Proactive hybrid HVDC breakers—A key innovation for reliable HVDC grids. *CIGRE 2011 Bol. Symp.—Electr. Power Syst. Futur. Integr. Supergrids Microgrids*, 2011.

[112] R. Zeng, and L. Xu. Design and operation of a hybrid modular multilevel converter. *IEEE Transactions on Power Electronics*, vol. 8993, no. c, pp. 1–10, 2014, doi: 10.1109/TPEL.2014.2320822.

[113] X. Yu, Y. Wei, and Q. Jiang. STATCOM Operation Scheme of the CDSM-MMC during a Pole-to-Pole DC Fault. *IEEE Trans. Power Deliv.*, vol. 31, no. 3, pp. 1150–1159, 2016, doi: 10.1109/TPWRD.2015.2464320.

[114] Z. Ou, G. Wang, and L. Zhang. Modular multilevel converter control strategy based on arm current control under unbalanced grid condition. *IEEE Trans. Power Electron.*, vol. 33, no. 5, pp. 3826–3836, 2018, doi: 10.1109/TPEL.2017.2717541.

[115] A. E. Leon, and S. J. Amodeo. Energy balancing improvement of modular multilevel converters under unbalanced grid conditions. *IEEE Trans. Power Electron.*, vol. 32, no. 8, pp. 6628–6637, 2017, doi: 10.1109/TPEL.2016.2621000.

[116] C. Zhao, F. Gao, Z. Li, P. Wang, and Y. Li. Capacitor Voltage ripples reduction of full-bridge mmc with circulating current injection under single-line-to-ground fault conditions. *IECON Proc. Industrial Electron. Conf.*, vol. 2019, pp. 6084–6089, 2019, doi: 10.1109/IECON.2019.8926930.

10 Review Analysis of Cascaded H-Bridge and Modular Multilevel Inverters

Ashutosh Kumar Singh, Rajib Kumar Mandal, and Ravi Anand

CONTENTS

10.1   INTRODUCTION

Multi-Level Inverters (MLIs) allow us to discover a suitable solution for DC to AC converters. These inverter play an important role in recent technology development [1], [2]. In power system applications, it is widely used (such as power quality improvement by Flexible AC Transmission System (FACTS) devices [3]–[6], Static-Compensators (STATCOM) [7]–[10], High Voltage Direct Current (HVDC) [11]–[14], dynamic voltage restorers [15]–[18], unified-power quality-conditioners [19]–[22], active-filters [23]–[26], Solid-Sate-Transformers (SST) [27]–[30], industrial drives used in transportation and traction [31]–[38], conveyors [39]–[40], marine propulsion [41]–[45], mine hoists [46], industrial applications as heating dissipates power supply [47] and Electro Magnetic Resonance Imaging (MRI) system [48]. The following are advantages of MLIs over two-level converters [49]–[51].

Due to its modular structure, it can be easily extended.

(1) It generates close to sinusoidal waveform as it has a good harmonic specification and so reduces filter size.

DOI: 10.1201/9781003323471-10

(2) Using redundant switching states, it enables fault-tolerant operation capability.
(3) Less voltage stress on the devices and lower switching frequency because of lower switching losses.
(4) Electro Magnetic Interference (EMI) related problems can be avoided due to low voltage stress (dv/dt).
(5) Improve in power harvesting and sharing of load can be done with combination of various MLIs with different sources (like wind, fuel cells, and photovoltaic).

To calculates the dependability of power electronics converters, MLIs must be fault-tolerant. Fault tolerance is influenced by the MLIs' design process. Modularity, number of total power electronics switches, appearance of each switch at various levels, and redundant pathways are all design parameters to consider when analysing fault-tolerant systems. The authors of [52]–[56], examined many MLIs topologies from a fault-tolerant perspective, and this chapter summarises their findings. By and large, duplicate states aid in improving fault tolerance. The pathways designated to each level give a redundant mechanism of generating the level in the case of a switch failure. Alternatively, if a switch fails, other methods for generating the same level exist. In this scenario, symmetric MLIs outperform asymmetric MLIs owing to their fewer transitions. Additionally, unipolar MLIs (which rely on the H-bridge to generate polarity) pose a significant risk in fault-tolerant capability applications, since if a single switch in the H-bridge fails, the MLIs will lose all negative or positive levels. When we can see, as the quantity of power electronics switches rises, the cost, volume, quantity of components, and complexity increase as well. Additionally, the greater degree of modularity of MLIs enables the creation of redundant pathways at multiple levels, which is a significant benefit of modular MLIs from a fault-tolerant standpoint. In fault-tolerant capability analysis, the behaviour of each switch represents a crucial condition. During the design process, it is important to consider the order in which the switches on each level should be activated. The switch that contributes to the majority of levels is at danger of failure, and if the switch fails, many levels in the power converter's output waveform will be lost. It is better if the shifts occur at the same pace as the levels. Despite the fact that MLIs are suitable for medium and high voltage applications, the increased number of switching components increases the overall cost and complexity of the overall system design [57]. The concept of MLIs with (CHB) topology came in 1970s [58]–[60], after that NPC structure came in 1980s [61]–[63], and flying capacitor (FC) topologies in 1990s [64]. The preceding topologies are classic topologies and these were considered to be the base of different multi-level inverters that existed in recent years. The invention of Modular Multilevel Converters (MMCs) presented a great development in many industrial drives applications [65]–[66]. P2 introduced in 2000, which is a generalised MLI that encompasses FC topologies and NPC topology [67]. With advancements in MLI technology, particularly in NPC structure, the Active NPC (ANPC) structure is explored in [68]. Other various multilevel structure with an application-oriented based have been published and researched throughout the last decade. Figure 10.1 illustrates the development of MLI topologies. Based on equal sources, the multilevel converters can be considered as symmetric [69]–[132], and based on unequal sources (binary, trinary, etc.), the multilevel converters can be considered as asymmetric [133]–[170] with inherent negative polarity level [67], [69]–[75], [77]–[92], [133]–[177] or without H-Bridge negative level [93]–[132], DC sources by capacitor buses link [77]–[79], [111], [114]–[118], [143]–[159], [176], [177], regenerative configurations to work as an inverter or a rectifier [104]–[110] and hybrid model based on the combination of NPC, CHB, and FC [140]–[142]. Most of the aforementioned strictures have same construction, which will be discussed in next sections. In recent times, a different type of MLIs topology have been analysed and discussed in literature review. Scientists' may get fascinated by some of these topologies due to their similar structure. Due to the behaviour of sub-structure of existing topologies, it needs to be deeply analysed. This chapter does a comprehensive structural analysis of both classic and current MLIs. As a result, the chapter may serve as a guidebook for developing many new MLIs for readers' specialised applications.

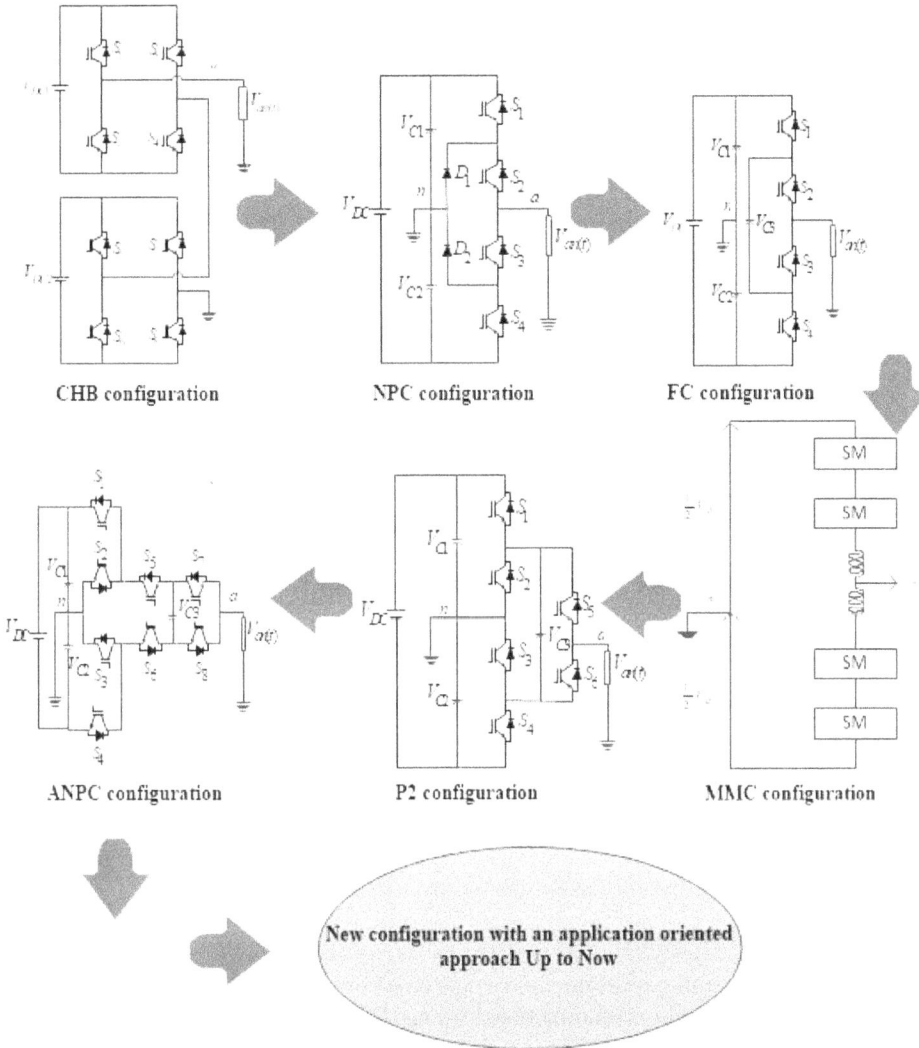

FIGURE 10.1   Point of view on advancement of MLIs configuration.

On the other hand, researchers may install the supplied submodules (SMS) according to their desired MLI structure and depending on the more DC power supply, levels, capacitors, and semiconductors. This chapter demonstrates how conventional inverters developed in response to the offered SMs. It may serve as a useful guideline for readers' comprehension of future MLIs. Section 10.2 discusses the working principles discussion and configuration of the five primary SMs that serve as the basis for MLI topologies. Then, in section 10.3 based on these main SMs, many other topologies are studied. Important features of these structures are explained and classified in different topologies. Advantages and disadvantages of each topology have been discussed in detail analysis. In this work, the strictures that are examined and evaluated are divided into two types [2]: Various strictures with inherent negative polarity voltage levels generated in [67], [69]–[75], [77]–[92], [133]–[177]. In this section 10.4 mainly focuses on the topologies features of the various MLIs topology. Aside from that, section 10.5 discusses a comparative examination of evaluated and studied topologies. In the final portion, conclusions are presented in section 10.6.

10.2 MLI-SPECIFIC INDICES AND TERMINOLOGY

Here are the MLI topologies, all the terms define which are used in this chapter.

10.2.1 TOPOLOGIES WITH DECREASED PART COUNT

MLI topologies that are suggested with the idea of decreasing the number of elements that are used while enhancing the power quality by improving other parts are called reduced number of component count construction [2].

10.2.2 MODULAR MLIs

It employs voltage sources of about same magnitude and semiconductor devices and capacitors of nearly identical rating in each standard cell of MLI. Output cells are typically linked in series one after the other to provide high voltage levels. The structure of a typical example of symmetric CHB MLI topology is novel. Section 10.2 deliberates the working principles and configuration of the five primary SMs that serve as the basis for MLI topologies.

10.2.3 DIFFERENT MAGNITUDE DC SUPPLY OF TOPOLOGIES

MLIs may have a symmetric/asymmetric configuration. Symmetric topologies strive to maximise modularity, while asymmetric topologies prioritise size reduction and cost minimisation while maintaining minimal THD in the output voltage waveform due to modularity.

10.2.4 CONTROL COMPLEXITY

The distribution here is about the hardware implementation problem of the control scheme in MLIs [132]. When talking about asymmetric topologies, regulation is also extremely dynamic.

10.2.5 POWER ELECTRONICS SWITCHES

This term means, with an anti-parallel diode, a single IGBT or MOSFET. It has the power of holding bidirectional current and blocking unidirectional voltage [88].

10.2.6 COST FACTOR

The expense of an MLI topology is calculated using this parameter. It is defined as follows:

$$G = N_{variety} * N_{source} * N_{switch} \qquad (1)$$

10.2.7 GENERATING LEVELS AND GENERATING POLARITY

MLI topologies also have their own level generation part and polarity generating part component of the reduced switch count. In level generation, the output produces positive and zero voltage values, while the portion of polarity generation transforms the unipolar voltage waveform of the staircase into the bipolar voltage waveform of the staircase. H-bridge works in MLI [94], and [132] as a portion of polarity generation.

10.2.8 Conduction and Switching Losses

The most prevalent forms of power losses in semiconductor devices are conduction losses, switching losses, and blocking losses. Conductor losses are the most common type of power loss. Because the quantity of leakage current is so small, these losses are often missed. When computing conductivity losses, the on-state voltage drop as well as the equal resistance of the semiconductor device are taken into account. Eqs. (2) and (3) respectively represent the instantaneous power conduction losses of transistors and power diodes, respectively.

$$P_{c,T}(t) = \left[V + R \cdot i^{\beta}(t) \right] \cdot i(t) \tag{2}$$

$$P_{c,D}(t) = i(t) \cdot [V + R \cdot i(t)] \tag{3}$$

Conduction losses in an MLI topology are depending on the components counts of switches and diodes in the current carrying direction. Switching power losses depend on the frequency of system switching when eqs. occur (4), and (5) are

$$E_{\mathrm{off}}, k = \frac{\left(V_{\mathrm{off}}, k \cdot I \cdot \mathrm{to}\, ff \right)}{6} \tag{4}$$

$$E_{\mathrm{on}}, k = \frac{\left(V_{\mathrm{off}}, k \cdot I' \cdot \mathrm{ton} \right)}{6} \tag{5}$$

Where V_{off}, k is off the k switch state voltage, I is the current flowing through the switch before switching off, I' is the current flowing through the switch after switching on, (t_{on} and t_{off}) are the switch's turn on/off time. E_{on}, k and E_{off}, k is the energy depletion switches of k_{th}.

10.2.9 The Redundancy

In some instances of MLI topology, without using a large number of switching combinations, the voltage level can be established. For a certain voltage standard, these alternative switching summations are considered redundant switching conduction states. In the meantime, redundancy has many applications, such as capacitor charge balancing, MLIs programming also includes fault tolerance capability operating, as well as power utilisation of the input DC power source [51].

10.2.10 Multilevel Waveform

A sinusoidal waveform in MLIs is similar to a staircase waveform in that it has the lowest total harmonic distortion (THD) in the output voltage and the least voltage stress on power electronic switches [85], [94].

10.2.11 Total Voltage Blocking (TSV)

It is the sum of all the switches used in the inverter topology for peak inverse voltage (PIV) ranking.

10.3 PROPOSED MODULE

Usually with the help of main SMs, MLIs can be designed. MLIs may usually be created with the assistance of the primary SMs. SM configuration and operation guidelines are outlined in Table 10.1. It is possible to link these basic elements in a variety of ways to meet the needs of a specific application. The advantages of each SM making them appropriate for a variety of mixed topologies.

(1) The right and left sides of SM-I have three terminals and two options for connecting it to other modules in parallel and series, respectively. It may also be used with a single DC supply. Each switch must be able to withstand the DC voltage stress of V_{DC}. The half bridge arrangement is somewhat similar to this one.

(2) The SM-II has four ports terminals and six states of operation, allowing it to be connected to other modules in parallel on the right side and in series on the left side of the module. It may also be powered by a single dc source. Each switch should have the ability to prevent V_{DC} from being executed. S_2 and S_3 may be used as bidirectional switches when the output terminals b and c are used as input terminals.

(3) The SM-III has four port terminals, and four ways of connection, allowing it to be linked in series to other modules on the left, right, up, and down sides, as well as in parallel to other modules through the various configurations of four terminals. It may also be used with a single DC supply. H-bridge with two additional terminals on top and bottom is a common layout. Each switch should prevent access to V_{DC}.

(4) The right and left sides of SM-IV each include four terminals and three options for connecting to other modules in parallel and series, respectively. With the present bidirectional switch design, it can run two dc sources. T-type arrangement is often referred to when there is an additional terminal on the right. Unidirectional, and bidirectional switches should be able to resist voltages of $2V_{DC}$ and V_{DC}, respectively, in order to function properly. To add insult to injury compared to the SM-II, it creates higher voltage levels with more blocking voltage stress on the power electronics switches.

(5) SM-V features four terminals ports and six states, with modules in series connection circuits on the left and parallel connection circuits on the right. With the setup of the bidirectional switch, two dc sources may be controlled in two different ways. With two extra terminals on the top and bottom of the board, T-type and half-bridge configurations are combined. In comparison to SM-III, it creates more output voltage-level while putting greater strain on the switches. Unidirectional and bidirectional switches must be able to bear the voltage stress of $2V_{DC}$ and V_{DC}, respectively.

SMs are clearly regulated in one directional current flowing, whereas diodes are un-controlled in the other way. The points "a" and "d" in SMs-III and SMs-V may be separated to generate three more modules (see Figure 10.2). Some switches are connected in series or parallel to handle power flow in all four quadrants while also handling bidirectional current and blocking positive/negative voltage (bipolar voltages).

Bidirectional switches refer to switches that are both bidirectional and bipolar. Several unidirectional switches may be connected in various configurations to generate bidirectional switches (see Figure 10.3). Figure 10.3 shows a typical VSI configuration. As an example, connecting two SMs results in the creation of a bidirectional switch, as illustrated in Figure 10.4, which illustrates how a T-type module may be created by connecting two half-bridge SMs in a vertical parallel configuration (SM-1). S_1 and S_2 must be able to handle both positive and negative polarity voltages from DC sources in order to work properly in this instance (V_{DC1} and V_{DC2}). As a result, switches might be considered to be bidirectional switches.

TABLE 10.1
Data Analysis

1) Main SM

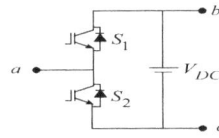

Switching Sequence

S_1	S_2
ON	OFF
OFF	ON

Voltage Polarities

V_{ba}	V_{ca}
0	$-V_{DC}$
V_{DC}	0

2) Main SM

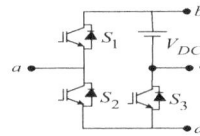

Switching Sequence

S_1	S_2	S_3
OFF	ON	ON
ON	OFF	ON
ON	ON	OFF
OFF	ON	OFF
ON	OFF	OFF

Voltage Polarities

V_{ba}	V_{ca}	V_{da}
$-V_{DC}$	0	0
0	$-V_{DC}$	$-V_{DC}$
0	$-V_{DC}$	0
0	0	0
0	$-V_{DC}$	0

3) Main SM

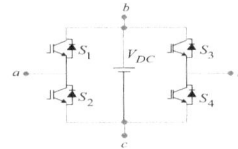

Switching Sequence

S_1	S_2	S_3	S_4
ON	OFF	OFF	ON
OFF	ON	ON	OFF
OFF	ON	OFF	ON
OFF	ON	OFF	ON
ON	OFF	ON	OFF

Voltage Polarities

V_{ba}	V_{ca}	V_{da}
0	0	$-V_{DC}$
0	0	V_{DC}
0	0	0
0	0	0
0	0	0

4) Main SM

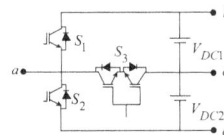

Switching Sequence

S_1	S_2	S_3
OFF	OFF	ON
OFF	ON	OFF
ON	OFF	OFF

Voltage Polarities

V_{ba}	V_{ca}	V_{da}
V_{DC}	0	0
$2V_{DC}$	V_{DC}	0
0	$-V_{DC}$	$-2V_{DC}$

(Continued)

TABLE 10.1
Continued

5) Main SM

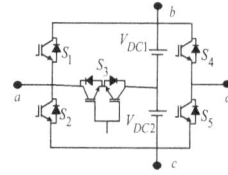

Switching Sequence					Voltage Polarities		
S_1	S_2	S_3	S_4	S_5	V_{ba}	V_{ca}	V_{da}
OFF	OFF	ON	ON	OFF	0	0	V_{DC}
OFF	OFF	ON	OFF	ON	0	0	$-V_{DC}$
OFF	ON	OFF	OFF	ON	0	0	0
ON	OFF	OFF	ON	OFF	0	0	0
OFF	ON	OFF	ON	OFF	0	0	$2V_{DC}$
ON	OFF	OFF	OFF	ON	0	0	$-2V_{DC}$

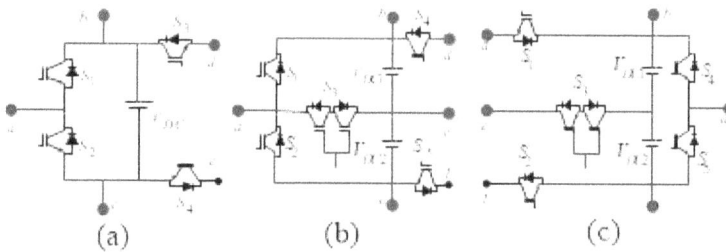

FIGURE 10.2   Point of view on advancement of MLIs: (a) SMs-III, (b) SMs-IV, and (c) SMs-V.

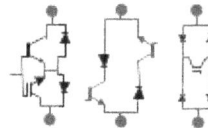

FIGURE 10.3   Bidirectional power electronic semiconductor switches.

10.4 MULTILEVEL TOPOLOGIES OVERVIEW

The SMs are shown in Table 10.1; based on that, this section contains a detailed discussion of MLIs. By the different connection of main SMs, that is, parallel, series, and anti-parallel, some derived inverters are studied in this section.

10.4.1 CONFIGURATIONS WITH INHERENT (-VE) POLARITY

As stated earlier, even without the use of H-bridge converter, some of MLI converters can produce both positive and negative output voltage levels. A review of these types of configurations have been referred as bipolar MLIs by the authors.

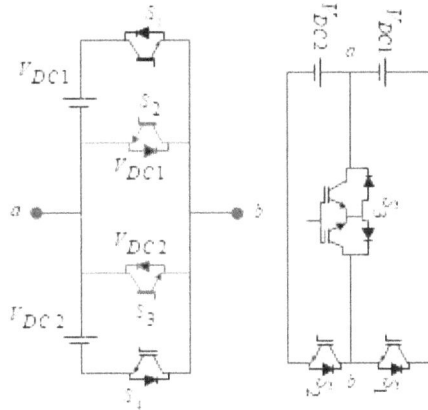

FIGURE 10.4 Bidirectional switch in T-type configuration (submodule–4).

The first patent was CHB that proposed a configuration which can produce multilevel output voltage from dc sources. Due to its modularity, it is good option to use in high voltage applications. However, each module required individual dc voltage sources that restrict it uses in higher output levels [69]. Moreover, due to its more redundancy conduction states and modularity, it is very suitable to fault-tolerant applications [71]. By connecting SM-III in series, CHB can be formed as in Figure 10.5. Symmetrical and asymmetrical sources are discussed in [83]. Table 10.2 depicts the number of output voltage-level and pattern of DC power supply for Figure 10.5 based on the patterns.

A five-level active neutral point clamp (also known as a 5L-ANPC) is shown in Figure 10.6(a), when two series SM-Is are being connected in parallel with a series SM-III [70], [134]–[136]. Let $V_{C1} = V_{C2} = V_{DC}$ and $V_{C3} = 1/2\ V_{DC}$, then 5L-ANPC substantially improves the output voltage level and provides five distinct load voltage levels $(0, \pm\ V_{DC}, 3/2V_{DC}, \pm, 2V_{DC})$. Three distinct levels $(0, V_{DC}, 2V_{DC})$ may be formed from the 5L-ANPC by dismantling the switches S_5, S_6 and dc link (V_{C3}) [70]–[71]. It is now possible to implement the NPC topology by simply deleting the neutral clamping switches (see Figure 10.6(b)) [72]. Instead of employing clamping diodes, 3L-ANPC employs neutral clamping switches. In NPC inverters, the unequal distribution of losses between the inner and outer switches, the high necessary number of diodes, and difficult voltage balancing are the key drawbacks, and neutral clamping switches should be discussed as a solution [172]. A new simplified configuration can be obtained by using SM-I in place of SM-III (see Figure 10.6(c)) [67]. The simplified inverter topology is a pyramid stricture of SM-I with its 3-L configuration [(see Figure 10.6(d)). In the configuration of symmetrical DC power supply, each switch needs to be able to block voltage V_{DC}. Additionally, some of the MLIS can be inferred from this simplified topology. For an example, an 3-L asymmetrical NPC can be formed by omitting dc link (V_{C3}). 3L-NPC can be obtained by reduced the clamping diodes (see Figure 10.6(c)). Using pyramid structure, new configurations are presented in [173], [174]. Table 10.3 depicts the number of output voltage levels and pattern of dc sources on the basis of patterns for Figure 10.6. Under different fault situation, a thorough study of the performance of a 3L-NPC has been presented in [175] and illustrates the fault tolerance capability, but it is controlled. Parallel circuits connection of SMs-I and SMs-III, FC topology can be realised as in Figure 10.7. As FC require large number of capacitors and have the problem of voltage balancing due to capacitor, FC has limited use in industrial application [137], [138]. Further, fault-tolerant operation can be done due to their redundancy states [139]. Steps of load voltage-level and the pattern of dc sources for Figure 10.7 are represented in Table 10.4.

A variety of topologies may now be generated using these configurations. Figure 10.8 illustrates three different circuit topologies that may be implemented using ordinary circuits. Figure 10.8 [140]

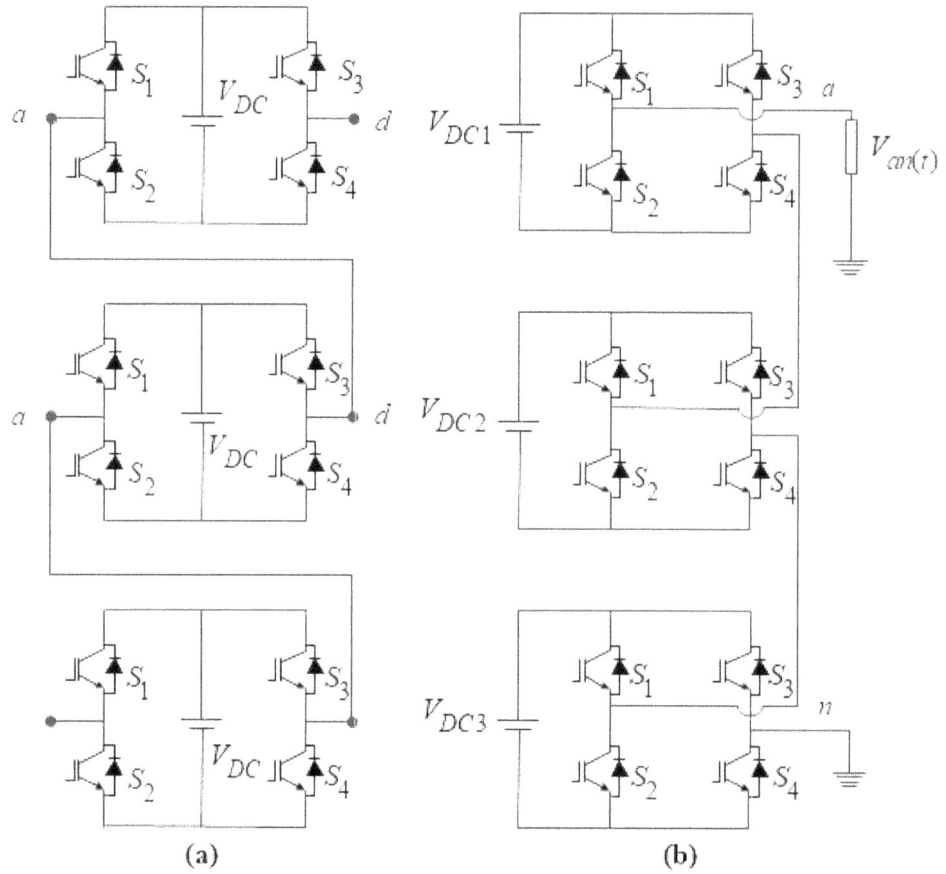

FIGURE 10.5 CHB inverter: (a) mixed connection of submodule-1 and -3, and (b) five-level CHB [69].

TABLE 10.2

Type of Source Configuration of the CHB [69], [133]

Type of source	Value of DC sources	N_{level}
Sym. sources	$V_{DC1} = V_{dc}$, $V_{DC2} = V_{dc}$ and $V_{DC3} = V_{dc}$	7
Asym. sources	$V_{DC1} = V_{dc}$, $V_{DC2} = 2V_{dc}$ and $V_{DC3} = 4V_{dc}$	15
	$V_{DC1} = V_{dc}$, $V_{DC2} = 3V_{dc}$ and $V_{DC3} = 9V_{dc}$	27

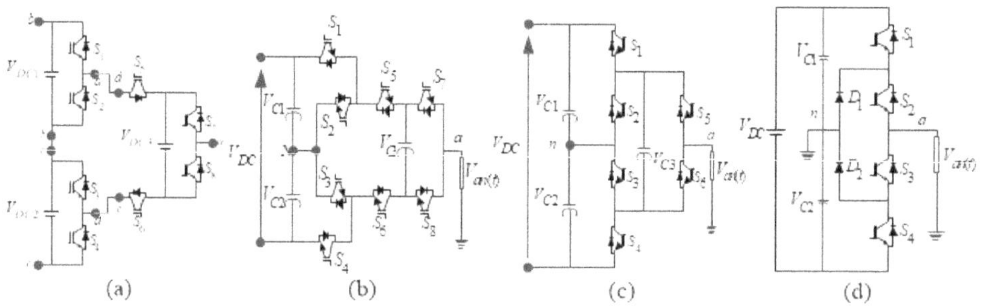

FIGURE 10.6 NPC converter: (a) mixed connection, (b) submodule-1, (c) submodule-2, and (d) submodule-3.

TABLE 10.3

The Three-Level NPC's Source Arrangement [72]

Type of source	Value of DC sources	N_{level}
Sym. Sources	$V_{C1} = V_{C2} = V_{dc}/2$	3

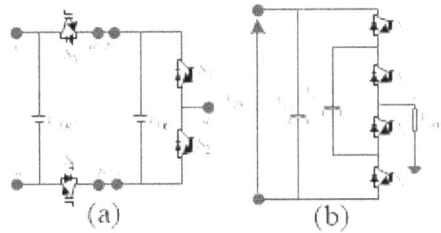

FIGURE 10.7   FC inverter: (a) parallel connection of submodule-1, and (b) three-level discussed in [96], [97].

TABLE 10.4

The three-level FC's source arrangement [137], [138].

Type of source	Value of DC sources	N_{level}
Sym. Sources	$V_{C1} = V_{dc}, V_{C2} = V_{dc}/2$	3

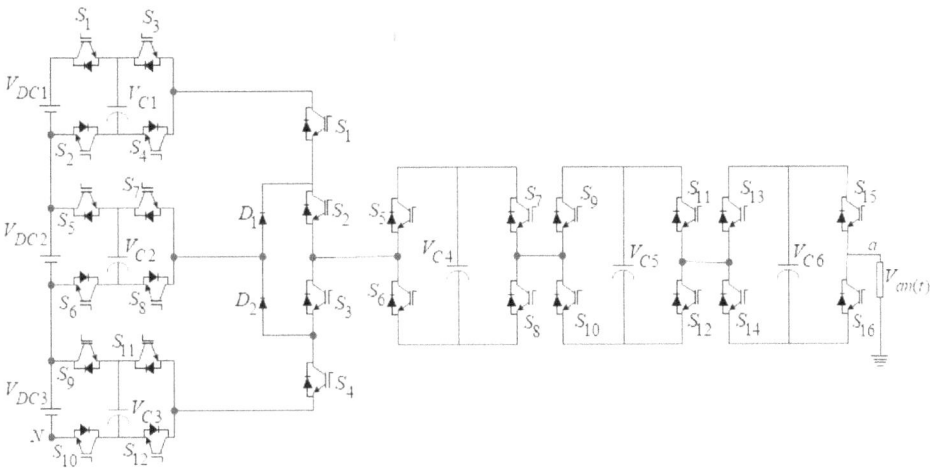

FIGURE 10.8   Combination of CHB, NPC, and FC (submodules-3) [142].

shows an FC inverter with an integrated three-level NPC. For example, this design has a low number of switching devices and can operate across a wide range of input voltage without the need for additional switches in series. It's also important to note that each power electronics switch does not contain to be able to withstand maximum blocking-voltage more than $1/4V_{DC}$ ($V_{DC} = V_{DC1} + V_{DC2}$).

[141] presented a six-level inverter with fewer components and a single DC power source, which helps minimise the weight, dimensions, and cost of the inverter. As seen in Figure 10.8, the

structures SMs-I and the inner three-level FC unit are connected to realise the topology. When $V_{C1} = 1/5V_{DC}$, $V_{C2} = 3/5V_{DC}$, $V_{C3} = 1/5V_{DC}$, and $V_{C4} = 2/5V_{DC}$, the switches have the following voltage ratings: $1/5V_{DC}$ is required for (S_1 and S_2) and $2/5V_{DC}$ is required for the remaining switches. Additionally, all power electronics switches are rated for the same stress current. Numerous SMs may be used in MMCs. The SM-I is the most basic bidirectional ac-dc converter module available in Figure 10.9 [73]–[76]. As shown in Figure 10.9, two structural SMs-III are joined in parallel to construct a new three-level SM, as proposed in [77] and [78]. The recommended 3L-SM, which may be replaced by 2 SMs-I, is known as double-SM. A two SM (8 switches) has more switches than 2 SMs-I. The rating of switches is half of the current rating, and two switches are always connected in parallel. As a result, the total power stress of the switches in twice SM is equivalent to SMs-I. Many power electronics switches are used in circuits that are linked together at the same time to improve the current rating of the SMs in practise. Double-SM may be produced in this situation by using the same components. Table 10.5 represents the number of output voltage levels and the layout of the dc sources for Figure 10.9.

Series-connection circuits of SM-III with parallel connection SM-I (see Figure 10.10) helps in realising a new MLI with asymmetrical source [143]. It can generate all required load levels by using four floating capacitors and single DC source. The upper, and lower, circuits switch in each module should block voltage related to their respective capacitors (dc link), excluding S_1 and S_2 that should block voltage equal to subtraction of V_{DC1} and V_{C1} and also these lower and upper switches

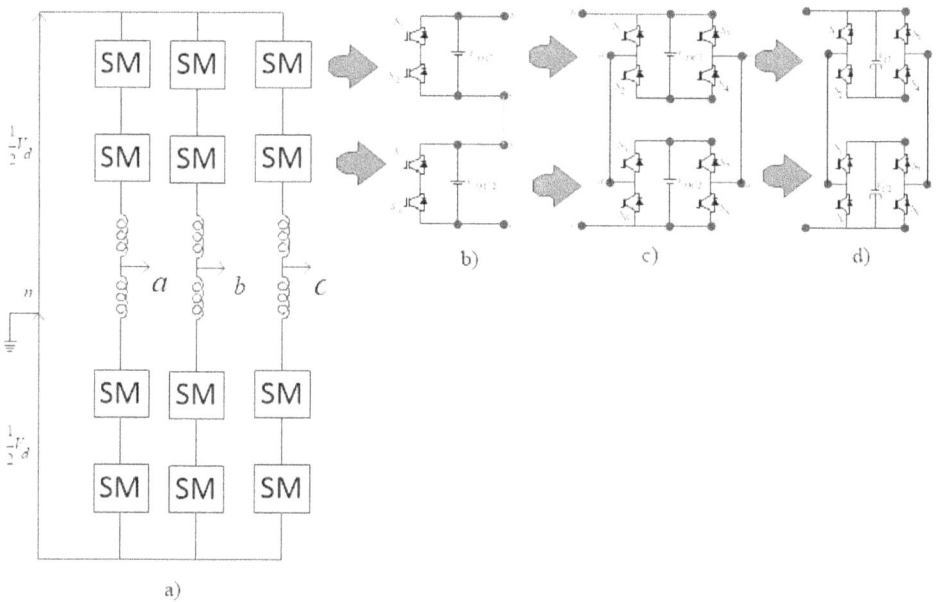

FIGURE 10.9   (a) MMC configuration presented in [73]–[78], (b) submodule in [103]–[106], (c) submodule in [107], [108], and (d) submodule in [107], [108].

TABLE 10.5
The Three-Level Source Arrangement [77] and [78]

Type of source	Value of DC sources	N_{level}
Sym. Sources	$V_{C1} = V_{C2} = 1/2\ V_{dc}$	3

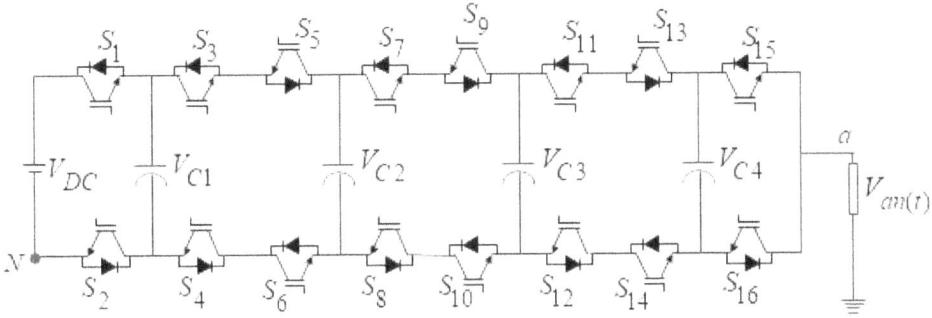

FIGURE 10.10 Presented in [143].

TABLE 10.6
The 17-Level Arrangement in [143]

Type of source	Value of DC sources	N_{level}
Asym. Sources	$V_{C1} = V_{DC}$, $V_{C2} = 1/2V_{DC}$ $V_{C3} = 1/4V_{DC}$, $V_{C4} = 1/8V_{DC}$, $V_{C5} = 1/16V_{DC}$	17

should work in complementary way. However, by using one/more than one switching states, each voltage level can be produced. The current flowing through capacitors can be negative and their voltages can be balanced by power electronics switching through the reducing switching combinations. Due to modularity and symmetrical structure of the suggested configuration enabling the inverter to be modified to a greater 5-phase and 6-phase configuration using single dc link with same control scheme for three-phase structure. Moreover, due to its high redundancy conduction states, it can be used for fault-tolerant operation. The number of load voltage levels and pattern of the dc sources for Figure 10.10 are represented in Table 10.6.

In [144]–[158], the Packed U-Cell (PUC) configuration is discussed. It can be designed either by parallel connection circuits of two SMs-III or SM-I and SM-III and removing spare switches and it uses a smaller number of components, as shown in Figure 10.11(a)–(c). The structure is achieved from easy and comprehensible design of components. To generate output voltage levels, one capacitor, six switches, and one dc source are used. Symmetric source configuration can't be utilised to get the most output from this topology. The amplitude of the capacitor voltage is held constant at one-second of DC power supply, allowing the PUC architecture to create five levels at the output terminals [155]–[156]. If the capacitor voltage is maintained at one-third of the dc voltage sources (trinary), this architecture may provide seven levels at the load terminals [157], [158]. As seen in Figure 10.11(f), the number of load voltages from specified input DC sources may be increased by adding two cross-over switches [159]. In addition, it should be noted that this design is comprised of two SMs-III, and all switches are required to carry the load current.

The advantages of this topology include lower design costs, less voltage blockage by switches, and good energy conversion quality with a lesser number of components. [80]–[86] may be achieved by eliminating additional switches (marked ones) and cross connection of the previously stated SMs (see Figure 10.11(d) and (e)) provided in Figure 10.11(g). It is made up of direct current sources and unidirectional switches. It may be utilised with both symmetric and asymmetric sources. Varying switches in symmetrical and asymmetrical source configurations need different voltage blocking capabilities, for example, switches S_1 and S_2, switches S_3 and S_4, and switches S_7 and S_8 should have voltage blocking ability of V_{DC1}, $V_{DC1} + V_{DC2}$ and V_{DC2} respectively.

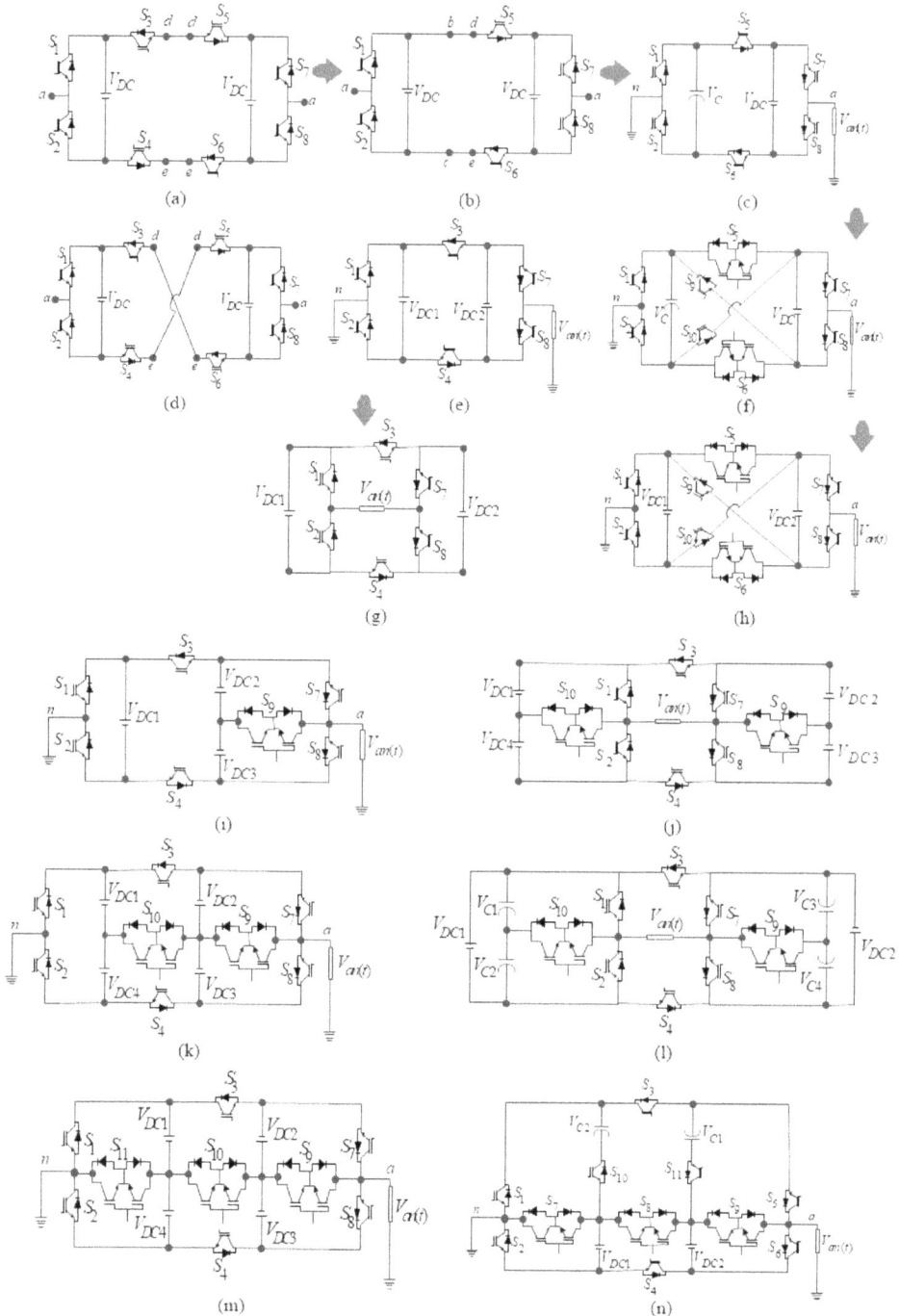

FIGURE 10.11 (a) Connection circuits of SMs-III in parallel, (b) connection circuits of SMs-I and -III in parallel, (c) introduced in [144]–[158], (d) cross-connection circuits of SMs-III, (e) cross-connection circuits of SMs-I, and III, presented in (f) [159], (g) [80]–[86], (h) [87], (i) [88], [89], (j) [90], [91], (k) [160]–[162], (l) [163], [164], (m) [165], and (n) [166].

On conducting, each switch should tolerate the load current. In [83] and [86], asymmetrical source configuration is discussed. It is possible to expand the number of load voltages from specified input DC sources by adding two-circuit-crossover switches, as shown in Figure 10.11(h) [87]. In comparison to the CHB, the aforementioned architecture requires a less quantity of switches and gate drivers to achieve the same number of output voltage levels as the CHB. The topology of [88], which is a new arrangement of Figure 10.11(i), is shown in Figure 10.11(g). A revised version of the aforementioned design (see Figure 10.11(i)) is presented in [89], which enhances the voltage level while lowering the number of power electronics components and blocking voltages stress through switches. The suggested topology in [90] and [91] may be constructed with little modifications by substituting one SM-IV for one SM-I, as depicted in Figure 10.11. This design is shown and investigated utilising symmetrical and asymmetrical DC sources configurations, as well as dc power supply with unidirectional and bidirectional switches connection. The primary disadvantages of the previous design are that switches S_3 and S_4 must be capable of resisting the full voltage of the input DC sources and that it does not support trinary source combinations due to its inability to generate all additive and subtractive signals. By swapping two SM-I and two SM-IV in Figure 10.11(i), the topology presented in [160] may be obtained as depicted in Figure 10.11(k). $V_{DC1} = V_{DC2} = V_{DC}$ and $V_{DC3} = V_{DC4} = nV_{DC}$, where n = 2 or 3. In [161], [162] and [177], many extension kinds of topology are examined and suggested. Furthermore, as indicated in [166], the number of dc sources in the aforementioned topology may be cut in half (see Figure 10.11(m)). The recommended topology in [163] and [164] may be developed as shown in Figure 10.11(j) by connecting SM-I with SM-IV (Figure 10.11(i)). It necessitated the use of bidirectional switches, unidirectional switches, and a mix of DC sources. It is obvious that all switches must be able to withstand uneven voltage. It is proposed and investigated in the context of an asymmetrical source arrangement (binary). S_3 and S_5 switches must be able to block the whole voltage of the input sources. As the preceding topology is not capable to generate all step levels and subtractive arrangement of the input voltage, the preceding topology does not aid trinary source configuration. By connecting a bidirectional switch as in Figure 10.11(n), the preceding demerit is solved in [165]. Furthermore, because of the high redundancy states, the topology shown in Figure 10.11(h) may be employed for fault-tolerant operation. If the recommended system in [167] is utilised, alternative suggested topologies may be better appropriate for fault-tolerant operations. In reality, an additional switch is recommended, which provides another state for the missing level and allows fault tolerance to operate if any one switch fails. Table 10.7 shows the number of output voltage levels and the layout of the dc sources for Figure 10.11. As illustrated in Figure 10.12(b), the proposed topology for MLI in [168] utilises cross switches and parallel circuits to connect SM-I and SM-II. The topology also includes four DC power supply sources, eight unidirectional switches, and two bidirectional power electronics switches, with the DC power supply sources on the left having the same voltage value. Furthermore, sources on the right-hand side have the same voltage stress value as those on the left, but they are not the same. It is obvious that switches must withstand varying amounts of voltage, especially S_6 and S_7, since the blocking voltage must be proportionate to the maximum output voltage. It is not cost effective to utilise the topology for high voltage applications since the number of components grows as the number of voltage levels increases, however this constraint is overcome by the cascade design of the recommended topology. Because of the limited number of redundant states, it is not well suited for fault-tolerant applications. Table 10.8 shows the number of load voltage levels and the layout of the dc sources for Figure 10.12. If the SM-I are connected together in series vertically and then reconnected in parallel horizontally and spare dc sources are omitted as in Figure 10.13, the given topology in [92] is made. To produce necessary voltage levels at the output terminals, the preceding topology needs to have dc sources and bidirectional switches. For a symmetrical DC power supply source arrangement, switches S_1, S_2, S_9, and S_{10} must resist the voltage stress of $4V_{DC}$, switches S_3, S_4, S_7, and S_8 need to withstand the voltage stress of $3V_{DC}$, switches S_5 and S_6 must be

TABLE 10.7
Source Configuration of Figure 10.11

	Source configuration	Value of DC sources	N_{level}
c	Asym. Sources	$V_{DC1} = V_{DC}, V_{DC2} = 2V_{DC}$	5
c	Asym. Sources	$V_{DC1} = V_{DC}, V_{DC2} = 3V_{DC}$	7
f	Asym. Sources	$V_{DC1} = V_{DC}, V_{DC2} = 3V_{DC}$	9
g	Sym. Sources	$V_{DC1} = V_{DC2} = V_{DC}$	5
g	Asym. Sources	$V_{DC1} = V_{DC}, V_{DC2} = 2V_{DC}$	7
h	Sym. Sources	$V_{DC1} = V_{DC2} = V_{DC}$	5
h	Asym. Sources	$V_{DC1} = V_{DC}, V_{DC2} = 3V_{DC}$	9
i	Sym. Sources	$V_{DC1} = V_{DC2} = V_{DC}$	7
i	Asym. Sources	$V_{DC1} = V_{DC}, V_{DC2} = 2V_{DC}$	7
j	Sym. Sources	$V_{DC1} = V_{DC2} = V_{DC3} = V_{DC}$	7
j	Asym. Sources	$V_{DC1} = V_{DC}, V_{DC2} = V_{DC3} = 4V_{DC}$	11
k	Sym. Sources	$V_{DC1} = V_{DC2} = V_{DC3} = V_{DC4} = V_{DC}$	9
k	Asym. Sources	$V_{DC1} = V_{DC4} = V_{DC}, V_{DC2} = V_{DC3} = 2V_{DC}$	13
k	Asym. Sources	$V_{DC1} = V_{DC4} = V_{DC}, V_{DC2} = V_{DC3} = 3V_{DC}$	17
l	Asym. Sources	$V_{DC1} = V_{DC4} = V_{DC}, V_{DC2} = V_{DC3} = 2V_{DC}$	13
m	Asym. Sources	$V_{DC1} = V_{DC4} = V_{DC}, V_{DC2} = V_{DC3} = 2V_{DC}$	13
m	Asym. Sources	$V_{DC1} = V_{DC4} = V_{DC}, V_{DC2} = V_{DC3} = 3V_{DC}$	17

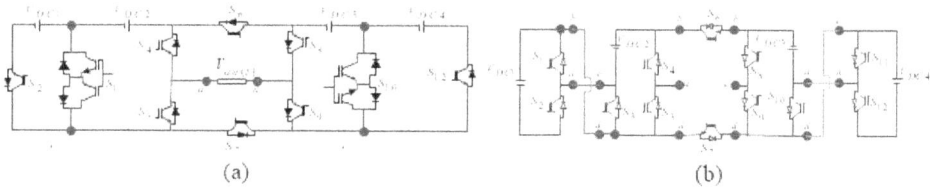

FIGURE 10.12 (a) Presented in [168], and (b) mixed connection of submodules-1 and -2.

TABLE 10.8
Source Arrangement of the Introduced Topology in [168]

Type of source	Value of DC sources	N_{level}
Asym. Sources	$V_{DC1} = V_{DC2} = V_{DC},$ $V_{DC3} = V_{DC4} = 2V_{DC}$	13

able to block $2V_{DC}$. Additionally, due to its low redundancy, the preceding topology is not suitable for fault-tolerant applications.

The number of output voltage levels and pattern of the dc sources for Figure 10.13 are represented in Table 10.9. By parallel connection of SMs-III and SMs-V as in Figure 10.14, formation different ANPC is done as suggested in [170].

The suggested topology consists of two dc sources, one capacitor, and eight switches. On comparison with 5L-ANPC, voltage ratings of the switches are unequal but the number of components

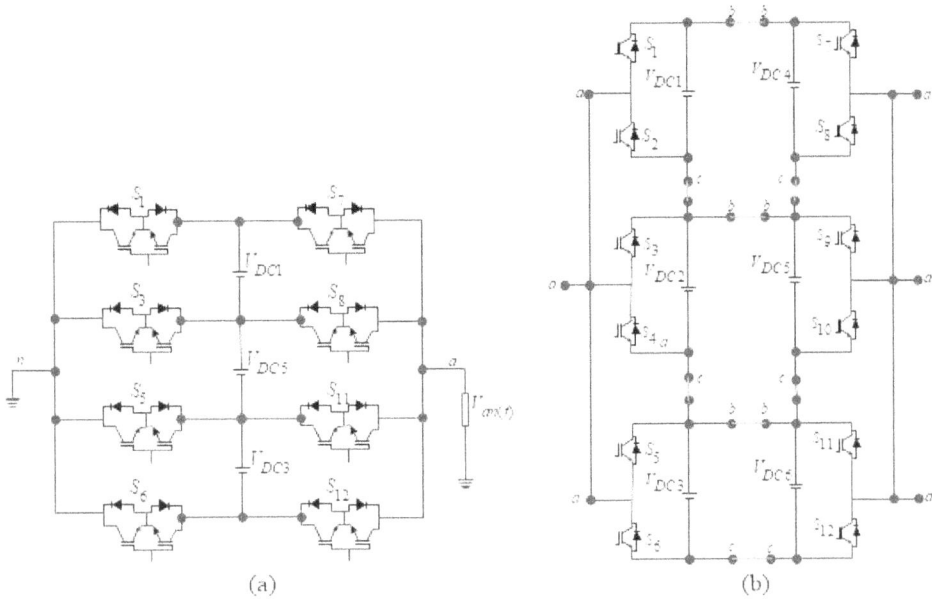

FIGURE 10.13   (a) Presented in [169], and (b) vertical series and horizontal parallel connection of submodule-1.

TABLE 10.9
Source Arrangement of the Introduced Topology in [169]

Type of Source	Value of DC sources	N_{level}
Sym. Sources	$V_{DC1} = V_{DC2} = V_{DC3} = V_{DC4} = V_{DC}$	7

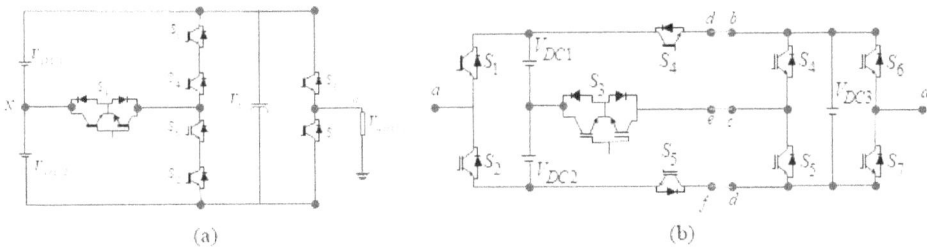

FIGURE 10.14   (a) Suggested in [170], and (b) parallel connection of submodules-1 and -5.

is equal. When $V_{DC1} = V_{DC2} = 2V_C = 1/2V_{DC}$, then the voltage ratings of S_1 and S_2 is $3/4V_{DC}$ and $1/4V_{DC}$ for rest of the switches. The number of load voltage levels and pattern of the dc sources for Figure 10.14 are represented in Table 10.10.

10.4.2   H-Bridge Topologies (Unipolar)

In general, there are two forms of MLI, each with a distinct purpose of producing levels either naturally or through an H-bridge (bipolar). As mentioned before in the description of MLI, they

TABLE 10.10

Source Arrangement of the Introduced Topology in [166]

Type of source	Value of DC sources	N_{level}
Asym. Sources	$V_{DC1} = V_{DC2} = 1/2VDC$	5

TABLE 10.11

Source Arrangement of the Introduced Topology in [93]–[110]

Type of source	Value of DC sources	N_{level}
Sym. Sources	$V_{DC1} = V_{DC2} = V_{DC3} = V_{DC}$	7

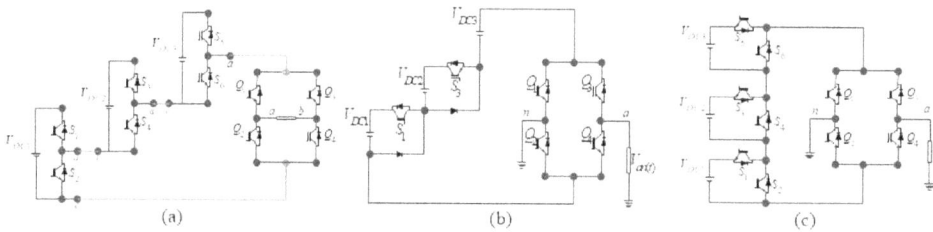

FIGURE 10.15 (a) Series connection of submodule-1, (b) proposed in [166]–[172], and (c) presented in [154]–[158].

are capable of producing both positive and negative load voltage levels. Other MLIs create just positive polarity in the first section and then convert the polarity waveform to generate ac voltage using an H-bridge in the second section. The first section denotes the "level-generation section," while the second section denotes the "polarity-generation section." Additionally, H-bridge topologies have fewer components yet their associated circuits can withstand high switch stress. The unipolar MLIs will lose all positive or negative levels if one of the switches in the H-bridge fails, and they also have a significant chance of failure in fault-tolerant applications. These MLIs are presented and studied in this part. They can be constructed from the major SMs. The suggested topology in [154]–[158] may be constructed (see Figure 10.15(b)) by sequentially linking SMs-I (half H-bridge cells). An H-bridge module is put at the finish of the series connection circuits for the polarity pats part. Level generation and polarity generation pats necessitate that each switch be capable of transporting and blocking the voltage of the corresponding dc source(s) respectively, and each switch must be capable of preventing polarity generation. As a result, H-bridge switches have a higher rating than level-generation switches. Although uneven source configurations are not assessed in [93]–[102], the binary source may be used [103]. In compared to CHB, binary topology requires fewer gate drivers and switches to implement the same number of load voltage steps. Table 10.11 illustrates the number of steps and pattern for symmetric, and asymmetric topologies. It may be extended as a new architecture, as seen in Figure 10.15(c), where switches are replaced with diodes [104]–[110]. Due to the fact that diodes cannot carry negative currents, the aforementioned architecture is incapable of powering inductive loads. These are referred to as none-regeneration topologies.

10.5 CONCLUSION

This chapter presents an in-depth assessment of MLIs and their development, as well as some ideas and recommendations for future scholars in their work. Additionally, the document may be exploited as an educational paper. In this chapter, the similarity of structural parts is systematically examined. In actuality, it turns out that specific SMs are discovered in all topologies on a regular basis, and all of these SMs have been described. A total of five main SMs are identified, linked, and generalised from these SMs. Multilevel topologies (MLIs) are constructed from these core SMs that have the capacity to connect and may be classed as either traditional or current MLIs. The chapter has undertaken a detailed examination of practically all of the MLIs in paper based on the suggested SMs with distinct links that have been provided. There is also a full description of the qualitative and quantitative capabilities of MLIs, as well as their whole potential operation, including symmetric, asymmetric, single-source, and regeneration modes, as well as the ability to produce negative levels.

REFERENCES

[1] A. Kumar Singh, R. Kumar Mandal, R. Raushan, and R. Anand. Five-level switched capacitor inverter for photovoltaic applications. *IETE Technical Review*, pp. 1–8, 2022.

[2] A. K. Singh, and R. K. Mandal. A Novel 17-level Reduced component single DC switched-capacitor-based inverter with reduced input spike current. *IEEE Journal of Emerging and Selected Topics in Power Electronics*, doi: 10.1109/JESTPE.2022.3166222.

[3] A. K. Singh, R. Raushan, R. K. Mandal, and M. W. Ahmad. A new single-source nine-level quadruple boost inverter (NQBI) for PV application. *IEEE Access*, vol. 10, pp. 36246–36253, 2022, doi: 10.1109/ACCESS.2022.3163262.

[4] A. Lashkar Ara, A. Kazemi, and S. A. Nabavi Niaki. Multiobjective optimal location of facts shunt-series controllers for power system operation planning. *IEEE Transactions on Power Delivery*, vol. 27(2), pp. 481–490, 2012.

[5] P. Neeraj, B. Mahajan Sagar, P. Sanjeevikumar, B. Frede Azam. Farooque: 'New CUK—SEPIC converter based photovoltaic power system with hybrid GSA—PSO algorithm employing MPPT for water pumping applications. *IET Power Electronics*, vol. 13, no. 13, pp. 2824–2830, 2020, doi: 10.1049/iet-pel.2019.1154.

[6] D. Soto, and T. C. Green. A comparison of high-power converter topologies for the implementation of facts controllers. *IEEE Transactions on Industrial Electronics*, vol. 49, no. 5, pp. 1072–1080, 2002.

[7] R. Sajadi, H. Iman-Eini, M. K. Bakhshizadeh, Y. Neyshabouri, and S. Farhangi. Selective harmonic elimination technique with control of capacitive dc-link voltages in an asymmetric cascaded h-bridge inverter for statcom application. *IEEE Transactions on Industrial Electronics*, vol. 65, no. 11, pp. 8788–8796, 2018.

[8] N. N. V. Surendra Babu, and B. G. Fernandes. Cascaded two-level inverter-based multilevel statcom for high-power applications. *IEEE Transactions on Power Delivery*, vol. 29, no. 3, pp. 993–1001, 2014.

[9] A. K. Singh, and R. K. Mandal. A grid-connected single-stage closed-loop controlled five-level boosting inverter for solar pv systems. *Australian Journal of Electrical and Electronics Engineering*, pp. 1–9, 2022.

[10] Y. Liu, and F. L. Luo. Trinary hybrid multilevel inverter used in statcom with unbalanced voltages. *IEE Proceedings—Electric Power Applications*, vol. 152, no. 5, pp. 1203–1222, 2005.

[11] A. Nami, J. Liang, F. Dijkhuizen, and G. D. Demetriades. Modular multilevel converters for hvdc applications: Review on converter cells and functionalities. *IEEE Transactions on Power Electronics*, vol. 30, no. 1, pp. 18–36, 2015.

[12] N. Priyadarshi, S. Padmanaban, P. Kiran Maroti, and A. Sharma. An extensive practical investigation of FPSO-based MPPT for grid integrated PV system under variable operating conditions with anti-islanding protection. *IEEE Systems Journal*, vol. 13, no. 2, pp. 1861–1871, 2019, doi: 10.1109/JSYST.2018.2817584.

[13] J. Jung, S. Cui, J. Lee, and S. Sul. A new topology of multilevel vsc converter for a hybrid hvdc transmission system. *IEEE Transactions on Power Electronics*, vol. 32, no. 6, pp. 4199–4209, 2017.

[14] R. Feldman, M. Tomasini, E. Amankwah, J. C. Clare, P. W. Wheeler, D. R. Trainer, and R. S. Whitehouse. A hybrid modular multilevel voltage source converter for hvdc power transmission. *IEEE Transactions on Industry Applications*, vol. 49, no. 4, pp. 1577–1588, 2013.

[15] H. Hafezi, and R. Faranda. Dynamic voltage conditioner: A new concept for smart lowvoltage distribution systems. *IEEE Transactions on Power Electronics*, vol. 33, no. 9, pp. 7582–7590, 2018.

[16] S. Galeshi, and H. Iman-Eini. Dynamic voltage restorer employing multilevel cascaded h-bridge inverter. *IET Power Electronics*, vol. 9, no. 11, pp. 2196–2204, 2016.

[17] F. Jiang, C. Tu, Z. Shuai, M. Cheng, Z. Lan, and F. Xiao. Multilevel cascaded-type dynamic voltage restorer with fault current-limiting function. *IEEE Transactions on Power Delivery*, vol. 31, no. 3, pp. 1261–1269, 2016.

[18] E. Babaei, M. F. Kangarlu, and M. Sabahi. Dynamic voltage restorer based on multilevel inverter with adjustable dc-link voltage. *IET Power Electronics*, vol. 7, no. 3, pp. 576–590, 2014.

[19] V. S. Cheung, R. S. Yeung, H. S. Chung, A. W. Lo, and W. Wu. A transformer-less unified power quality conditioner having fast dynamic control. *2017 IEEE Energy Conversion Congress and Exposition (ECCE)*, pp. 2962–2968, 2017.

[20] S. Lakshmi, and S. Ganguly. Modelling and allocation of open-upqc-integrated pv generation system to improve the energy efficiency and power quality of radial distribution networks. *IET Renewable Power Generation*, vol. 12, no. 5, pp. 605–613, 2018.

[21] A. M. Rauf, A. V. Sant, V. Khadkikar, and H. H. Zeineldin. A novel ten-switch topology for unified power quality conditioner. *IEEE Transactions on Power Electronics*, vol. 31, no. 10, pp. 6937–6946, 2016.

[22] G. J. Li, F. Ma, S. S. Choi, and X. P. Zhang. Control strategy of a cross-phase-connected unified power quality conditioner. *IET Power Electronics*, vol. 5, no. 5, pp. 600–608, 2012.

[23] Priyadarshi, Neeraj, Sanjeevikumar Padmanaban, Dan M. Ionel, Lucian Mihet-Popa, and Farooque Azam. 2018. "Hybrid PV-wind, micro-grid development using quasi-z-source inverter modeling and control—experimental investigation. *Energies*, vol. 11, no. 9: p. 2277, doi: 10.3390/en11092277.

[24] Z. Shu, H. Lin, Z. Ziwei, X. Yin, and Q. Zhou. Specific order harmonics compensation algorithm and digital implementation for multi-level active power filter. *IET Power Electronics*, vol. 10, no. 5, pp. 525–535, 2017.

[25] P. Acuña, L. Morán, M. Rivera, J. Dixon, and J. Rodriguez. Improved active power filter performance for renewable power generation systems. *IEEE Transactions on Power Electronics*, vol. 29, no. 2, pp. 687–694, 2014.

[26] A. Varschavsky, J. Dixon, M. Rotella, and L. Morán. Cascaded nine-level inverter for hybrid-series active power filter, using industrial controller. *IEEE Transactions on Industrial Electronics*, vol. 57, no. 8, pp. 2761–2767, 2010.

[27] L. Wang, D. Zhang, Y. Wang, B. Wu, and H. S. Athab. Power and voltage balance control of a novel three-phase solid-state transformer using multilevel cascaded h-bridge inverters for microgrid applications. *IEEE Transactions on Power Electronics*, vol. 31, no. 4, pp. 3289–3301, 2016.

[28] L. Tarisciotti, P. Zanchetta, A. Watson, P. Wheeler, J. C. Clare, and S. Bifaretti. Multiobjective modulated model predictive control for a multilevel solid-state transformer. *IEEE Transactions on Industry Applications*, vol. 51, no. 5, pp. 4051–4060, 2015.

[29] S. Madhusoodhanan, A. Tripathi, D. Patel, K. Mainali, A. Kadavelugu, S. Hazra, S. Bhattacharya, and K. Hatua. Solid-state transformer and mv grid tie applications enabled by 15 kv sic igbts and 10 kv sic mosfets based multilevel converters. *IEEE Transactions on Industry Applications*, vol. 51, no. 4, pp. 3343–3360, 2015.

[30] T. Zhao, G. Wang, S. Bhattacharya, and A. Q. Huang. Voltage and power balance control for a cascaded h-bridge converter-based solid-state transformer. *IEEE Transactions on Power Electronics*, vol. 28, no. 4, pp. 1523–1532, 2013.

[31] M. Quraan, P. Tricoli, S. D'Arco, and L. Piegari. Efficiency assessment of modular multilevel converters for battery electric vehicles. *IEEE Transactions on Power Electronics*, vol. 32, no. 3, pp. 2041–2051, 2017.

[32] N. Priyadarshi, S. Padmanaban, J. B. Holm-Nielsen, F. Blaabjerg, and M. S. Bhaskar. An experimental estimation of hybrid ANFIS—PSO-based MPPT for PV grid integration under fluctuating sun irradiance. *IEEE Systems Journal*, vol. 14, no. 1, pp. 1218–1229, 2020, doi: 10.1109/JSYST.2019.2949083.

[33] M. Z. Youssef, K. Woronowicz, K. Aditya, N. A. Azeez, and S. S. Williamson. Design and development of an efficient multilevel dc/ac traction inverter for railway transportation electrification. *IEEE Transactions on Power Electronics*, vol. 31, no. 4, pp. 3036–3042, 2016.

[34] F. Azam, N. Priyadarshi, H. Nagar, S. Kumar, and A. K. Bhoi. An overview of solar-powered electric vehicle charging in vehicular adhoc network. In: Patel, N., Bhoi, A. K., Padmanaban, S., Holm-Nielsen, J. B. (eds) *Electric Vehicles. Green Energy and Technology*. Singapore: Springer, 2021, doi: 10.1007/978-981-15-9251-5_5.

[35] F. Azam, S. K. Yadav, N. Priyadarshi, S. Padmanaban, and R. C. Bansal. A comprehensive review of authentication schemes in vehicular ad-hoc network. *IEEE Access*, vol. 9, pp. 31309–31321, 2021, doi: 10.1109/ACCESS.2021.3060046.

[36] P. Neeraj, S. Padmanaban, L. Mihet-Popa, F. Blaabjerg, and F. Azam. 2018. "Maximum power point tracking for brushless DC motor-driven photovoltaic pumping systems using a hybrid ANFIS-FLOWER pollination optimization algorithm. Energies, vol. 11, no. 5, p. 1067, doi: 10.3390/en11051067.

[37] M. Carpita, M. Marchesoni, M. Pellerin, and D. Moser. Multilevel converter for traction applications: Small-scale prototype tests results. *IEEE Transactions on Industrial Electronics*, vol. 55, no. 5, pp. 2203–2212, 2008.

[38] S. Padmanaban et al. A novel modified sine-cosine optimized MPPT algorithm for grid integrated PV system under real operating conditions. *IEEE Access*, vol. 7, pp. 10467–10477, 2019, doi: 10.1109/ACCESS.2018.2890533.

[39] J. Rodriguez, Jih-Sheng Lai, and Fang Zheng Peng. Multilevel inverters: a survey of topologies, controls, and applications. *IEEE Transactions on Industrial Electronics*, vol. 49, no. 4, pp. 724–738, 2002.

[40] F. Azam, A. Biradar, N. Priyadarshi, S. Kumari, D. Almakhles, and S. Tangade. A framework for secured dissemination of messages in Internet of Vehicle (IoV) using blockchain approach. *2021 IEEE International Conference on Mobile Networks and Wireless Communications (ICMNWC)*, 2021, pp. 1–6, doi: 10.1109/ICMNWC52512.2021.9688397.

[41] P. Neeraj, and K. Vigna Ramachandaramurthy, S. Padmanaban, and F. Azam. 2019. "An ant colony optimized MPPT for standalone hybrid PV-wind power system with single cuk converter. Energies, vol. 12, no. 1, p. 167, doi: 10.3390/en12010167.

[42] S. Padmanaban, N. Priyadarshi, M. S. Bhaskar, J. B. Holm-Nielsen, E. Hossain, and F. Azam. A hybrid photovoltaic-fuel cell for grid integration with jaya-based maximum power point tracking: Experimental performance evaluation. *IEEE Access*, vol. 7, pp. 82978–82990, 2019, doi: 10.1109/ACCESS.2019.2924264.

[43] F. Azam, A. Biradar, N. Priyadarshi, S. Kumari, and S. Tangade. A review of blockchain based approach for secured communication in Internet of Vehicle (IoV) scenario. *2021 Second International Conference on Smart Technologies in Computing, Electrical and Electronics (ICSTCEE)*, 2021, pp. 1–6, doi: 10.1109/ICSTCEE54422.2021.9708555.

[44] K. A. Corzine, and S. Lu. Comparison of hybrid propulsion drive schemes. *IEEE Electric Ship Technologies Symposium*, vol. 2005, pp. 355–362, 2005.

[45] S. Lu, and K. Corzine. Multilevel multi-phase propulsion drives. *IEEE Electric Ship Technologies Symposium*, vol. 2005, pp. 363–370, 2005.

[46] P. Neeraj, P. Sanjeevikumar, H.-N., Jens Bo, B. Mahajan Sagar, and A. Farooque, 'Internet of things augmented a novel PSO-employed modified zeta converter-based photovoltaic maximum power tracking system: hardware realisation', *IET Power Electronics*, vol. 13, no. 13, pp. 2775–2781, 2020, doi: 10.1049/iet-pel.2019.1121.

[47] Y. Yue, Q. Xu, A. Luo, P. Guo, Z. He, and Y. Li. Analysis and control of tundish induction heating power supply using modular multilevel converter. *IET Generation, Transmission Distribution*, vol. 12, no. 14, pp. 3452–3460, 2018.

[48] J. Sabate, L. J. Garces, P. M. Szczesny, Qiming Li, and W. F. Wirth. High-power highfidelity switching amplifier driving gradient coils for mri systems. In *2004 IEEE 35th annual power electronics specialists conference (IEEE Cat. No.04CH37551)*, vol. 1, pp. 261–266, 2004.

[49] J. Venkataramanaiah, Y. Suresh, and Anup Kumar Panda. A review on symmetric, asymmetric, hybrid and single DC sources based multilevel inverter topologies. *Renewable and Sustainable Energy Reviews*, vol. 76, pp. 788–812, 2017.

[50] M. J. Mojibian, and M. Tavakoli Bina. Classification of multilevel converters with a modular reduced structure: implementing a prominent 31-level 5 kva class b converter. *IET Power Electronics*, vol. 8, no. 1, pp. 20–32, 2015.

[51] J. Rodriguez, S. Bernet, B. Wu, J. O. Pontt, and S. Kouro. Multilevel voltage-sourceconverter topologies for industrial medium-voltage drives. *IEEE Transactions on Indus-trial Electronics*, vol. 54, no. 6, pp. 2930–2945, 2007.

[52] P. Lezana, J. Pou, T. A. Meynard, J. Rodriguez, S. Ceballos, and F. Richardeau. Survey on fault operation on multilevel inverters. *IEEE Transactions on Industrial Electronics*, vol. 57, no. 7, pp. 2207–2218, 2010.

[53] M. A. Parker, L. Ran, and S. J. Finney. Distributed control of a fault-tolerant modular multilevel inverter for direct-drive wind turbine grid interfacing. *IEEE Transactions on Industrial Electronics*, vol. 60, no. 2, pp. 509–522, 2013.

[54] A. Chen, L. Hu, L. Chen, Y. Deng, and X. He. A multilevel converter topology with fault-tolerant ability. *IEEE Transactions on Power Electronics*, vol. 20, no. 2, pp. 405–415, 2005.

[55] S. P. Gautam, L. Kumar, S. Gupta, and N. Agrawal. A single-phase five-level inverter topology with switch fault-tolerance capabilities. *IEEE Transactions on Industrial Electronics*, vol. 64, no. 3, pp. 2004–2014, 2017.

[56] M. Aly, E. M. Ahmed, and M. Shoyama. A new single-phase five-level inverter topology for single and multiple switches fault tolerance. *IEEE Transactions on Power Electronics*, vol. 33, no. 11, pp. 9198–9208, 2018.

[57] H. Abu-Rub, J. Holtz, J. Rodriguez, and G. Baoming. Medium-voltage multilevel converters—state of the art, challenges, and requirements in industrial applications. *IEEE Transactions on Industrial Electronics*, vol. 57, no. 8, pp. 2581–2596, 2010.

[58] W. McMurray. Fast response stepped-wave switching power converter circuit. *U.S. Patent 3 581 212*, 1971.

[59] J. A. Dickerson, and G. H. Ottaway. Transformerless power supply with line to load isolation. *U.S.Patent 3 596 369*, Aug. 1971.

[60] Fang Zheng Peng, Jih-Sheng Lai, J. McKeever, and J. VanCoevering. A multilevel voltage-source inverter with separate dc sources for static var generation. *IAS'95. Conference Record of the 1995 IEEE Industry Applications Conference Thirtieth IAS Annual Meeting*, vol. 3, pp. 2541–2548, 1995.

[61] R. H. Baker. High-voltage converter circuit. *U.S.Patent 4 203 151*, 1980.

[62] R. H. Baker. Bridge converter circuit. *U.S.Patent 4 270 163*, 1981.

[63] J. Rodriguez, S. Bernet, P. K. Steimer, and I. E. Lizama. A survey on neutral-pointclamped inverters. *IEEE Transactions on Industrial Electronics*, vol. 57, no. 7, pp. 2219–2230, 2010.

[64] T. A. Meynard, and H. Foch. Multi-level choppers for high voltage applications. *EPE Journal*, vol. 2, no. 1, pp. 45–50, 1992.

[65] R. Marquardt. tromrichterschaltungen mit verteilten energiespeichern. *German Patent DE10103031A1*, Jan. 2001.

[66] R. Marquardt, A. Lesnicar, and J. Hildinger. Modulares stromrichterkonzept fu¨r netzkupplungsanwendungen bei hohen spannungen. *ETG-Conference*, 2002.

[67] Fang Zheng Peng. A generalized multilevel inverter topology with self-voltage balancing. *IEEE Transactions on Industry Applications*, vol. 37, no. 2, pp. 611–618, 2001.

[68] P. Barbosa, P. Steimer, L. Meysenc, M. Winkelnkemper, J. Steinke, and N. Celanovic. Active neutral-point-clamped multilevel converters. In *2005 IEEE 36th Power Electronics Specialists Conference*, pp. 2296–2301, 2005.

[69] M. Malinowski, K. Gopakumar, J. Rodriguez, and M. A. P´erez. A survey on cascaded multilevel inverters. *IEEE Transactions on Industrial Electronics*, vol. 57, no. 7, pp. 2197–2206, 2010.

[70] T. Bruckner, S. Bernet, and H. Guldner. The active npc converter and its loss-balancing control. *IEEE Transactions on Industrial Electronics*, vol. 52, no. 3, pp. 855–868, 2005.

[71] O. Apeldoorn, B. Odegard, P. Steimer, and S. Bernet. A 16 mva anpc-pebb with 6 ka igcts. In *Fourtieth IAS Annual Meeting. Conference Record of the 2005 Industry Applications Conference, 2005*, vol. 2, pp. 818–824, 2005.

[72] A. Nabae, I. Takahashi, and H. Akagi. A new neutral-point-clamped pwm inverter. *IEEE Transactions on Industry Applications*, vol. IA–17, no. 5, pp. 518–523, 1981.

[73] A. Lesnicar, and R. Marquardt. An innovative modular multilevel converter topology suitable for a wide power range. In *2003 IEEE Bologna Power Tech Conference Proceedings*, vol. 3, p. 6, 2003.

[74] M. Hagiwara, K. Nishimura, and H. Akagi. A modular multilevel pwm inverter for medium-voltage motor drives. In *2009 IEEE Energy Conversion Congress and Exposition*, pp. 2557–2564, 2009.

[75] M. Hagiwara, K. Nishimura, and H. Akagi. A medium-voltage motor drive with a modular multilevel pwm inverter. *IEEE Transactions on Power Electronics*, vol. 25, no. 7, pp. 1786–1799, 2010.

[76] B. Jacobson, P. Karlsson, G. Asplund, L. Harnefors, and T. Jonsson. VSC-HVDC transmission with cascaded two-level converters. In *Proceedings Cigre*, pp. B4–B110, 2010.

[77] K. Ilves, F. Taffner, S. Norrga, A. Antonopoulos, L. Harnefors, and H. Nee. A submodule implementation for parallel connection of capacitors in modular multilevel converters. In *2013 15th European Conference on Power Electronics and Applications (EPE)*, pp. 1–10, 2013.

[78] K. Ilves, F. Taffner, S. Norrga, A. Antonopoulos, L. Harnefors, and H. Nee. A submodule implementation for parallel connection of capacitors in modular multilevel converters. *IEEE Transactions on Power Electronics*, vol. 30, no. 7, pp. 3518–3527, 2015.

[79] M. Vizheh, M. Rezanejad, and E. Samadaei. New asymmetrical commutation cell for multilevel inverters with reduced number of components. *2016 7th Power Electronics and Drive Systems Technologies Conference (PEDSTC)*, 2016, pp. 153–158.

[80] K. K. Gupta, and S. Jain. A novel multilevel inverter based on switched dc sources. *IEEE Transactions on Industrial Electronics*, vol. 61, no. 7, pp. 3269–3278, 2014.

[81] K. K. Gupta, and S. Jain. Multilevel inverter topology based on series connected switched sources. *IET Power Electronics*, vol. 6, no. 1, pp. 164–174, 2013.

[82] S. Thamizharasan, J. Baskaran, S. Ramkumar, and S. Jeevananthan. Cross-switched multilevel inverter using auxiliary reverse-connected voltage sources. *IET Power Electronics*, vol. 7, no. 6, pp. 1519–1526, 2014.

[83] K. K. Gupta, and S. Jain. Comprehensive review of a recently proposed multilevel inverter. *IET Power Electronics*, vol. 7, no. 3, pp. 467–479, 2014.

[84] A. Nami, L. Wang, F. Dijkhuizen, and A. Shukla. Five level cross connected cell for cascaded converters. In *2013 15th European Conference on Power Electronics and Applications (EPE)*, pp. 1–9, 2013.

[85] M. F. Kangarlu, and E. Babaei. Cross-switched multilevel inverter: An innovative topology. *IET Power Electronics*, vol. 6, no. 4, pp. 642–651, 2013.

[86] M. F. Kangarlu, E. Babaei, and M. Sabahi. Cascaded cross-switched multilevel inverter in symmetric and asymmetric conditions. *IET Power Electronics*, vol. 6, no. 6, pp. 1041–1050, 2013.

[87] K. K. Gupta, and S. Jain. Topology for multilevel inverters to attain maximum number of levels from given dc sources. *IET Power Electronics*, vol. 5, no. 4, pp. 435–446, 2012.

[88] E. Babaei, S. Laali, and S. Alilu. Cascaded multilevel inverter with series connection of novel h-bridge basic units. *IEEE Transactions on Industrial Electronics*, vol. 61, no. 12, pp. 6664–6671, 2014.

[89] E. Babaei, and S. Laali. Optimum structures of proposed new cascaded multilevel inverter with reduced number of components. *IEEE Transactions on Industrial Electronics*, vol. 62, no. 11, pp. 6887–6895, 2015.

[90] M. Vijeh, E. Samadaei, M. Rezanejad, H. Vahedi, and K. Al-Haddad. Design and implementation of a new three source topology of multilevel inverters with reduced number of switches. In *IECON 2016–42nd Annual Conference of the IEEE Industrial Electronics Society*, pp. 6500–6505, 2016.

[91] C. I. Odeh, E. S. Obe, and O. Ojo. Topology for cascaded multilevel inverter. *IET Power Electronics*, vol. 9, no. 5, pp. 921–929, 2016.

[92] E. Babaei. A cascade multilevel converter topology with reduced number of switches. *IEEE Transactions on Power Electronics*, vol. 23, no. 6, pp. 2657–2664, 2008.

[93] Gui-Jia Su. Multilevel dc link inverter. *Conference Record of the 2004 IEEE Industry Applications Conference, 2004.39th IAS Annual Meeting*, vol. 2, pp. 806–812, 2004.

[94] Gui-Jia Su. Multilevel dc-link inverter. *IEEE Transactions on Industry Applications*, vol. 41, no. 3, pp. 848–854, 2005.

[95] G. Waltrich, and I. Barbi. Three-phase cascaded multilevel inverter using power cells with two inverter legs in series. In *2009 IEEE Energy Conversion Congress and Exposition*, pp. 3085–3092, 2009.

[96] A. Salem, E. M. Ahmed, M. Orabi, and M. Ahmed. New three-phase symmetrical multilevel voltage source inverter. *IEEE Journal on Emerging and Selected Topics in Circuits and Systems*, vol. 5, no. 3, pp. 430–442, 2015.

[97] M. M. Hasan, S. Mekhilef, and M. Ahmed. Three-phase hybrid multilevel inverter with less power electronic components using space vector modulation. *IET Power Electronics*, vol. 7, no. 5, pp. 1256–1265, 2014.

[98] H. Belkamel, S. Mekhilef, A. Masaoud, and M. A. Naeim. Novel three-phase asymmetrical cascaded multilevel voltage source inverter. *IET Power Electronics*, vol. 6, no. 8, pp. 1696–1706, 2013.

[99] N. Booma, and N. Sridhar. Nine level cascaded h-bridge multilevel dc-link inverter. In *2011 International Conference on Emerging Trends in Electrical and Computer Technology*, pp. 315–320, 2011.

[100] M. S. Varna, and J. Jose. Seven level inverter with nearest level control. In *2014 International Conference on Green Computing Communication and Electrical Engineering (ICGCCEE)*, pp. 1–7, 2014.

[101] M. S. Varna, and J. Jose. A novel seven level inverter with reduced number of switches. In *2014 IEEE 2nd International Conference on Electrical Energy Systems (ICEES)*, pp. 294–299, 2014.

[102] S. N. Rao, D. V. A. Kumar, and C. S. Babu. New multilevel inverter topology with reduced number of switches using advanced modulation strategies. In *2013 International Conference on Power, Energy and Control (ICPEC)*, pp. 693–699, 2013.

[103] Ebrahim Babaei and Seyed Hossein Hosseini. New cascaded multilevel inverter topology with minimum number of switches. *Energy Conversion and Management*, vol. 50, no. 11, pp. 2761–2767, 2009.

[104] D. Mudadla, Sandeep N, and G. R. Rao. Novel asymmetrical multilevel inverter topology with reduced number of switches for photovoltaic applications. In *2015 International Conference on Computation of Power, Energy, Information and Communication (ICCPEIC)*, pp. 0123–0128, 2015.

[105] R. Shalchi Alishah, D. Nazarpour, S. H. Hosseini, and M. Sabahi. Novel topologies for symmetric, asymmetric, and cascade switched-diode multilevel converter with minimum number of power electronic components. *IEEE Transactions on Industrial Electronics*, vol. 61, no. 10, pp. 5300–5310, 2014.

[106] R. Shalchi Alishah, D. Nazarpour, S. H. Hosseini, and M. Sabahi. New hybrid structure for multilevel inverter with fewer number of components for high-voltage levels. *IET Power Electronics*, vol. 7, no. 1, pp. 96–104, 2014.

[107] S. P. Gautam, L. Kumar, and S. Gupta. A modified structure for symmetrical and asymmetrical configuration of multilevel inverter. In *IECON 2015–41st Annual Conference of the IEEE Industrial Electronics Society*, pp. 001430–001435, 2015.

[108] K. Dhanalakshmi, and K. S. Kavin. Single phase thirteen-level inverter using seven switches for photovoltaic systems. In *2014 International Conference on Circuits, Power and Computing Technologies [ICCPCT-2014]*, pp. 1002–1005, 2014.

[109] M. F. Kashif, and A. K. Rashid. A multilevel inverter topology with reduced number of switches. In *2016 International Conference on Intelligent Systems Engineering (ICISE)*, pp. 268–271, 2016.

[110] S. Thamizharasan, J. Baskaran, S. Ramkumar, and S. Jeevananthan. A new dual bridge multilevel dc-link inverter topology. *International Journal of Electrical Power Energy Systems*, vol. 45, no. 1, pp. 376–383, 2013.

[111] E. Babaei, M. F. Kangarlu, and M. Sabahi. Extended multilevel converters: an attempt to reduce the number of independent dc voltage sources in cascaded multilevel converters. *IET Power Electronics*, vol. 7, no. 1, pp. 157–166, 2014.

[112] Y. Hinago, and H. Koizumi. A single phase multilevel inverter using switched series/parallel dc voltage sources. In *2009 IEEE Energy Conversion Congress and Exposition*, pp. 1962–1967, 2009.

[113] Y. Hinago, and H. Koizumi. A single-phase multilevel inverter using switched series/parallel dc voltage sources. *IEEE Transactions on Industrial Electronics*, vol. 57, no. 8, pp. 2643–2650, 2010.

[114] Y. Hinago, and H. Koizumi. A switched-capacitor inverter using series/parallel conversion with inductive load. *IEEE Transactions on Industrial Electronics*, vol. 59, no. 2, pp. 878–887, 2012.

[115] R. Barzegarkhoo, H. M. Kojabadi, E. Zamiry, N. Vosoughi, and L. Chang. Generalized structure for a single phase switched-capacitor multilevel inverter using a new multiple dc link producer with reduced number of switches. *IEEE Transactions on Power Electronics*, vol. 31, no. 8, pp. 5604–5617, 2016.

[116] E. Zamiri, N. Vosoughi, S. H. Hosseini, R. Barzegarkhoo, and M. Sabahi. A new cascaded switched-capacitor multilevel inverter based on improved series—parallel conversion with less number of components. *IEEE Transactions on Industrial Electronics*, vol. 63, no. 6, pp. 3582–3594, 2016.

[117] J. Liu, J. Wu, J. Zeng, and H. Guo. A novel nine-level inverter employing one voltage source and reduced components as high-frequency ac power source. *IEEE Transactions on Power Electronics*, vol. 32, no. 4, pp. 2939–2947, 2017.

[118] Y. Ye, K. W. E. Cheng, J. Liu, and K. Ding. A step-up switched-capacitor multilevel inverter with self-voltage balancing. *IEEE Transactions on Industrial Electronics*, vol. 61, no. 12, pp. 6672–6680, 2014.

[119] W. Choi, and F. Kang. H-bridge based multilevel inverter using pwm switching function. In *INTELEC 2009–31st International Telecommunications Energy Conference*, pp. 1–5, 2009.

[120] E. Najafi, A. H. M. Yatim, and A. S. Samosir. A new topology -reversing voltage (rv) for multi-level inverters. In *2008 IEEE 2nd International Power and Energy Conference*, pp. 604–608, 2008.

[121] E. Najafi, and A. H. M. Yatim. Design and implementation of a new multilevel inverter topology. *IEEE Transactions on Industrial Electronics*, vol. 59, no. 11, pp. 4148–4154, 2012.

[122] R. Yadav, P. Bansal, and A. R. Saxena. A three-phase 9-level inverter with reduced switching devices for different pwm techniques. In *2014 6th IEEE Power India International Conference (PIICON)*, pp. 1–6, 2014.

[123] E. Babaei, M. Sarbanzadeh, M. A. Hosseinzadeh, and C. Cecati. A new basic unit for symmetric and asymmetric cascaded multilevel inverter with reduced number of components. In *IECON 2016–42nd Annual Conference of the IEEE Industrial Electronics Society*, pp. 3147–3152, 2016.

[124] J. Ebrahimi, E. Babaei, and G. B. Gharehpetian. A new multilevel converter topology with reduced number of power electronic components. *IEEE Transactions on Industrial Electronics*, vol. 59, no. 2, pp. 655–667, 2012.

[125] G. Ceglia, V. Guzman, C. Sanchez, F. Ibanez, J. Walter, and M. I. Gimenez. A new simplified multilevel inverter topology for dc—ac conversion. *IEEE Transactions on Power Electronics*, vol. 21, no. 5, pp. 1311–1319, 2006.

[126] J. Antenor Pomilio, G. M. Martins, S. Buso, and G. Spiazzi. Three-phase low-frequency commutation inverter for renewables. *IEEE 2002 28th Annual Conference of the Industrial Electronics Society. IECON 02*, vol. 2, pp. 1119–1124, 2002.

[127] G. M. Martins, J. A. Pomilio, S. Buso, and G. Spiazzi. Three-phase low-frequency commutation inverter for renewable energy systems. *IEEE Transactions on Industrial Electronics*, vol. 53, no. 5, pp. 1522–1528, 2006.

[128] N. A. Rahim, K. Chaniago, and J. Selvaraj. Single-phase seven-level grid-connected inverter for photovoltaic system. *IEEE Transactions on Industrial Electronics*, vol. 58, no. 6, pp. 2435–2443, 2011.

[129] L. Mohammadalibeigy, and N. A. Azli. A new symmetric multilevel inverter structure with less number of power switches. In *2014 IEEE Conference on Energy Conversion (CENCON)*, pp. 321–324, 2014.

[130] R. Shalchi Alishah, D. Nazarpour, S. H. Hosseini, and M. Sabahi. Reduction of power electronic elements in multilevel converters using a new cascade structure. *IEEE Transactions on Industrial Electronics*, vol. 62, no. 1, pp. 256–269, 2015.

[131] M. F. Kangarlu, E. Babaei, and S. Laali. Symmetric multilevel inverter with reduced components based on non-insulated dc voltage sources. *IET Power Electronics*, vol. 5, no. 5, pp. 571–581, 2012.

[132] M. F. Kangarlu, and E. Babaei. A generalized cascaded multilevel inverter using series connection of submultilevel inverters. *IEEE Transactions on Power Electronics*, vol. 28, no. 2, pp. 625–636, 2013.

[133] Y. Lai, and F. Shyu. Topology for hybrid multilevel inverter. *IEE Proceedings Electric Power Applications*, vol. 149, no. 6, pp. 449–458, 2002.

[134] J. Meili, S. Ponnaluri, L. Serpa, P. K. Steimer, and J. W. Kolar. Optimized pulse patterns for the 5-level anpc converter for high speed high power applications. *IECON 2006 32nd Annual Conference on IEEE Industrial Electronics*, pp. 2587–2592, 2006.

[135] L. A. Serpa, P. M. Barbosa, P. K. Steimer, and J. W. Kolar. Five-level virtual-flux direct power control for the active neutral-point clamped multilevel inverter. *2008 IEEE Power Electronics Specialists Conference*, pp. 1668–1674, 2008.

[136] F. Kieferndorf, M. Basler, L. A. Serpa, J. Fabian, A. Coccia, and G. A. Scheuer. A new medium voltage drive system based on anpc-5l technology. *2010 IEEE International Conference on Industrial Technology*, pp. 643–649, 2010.

[137] B. P. McGrath, and D. G. Holmes. Analytical modelling of voltage balance dynamics for a flying capacitor multilevel converter. *IEEE Transactions on Power Electronics*, vol. 23, no. 2, pp. 543–550, 2008.

[138] Jing Huang, and K. A. Corzine. Extended operation of flying capacitor multilevel inverters. *IEEE Transactions on Power Electronics*, vol. 21, no. 1, pp. 140–147, 2006.

[139] Xiaomin Kou, K. A. Corzine, and Y. L. Familiant. A unique fault-tolerant design for flying capacitor multilevel inverter. *IEEE Transactions on Power Electronics*, vol. 19, no. 4, pp. 979–987, 2004.

[140] M. Narimani, B. Wu, and N. R. Zargari. A novel five-level voltage source inverter with sinusoidal pulse width modulator for medium-voltage applications. *IEEE Transactions on Power Electronics*, vol. 31, no. 3, pp. 1959–1967, 2016.

[141] Q. A. Le, and D. Lee. A novel six-level inverter topology for medium-voltage applications. *IEEE Transactions on Industrial Electronics*, vol. 63, no. 11, pp. 7195–7203, 2016.

[142] V. Nair R., A. Rahul S., R. S. Kaarthik, A. Kshirsagar, and K. Gopakumar. Generation of higher number of voltage levels by stacking inverters of lower multilevel structures with low voltage devices for drives. *IEEE Transactions on Power Electronics*, vol. 32, no. 1, pp. 52–59, 2017.

[143] P. R. Kumar, R. S. Kaarthik, K. Gopakumar, J. I. Leon, and L. G. Franquelo. Seventeenlevel inverter formed by cascading flying capacitor and floating capacitor h-bridges. *IEEE Transactions on Power Electronics*, vol. 30, no. 7, pp. 3471–3478, 2015.

[144] Y. Ounejjar, and K. Al-Haddad. A novel high energetic efficiency multilevel topology with reduced impact on supply network. *2008 34th Annual Conference of IEEE Industrial Electronics*, pp. 489–494, 2008.

[145] Y. Ounejjar, and K. Al-Haddad. A new high power efficiency cascaded u cells multilevel converter. *2009 IEEE International Symposium on Industrial Electronics*, pp. 483–488, 2009.

[146] Y. Ounejjar, K. Al-Haddad, and L. Grégoire. Novel three phase seven level pwm converter. *2009 IEEE Electrical Power Energy Conference (EPEC)*, pp. 1–6, 2009.

[147] Y. Ounejjar, and K. Al-haddad. A novel high efficient fifteen level power converter. *2009 IEEE Energy Conversion Congress and Exposition*, pp. 2139–2144, 2009.

[148] Y. Ounejjar, and K. Al-Haddad. Multilevel hysteresis controller of the novel seven-level packed u cells converter. *SPEEDAM 2010*, pp. 186–191, 2010.

[149] Y. Ounejjar, and K. Al-Haddad. A novel six-band hysteresis control of the packed u cells seven-level converter. *2010 IEEE International Symposium on Industrial Electronics*, pp. 3199–3204, 2010.

[150] Y. Ounejjar, K. Al-Haddad, and L. A. Dessaint. A novel six-band hysteresis control for the packed u cells seven-level converter: Experimental validation. *IEEE Transactions on Industrial Electronics*, vol. 59, no. 10, pp. 3808–3816, 2012.

[151] Y. Ounejjar, K. Al-Haddad, and L. Gregoire. Packed u cells multilevel converter topology: Theoretical study and experimental validation. *IEEE Transactions on Industrial Electronics*, vol. 58, no. 4, pp. 1294–1306, 2011.

[152] H. Vahedi, H. Y. Kanaan, and K. Al-Haddad. Puc converter review: Topology, control and applications. In *IECON 2015–41st Annual Conference of the IEEE Industrial Electronics Society*, pp. 004334–004339, 2015.

[153] P. Qashqai, A. Sheikholeslami, H. Vahedi, and K. Al-Haddad. A review on multilevel converter topologies for electric transportation applications. *2015 IEEE Vehicle Power and Propulsion Conference (VPPC)*, pp. 1–6, 2015.

[154] H. Vahedi, K. Al-Haddad, and H. Y. Kanaan. A new voltage balancing controller applied on 7-level puc inverter. *IECON 2014–40th Annual Conference of the IEEE Industrial Electronics Society*, pp. 5082–5087, 2014.

[155] H. Vahedi, and K. Al-Haddad. Puc5 inverter—a promising topology for single-phase and three-phase applications. In *IECON 2016–42nd Annual Conference of the IEEE Industrial Electronics Society*, pp. 6522–6527, 2016.

[156] H. Vahedi, P. Labb´e, and K. Al-Haddad. Corrections to "sensor-less five-level packed ucell (puc5) inverter operating in stand-alone and grid-connected modes" [feb 16 361–370]. *IEEE Transactions on Industrial Informatics*, vol. 12, no. 4, pp. 1298–1298, 2016.

[157] H. Vahedi, and K. Al-Haddad. Real-time implementation of a seven-level packed u-cell inverter with a low-switching-frequency voltage regulator. *IEEE Transactions on Power Electronics*, vol. 31, no. 8, pp. 5967–5973, 2016.

[158] J. I. Metri, H. Vahedi, H. Y. Kanaan, and K. Al-Haddad. Real-time implementation of model-predictive control on seven-level packed u-cell inverter. *IEEE Transactions on Industrial Electronics*, vol. 63, no. 7, pp. 4180–4186, 2016.

[159] H. Vahedi, K. Al-Haddad, Y. Ounejjar, and K. Addoweesh. Crossover switches cell (csc): A new multilevel inverter topology with maximum voltage levels and minimum dc sources. *IECON 2013–39th Annual Conference of the IEEE Industrial Electronics Society*, pp. 54–59, 2013.

[160] M. Sarbanzadeh, E. Babaei, M. A. Hosseinzadeh, and C. Cecati. A new sub-multilevel inverter with reduced number of components. *IECON 2016–42nd Annual Conference of the IEEE Industrial Electronics Society*, pp. 3166–3171, 2016.

[161] R. S. Alishah, S. H. Hosseini, E. Babaei, and M. Sabahi. Optimal design of new cascaded switch-ladder multilevel inverter structure. *IEEE Transactions on Industrial Electronics*, vol. 64, no. 3, pp. 2072–2080, 2017.

[162] Reza Barzegarkhoo, Naser Vosoughi, Elyas Zamiri, Hossein Madadi Kojabadi, and Liuchen Chang. A cascaded modular multilevel inverter topology using novel series basic units with a reduced number of power electronic elements. *Journal of Power Electronics*, vol. 16, no. 6, pp. 2139–2149, 2016.

[163] E. Samadaei, S. A. Gholamian, A. Sheikholeslami, and J. Adabi. An envelope type (etype) module: Asymmetric multilevel inverters with reduced components. *IEEE Transactions on Industrial Electronics*, vol. 63, no. 11, pp. 7148–7156, 2016.

[164] E. Samadaei, S. A. Gholamian, A. Sheikholeslami, and J. Adabi. Cascade topologies for the asymmetric multilevel inverter by new module to achieve maximum number of levels. *Int. Islamic Univ. Malaysia Eng. J.*, vol. 17, no. 2, pp. 83–93, 2016.

[165] E. Samadaei, A. Sheikholeslami, S. A. Gholamian, and J. Adabi. A square t-type (sttype) module for asymmetrical multilevel inverters. *IEEE Transactions on Power Electronics*, vol. 33, no. 2, pp. 987–996, 2018.

[166] E. Samadaei, M. Kaviani, and K. Bertilsson. A 13-levels module (k-type) with two dc sources for multilevel inverters. *IEEE Transactions on Industrial Electronics*, vol. 66, no. 7, pp. 5186–5196, 2019.

[167] Niraj Kumar Dewangan, Shubhrata Gupta, and Krishna Kumar Gupta. Fault-tolerant operation of some reduced-device-count multilevel inverters with improved performance. *International Transactions on Electrical Energy Systems*, vol. 29, no. 2, p. e2731, 2019.

[168] R. S. Alishah, S. H. Hosseini, E. Babaei, and M. Sabahi. A new general multilevel converter topology based on cascaded connection of submultilevel units with reduced switching components, dc sources, and blocked voltage by switches. *IEEE Transactions on Industrial Electronics*, vol. 63, no. 11, pp. 7157–7164, 2016.

[169] E. Babaei, S. H. Hosseini, G. B. Gharehpetian, M. Tarafdar Haque, and M. Sabahi. Reduction of dc voltage sources and switches in asymmetrical multilevel converters using a novel topology. *Electric Power Systems Research*, vol. 77, no. 8, pp. 1073–1085, 2007.

[170] J. Korhonen, A. Sankala, J. Str¨om, and P. Silventoinen. Hybrid five-level t-type inverter. In *IECON 2014–40th Annual Conference of the IEEE Industrial Electronics Society*, pp. 1506–1511, 2014.

[171] W. Song, and A. Q. Huang. Fault-tolerant design and control strategy for cascaded hbridge multilevel converter-based statcom. *IEEE Transactions on Industrial Electronics*, vol. 57, no. 8, pp. 2700–2708, 2010.

[172] S. Kouro, M. Malinowski, K. Gopakumar, J. Pou, L. G. Franquelo, B. Wu, J. Rodriguez, M. A. P´erez, and J. I. Leon. Recent advances and industrial applications of multilevel converters. *IEEE Transactions on Industrial Electronics*, vol. 57, no. 8, pp. 2553–2580, 2010.

[173] A. Taghvaie, J. Adabi, and M. Rezanejad. Circuit topology and operation of a step-up multilevel inverter with a single dc source. *IEEE Transactions on Industrial Electronics*, vol. 63, no. 11, pp. 6643–6652, 2016.

[174] A. Taghvaie, J. Adabi, and M. Rezanejad. A multilevel inverter structure based on a combination of switched-capacitors and dc sources. *IEEE Transactions on Industrial Informatics*, vol. 13, no. 5, pp. 2162–2171, 2017.

[175] S. Ceballos, J. Pou, E. Robles, J. Zaragoza, and J. L. Martin. Performance evaluation of fault-tolerant neutral-point-clamped converters. *IEEE Transactions on Industrial Electronics*, vol. 57, no. 8, pp. 2709–2718, 2010.

[176] X. Sun, B. Wang, Y. Zhou, W. Wang, H. Du, and Z. Lu. A single dc source cascaded seven-level inverter integrating switched-capacitor techniques. *IEEE Transactions on Industrial Electronics*, vol. 63, no. 11, pp. 7184–7194, 2016.

[177] E. Samadaei, A. Sheikholeslami, S. A. Gholamian, and J. Adabi. A square T-type (ST-type) module for asymmetrical multilevel inverters. *IEEE Transactions on Power Electronics*, vol. 33, no. 2, pp. 987–996, 2018.

11 Development of BESS with PV Output Characteristics for Direct PV String DC Coupling Using Half-Bridge Bidirectional DC-DC Converter

Jordan S. Z. Lee, Rodney H. G. Tan,
T. Sudhakar Babu, and Nadia M. L. Tan

CONTENTS

DOI: 10.1201/9781003323471-11

11.1 INTRODUCTION

Rapid development in technology and the rise in world population growth leading to upsurge in energy demand. The high energy demand causes the increase of depletion rate of non-renewable energy resources. Besides, the combustion of non-renewable energy resources produces carbon dioxide as by-product which causes air pollution, global warming and climate change. Therefore, renewable energy has been the focus in energy generation industry for the pass years. Solar PV system is the most installed type of renewable energy system in recent years. This leads to the demand in the advancement of solar technology. Nevertheless, the energy generation of PV system highly relies on the availability of sunlight throughout the day. Commercial PV system installation supply energy to load as priority and exporting excess energy to utility grid depending on the involvement in energy exporting scheme provide by the local government. PV system installation for self-consumption is not allowed to export energy to the utility grid. Situation where solar energy generation more than load demand required the curtailment of excess energy. Integration of BESS with PV system, also known as PV-BESS system, is used to avoid the curtailment of excess energy by storing them. Besides, this system combination also provides functions such as power quality control for the utility grid. This is because the injection of utility scale PV system may cause power quality issue in the utility grid. In 2020, the total of PV system installed in a single achieved up to 134GW. Thus, numerous countries PV system to include 10% to 20% of BESS [1].

There are various types of coupling topology for the PV-BESS system such as alternating current (AC) coupling, direct current (DC) coupling and direct PV string DC coupling. Many studies were carried out to evaluate these topologies in terms of efficiency, reliability, cost, flexibility and complexity [2]–[6]. Figure 11.1 shows the three common types of coupling topology for PV-BESS.

All of the coupling topologies required fast communication link to facilitate the application, charging and discharging of BESS. AC coupling topology provides easy retrofitting to existing PV system. DC coupling and direct PV string DC coupling provides better efficiency for BESS charging due to lesser power conversion stage. A newly emerged hybrid inverter for PV-BESS system where the PV system and BESS is connected in the DC link within the hybrid inverter. However, due to the limitation in technology, only small capacity of hybrid inverter is available for residential PV-BESS system. Conventional direct PV string DC coupling regulate the DC bus voltage or current to allow the PV string to operate at maximum power point [6]. Method of regulating the bus voltage or current in this topology is rarely discussed. PV emulating function is one of the methods to regulate the DC bus voltage and current. In this chapter, PV emulating function is implemented into the BESS for direct PV string DC coupling.

11.2 RELATED WORKS

This section is divided into two main parts consisting the PV-BESS system coupling topology and the PV emulating function. Different coupling topology used in PV-BESS system and the application of PV emulating function are reviewed.

11.2.1 PV-BESS System Coupling Topology

The PV-BESS system coupling topology has been studied throughout the last decades. The object of these studies is mainly to evaluate the efficiency, reliability and cost by proposing different coupling topology and the power converter topology used for integrating the BESS in PV system.

In 2010, [7] proposed two model of BESS with the needed power electronics for grid integration to store energy from renewable energy such as PV energy and wind energy. The important parameters used for the investigation of the method used for addition of renewable power at medium voltage (MV) and low voltage (LV) grid is such as the efficiency and cost. DC coupling topology for integrating the PV system and BESS is used. A 5kWp rooftop solar panel was used as reference.

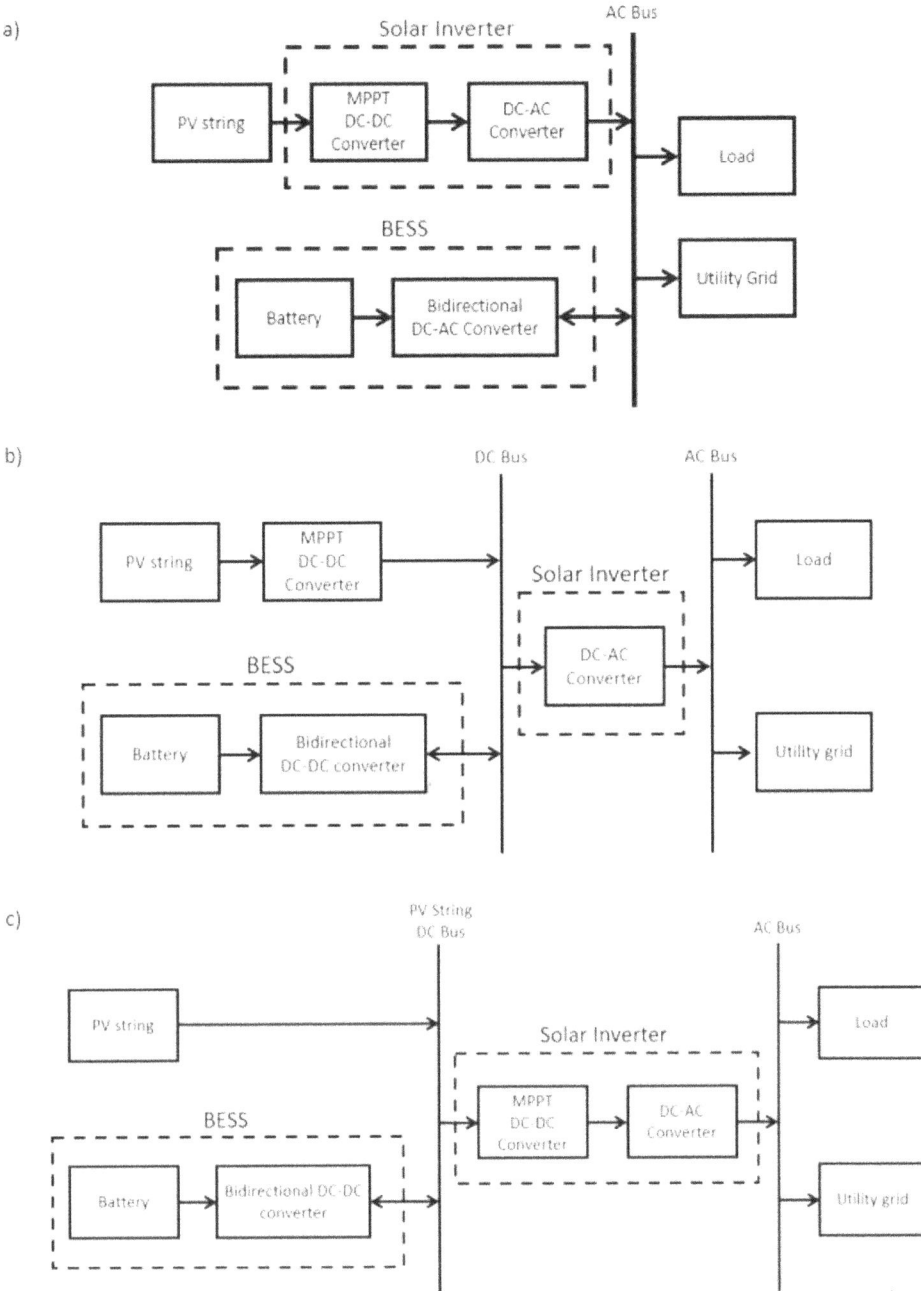

FIGURE 11.1 PV-BESS system coupling topology: (a) AC coupling topology, (b) DC coupling topology, (c) direct PV string DC coupling.

The proposed battery converter is a two-phase synchronous converter with double loop control. The battery converter is placed between the battery and the DC bus. The simulation shows that the battery converter efficiency is more than 94% at different battery voltage. However, the overall system efficiency is not presented. With PV converter efficiency considered, the efficiency of the DC coupling method may be further reduced.

A study to enhance the battery life span in a PV-BESS system using DC coupling method is presented by [8]. This studies mainly focus on the operating approach for PV-BESS system to improve the life span of lithium-ion batteries. Forecasting based strategy is used to predict the state of charge (SOC) in various period. For the DC coupling method, two DC-DC converters and a DC-AC converter is used. The overall system efficiency in terms of power conversion was not considered for the forecasting strategy.

PV-BESS system with maximum power point tracking (MPPT) for voltage and frequency (V-f) control in micro-grid is proposed by [9]. Single diode model is used for the modelling of a 125kW PV system. The battery model in MATLAB SimPower Systems library is used. The type of battery used is lead-acid. DC coupling method is used for the PV system integration with BESS. A DC-DC boost converter with MPPT is used for the PV inverter. The battery converter is developed using DC-DC buck-boost converter. A double loop control with PI controller is used to determine the charging or discharging mode of the battery converter. Buck converter mode is used for charging while boost converter is used for discharging. The efficiency of the power conversion of the system is not evaluated as the studies focus on the V-f control, MPPT control and energy storage control.

Tesla's Powerwall was analysed by [10] in terms of various influencing parameter such as method of BESS coupling, aging of battery and retrofitting of existing PV systems. It is mentioned that the Powerwall can be connected either using AC coupling or DC coupling topology. For AC coupling, an additional bidirectional DC-AC inverter is required. The system simulation is carried out with the consideration of efficiency losses. Retrofitting for DC coupling topology required rewiring of the system. AC coupling is easier for retrofitting as separate inverter is used but the cost of the system is higher than DC coupling. The ROI in terms of coupling topology varies depending on the size of the system required. Some scenario shows that AC coupling is more energy efficiency than DC coupling when the load demand is low. Nonetheless, the evaluation is based on residential load demand.

A review on the BESS for applications in modern power grids is presented by [2]. Three types of coupling topologies were mentioned such as AC coupling, DC coupling and direct PV string DC coupling. For AC coupling, the battery is connected to the battery inverter through a DC-DC converter. The PV inverter is connected to the load and utility grid. For DC coupling, the BESS is connected to the DC bus between the PV system MPPT DC-DC converter and the solar inverter. The solar inverter is then connected to the utility grid. Lesser conversion is required for power flow from PV to BESS as compare to AC coupling. The battery in direct PV string DC coupling is connected to the PV system MPPT DC-DC converter. This topology allows higher conversion efficiency but required rapid and accurate communication signal between the BESS and the AC grid.

DC coupling topology for PV-BESS system modelling is proposed by [11]. The DC coupling topology proposed is to allow the PV system to supply power to battery through a DC-DC converter and also use the same bidirectional inverter to supply power to the load or charge battery from the utility grid. For AC coupling topology, the PV system is connected to a DC-AC inverter, or also known as solar inverter. The battery is using a separate bidirectional DC-AC inverter. It was mentioned that the additional DC-DC converter loss in DC coupling should be considered for the battery. This is the concern when battery is required to discharge to supply power to the load. However, for charging of the battery, there is additional DC-AC power conversion from PV system to battery which leads to more losses. Besides, conventional solar inverter includes a DC-DC converter for MPPT tracking purposes.

A study is carried by [3] to evaluate the reliability of coupling topologies used for combining BESS with PV systems. The system configuration will directly influence the operation, efficiency, cost and reliability of the PV-BESS system. The DC coupling topology has lower flexibility because of the amount of power supply to the load depends on the inverter capacity. The PV system and battery in AC coupling topology are using separate inverter and able to supply to the load simultaneously. However, the cost of AC coupling topology is higher because of the extra DC-AC power conversion which also leads to complex system designs and balance. Besides, the efficiency of the AC coupling is lesser than DC coupling because it has higher number of power conversion units. DC coupling topology consists of a PV converter, battery converter and a solar inverter. In this study, a boost converter with MPPT control is used for the PV converter. A bidirectional DC-DC converter is used for the battery converter where it operates in boost converter mode when discharging and buck converter mode when charging. The solar inverter is developed using a full-bridge single phase inverter. As for AC coupling, the solar inverter is identical to that used in DC coupling. For battery inverter, it operates in inverter mode when discharging and in rectifier mode when charging. A 6kWp PV system is used as the reference for simulation. The results show that the AC coupling topology has lower reliability where the main reason that affects its reliability is the battery inverter.

An adaptive command-filtered backstepping sliding mode control for grid-connected PV system with BESS is proposed by [12]. The study was carried out to resolve the power fluctuation in the PV system by adjusting the DC bus voltage and the AC bus current in the grid-connected PV system. DC coupling topology is used to integrate the BESS with PV system for PV smoothing which is required because of the changes in irradiation and temperature that leads to fluctuation of PV system output power. Single diode model was used to model the PV array characteristics. A 300W PV system is used for simulation. Boost converter with MPPT is used to develop the PV converter. A bidirectional DC-DC converter is used to develop the battery converter.

An assessment of performance methods for residential PV-BESS system is presented by [4]. Three types of PV-BESS system coupling topology which includes AC coupling, DC coupling and direct PV string DC coupling are mentioned in the study. Three cases of power conversion were simulated namely PV system to AC grid, PV system to BESS and BESS to AC grid. For PV systems to AC grid, the efficiency for AC coupling system remains over 95% when the rated power of the system is over 200W. For DC coupling, the efficiency only started to increase to over 95% when the rated power is high. For PV system to BESS, the DC coupling is showing better efficiency than the AC coupling at various rated power. For BESS to AC grid, AC coupling shows better efficiency at low rated power while DC coupling has higher efficiency at high rated power.

A study is carried out by [5] to evaluate the reliability of DC and AC coupling topology for BESS integration with 1500V PV system. The evaluation is performed using a case study on a 160kW PV-BESS system for PV smoothing and ramp-rate regulation. According to the results, the PV inverter in DC coupling topology has better reliability. However, from the perspective of system-level reliability, both DC and AC coupling topology will reduce the general system reliability. The AC coupling has longer lifetime and higher reliability than DC coupling in a 1500V PV system.

An efficiency comparison is carried out for the DC and AC coupling topology for large scale PV-BESS system by [6]. The study is carried out using a 288MWp PV system with 275MWh BESS. The charging and discharging for BESS are proposed. The direction of power flow of all the components in the system are acquired based on practicable operation condition. The overall power losses and efficiency are calculated. Based on the results, the DC coupling shows better efficiency than the AC coupling topology. For DC coupling, the direct PV string DC coupling has better efficiency than the conventional DC coupling at PV converter.

In summary, most of the study focus on the efficiency, reliability and cost of the coupling topology. The methodology used to model each coupling topology was not discussed in depth. Besides, the development of direct PV string DC coupling was not discussed in terms of battery converter modelling and BESS control. Therefore, this chapter presents the half-bridge bidirectional converter for the BESS with PV emulated output characteristics, which has not been used for direct PV string coupling. The BESS control system for charging and discharging is presented as well.

11.2.2   PV EMULATING FUNCTION

PV emulating function also known as PV emulator where its primary purpose is to emulate the characteristics of a PV module output. Numerous studies are carried out to enhance the performance of the PV emulating function by using various PV model modelling method, type of power converter and control strategy.

In 2007, [13] proposed a DC power supply used as PV emulator to test MPPT algorithm. A basic circuit which consists of a DC power supply, a series resistor and a load resistor. The maximum power supplied to the load when its resistance value is equal to the source internal resistance according to the maximum power transfer theorem. The results show that the maximum power point is tracked for the single peak power curve. However, the practical use MPPT is used to track the maximum power point when there is multiple peak due to shading.

A programmable energy source emulator with partial shadow effect is presented by [14]. Single diode model is used to develop the PV module model. A partial shading model with two PV modules connected in series and another configuration where two PV modules connected parallelly are proposed in this study. DC buck converter is used to control the output of the emulator depending on the operating point. The feedback control loop consists of a digital signal processor. To reduce the computation burden, a characteristic curve segmentation method is proposed. A 300W PV emulating function was successfully developed under uniform solar irradiance and partial shading condition. Power curve with multiple peak was produced, showing the partial shading effect.

PV emulator using buck converter using fast convergence resistance feedback method was proposed by [15]. This method was presented to solve the oscillating output voltage that leads to high number of iterations. Current-resistance (I-R) PV model developed based on single diode model is used and is combined with the PI controller in the closed-loop control of the buck converter. A boost converter is developed to test the hill-climbing MPPT algorithm.

A study is carried out by [16] to develop a low-cost PV emulator for validating the MPPT algorithm under various irradiance and temperature changes. The PV emulator consists of a DC source in series with a variable resistor. To test the P&O MPPT algorithm, a boost converter is used. This is also to validate the function of the PV emulator. Based on the result, the power curve is produced but it is not smooth after the maximum power point.

PV emulator is modelled by [17] using buck converter and Z-source converter and is compared in this study. According to the results, buck converter is better than Z-source converter in terms of control and cost. It also can be easily modified to produce the I-V curve at different irradiance. The voltage and current control are used in buck converter feedback control loop. The results show that the current mode control is better than voltage mode control such as it has better reliability.

The design of a low-cost PV emulator applied for PV energy conversion system is proposed by [18]. Single diode model is implemented using analog electronic components. PI controller is used in the feedback control loop of the buck converter. A hardware prototype is developed to validate the functionality of the PV emulator proposed. The emulated I-V and P-V curve are validated using resistive load and batteries. Its performance is assessed by its ability to recharge two 12V 7Ah battery. Based on the results, the hardware PV emulator is not as accurate when using resistor load. It is more accurate when a battery is used.

A new control method using shift controller for the PV emulator is proposed by [19]. PI controller and fuzzy logic controller are used for comparison. The current-resistance PV module model is used because the resistance feedback control strategy is utilised. It is based on a single diode model because it is simple and accurate. Buck converter is used as the power converter. Based on the results, the shift controller shows better response than PI controller during load variation. Besides, it also computes faster than the fuzzy logic controller.

PV emulator developed using four-switch buck-boost converter is presented by [20]. Linear approach and Newton's method are used to find the voltage reference which matches the I-V curve. The power converter operates at different mode at each part of the I-V curve. It operates in buck mode and boost mode at constant current part and constant voltage part, respectively. It operates in buck-boost mode at the knee part of the curve. The PV emulator shows good accuracy but required high computational burden as two PI controllers are used in the control loop. The results also show that the emulator has good performance under load variations.

Hybrid damping injection controller was proposed by [21] for developing the PV emulator. Partial shading emulation is included in the design of the PV emulator. Look-up table is used as the

PV model implementation method. Buck converter is selected to develop the PV emulator. Voltage and current referencing were used at constant voltage part and constant current part, respectively. The results show that the PV emulator has low ripple under load variation and have fast transient response. P-V curve with up to four peaks was generated under partial shading condition.

PV module modelling is required for emulating the PV module output characteristics in simulation to validate the functionality of proposed MPPT algorithm. The MPPT algorithms proposed by [22] and [23] are implemented into the feedback control system of a zeta converter to achieve MPPT function. Single diode model was used for modelling th PV module model because of its simplicity and accuracy. A Luo converter with MPPT algorithm integrated into its feedback control loop to track the maximum power of the PV module is presented by [24]. A double diode model was selected to model the PV module model because it has better accuracy than single diode model. However, double diode model has higher computation burden because it required more parameters for the reference current computation.

A MPPT with ant colony optimisation (ACO) utilising cuk converter with fuzzy logic control is developed by [25]. Single diode model is used for the PV module modelling and combined with wind energy system model to form a hybrid solar-wind renewable energy system. The ACO based MPPT is implemented into the cuk converter feedback control system to charge to battery and supply power to load. The battery equivalent electric circuit is used for modelling the battery because of its adaptability in approximating the state of charge of the battery.

PV emulator was developed by [26] to test MPPT algorithm and BESS charging controller. A dual loop PI controller DC-DC buck converter is selected for the development. Single diode is used again for the modelling of PV module model. Various simulation condition was carried with varied irradiance, temperature and wind speed. A boost converter is developed to test the P&O MPPT algorithm while buck converter is used as battery charge controller. Real time evaluation is also carried out for the PV emulator. The simulation results show the PV emulator has good performance and efficiency.

A PV emulator was modelled by [27] for the testing of an adaptive TS-fuzzy based RBF neural network alogoritm-based MPPT. Single diode model was employed as the mathematical model. The topology DC-DC converter used is boost converter while PI controller was used in the feedback control-loop. Similarly, a PV emulator was modelled by [28] to test the proposed hybrid simplified firefly and neigbourhood attraction firefly algorithm-based MPPT. Single diode model is used as well while an isolated unifirectional dual-brigde converter is developed for the charging of battery in electric vehicle. Industrial PV simulator is used for experimental validation. Both the author has used dSpace as the hardware controller.

In summary, most of the studies for PV emulator are focussing on the improvement of the PV emulator is terms of accuracy, response and robustness. Application of PV emulator discussed are mainly to test to MPPT algorithm, the charging of battery and partial shading emulation. This chapter proposed a new application for the PV emulating function by utilising it in direct PV string DC coupling for PV-BESS system.

11.3 METHODS

A conventional solar PV system with BESS typically consists of PV string, BESS and solar inverter connected to the utility grid. The PV string produces DC power from the sun and BESS is integrated into solar PV system through AC or DC bus coupling to store excess solar energy generation. The energy stored is used for various applications such as load management, demand respond, PV smoothing and frequency control. The solar inverter will convert the DC power to AC power for load consumption and supply to electrical grid for utilisation.

This section will demonstrate the development of BESS with PV output characteristic for direct PV string DC coupling. The overview of direct PV string DC coupling method for BESS and solar PV system integration is shown in Figure 11.2. It consists of PV string, Battery System, PVE BESS and Inverter load.

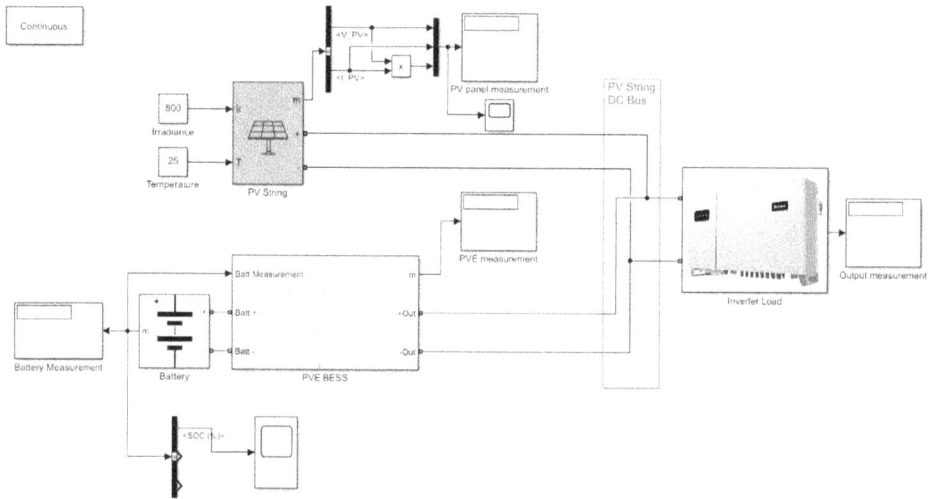

FIGURE 11.2   Direct PV string DC coupling for BESS and solar PV system integration.

11.3.1 PV STRING

A PV string consists of multiple PV modules connected in series to increase the PV system voltage and output power. The PV system voltage is increased to achieve the MPPT operating voltage of the solar inverter. The operating point of the PV string is computed by dividing the output voltage with the output current. The operating voltage and current of the maximum power can be obtained through PV module datasheet. The inverter load value is set to the maximum power operating point to obtain maximum power from the PV string. There are various factors affecting the solar energy generation such as irradiance and PV cell temperature. When the solar irradiance increases, the short circuit current and maximum output power of PV string increases significantly. When the solar cell temperature increases, the open circuit voltage and maximum output power of PV string decreases. The standard testing condition (STC) for irradiance and cell temperature are usually at 1000 W/m² and 25°C, respectively. A PV string with four 250W PV modules connected in series is used as reference. Figure 11.3 shows the PV characteristics curve of the PV string at different irradiance. Table 11.1 shows the parameters of a 250W PV module.

11.3.2 BATTERY ENERGY STORAGE SYSTEM (BESS)

A BESS consists of a battery and a half-bridge bidirectional DC-DC converter. When the load usage is higher than the PV system output, BESS will discharge to accommodate the high load demand. When the solar generation is higher than the load demand, the excess solar energy generation is use to charge the BESS.

11.3.2.1 Battery

A battery can act as a source to supply power and a storage to store energy. There are various types of battery technology used in BESS such as lead-acid, lithium ion, and nickel-cadmium. The most commonly used battery technology in industry in recent years is lithium-ion based battery because of its long-life span, high energy density, high depth of discharge (DOD) and safety aspect. The general parameter of a battery includes nominal voltage, capacity, and state of charge (SOC). The nominal voltage of a battery is the average output voltage of a charged battery at a specific discharge rate. The capacity of a battery with the unit of Ah or Wh is the amount of energy can be

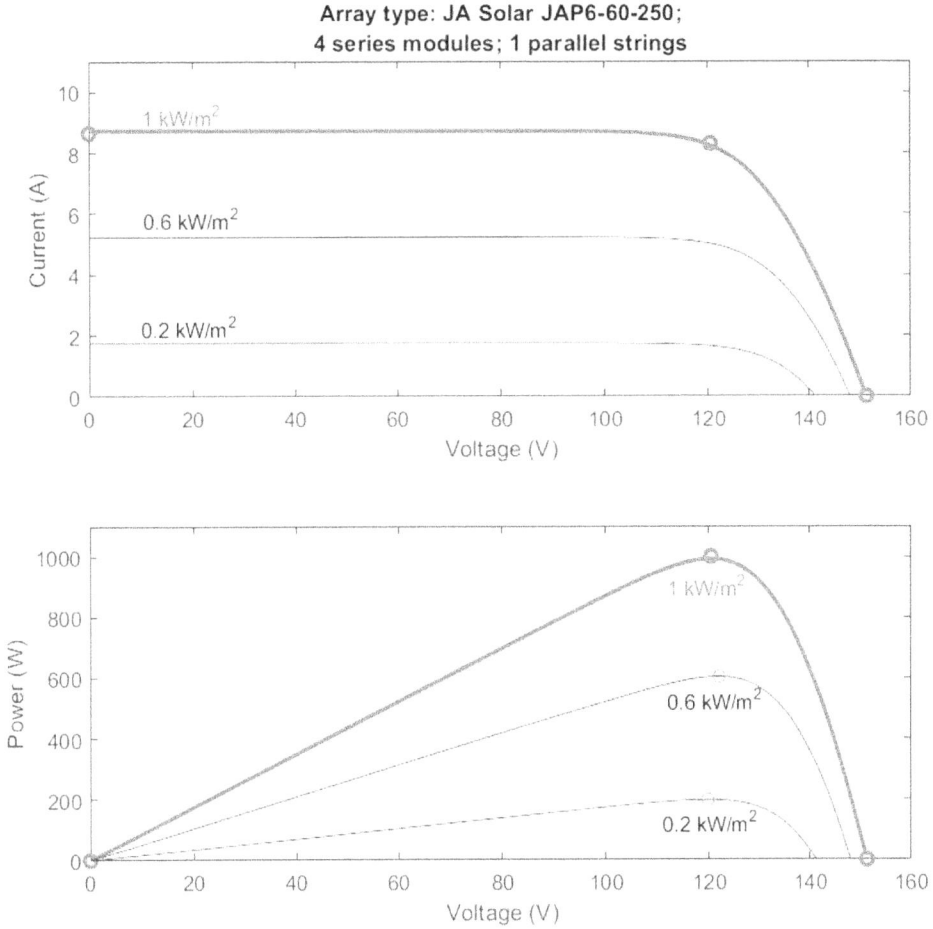

FIGURE 11.3 (a) PV string output characteristics curve with various solar irradiance; (b) PV string output characteristics with various solar cell temperature.

TABLE 11.1
PV Module Parameters

PV module model	JA Solar JAP6–60–250
Maximum power	250 W
Cells per module	60
Open circuit voltage (V_{OC})	37.85 V
Short-circuit current (I_{SC})	8.65 A
Voltage at maximum power point (V_{mp})	30.12 V
Current at maximum power point (I_{mp})	8.3 A
Temperature coefficient of Voc	−0.3642 %/degC
Temperature coefficient of Isc	0.064405 %/degC
Diode ideality factor	1.0251
Shunt resistance (R_{sh})	1372.43 Ω
Series resistance (R_s)	0.38179 Ω

stored. The SOC of a battery is the stored energy level in relative to the battery rated capacity. The higher the stored energy level, the higher the SOC. When the battery is discharging, the SOC will decrease while the battery is charging, the SOC will increase. For this chapter, a lithium-ion battery with nominal voltage of 80V and capacity of 10Ah is used.

11.3.2.2 Bidirectional Power Flow for Charging and Discharging

Bidirectional power flow is required for the charging and discharging of the BESS. When the BESS discharge, the power will flow from BESS to the load and the utility grid. To charge the BESS, the power will flow from the PV string or the utility grid to the BESS. BESS is a DC source and required DC power to charge. Since PV string is DC source, it can be used to charge the BESS without power conversion. However, utility grid supply AC power and power conversion is required to charge the BESS. A half-bridge bidirectional DC-DC converter topology is used to achieve the bidirectional power flow.

11.3.3 HALF-BRIDGE BIDIRECTIONAL DC-DC CONVERTER

The main purpose of a half-bridge bidirectional DC-DC converter is to allow the control of output voltage or current while maintaining the output power to the load. This function is achieved through the control of switching frequency and duty cycle to the MOSFET. Referring to Figure 11.4, it consists of a battery source, an inductor, two MOSFETs (Q1 and Q2) and two capacitors. Sizing of LC filter and switching frequency is required to achieve good performance for the converter. It can operate in either buck mode or boost mode. Boost mode is used for the discharging of BESS while buck mode is used for the charging of the BESS. Since PV module is a current source, a current-controlled half-bridge bidirectional DC-DC converter is used.

11.3.3.1 LC Filter and Switching Frequency Sizing

The ripple of the output should be kept as low as possible at the design phase. Ripple current and voltage is affected by the parameters such as the inductor, capacitor and the switching frequency value. The inductor and capacitor value are calculated based on Equations (1) and (2), respectively.

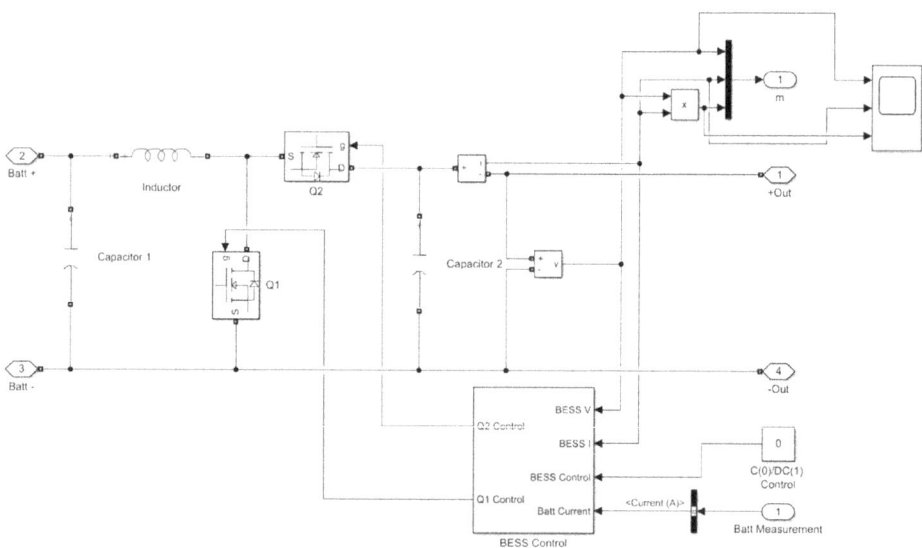

FIGURE 11.4 Half-bridge bidirectional DC-DC converter.

$$L = \frac{V_{in}D}{\Delta I \times f} \tag{1}$$

$$C = \frac{I_{out}D}{\Delta V \times f} \tag{2}$$

Where L is the inductance of inductor, C is the capacitance of capacitor, V_{in} is the source voltage, ΔI is the allowable ripple current, ΔV is the allowable ripple voltage magnitude, D is the duty cycle and f is the switching frequency. Based on Equation (1), the inductor value will affect the ripple current. According to Equation (2), the capacitor value will affect the ripple voltage. The ripple current and voltage is affected by the switching frequency according to Equations (1) and (2). The switching frequency is set to 10kHz. The ripple voltage and current magnitude is set to 2V and 40A, respectively. Therefore, 100μH and 300μF is selected for the value of inductor and capacitor, respectively.

11.3.3.2   Charging Mode
During charging, the bidirectional converter is operating in buck converter mode, the input voltage is step-down to a lower output voltage. In this mode, the PV string will act as a source to supply power to the load and charging the battery simultaneously. Referring to Figure 11.4, the PV string is connected to the inductor through Q2, which act a switch. The operating condition of MOSFET can be divided into two states which are on-state and off-state. During on-state, the PV string will supply current to load and battery. The inductor is charged concurrently. Q1 act as a diode in this mode and works in reverse-biased in this state. During off-state, the PV string is cut-off from the circuit, and the inductor will become the source and discharge to ensure current flow in the circuit. The switch is closed again before the inductor is fully discharged to allow the buck mode to operate in continuous current mode (CCM). The duty cycle that decides the Q2 on and off-state switching can be determined by Equation (3), where V_{out} is the output voltage, V_{in} is the input voltage, I_{out} is the output current and I_{in} is the input current. A reference current is used to control the duty cycle in this mode.

$$D = \frac{V_{out}}{V_{in}} \ or \ \frac{I_{out}}{I_{in}} \tag{3}$$

11.3.3.3   Discharging Mode
During discharging, the bidirectional converter is operating in boost converter mode. The battery and PV string become the source when operating in boost mode. The battery is connected to the Q1 through the inductor. During on-state, the battery will supply current to the load while charging the inductor concurrently. Q2 will act as a diode in this mode and work in reverse-biased in this state. During off-state, both battery and inductor is the source and supply current in the circuit. Similar to buck mode, the switch is closed again before the inductor is completely discharged to allow CCM operation. The duty cycle that decides the Q1 on and off-state switching can be determined by Equation (4). The reference current is computed by the PV emulating function to control the duty cycle.

$$D = 1 - \frac{V_{in}}{V_{out}} \ or \ 1 - \frac{I_{in}}{I_{out}} \tag{4}$$

11.3.4   BESS Control System

The BESS control system consists of a feedback control loop of the half-bridge bidirectional DC-DC converter and a charging/discharging mode control for the BESS as shown in Figure 11.5.

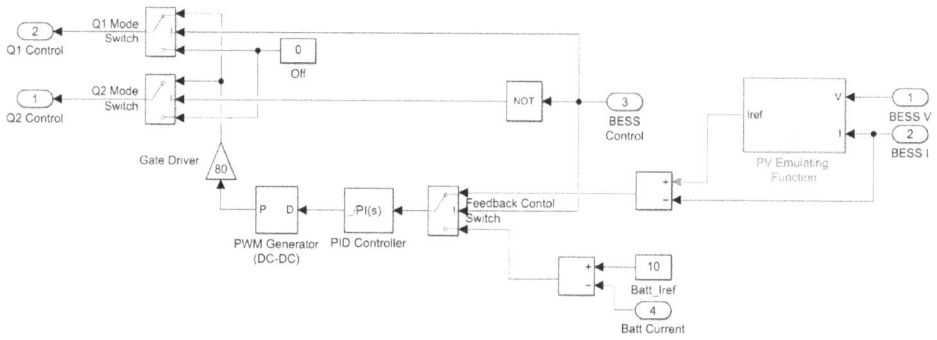

FIGURE 11.5   BESS control system.

11.3.4.1   Feedback Control Loop

The feedback control loop consists of PV emulating function block to generates reference current for discharging mode, a reference current for charging mode, a PI controller and a pulse width modulation (PWM) generator. While operating in discharging mode, a new corresponding reference current is calculated by the PV emulating function block when the voltage and current changes. For charging mode, the reference current remains constant regardless of the changes in load resistance. The PI controller will generate the duty cycle signal based on the difference of the reference current and the converter current. The converter current for charging mode is the battery output current while the output current for discharging mode is the BESS output current. The PWM generator will transmit a PWM duty cycle to the gate of MOSFET based on the PI controller signal. A gain block is used to act as the gate driver to drive the gate of the MOSFET. The value of gate driver is set based on the nominal voltage of the battery.

11.3.4.2   Charging/Discharging Mode Control System

A charging/discharging mode control system is developed to control the BESS operating mode. Referring to the BESS control flowchart as shown in Figure 11.6, when the BESS control is set to 0, BESS will operate in charging mode. The feedback control switch is switched to input 2 and a constant reference current is used to determine the duty cycle. Q2 mode switch is switched to input 1 for Q2 to receive PWM duty cycle. Q1 mode switch is switched to input 2 for Q1 to work as a diode. When the BESS control is set to 1, it will operate in discharging mode. The feedback control switch is switched to input 1 to activate the PV emulating function for discharging mode. Q1 mode switch is switched to input 1 to receive PWM duty cycle while Q2 mode switch is switched to input 2 for Q2 to act as a diode.

11.3.5   PV MODULE MATHEMATICAL MODEL

The PV string output characteristics can be represented by diode equivalent circuit such as single diode circuit and double diode circuit. For this chapter, single diode model is used to model the PV module mathematical model. It is selected because of its simplicity and good accuracy. Referring to Figure 11.7, the single diode model equivalent circuit consists of a current source that represents the photocurrent produced by the solar irradiance, a diode which represents the semiconductor substance characteristics of solar cell, a series resistor (R_s) which represents the power losses and shunt resistor (R_{sh}) which represents the leakage current for the diode.

Based on Kirchhoff's current law, the output current equation of the single diode equivalent circuit is derived as Equation (5).

$$I_{ref} = I_{ph} - \left(I_d \times e^{\left(\frac{V_{out} + \left(I_{out} \times R_s \right)}{V_t} \right)} - 1 \right) - \left(\frac{V_{out} + \left(I_{out} \times R_s \right)}{R_{sh}} \right) \tag{5}$$

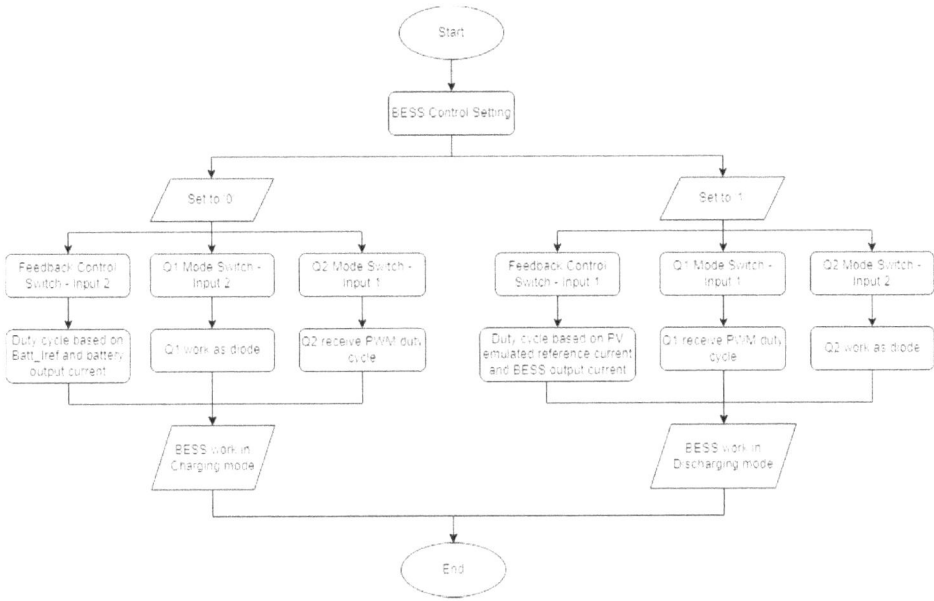

FIGURE 11.6   Flowchart of BESS control.

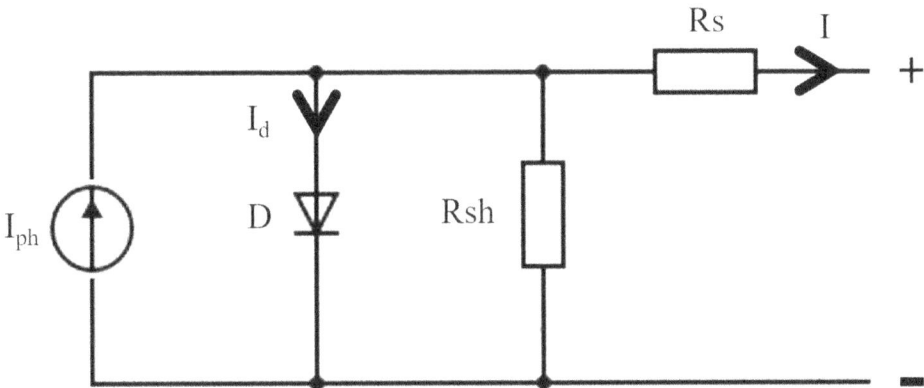

FIGURE 11.7   Single diode model equivalent circuit.

Where I_{ph} is the photocurrent, I_d is the diode reverse saturation current, V_{out} is the output voltage, I_{out} is the output current, v_t is the thermal voltage, R_s is the series resistance, and R_{sh} is the shunt resistance. The higher the photocurrent, the higher the irradiance according to Equation (6).

$$I_{ph} = I_{sc\_T} \times \left(\frac{G}{G_{stc}} \right) \tag{6}$$

Where G is the irradiance in W/m^{2,} and G_{stc} is the irradiance at STC. Equation (7) shows the computation of thermal voltage v_t.

$$v_t = \frac{A \times k \times T_{cell} \times N_s}{q} \tag{7}$$

Where A is the diode ideality factor, k is the Boltzmann constant (1.38×10^{-23}), T_{cell} is the PV cell temperature in Kelvin, N_s is the number of PV cells in a string, and q is the electron charge (1.6×10^{-19}). The temperature of the PV cell will alter the open-circuit voltage (V_{oc}) and short circuit current (I_{sc}) of a PV panel. The cell temperature corrected open-circuit voltage (V_{oc_T}) and short circuit current (I_{sc_T}) can be calculated using Equations (8) and (9), respectively.

$$I_{sc\_T} = I_{sc} + \left(\left(I_{sc} \times \left(\frac{K_I}{100} \right) \right) \times \left(T - T_{stc} \right) \right) \tag{8}$$

$$V_{oc\_T} = V_{oc} + \left(\left(V_{oc} \times \left(\frac{K_V}{100} \right) \right) \times \left(T - T_{stc} \right) \right) \tag{9}$$

Where K_I is the temperature coefficient of I_{sc}, K_V is the temperature coefficient V_{oc}, T is the temperature of the PV cell in degree Celsius, and T_{stc} is the temperature of the STC at 25°C. Equation (10) shows the calculation of diode reverse saturation current using V_{oc_T} and I_{sc_T}.

$$I_d = \frac{I_{sc\_T}}{e^{\frac{V_{ocT}}{v_t}} - 1} \tag{10}$$

The PV emulating function subsystem is implemented using MATLAB function block as shown in Figure 11.8. It will receive two inputs which are output voltage and output current of the DC-DC converter and produce one output which is the reference current. The remaining inputs are the required PV module mathematical model parameters based on Equation (7) which are V_{oc}, I_{sc}, A, R_s, R_{sh}, N_s, K_I, and K_V. However, there are several parameters such as R_s, R_{sh}, and A are not accessible from the PV module datasheet. Therefore, these parameters are acquired using DeSoto Algorithm from NREL.

According to Figure 11.8, V_{oc}, R_s, R_{sh}, and N_s are multiplied by four as compared to the value shown in Table 11.1. This is because the four PV modules are connected in series. Thus, these four parameters increase in accordance to single diode model equivalent circuit. The PV module mathematical model Equations (5) to (10) are coded in MATLAB function block using MATLAB script as shown in Figure 11.9.

11.4 RESULTS AND DISCUSSION

The direct PV string DC coupling method for discharging and charging of BESS are evaluated here. BESS without and with PV output characteristics are simulated to observe its effect on the PV string output characteristics. The parameter of a 1000W PV string is used as reference for this chapter where the string voltage and current at maximum power operating point is 120.48V and 8.3A, respectively.

11.4.1 PV String Characteristic Curve

Simulation of PV String is required to assess the potential of the proposed BESS with PV emulated characteristics. A basic circuit consists of a PV string and a load resistor as shown in Figure 11.10 is established to produce a reference PV characteristic curve.

The load resistance is set to run from 10Ω to 24Ω in the span of 1Ω generating up to 15 measurement points. Every load resistance is simulated for 1 second. The voltage, current and power is recorded to plot the I-V and P-V characteristics curve as shown in Figure 11.11.

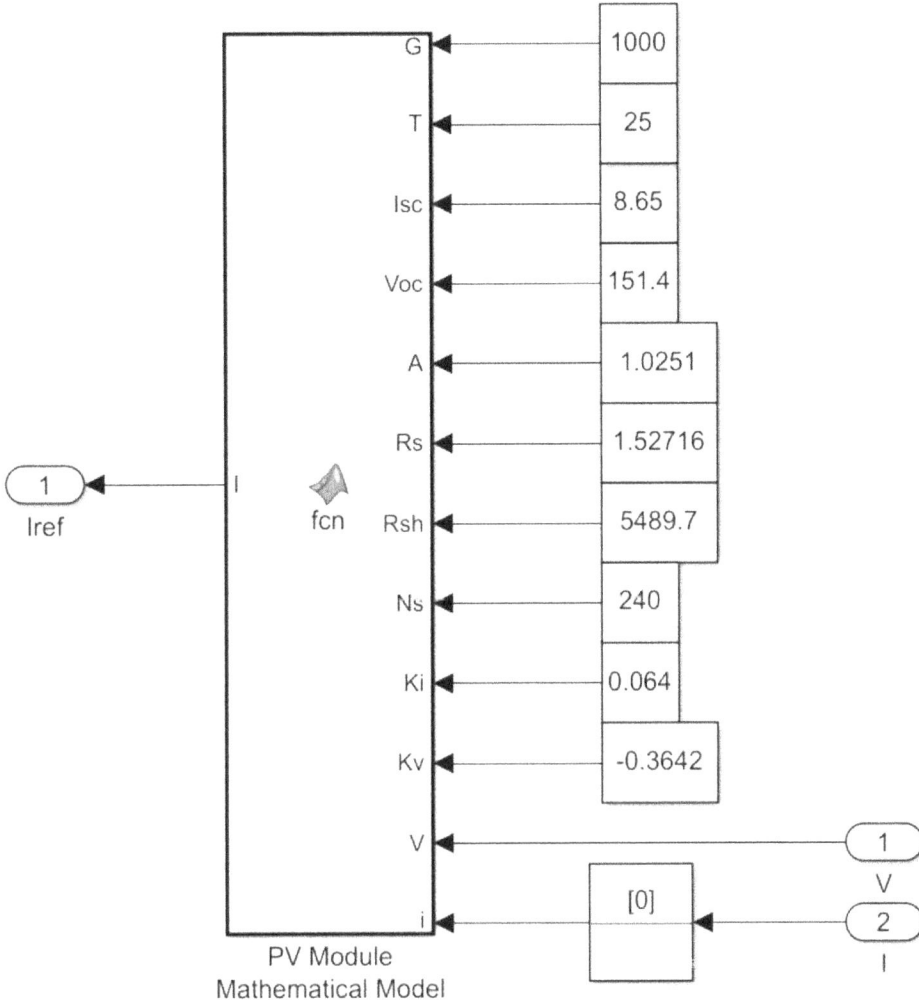

FIGURE 11.8 PV module mathematical model.

```
1       function I = fcn(G,T,Isc,Voc,A,Rs,Rsh,Ns,Ki,Kv,V,i)
2
3           q = 1.6e-19;                           % Electron charge
4           k = 1.38e-23;                          % Boltzmann constant
5           Tcell = 273+T;                         % Cell temperature in Kelvin
6           vt=(A*k*Tcell*Ns)/q;                   % Thermal voltage Equation(5)
7
8           Isc_T = Isc+((Isc*(Ki/100))*(T-25));   % Isc temperature coef Equation(6)
9           Voc_T = Voc+((Voc*(Kv/100))*(T-25));   % Voc temperature coef Equation(7)
10          Id = Isc_T/(exp(Voc_T/vt)-1);          % Reverse saturation current Equation(8)
11          Iph = Isc_T*(G/1000);                  % Photon current Equation(9)
12
13          % Single diode model output current Equation(4)
14          I = Iph-(Id*(exp((V+(i*Rs))/vt)-1))-((V+(i*Rs))/Rsh);
15      end
```

FIGURE 11.9 PV module mathematical model code in MATLAB function block.

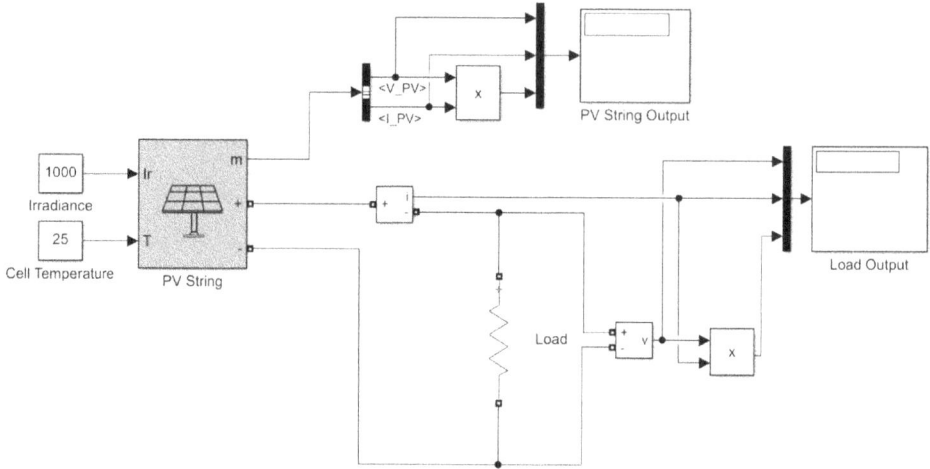

FIGURE 11.10   Circuit model to generate reference PV characteristics curve.

It can be observed that the peak power of the PV string is 990.64W at 121.9V and 8.126A which is similar with the rated power of 1000W. Based on the voltage and current, the maximum power operating point of the PV string is 15Ω. These I-V and P-V characteristics curve are use as reference to assess the BESS with PV emulated characteristics. Mean Absolute Percentage Error (MAPE) is used for the performance evaluation to assess the accuracy of the PV emulated BESS output by comparing with the reference PV characteristics curves. The lower the MAPE, the better the performance of the BESS. The MAPE is calculated using Equation (11).

$$MAPE = \frac{1}{n}\left(\sum\nolimits_{i=1}^{n} \left| \frac{P_{PVS}\left(V_i\right) - P_{BESS}\left(V_i\right)}{P_{PVM}\left(V_i\right)} \right| \right) \times 100\% \tag{11}$$

Where $P_{PVS}(V_i)$ is the power of the PV string at an operating voltage, $P_{BESS}(V_i)$ is the power of the PV emulator at an operating voltage, and n is the number of the operating voltage point.

11.4.2   BESS WITH PV EMULATED CHARACTERISTICS SIMULATION

Referring to Figure 3.8, the PV emulating function is added to the feedback control path of the BESS. Therefore, the BESS output will be converter to output with PV characteristics. Similar method used in PV string simulation is used for the setting of load resistance and simulation run time. A total of 15 measurement points is recorded to plot the PV characteristics curves. The I-V and P-V characteristics curve are as shown in Figure 11.12 and Figure 11.13.

It is noticed that the BESS output characteristics curves overlapped on the reference PV string. The maximum power of 997.153W is located at 122.3V and 8.153A with the MAPE of 0.66%. The average MAPE is 0.85%. Since the MAPE is so low, the BESS is operating with high accuracy.

11.4.3   BESS DISCHARGING MODE

Direct PV string DC coupling is similar to connecting two PV string parallelly. When one of the strings behave differently from the other string, it may cause output mismatch leading to operating point of the PV string being shifted. Two cases are carried out where is BESS with and without PV emulated output characteristics are simulated to observe the effect on the PV string output.

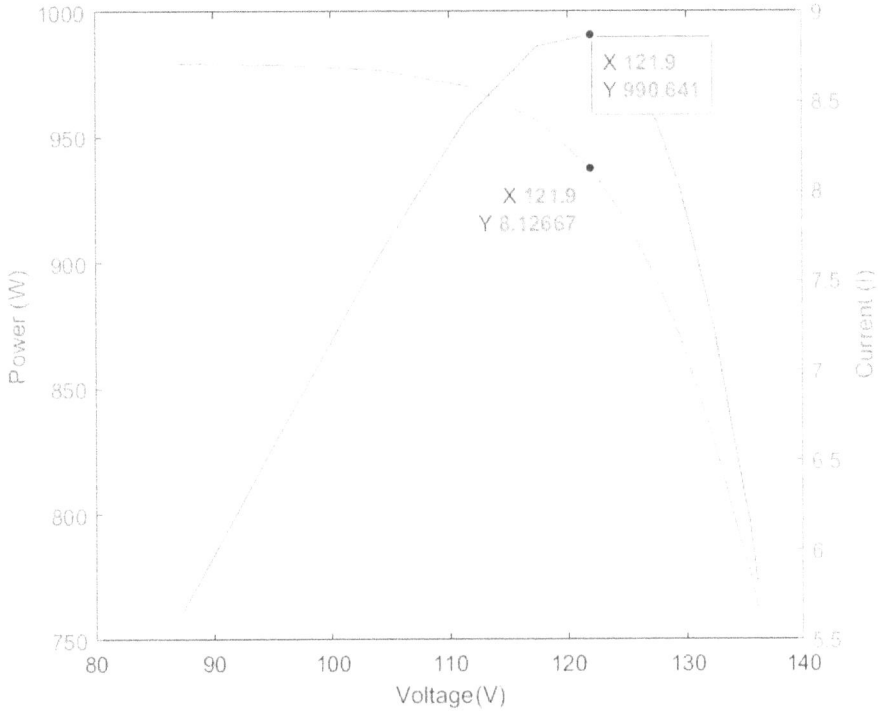

FIGURE 11.11   1000W PV string reference current and power characteristics curve.

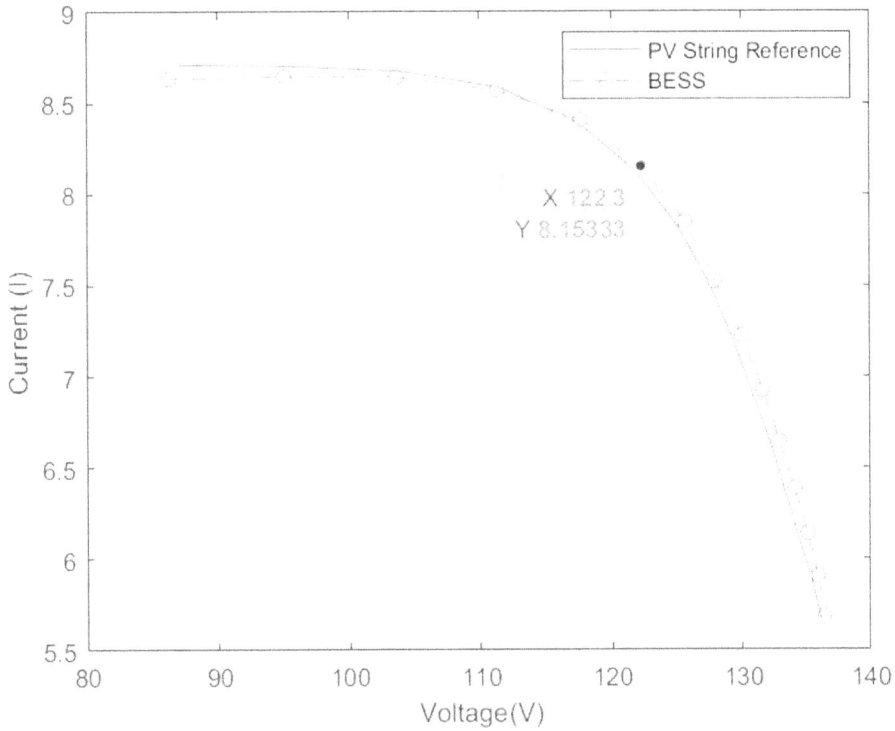

FIGURE 11.12   BESS with PV emulated characteristics simulation I-V curve performance.

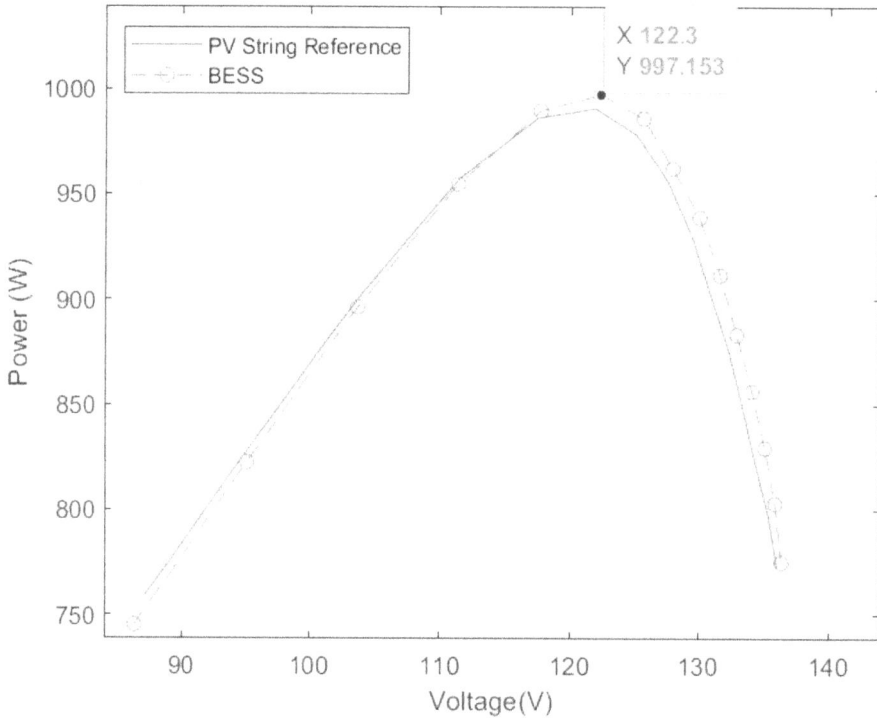

FIGURE 11.13 BESS with PV emulated characteristics simulation P-V curve performance.

TABLE 11.2
Simulation Results of BESS without PV Emulated Characteristics

Reference current	DC bus voltage	PV string current	BESS current	Inverter load current	Inverter load power
4 A	95.4 V	8.702 A	4.017 A	12.719 A	1213.39 W
6 A	110.0 V	8.610 A	6.055 A	14.655 A	1613.15 W
8.3 A	123.6 V	7.970 A	8.513 A	16.483 A	2037.30 W
10 A	130.6 V	6.974 A	10.430 A	17.404 A	2272.96 W
12 A	18.84 V	5.437 A	12.820 A	18.257 A	2499.38 W

11.4.3.1   BESS Discharging without PV Emulated Output Characteristics

The simulation is performed to verify that the BESS without PV emulate output characteristics will affect the operating point of the PV string. Referring to Figure 11.5, the PV emulating function block is replaced by a constant reference current for this simulation. When two 1000W PV strings are connected in parallel, the maximum operating point shifts to 7.5Ω. Therefore, the inverter load is set to 7.5Ω to allow both PV string and BESS to produce maximum power. The BESS reference current is set to various value to observe the effect on the PV string output. The result of the simulation is as shown in Table 11.2.

According to the results in Table 11.2, the PV string is not producing the maximum power when the reference current is not set to the current at maximum power operating point. This is because the PV string and BESS is sharing the load. When the BESS current increases, the PV string output

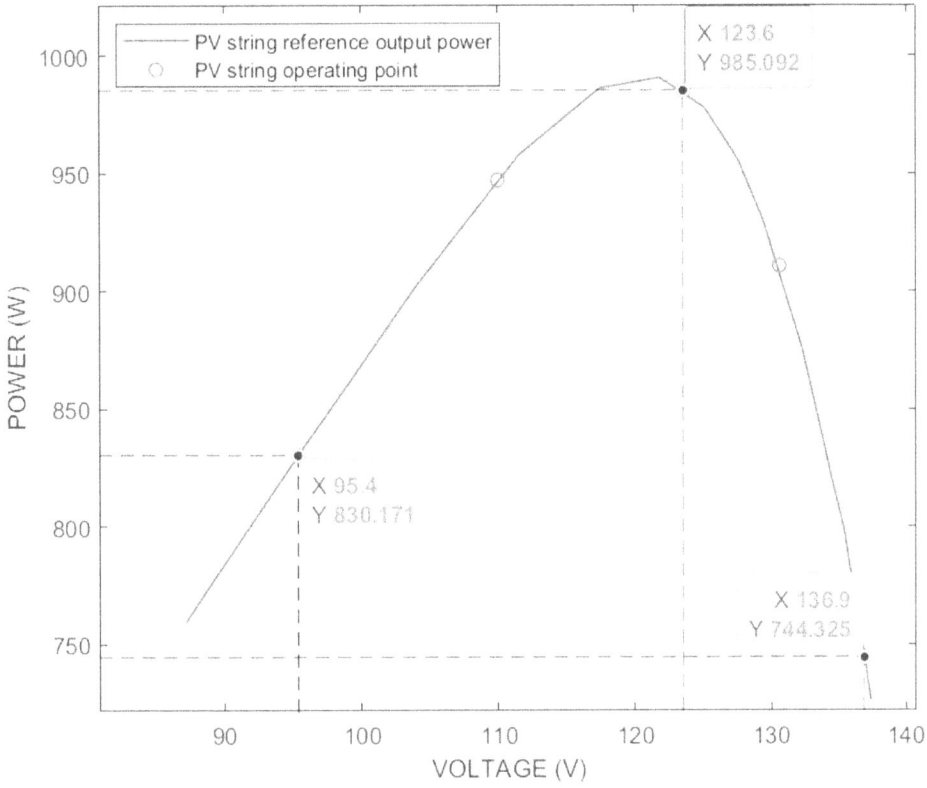

FIGURE 11.14   Operating point of PV string at each BESS reference current.

current decreases because the inverter load is constant. Besides, the inverter load power increases when the BESS current increases. This is because when the PV string output current decreases, the PV string voltage increases as refer to Figure 11.11. As a result, the DC bus voltage increases as well, causing the BESS to supply more power. The operating point and output current of PV string at each BESS reference current are as shown in Figure 11.14 and Figure 11.15, respectively.

It is observed that the PV string operating point is shifted away from the maximum power operating point when the reference current of BESS is not set to the current at maximum power operating point. When the reference current is set to value lesser than current at maximum power operating point, the PV string is operating at the constant current region. At this region, the current decrease gradually while the voltage increases significantly. When the BESS is supplying 4A, the PV string need to supply more current to accommodate the load demand, shifting the operating point of the PV string. When the reference current is set to value more than the current at maximum power operating point, the PV string is operating at the constant voltage region. At this region, the current decrease significantly while the voltage decreases gradually. When the BESS is supplying 12A, the PV string will supply lesser current to the load, shifting the operating point to constant voltage region. To allow the PV string to produce maximum power, a PV emulating function is used to convert the BESS output to have similar operating point as the PV string.

11.4.3.2   BESS Discharging with PV Emulated Output Characteristics

This case is carried out to verify that the PV string can operate at maximum power operating point when the BESS has PV emulated output characteristics. The irradiance and cell temperature of PV string and PV emulating function are set to STC. The inverter load is set to value ranging from 5Ω to

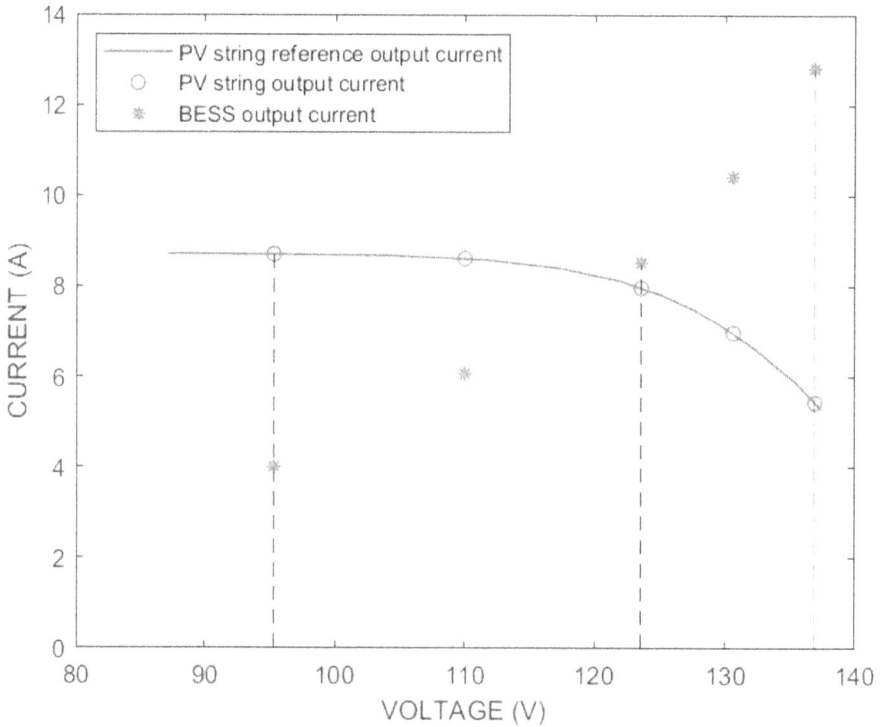

FIGURE 11.15 PV string output current at each BESS reference current.

12Ω with interval of 0.5Ω to cover the maximum power operating point. A total of 15 measurement points is recorded to plot the P-V curve. The simulation result is as shown in Figure 11.16.

Based on the simulation results, it is observed that the inverter load output power is the sum of BESS output power and PV string output power. Besides, it is noticed that the PV emulated BESS characteristics curve is similar to the PV string characteristics curve where the maximum power of BESS is similar to PV string. Thus, it can be concluded that the PV string operating point is not affected when PV emulating function is added into the feedback control path of the BESS. Different value of irradiance is use for simulation to verify that the BESS can operate at any irradiance. When the irradiance decreases, the operating point will change. Therefore, the inverter load is set to different range of value to cover the operating point of each irradiance. The simulation results with various irradiance value are as shown in Figure 11.17 and Figure 11.18.

It can be observed that the BESS output at different irradiance is similar to the PV string output at each operating point. Therefore, the BESS can operate with high accuracy at any irradiance value. Besides, the inverter load output is the sum of the BESS output power and PV string output power at each operating point.

11.4.4 BESS CHARGING MODE

When BESS is operating in charging mode, the BESS become a load and store energy from the PV string. The BESS and inverter load will share the power generated by the PV string. The simulation is performed with the inverter load ranging from 7Ω to 21Ω with the interval of 1Ω to observe how the inverter load will affect the charging of BESS. Based on the BESS output current results as shown in Figure 11.19, the BESS is discharging when the inverter load is set at 7.5Ω. This is because the PV string is operating at 85.11V and 8.712A where the operating point

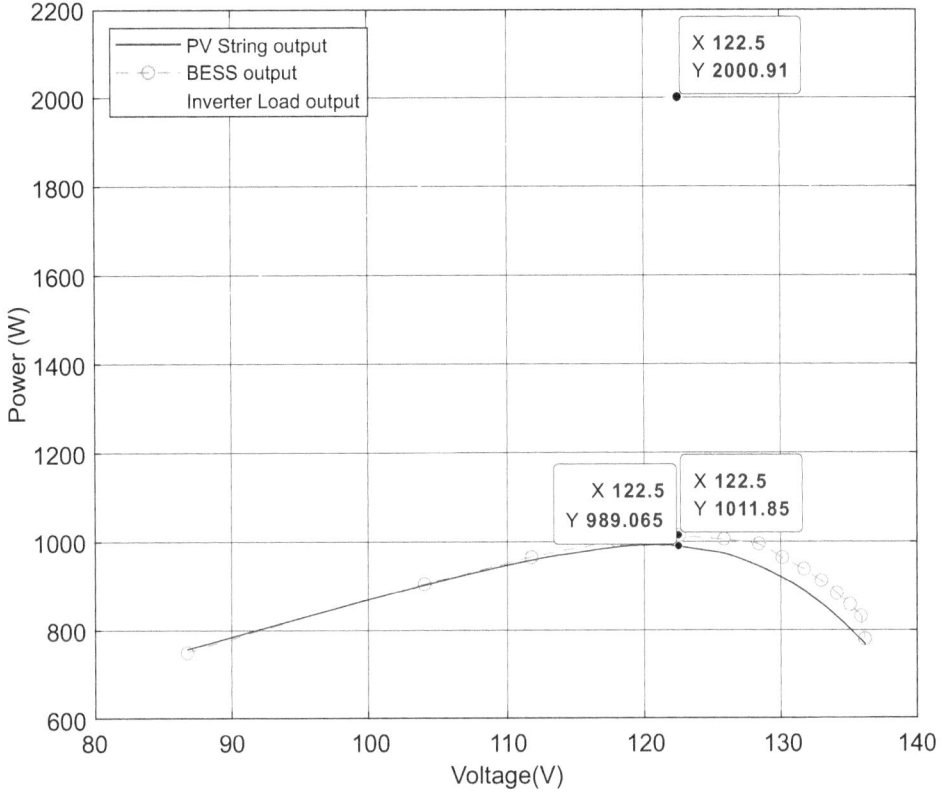

FIGURE 11.16   BESS with PV emulated output characteristics for direct PV string DC coupling.

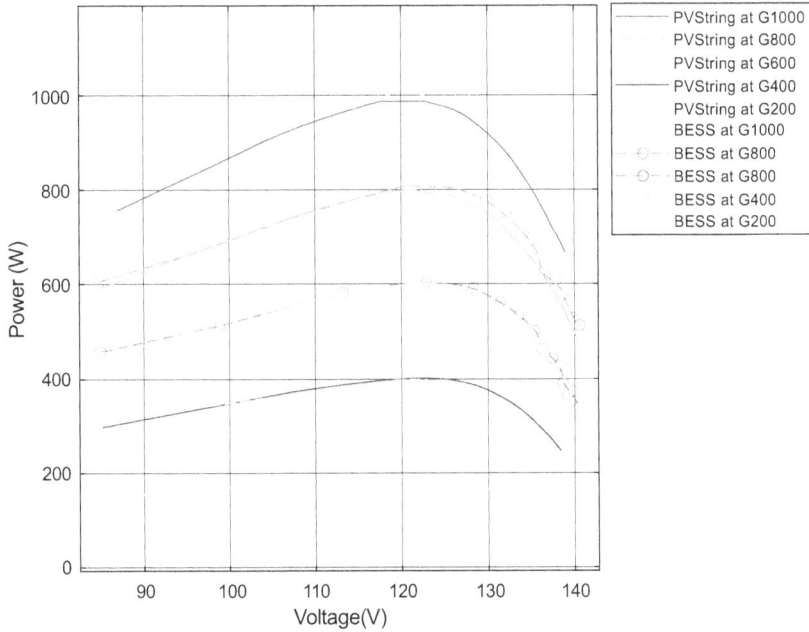

FIGURE 11.17   PV string output characteristics P-V curve and BESS with PV emulated output characteristics P-V curve at different irradiance value for direct PV string DC coupling.

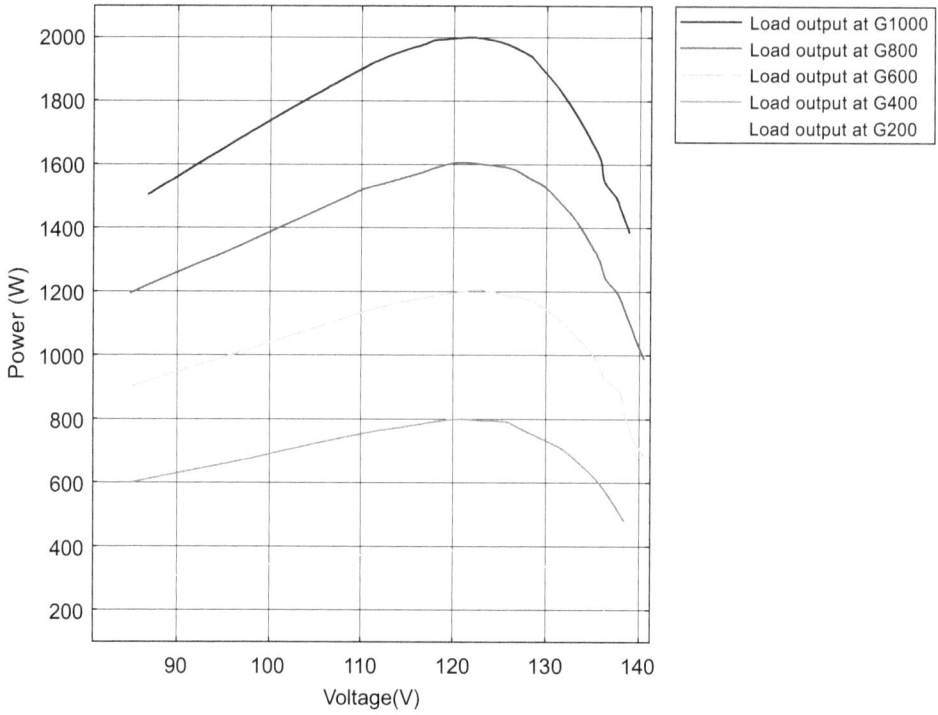

FIGURE 11.18   Inverter load P-V curve.

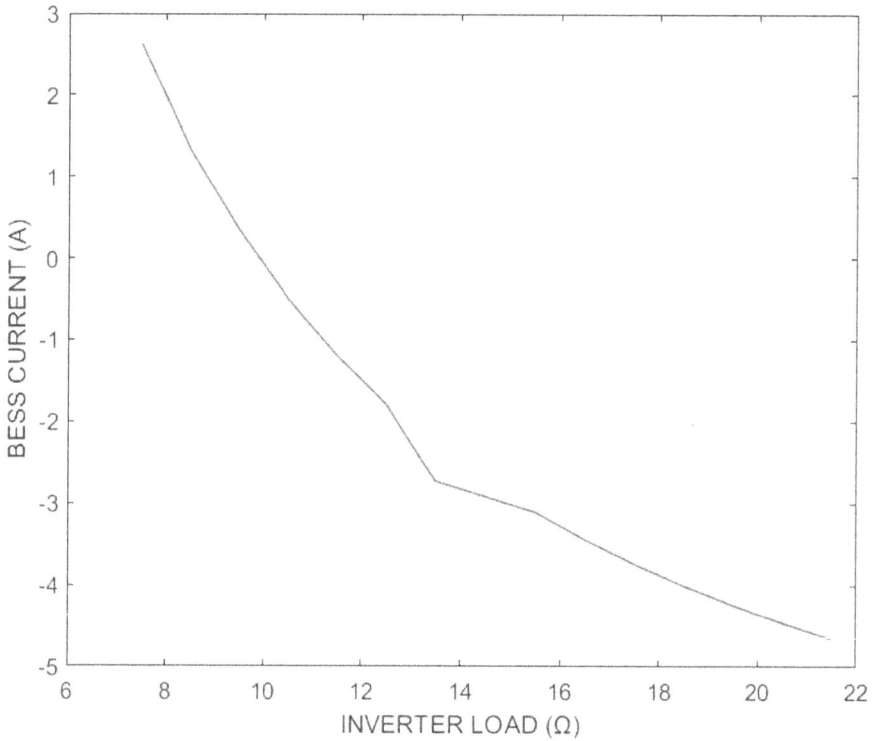

FIGURE 11.19   BESS current at various inverter load value when operating in charging mode.

is 9.77Ω. The operating point is higher than the inverter load value. The inverter load requires more current than the PV string is supplying. Thus, the BESS discharge to accommodate the load demand. When the inverter load value is set to 10.5Ω, the BESS start to charge. This is because the inverter load is now more than the PV operating point. As the inverter load value increases, the BESS charging current increases because the current draw by the inverter load decreases.

11.4.5 BESS State of Charge

The initial SOC of the battery is set to 50% and the simulation run time is set to 60s to observe the change in SOC level when the BESS is discharging and charging. For discharging mode, the inverter load value is set to the maximum power operating point of 7.5Ω. The discharging current of BESS is 8.26A. For charging mode, the inverter load value is set to 18.5Ω. The PV string is operating at 87.04V and 8.711A, generating 758.21W of power. The BESS charging current is 4.006A. The change is SOC during charging and discharging is shown in Figure 11.20.

It is observed that the SOC when discharging decreased from 50% to 48% and the SOC when charging increased from 50% to 50.7%. In summary, the proposed BESS with PV output emulated characteristics for direct PV string DC coupling is proved to be able to operate in both discharging and charging mode. The PV string able to operate at maximum power when PV emulating function is included in the BESS control system further proven that the output of BESS with PV characteristic is required to prevent the PV string operating point from being shifted.

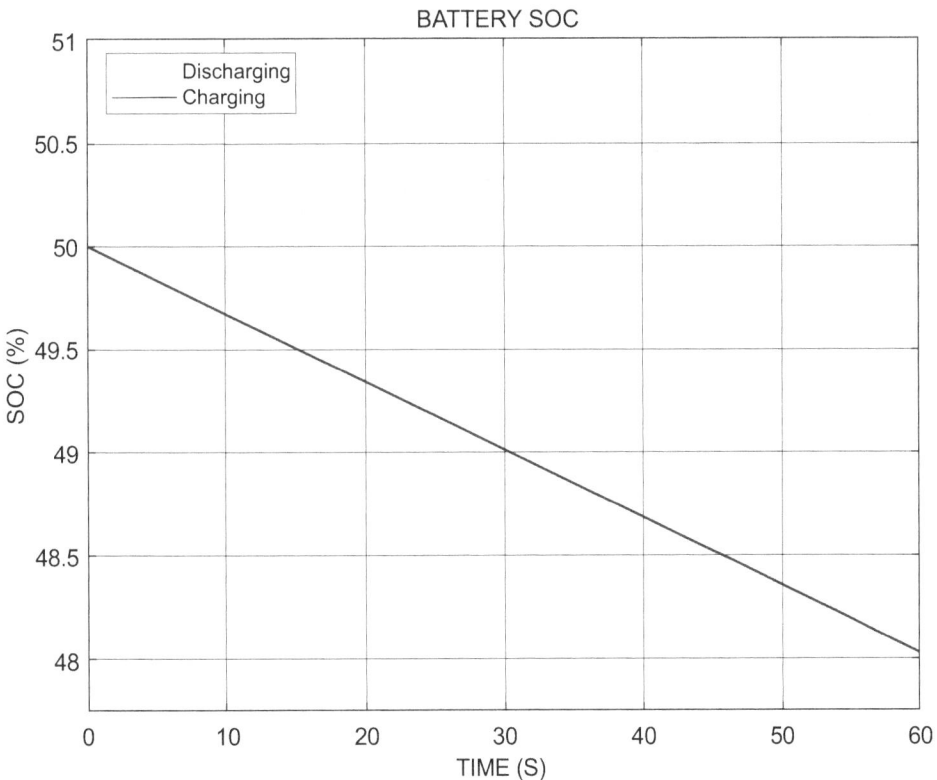

FIGURE 11.20 SOC when BESS operates in discharging mode.

11.5 CONCLUSION

A BESS with PV emulated output characteristics for direct PV string DC coupling using half-bridge bidirectional DC-DC converter is developed in MATLAB/Simulink. The design methodology of the half-bridge bidirectional DC-DC converter for the charging and discharging with PV emulated output characteristics of the BESS are presented in this study. The PV emulating accuracy is evaluated by comparing its PV characteristics curve with the reference PV characteristic curve produced by the PV string block in MATLAB/Simulink SimPowerSystems library. The average MAPE based on the 15 operating points data is 0.85%, which indicates that the PV emulating function have good accuracy. The effect on the PV string output when the direct PV string coupling is used with BESS without PV output characteristics was simulated. The result shows that the PV string operating point is shifted away from maximum power point when the BESS current is not at the maximum power point current. When the PV emulating function is integrated into the feedback control loop of the converter when operating in boost converter mode, the maximum power generation of the PV string is achieved. The SOC of the BESS is presented to verify the BESS charging and discharging operating mode. This chapter presents an inclusive method on how BESS with PV emulated output characteristics can be realised in direct PV string DC coupling for PV-BESS system.

In future research, the MPPT function can be integrated into the feedback control loop of the DC-DC converter operating in buck converter mode for charging of BESS to allow the PV string to operate at maximum power point. Besides, double loop control can be implemented into the control system of BESS for controlling the charging and discharging mode. Furthermore, larger PV-BESS system can be developed using direct PV string DC coupling topology to suit industry demands.

REFERENCES

[1] "Challenges and opportunities: The growing demand for BESS integration with PV. Energy Storage Association, Jul. 13, 2021. https://energystorage.org/challenges-on-growing-demand-of-bess-integration-with-pv/ (accessed Mar. 08, 2022).

[2] H. C. Hesse, M. Schimpe, D. Kucevic, and A. Jossen. Lithium-ion battery storage for the grid—A review of stationary battery storage system design tailored for applications in modern power grids. *Energies,* vol. 10, no. 12, Art. no. 12, 2017, doi: 10.3390/en10122107.

[3] M. Sandelic, A. Sangwongwanich, and F. Blaabjerg. Reliability evaluation of PV systems with integrated battery energy storage systems: DC-coupled and AC-coupled configurations. *Electronics*, vol. 8, no. 9, Art. no. 9, 2019, doi: 10.3390/electronics8091059.

[4] F. Niedermeyer, and M. Braun. Comparison of performance-assessment methods for residential PV battery systems. *Energies*, vol. 13, no. 21, Art. no. 21, 2020, doi: 10.3390/en13215529.

[5] J. He, Y. Yang, and D. Vinnikov. Energy storage for 1500 V photovoltaic systems: A comparative reliability analysis of DC- and AC-coupling. *Energies*, vol. 13, no. 13, Art. no. 13, 2020, doi: 10.3390/en13133355.

[6] F. Lo Franco, A. Morandi, P. Raboni, and G. Grandi. Efficiency comparison of DC and AC coupling solutions for large-scale PV+BESS power plants. *Energies*, vol. 14, no. 16, Art. no. 16, 2021, doi: 10.3390/en14164823.

[7] M. Bragard, N. Soltau, S. Thomas, and R. W. De Doncker. The balance of renewable sources and user demands in grids: power electronics for modular battery energy storage systems. *IEEE Transactions on Power Electronics*, vol. 25, no. 12, pp. 3049–3056, Dec. 2010, doi: 10.1109/TPEL.2010.2085455.

[8] G. Angenendt, S. Zurmühlen, R. Mir-Montazeri, D. Magnor, and D. U. Sauer. Enhancing battery lifetime in PV battery home storage system using forecast based operating strategies. *Energy Procedia*, vol. 99, pp. 80–88, 2016, doi: 10.1016/j.egypro.2016.10.100.

[9] B. V. Rajanna, S. Lalitha, G. J. Rao, and S. K. Shrivastava. Solar photovoltaic generators with MPPT and battery storage in microgrids. *International Journal of Power Electronics and Drive Systems (IJPEDS)*, vol. 7, no. 3, Art. no. 3, 2016, pp. 701–712, doi: 10.11591/ijpeds.v7.i3.

[10] C. N. Truong, M. Naumann, R. C. Karl, M. Müller, A. Jossen, and H. C. Hesse. Economics of residential photovoltaic battery systems in Germany: The case of Tesla's Powerwall. *Batteries*, vol. 2, no. 2, Art. no. 2, 2016, doi: 10.3390/batteries2020014.

[11] N. A. DiOrio, J. M. Freeman, and N. Blair. DC-connected solar plus storage modeling and analysis for behind-the-meter systems in the system advisor model. in *2018 IEEE 7th World Conference on Photo-voltaic Energy Conversion (WCPEC) (A Joint Conference of 45th IEEE PVSC, 28th PVSEC & 34th EU PVSEC)*, Waikoloa Village, HI, Jun. 2018, pp. 3777–3782, doi: 10.1109/PVSC.2018.8547329.

[12] D. Xu, G. Wang, W. Yan, and X. Yan. A novel adaptive command-filtered backstepping sliding mode control for PV grid-connected system with energy storage. *Solar Energy*, vol. 178, pp. 222–230, 2019, doi: 10.1016/j.solener.2018.12.033.

[13] A. K. Mukerjee, and N. Dasgupta. DC power supply used as photovoltaic simulator for testing MPPT algorithms. *Renewable Energy*, vol. 32, no. 4, pp. 587–592, 2007, doi: 10.1016/j.renene.2006.02.010.

[14] C.-C. Chen, H.-C. Chang, C.-C. Kuo, and C.-C. Lin. Programmable energy source emulator for photovoltaic panels considering partial shadow effect. *Energy*, vol. 54, pp. 174–183, 2013, doi: 10.1016/j.energy.2013.01.060.

[15] R. Ayop, and C. W. Tan. Rapid Prototyping of photovoltaic emulator using buck converter based on fast convergence resistance feedback method. *IEEE Transactions on Power Electronics*, vol. 34, no. 9, pp. 8715–8723, 2019, doi: 10.1109/TPEL.2018.2886927.

[16] A. Chalh, S. Motahhir, A. El Hammoumi, A. El Ghzizal, and A. Derouich. Study of a low-cost PV emulator for testing MPPT algorithm under fast irradiation and temperature change. *Technol Econ Smart Grids Sustain Energy*, vol. 3, no. 1, p. 11, 2018, doi: 10.1007/s40866-018-0047-8.

[17] M. T. Iqbal, M. Tariq, M. K. Ahmad, and M. S. B. Arif. Modeling, analysis and control of buck converter and Z-source converter for photo voltaic emulator. *2016 IEEE 1st International Conference on Power Electronics, Intelligent Control and Energy Systems (ICPEICES)*, 2016, pp. 1–6, doi: 10.1109/ICPEICES.2016.7853605.

[18] I. Moussa, A. Khedher, and A. Bouallegue. Design of a Low-Cost PV emulator applied for PVECS. *Electronics*, vol. 8, no. 2, Art. no. 2, 2019, doi: 10.3390/electronics8020232.

[19] R. Ayop, C. W. Tan, and A. L. Bukar. Simple and fast computation photovoltaic emulator using shift controller. *IET Renewable Power Generation*, vol. 14, no. 11, pp. 2017–2026, 2020, doi: 10.1049/iet-rpg.2019.1504.

[20] L. L. O. Carralero, G. S. Barbara da S. e Silva, F. F. Costa, and A. P. N. Tahim. PV emulator based on a four-switch buck-boost DC-DC converter. *2019 IEEE 15th Brazilian Power Electronics Conference and 5th IEEE Southern Power Electronics Conference (COBEP/SPEC)*, 2019, pp. 1–5, doi: 10.1109/COBEP/SPEC44138.2019.9065364.

[21] M. Alaoui, H. Maker, A. Mouhsen, and H. Hihi. Photovoltaic emulator of different solar array configurations under partial shading conditions using damping injection controller. *International Journal of Power Electronics and Drive Systems (IJPEDS)*, vol. 11, no. 2, Art. no. 2, 2020, doi: 10.11591/ijpeds.v11.i2.pp1019-1030.

[22] N. Priyadarshi, S. Padmanaban, P. Kiran Maroti, and A. Sharma. An extensive practical investigation of FPSO-based MPPT for grid integrated PV system under variable operating conditions with anti-islanding protection. *IEEE Systems Journal*, vol. 13, no. 2, pp. 1861–1871, Jun. 2019, doi: 10.1109/JSYST.2018.2817584.

[23] N. Priyadarshi, S. Padmanaban, J. B. Holm-Nielsen, F. Blaabjerg, and M. S. Bhaskar. An experimental estimation of hybrid ANFIS—PSO-based MPPT for PV grid integration under fluctuating sun irradiance. *IEEE Systems Journal*, vol. 14, no. 1, pp. 1218–1229, 2020, doi: 10.1109/JSYST.2019.2949083.

[24] N. Priyadarshi, S. Padmanaban, L. Mihet-Popa, F. Blaabjerg, and F. Azam. Maximum power point tracking for brushless DC motor-driven photovoltaic pumping systems using a hybrid ANFIS-FLOWER pollination optimization algorithm. *Energies*, vol. 11, no. 5, Art. no. 5, 2018, doi: 10.3390/en11051067.

[25] N. Priyadarshi, V. K. Ramachandaramurthy, S. Padmanaban, and F. Azam. An ant colony optimized MPPT for standalone hybrid PV-wind power system with single cuk converter. *Energies*, vol. 12, no. 1, Art. no. 1, 2019, doi: 10.3390/en12010167.

[26] A. Nazar Ali, K. Premkumar, M. Vishnupriya, B. V. Manikandan, and T. Thamizhselvan. Design and development of realistic PV emulator adaptable to the maximum power point tracking algorithm and battery charging controller. *Solar Energy*, vol. 220, pp. 473–490, 2021, doi: 10.1016/j.solener.2021.03.077.

[27] N. Priyadarshi, P. Sanjeevikumar, M. Bhaskar, F. Azam, I. B. M. Taha, and M. G. Hussien. An adaptive TS-fuzzy model based RBF neural network learning for grid integrated photovoltaic applications. *IET Renewable Power Generation*, vol. 16, no. 14, pp. 3149–3160, 2022, doi: 10.1049/rpg2.12505.

[28] N. Priyadarshi, M. S. Bhaskar, P. Sanjeevikumar, F. Azam, and B. Khan. High-power DC-DC converter with proposed HSFNA MPPT for photovoltaic based ultra-fast charging system of electric vehicles. *IET Renewable Power Generation*, doi: 10.1049/rpg2.12513.

12 A New Hybrid Islanding Detection Technique for Microgrid Using Virtual Synchronous Machine

Vikash Gurugubelli, Ankit Anand, and Arnab Ghosh

CONTENTS

DOI: 10.1201/9781003323471-12

The development of ecologically friendly distributed sources is becoming more common these days. Interest in installing solar, wind, and other renewable-based Distributed Energy Resources (DERs) and Energy Storage (ES) systems have increased in recent years [1]–[3] because of rising energy prices, an increase in power quality demand, and the need for more effective power transfer to prevent collapse and exhaustion. Small-scale DERs and ES are placed to provide auxiliary services at the distribution level like peak load sharing, Static-VAr Compensation (STATCOM), ES, and Uninterruptible Power Supply (UPS) capabilities via microgrid (MG) systems is a viable way to address all of these requirements. A controllable device in an MG system is thought to be a clump of loads and DERs/ESs units that supply the local area network with a power and stabilising system. To connect DER and ES systems to an MG system, grid-connected voltage source converters (VSCs) with control systems designed to provide these ancillary services are required; specifically, a control system with multiple operational modes, islanding detection algorithms, and the ability to quickly and accurately accompany with the grid for reconnection.

DG helps raise significant sums of money per year due to transmission congestion, which happens when there is insufficient capacity to satisfy all consumers' demands, impacting general power market prices. From a futuristic standpoint, this technology is critical to achieving today's vision of eco-friendly solutions by micro-grids. It is a future-proof solution because it allows communities to raise the total energy supply rapidly and effectively using renewable resources such as small local generators, solar panels, and wind turbines to meet known and unknown future needs. Furthermore, smart grids allow for the use of plug-in electric vehicles. This system would be more flexible because it will be able to use local renewable or natural gas energy production. Smart grids will repurpose the energy used to generate electricity for heating, hot water, and sterilisation. In the last few years, numerous safety and control tactics for MGs have been suggested in the literature. One significant difficulty is that MGs can be built to run in one of the two methods: grid-connected (GC) and islanding mode (IM). Power-sharing, synchronisation, and islanding detection are all important factors to consider [4], [5]. Area of a non-detection zone is derived and discussed for under/over voltage and frequency for IDT [6]. The communication-based IDT is more unfailing and it's not affecting the power aspect of the system but the installation cost is high [7]. Active method got attention these days due to high performances (detecting islanding condition is faster) [8]. There is no need to set up the threshold value of the parameter like voltage and frequency since they are using data-mining technology [9], [10].

In this work, various IDTs are discoursed in section 12.2. Modelling of the active and passive IDT is derived and analysed in section 12.3. The hybrid IDT is presented in section 12.4. The VSM operation and its implementation are presented in section 12.5. Results obtained from the simulation of active and passive IDT is discussed and compared in section 12.6. Section 12.7 gives the conclusions of this work.

12.1 ISLANDING CONDITION

Since islanding happens when the entire electricity distribution system becomes isolated or disconnected from the rest of the electrical power system but is still energised by a distributed resource (DR) [6], it was critical to comprehend how the islanding phenomenon works to prevent islanding problems. If not identified and configured on time, islanding may have negative consequences for device activity as well as safety concerns. Some of the key islanding problems, for example, are primarily related to power quality, protection, and operation. Technically, islanding can jeopardise power quality by causing harm to voltage and frequency, which cannot be bolstered within a normal allowable range. Furthermore, the DG interconnection may not sufficiently ground the islanded device. Furthermore, islanding can pose significant health and safety risks, especially for line workers. In terms of operation, instantaneous reclosing could cause out-of-phase DG reclosing, resulting of damage to utilities' and customers' equipment [11]. Figure 12.1 shows the islanding detection line diagram.

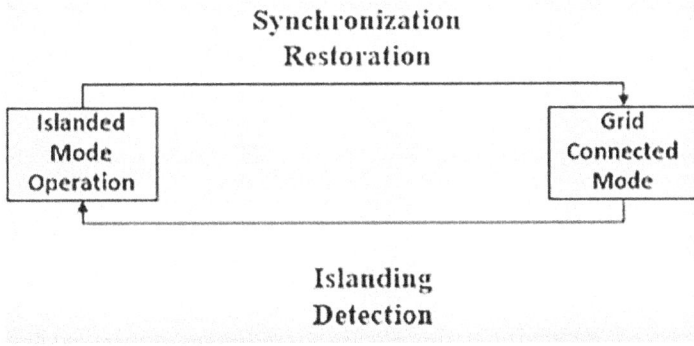

FIGURE 12.1 Block diagram of islanding condition.

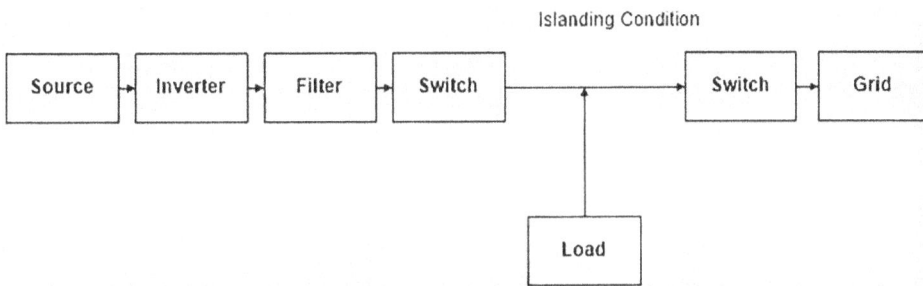

FIGURE 12.2 Overall system line diagram.

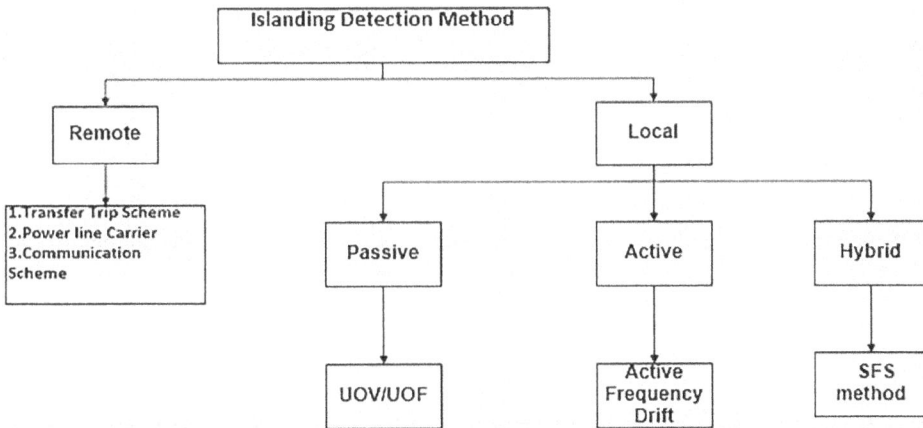

FIGURE 12.3 Different types of islanding detection methods.

12.2 ISLANDING DETECTION TECHNIQUES

The ability to link all distributed generations to the distribution system is made possible by islanding identification of distributed generations. This section provides a summary of power system islanding methods. The utility side is addressed by the remote islanding detection system, while the DG side is addressed by the local method. The three methods of islanding detection for the local approach are passive, active, and hybrid. Figure 12.2 shows the overall system and Figure 12.3 shows the different types of islanding detection methods.

12.2.1 REMOTE IDT

The remote system, primarily the transfer trip arrangement and the power line communication system, can be used to detect the islanded condition issue. They are dependent on the utilities' and the DG's contact. In certain cases, these remote methods have been more precise and more consistent than local methods, although these methods are very costly and uneconomical [12]. The remote process, for example, is primarily concerned with establishing a communication interface between utilities and DGs.

12.2.1.1 Transfer Trip Arrangement

A move trip arrangement is a method of tracking individual circuit breakers and re-closers in a delivery system to detect islands [13]–[20]. The transition trip system works with supervisory control and data acquisition (SCADA) to keep track the data. The utility and the DGs must communicate effectively for this approach to work.

12.2.1.2 Power Line Carrier Communication System

The PLCC is in charge of broadcasting a signal across the power line from the transmission grid to the distribution feeders. Each DG has a receiver that receives signals. When it detects no signal, it switches from GC mode to IM [21]–[25]. In the case of multiple DGs, the power line scheme is thought to be an efficient detection process.

12.2.2 LOCAL IDT

Local IDTs are focused on the measurement of device parameters on the DG side such as frequency, voltage, harmonics, and so on. There are three types of techniques: passive, active, and hybrid.

12.2.2.1 Passive IDT

Where there is a significant discrepancy among generation and demand in the standalone system, passive methods are used. These methods detect IM by watching various parameters such as frequency, voltage, harmonic distortion, and so on, and comparing them to a threshold value [26]. These methods are fast and do not disrupt the system and have lower harmonics, but they have a wide non-detectable zone (NDZ).

12.2.2.2 Active IDT

The successful IM detection method relies on disturbance signal into specific PCC parameters [27]. Unlike passive detection systems, the active method can be identified when the generation and load are perfectly matched. That is to say, when deducing the IM of operation using this tool, even a minor disturbance signal is clear and visible. The key advantage of the active technique is its high islanding detection response, which can cover the majority of the area and is defined as having a small NDZ, as opposed to the passive technique's large non-detection zone.

12.3 MODELLING OF UOV/UOF METHOD AND AFD METHOD

12.3.1 UNDER/OVER VOLTAGE (UOV) AND UNDER/OVER FREQUENCY (UOF)

The voltage and/or frequency at PCC are simply measured with this technique. Here, the IM is detected based on the threshold values [28]. The relays are mounted on the feeders to protect the distribution system. The UOV and UOF relays can detect irregular conditions. The threshold values for the relays can be measured using equations (1) and (2), as follows:

$$\left(\frac{v}{V_{max}}\right)^2 - 1 \le \frac{\Delta P}{P_{DG}} \le \left(\frac{v}{V_{min}}\right)^2 - 1 \tag{1}$$

$$Q_f \left(1 - \left(\frac{f}{f_{min}} \right)^2 \right) \leq \frac{\Delta Q}{P_{DG}} \leq Q_f \left(1 - \left(\frac{f}{f_{max}} \right)^2 \right) \tag{2}$$

Where V_{min}, V_{max}, f_{min} and f_{max} are the threshold values of the relays. There is no detection zone present in the passive method where the Islanding is not detectable. In reality, the region of the NDZ, specified in power discrepancy space at the PCC is used to classify the efficiency of IDT. P is the grid's active power (AP), Q is the grid's reactive power (RP). P_{DG}, Q_{DG}, P_{load} and Q_{load} are the distributed generation's and load real and reactive output power, respectively.

$$P_{load} = P_{DG} + \Delta P \tag{3}$$

$$Q_{load} = Q_{DG} + \Delta Q \tag{4}$$

The change in AP and RP because of the islanding cause's voltage and frequency change. Since the corresponding association between AP and voltage, and RP and frequency, a significant power imbalance causes a voltage and frequency drift that exceeds the NDZ's limits, allowing islanding to be determined. The logic of under/over voltage of frequency is shown in Figure 12.4. The actual frequency is compared with the threshold frequency for a small-time delay; if the actual frequency is crossing the threshold boundary then the IM is detected. A similar mechanism is shown in Figure 12.5. In which voltage is compared with the threshold value of the voltage, if the measured voltage excels the threshold limits, the IM is determined.

12.3.2 Active Frequency Drift (AFD)

While passive detection of islanded conditions appears to be easy, setting the threshold and detecting islanding conditions. Active methods will now be considered the key to resolving islanding issues. AFD is an active approach that can be easily implemented using a microprocessor-based controller [29]–[32]. Initially, disturbances signal is injected into the systems shown in Figure 12.6. Furthermore, the PV system's current waveform pumped into the utility grid is slightly skewed, indicating a phase lag between the inverter output voltage and the voltage at the PCC. The frequency at the PCC will drift up or down, augmenting the normal frequency, while islanding occurs [33]–[51].

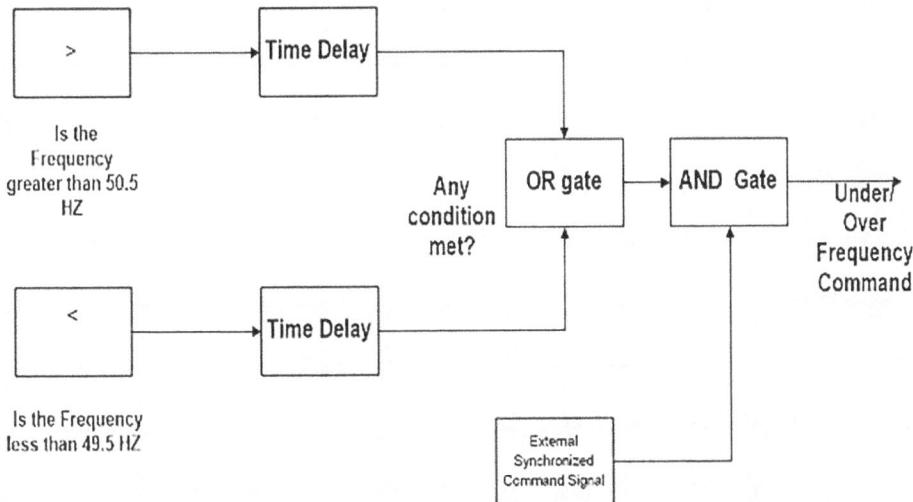

FIGURE 12.4 Block diagram of the under/over frequency passive IDM logic.

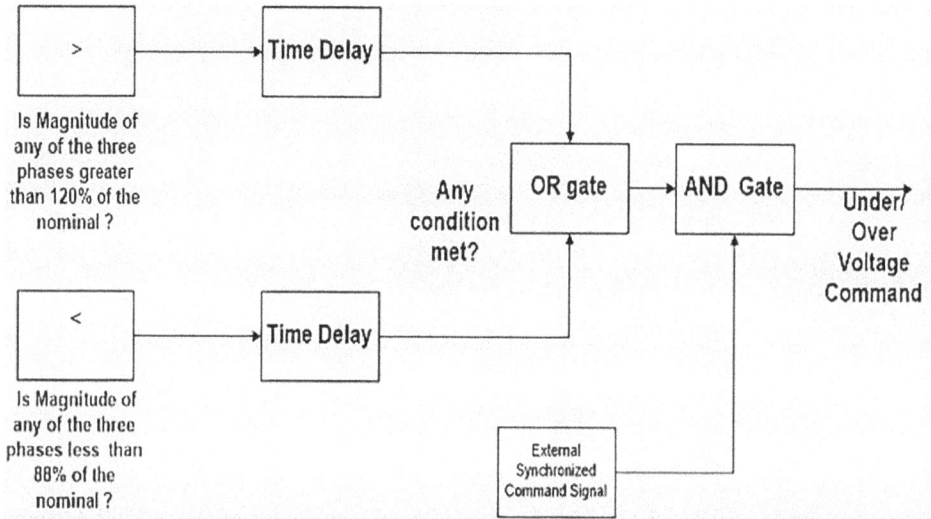

FIGURE 12.5   Block diagram of the under/over voltage passive IDM logic.

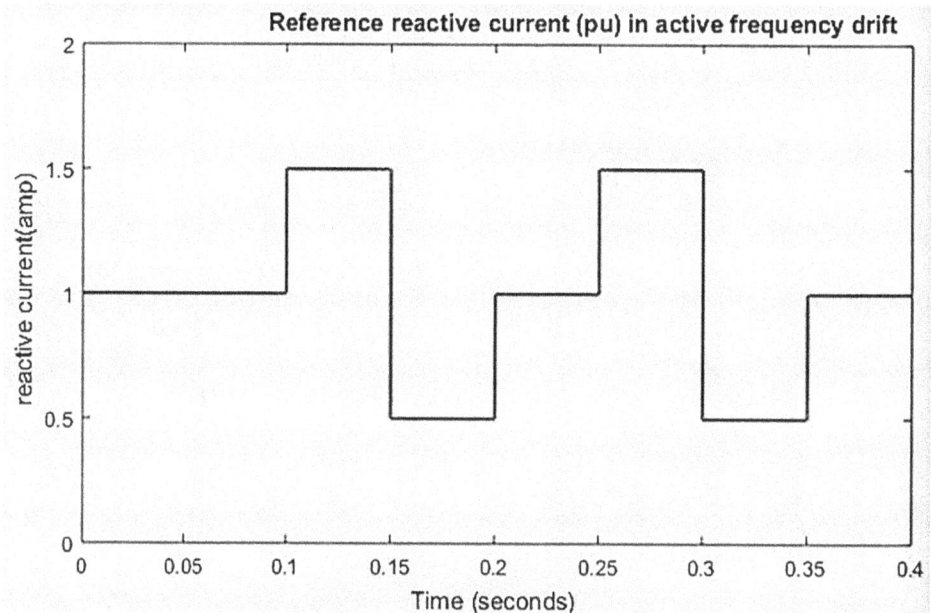

FIGURE 12.6   The waveform of reference reactive current (Pu) versus time.

12.4   HYBRID IDT

The hybrid IDT approach is based on AP, RP, and voltage discrepancy detection.

12.4.1   IMPLEMENTATION OF THE HYBRID IDT

According to its fixed power P_{set} and Q_{set}, the inverter sends AP, and RP to the grid, and the inverter current is used to measure the AP and RP.

$$P = P_{SET} = \frac{V_0^2}{R_T} \qquad (5)$$

$$Q = Q_{set} = V_0^2 \left(\frac{1}{2\pi f L_T} - 2\pi f C_T \right) \qquad (6)$$

The grid controls the voltage and frequency in GC mode, and the DG inverters transmit power to the grid based on their reference power, P_{set} and Q_{set}. The DG inverters have to maintain voltage and frequency in the IM and provide power to the essential or critical load. Between the set power and the actual power, there is a power discrepancy because the inverters do not transmit power according to their P_{set} and Q_{set} in the islanding state.

$$\Delta P = P_{set} - P \qquad (7)$$

$$\Delta Q = Q_{set} - Q \qquad (8)$$

Equations (5) and (6) show a shift in V_0, f, and connected load causes power oscillation from GC mode to IM.

$$\Delta P = -\frac{V_0^2}{R_T^2} \Delta R_T \qquad (9)$$

$$\Delta Q = -\frac{V_0^2}{2\pi f L_T^2} - 2\pi f V_0^2 \Delta C_T \qquad (10)$$

Δ is a term that describes the transition from a GC operation to an IM. For example, in a GC operation, the equivalent resistance is R_T, but in an IM, it is $R_T + \Delta R_T$. Since the voltage and frequency are controlled in both the GC and IMs with the help of AP-frequency droop and RP and voltage droop laws. When the inverter is tied with the grid system, the RP set point is generally fixed to zero such that the system will supply power at near to unity power factor. However, depending on the linked L_T and C_T, RP is supplied or absorbed by the inverter in an islanding state. In an islanding state, the set RP set point and the inverter's supplied RP must be out of sync.

12.4.2 Conditions for Hybrid IDM

In starting the condition is checked whether islanding is present or not in the network. In this approach, the variation of reactive power should be present within their limits, not cross their threshold limits value, as the threshold limit is about 0.1 pu.

$$|\Delta Q| \ge Q_L \qquad (11)$$

It is the secondary criteria but not compulsory criteria to be for islanding conditions. In this approach, the variation of active power should be present within their limits, not cross their threshold limits value, as the threshold limit is about 0.1 pu.

$$|\Delta P| \ge P_L \qquad (12)$$

In the IM, the DG system has to maintain the voltage and frequency by absorbing or delivering the power from the network. Voltage cannot be regulated when the inverter's current is cross its limits value, by increasing its power and the voltage magnitude will begin to exceed 0.95–1.05 Pu. The voltage incompatibility is described as follows:

$$\Delta V = E^* - V_0 \tag{13}$$

The rated voltage, E, maybe greater than the normal operating range. As a result, if V reaches the upper limit V_L, (4.10) is fulfilled, the voltage discrepancy has been taken into account. as one more condition for hybrid IDM.

$$|\Delta V| \geq V_L \tag{14}$$

The algorithm in the hybrid IDM is kept such that if condition (11) is met, and at least one of the other conditions (either (12) or (14)) is met. The hybrid IDM scheme senses the situation as an islanding state based on the aforementioned conditions.

12.4.3 CRITICAL CASE

In an islanding condition, the critical case is defined as when DG capacity matches the local load capacity. Hybrid IDM is a passive AIP scheme since it is focused on the monitoring of device parameter mismatches, as discussed in the previous paragraph. The passive method is simple to implement, but it has one drawback: it cannot find out an islanding situation, $P_{set} = P = P_{LOAD}$. In an islanding state, the inverter controls the voltage so that it will remains within the cap, and (14) is not satisfied. Since at least one of the conditions from (12) and (14) must be fulfilled in the proposed hybrid IDM in addition to (11) to find out an islanding situation when $P_{set} = P_{LOAD}$, this method cannot find out to detect an islanding situation when $P_{set} = P_{LOAD}$. When only (11) is satisfied, the hybrid IDM is changed to an active scheme to prevent this failure. An active power new setpoint is employed to avoid a potential instability issue. To change the active power set point, the hybrid IDM gives a command alert to the inverter.

$$|\Delta P| = |P_{set} - P_{load}| \approx 0 \leq P_L \tag{15}$$

The value of ΔP_{set} should be greater than the threshold limit. If the inverter is tied with the grid, changing ΔP_{set} will not result in any changes to the mode. Since P will obey the power relation, the threshold will be exceeded. The inverter will provide electricity in the event of an islanding incident. The inverter capacity equal to the local load capacity will not obey the order to modify the power reference, ΔP_{set}.

$$\Delta P = (P_{set} - \Delta P_{set}) - P_{load} = -\Delta P_{set} \tag{16}$$

In an islanding condition, there is a possibility of reference reactive power equals the complete reactive power load, $Q_{set} = Q = Q_{LOAD}$. To avoid an undetected condition in such critical case scenarios, a phase reduction of reference reactive power, Q_{set}, is implemented, resulting in a new reactive power setpoint of ($Q_{set} - \Delta Q_{set}$). As a result, $\Delta Q = (Q_{set} - Q_{set}) - Q_{LOAD} = -\Delta Q_{set}$ can be used to calculate the current reactive power mismatch. The value of ΔQ_{set} should be greater than the threshold limit. After that Giving the command to change the inverter's reactive power relation causes the reactive power to reach the threshold value.

12.4.4 CRITICAL FAULT TRIGGERING

For a brief amount of time, a grid failure, voltage drop, frequency drop, and different moments produced can all produce a power and voltage mismatch. In such situations, the hybrid IBM does not give a false signal. A time delay T_D with (11) and (12) is added to prevent a fault triggering in such situations. In the proposed IDM in an islanding design, the inverter is taken into account when (12) and/or (14) are fulfilled and the reactive power imbalance in (4.7) is satisfied for TD time.

12.4.5 LOGIC IMPLEMENTATION

P_{set} will be adjusted by a command signal sent by the hybrid IDM and then recalculate the P if they ΔQ exceed the upper limit Q_L for the length of the delay time T_D but the ΔP or ΔV does not go beyond the upper limit, P_L and V_L. If the updated power discrepancy crosses the P_L limit, the situation is identified as a standalone mode by Hybrid IDM. If the variation of reactive power crosses their threshold limits value but the active power's variation is present in their limits, the hybrid IDM will transmit an instructing signal to adjust Q_{set} and then recalculate the Q. After that updated power inconsistency reaches a cap Q_L, the situation is recognised as an islanding state by hybrid IDM. Figure 12.7 shows the logic diagram of the hybrid IDM.

12.5 VIRTUAL SYNCHRONOUS MACHINE (VSM)

The swing equation is the heart of the VSM and Figure 12.8 shows the block diagram of the VSM, which includes energy storage, an inverter, and a control system. The VSM control mechanism is implemented on the inverter to behave like a traditional SM. The damping power signifies the damping effect of the VSM like a traditional SM. It is characterised by D (damping factor). D is manipulated on the difference between the virtual angular speed of VSM and the angular frequency of the grid.

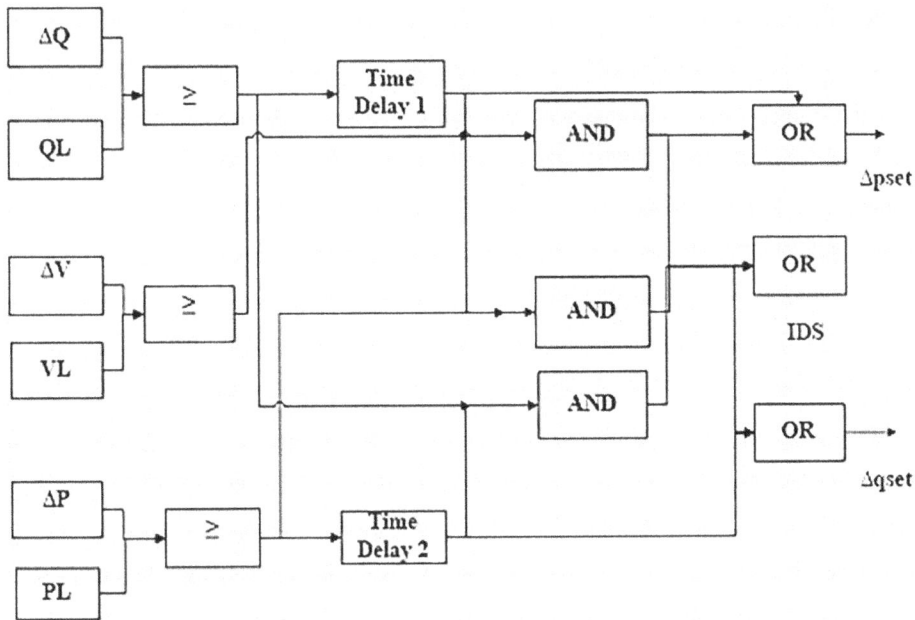

FIGURE 12.7 Logic diagram of the hybrid islanding detection method.

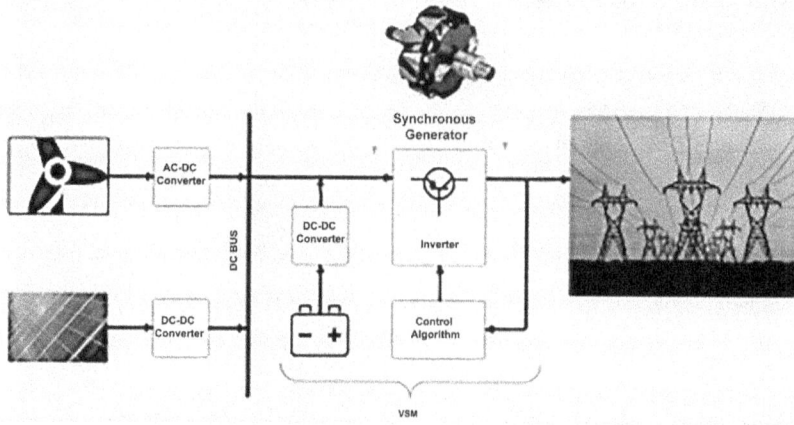

FIGURE 12.8 VSM block diagram.

This model, which is equal to the steady-state properties of the governor of SM, considers the power frequency (P_w) droop. Kw is a unit of measurement for power frequency (frequency droop constant). The divergence between the reference frequency and the genuine frequency is used to control KW. The time constant of the VSM unit is governed by the rotor's moment inertia and damping factor. After measuring voltage, the current, frequency at the VSM terminal then output power (P_{out}) is to be calculated. The equation (1) is to be solved after putting the value of P_{out}, P_{in}, and grid frequency. After solving the preceding equation, one can get W_m. W_m is integrated to get a virtual mechanical phase angle (θ_m). The phase angle controls the AP flowing through the inverter to the grid.

12.5.1 VIRTUAL INERTIA

In a traditional synchronous generator, there is a rotating mass. The rotating mass's moment of inertia affects the rotor's rate of change of speed. In a steady state, a little amount of kinetic energy is stored in the spinning mass. Whenever any type of fault, disturbance, and transient or sub-transient condition, the rotating mass whose kinetic energy is released in the system removes the disturbances and maintains synchronisation. But in VSM there is a rotating mass like traditional SM. So in replace of these rotating masses, flywheels and batteries act as energy storage devices. These energy devices act as rotating mass. The energy storage devices also deliver the power when there is unbalancing power or inconsistency power between the input and output power.

12.5.2 DIFFERENT CONCEPTS IN VSM

12.5.2.1 Active Power Frequency Droop Control

If consider the line impedance is inductance dominant so active power dominantly on power angle and RP is dominantly on voltage. P-f control is used to control active real power flowing through from the source side to the load side. For stability of the system, frequency control must have a drooping characteristic concerning the output of the generator.

f* has a frequency of 50 Hz. P stands for power measured on the grid. The active power frequency control droop coefficients are denoted by K_f. P* is the system's allotted power. Power generator if there is a change in power demand as result drop in frequency which suggests increasing the fuel supply. After that this operation frequency will be increased, this objective is implemented

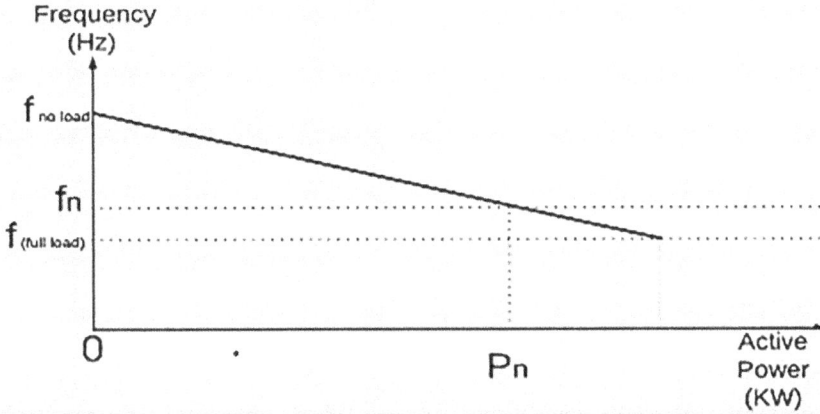

FIGURE 12.9 Active power frequency droop curve.

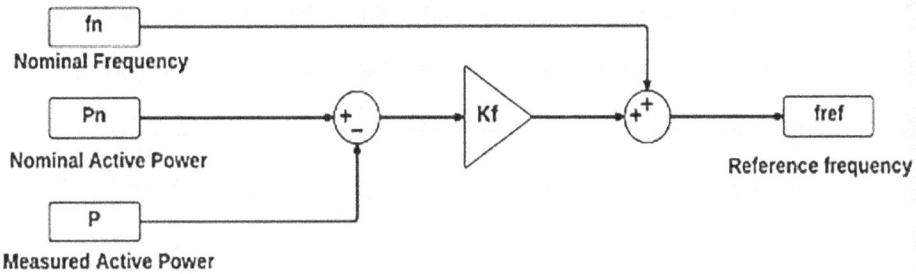

FIGURE 12.10 Block diagram of active power-frequency droop.

by active power frequency control. This droop plot is shown in Figure 12.9, the implementation diagram of this law is shown in Figure 12.10, and the droop laws are defined in equation (17).

$$F_{ref} = F_n + K_f \left(P_n - P \right) \tag{17}$$

12.5.2.2 Reactive Power Voltage Droop Control

The relation between RP and voltage is understood from the literature. The voltage and reactive power relation are shown in equation (18). The drooping curve between the voltage and reactive power is shown in Figure 12.11, and its implementation is shown in Figure 12.12.

$$V_{ref} = V_n + K_q (Q - Q_n) \tag{18}$$

Where K_q is droop coefficients of the reactive power voltage droop. V_{ref} is assigned voltage to the system. With the handling of the voltage and frequency, in this system real and reactive power is governed. The droop coefficients of this P-f and Q-V controller play an important role in governing the actual and reactive power in the system. So selecting the value of droop coefficients in this system is a little bit difficult.

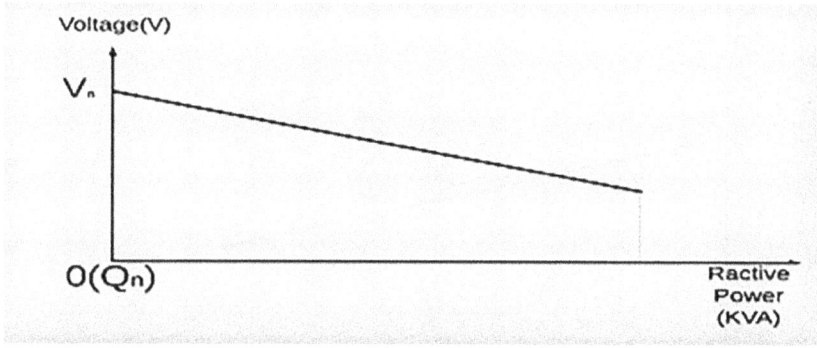

FIGURE 12.11   Reactive power—voltage droop curve.

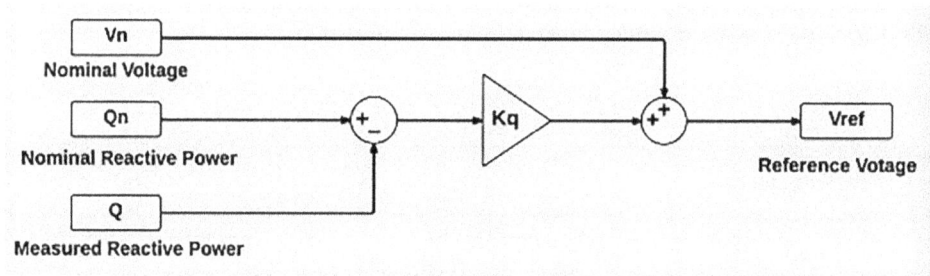

FIGURE 12.12   Block diagram of reactive power voltage droop.

12.5.2.3 Swing Equation

Adding swing equation as a solution to imitate virtual inertia is commonly regarded as a solution to imitate inertia property. The difficulty with using the swing equation is whether regardless of the load, the system's input/output power remains constant. These are power equations as given here.

$$P_{in} = P_{out} = P_{load} \tag{19}$$

$$f = f_n + K_f \left(P_{in} - P_{out} \right) \tag{20}$$

$$P_{in} = \left(f_n - f \right) / K_f + P_n \tag{21}$$

$$P_{in} - P_{out} = K_i \frac{df}{dt} \tag{22}$$

$$f = f_0 + \int K_i \left(P_{in} - P_{out} \right) \tag{23}$$

$$f = f_0 + \int_0^t K_i \left(\left(F_n - F_i \right) / K_f + P_{in} - P_{out} \right) dt \tag{24}$$

There is no controller to guide the BESS's operation with this function. It simply tracks the current value of demand power. The virtual power should be associated with the output frequency as a conventional synchronous system, there is a relationship between the power and the rotor inertia. The deviation of the frequency should be minimal due to the virtual inertia. After receiving P_{out} from a grid side power meter, it is possible to use the swing equation. The droop controller with the swing equation is shown in Figure 12.13.

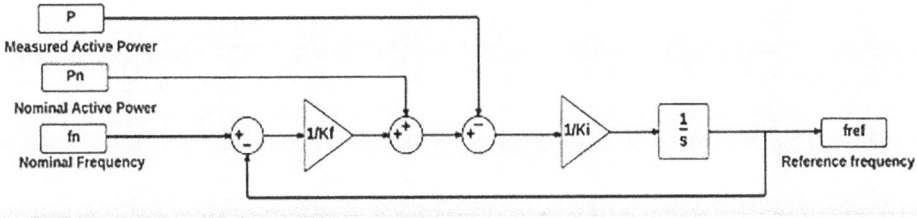

FIGURE 12.13   *P-f* droop scheme with swing equation.

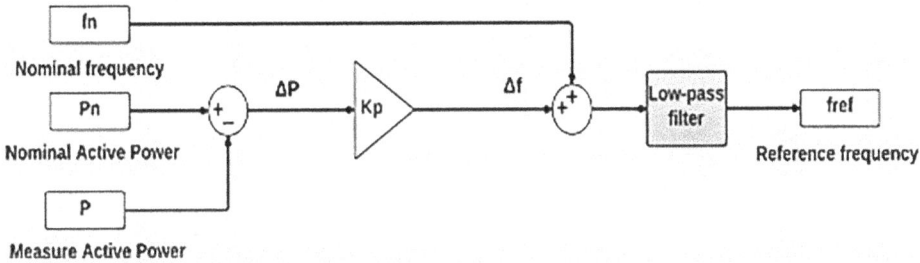

FIGURE 12.14   *P-f* droop control with low pass filter.

First harmonics must be removed such that the frequency of the system will be improved, mini-mising (or eliminating) overshoots, and smoothing its dynamic behaviour. The required impact could be achieved by using a low-pass filter for the inverter control reference frequency. The filter functions as a harmonic filter when coefficient k is tiny, and coefficient k denotes a frequency value that has been reduced down. The control law of the VSM with filter is shown in Figure 12.14. The frequency is regulated using the low pass filter's transfer function as shown in equation (25).

$$H(s) = \frac{1}{ks+1} \tag{25}$$

12.5.2.4   Damping Factor
The damping factor also contributes to system synchronisation and eliminates the oscillations of the rotor. The damping factor always tries to maintain synchronisation either rotor speed is more or less than synchronous speed. So damping windings are short-circuited coil. In steady-state opera-tion, damper winding doesn't come in action due to the relative velocity between the stator and rotor fields being zero. Whenever disturbances or any type of fault happens, damper winding comes into action. Oscillation of the rotor is known as hunting. If rotor oscillation is not removed within small intervals of time then results will be more dangerous. Oscillation is removed when emf is produced within the damper winding and after that flux will be generated to remove problems of hunting. This type of damping winding does not exist in a virtual synchronous generator. So the term damp-ing factor is added in the VSM. Here the value of the damping factor is considered a constant value. In VSM damping power is defined. K_d is a damping factor.

12.5.3   Current Controller Mechanism

Dq0 is a two-phase system in which d-axis (AP/torque) and q-axis (RP/flux) are controlled indepen-dently. In the dq0 reference frame, the AC signal is converted into dc quantities. If the DC signal is present, then the PI controller can easily manipulate and make the system with zero steady-state error. PLL is also used to track grid frequency and minimise the phase difference between the inverter and grid voltage such that the system will maintain the synchronisation. The current of the inverter is supervised, and the control mechanism is executed in a rotating dq-reference frame to ensure complete disassociating of the AP and RP control loops. The following equations are taken from the block diagram, which is shown in Figure 12.15.

$$V_{gk}(t) = V_{ik}(t) - \left[R_t i_{gk} + L_t \frac{di_{gk(t)}}{dt} \right]$$ (26)

Applying the Park transformation to (26), then the equation is:

$$V_{gd}(t) = V_{id}(t) - \left[R_t i_{gd}(t) + L_t \frac{di_{gd}(t)}{dt} - \omega_n L_t i_{gq}(t) \right]$$ (27)

$$V_{gq}(t) = V_{iq}(t) - \left[R_t i_{gq}(t) + L_t \frac{di_{gq}(t)}{dt} - \omega_n L_t i_{gd}(t) \right]$$ (28)

Where ω_n is system nominal angular frequency. Equations (26) and (27) are used to implement in Figure 12.15 to supervise AP and RP separately. The reference frame synced with the grid voltages, from the AP and RP credentials, as well as the computed grid voltages. Two PI regulators were chosen for the new controller to satisfy the system's criteria for reliability and to ensure that the steady-state error is zero. Because for the PWM logic, the control system's outputs must be modulating signals.

12.6 SIMULATION RESULTS AND DISCUSSION

12.6.1 UNDER AND OVER-VOLTAGE (UOV) AND UNDER AND OVER FREQUENCY (UOF) PASSIVE METHOD

In this method, the grid is disconnected at 0.25 seconds intentionally to create an islanding condition, which is shown in Figure 12.16. After 0.25 seconds the grid current flowing to the load is zero as shown in Figure 12.17. The system can identify the islanding condition based on voltage and

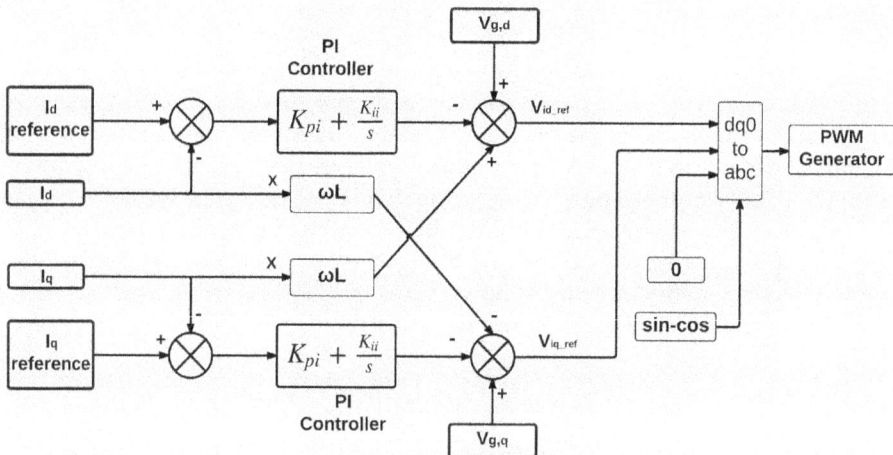

FIGURE 12.15 Block diagram of current control strategy.

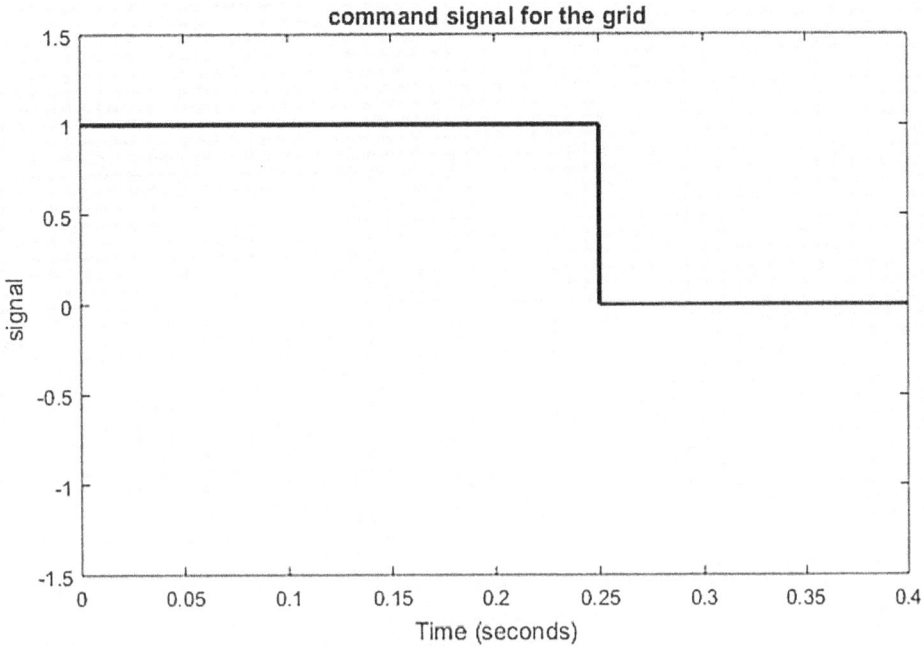

FIGURE 12.16   The waveform of a command signal to isolate the grid from the load.

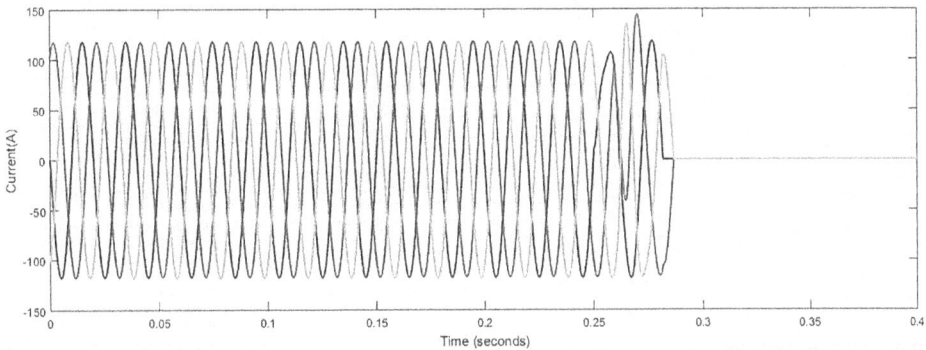

FIGURE 12.17   The load current from the grid.

frequency changes, but it takes some time to decouple the system from the grid. At 0.2806 seconds, the system is decoupled from the grid, indicating that there is a delay. The current form of the grid is shown in Figure 12.17, from which one can understand the delay. The voltage waveform of the grid is shown in Figure 12.18. The active power from the grid to load is shown in Figure 12.19. Here authors considered the weak grid, so the system is islanded from the load; there is an increase in the system frequency value shown in the Figure 12.20.

12.6.2   ACTIVE FREQUENCY DRIFT (AFD) ISLANDING METHOD

The inverter is controlled by the current controller with the reference current. In Figure 12.21, reactive currents is injected into the inverter side. The reference current is varied to the max or min to 10% of the rated current. When the microgrid is associated to the grid. System voltage is tied with grid voltage

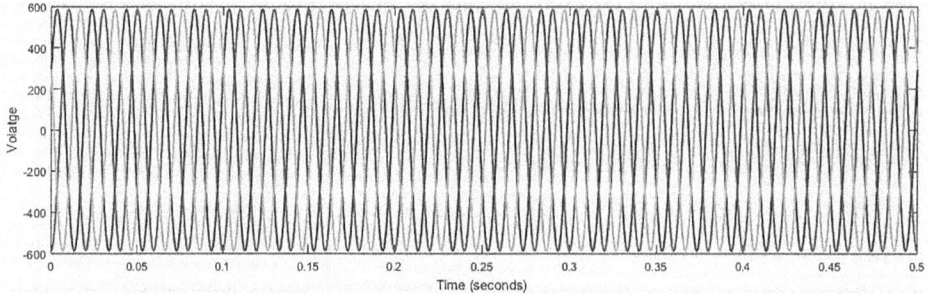

FIGURE 12.18   The grid voltage.

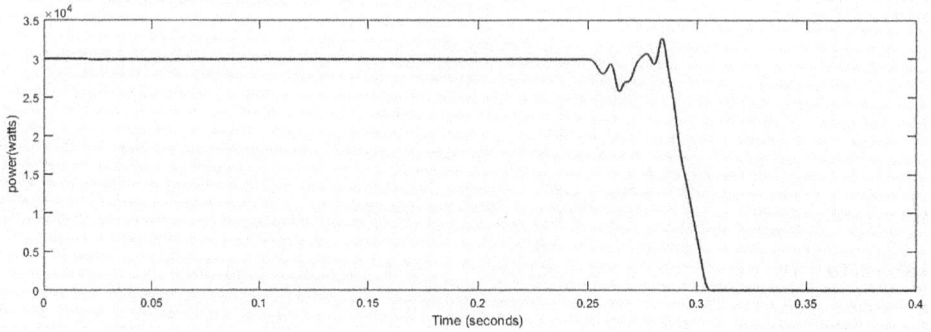

FIGURE 12.19   The active power drawn by the load from grid.

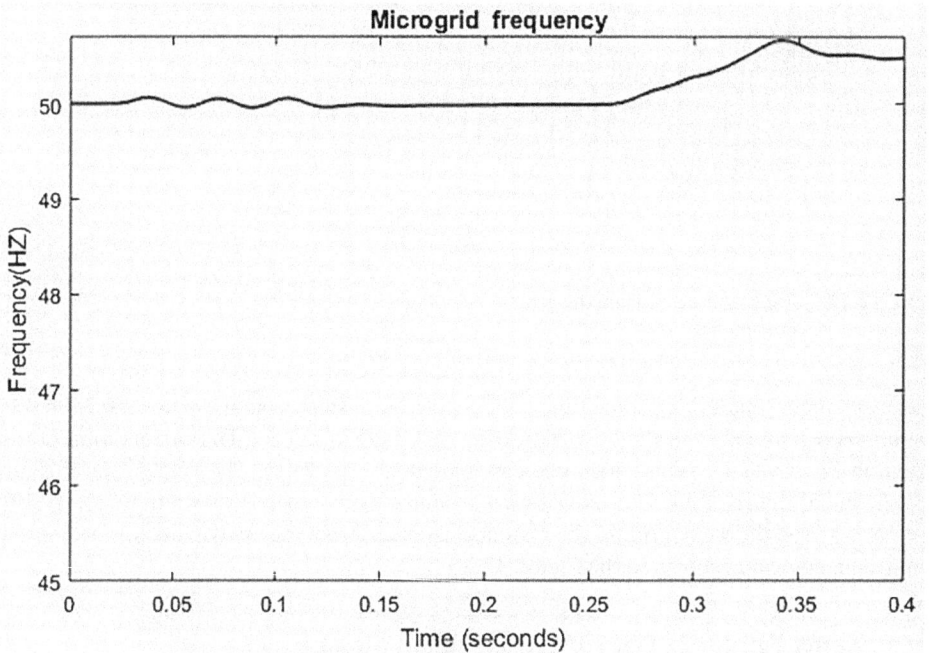

FIGURE 12.20   The grid frequency.

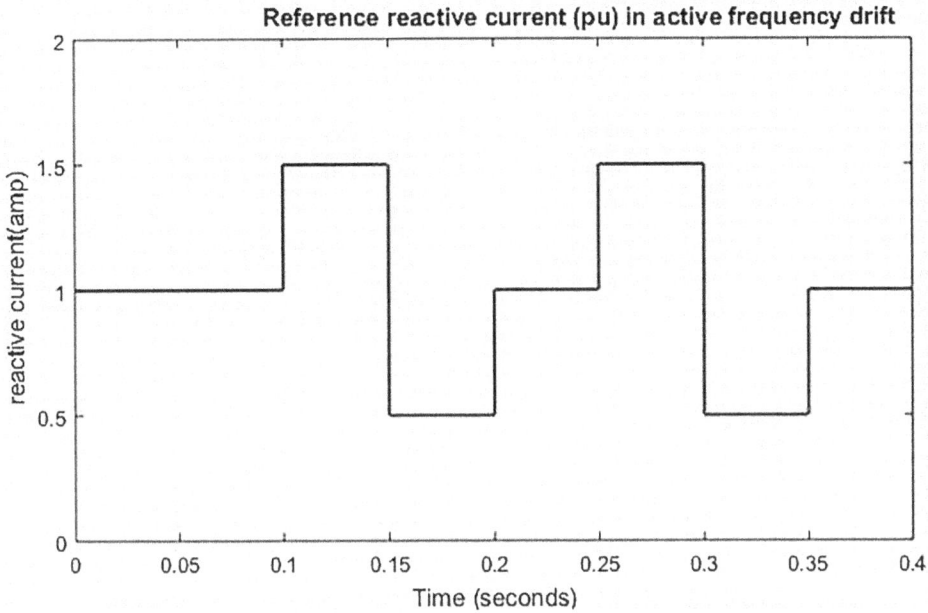

FIGURE 12.21   The waveform of reference reactive current (Pu) versus time.

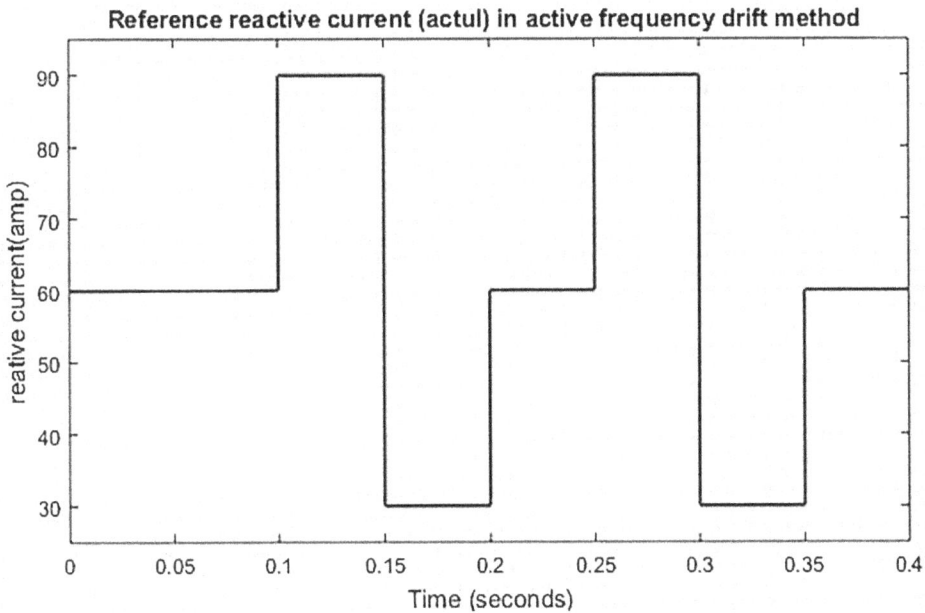

FIGURE 12.22   The waveform of reference reactive current (actual value) versus time for the controller.

so there will be no consequence on the voltage and frequency. When the grid is lost from the main network then voltage and frequency will be oscillating up somewhat. Microgrid side sensors (Relay) will sense the voltage and frequency. If the voltage and frequency will be cross their limits or their threshold values then the relay consigns the signal to the circuit breaker. The circuit breaker disconnects the microgrid from the major network and the inverter stops sending power to the network.

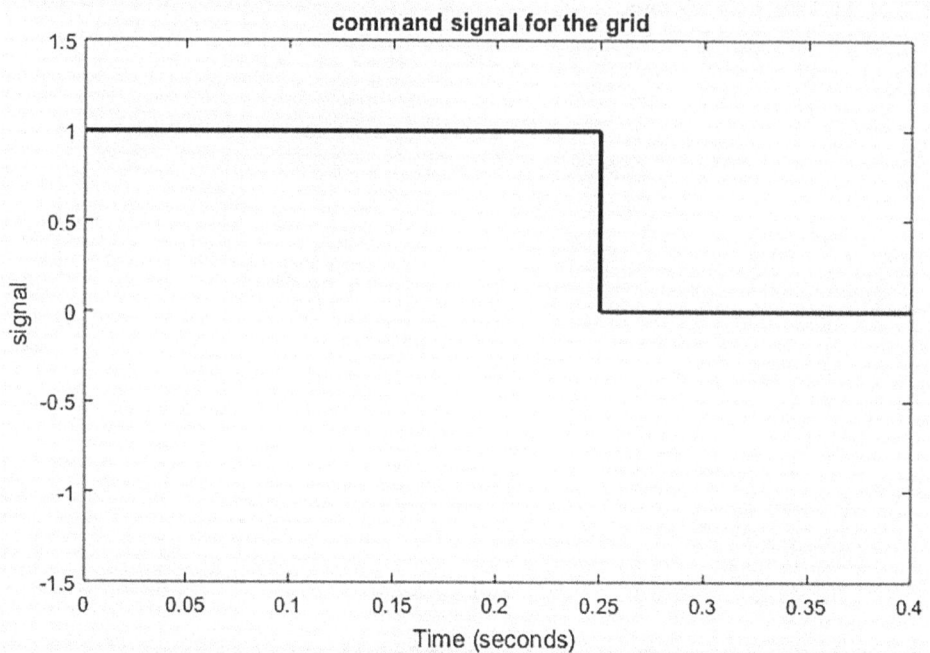

FIGURE 12.23   The waveform of command signal versus time for the grid.

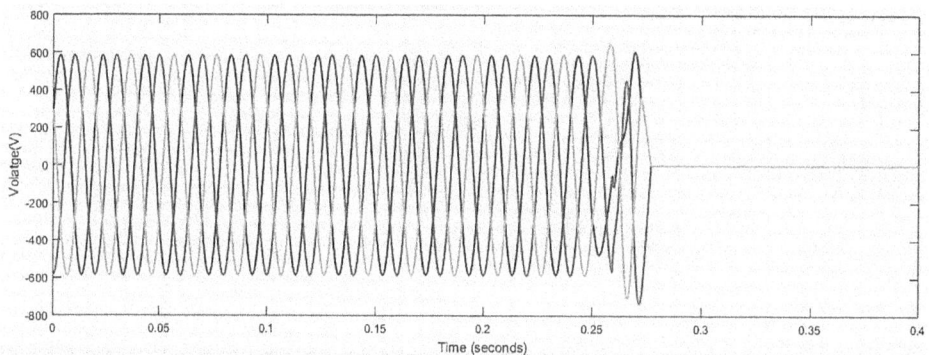

FIGURE 12.24   The waveform of grid voltage versus time.

In this method, the grid is disconnected at t = 0.25 seconds intentionally to create an islanding condition in Figure 12.23. Grid voltage and grid current is shown in Figures 12.24 and 12.25 respectively. After grid disconnection, the only microgrid is present to feed the local load. In the active drift frequency method frequency or phase, lag is absorbed after providing the signal. As a result frequency mismatch is absorbed in the point of common coupling. Relay sense frequency fluctuation, islanding condition is absorbed and circuit breaker sent islanding detection signal (IDS) to the microgrid at 0.2709 secs as shown in Figure 12.26. As microgrid would also stop feeding the power to load as shown in Figure 12.28. Microgrid frequency is shown in Figure 12.27. In this chapter, the DG system is replaced with equivalent DC input voltage as shown in Figure 12.29.

12.6.3   Comparison of UOV/UOF and AFD Islanding Detection Method

The active method is fast in detecting the islanding condition for the microgrid. Active detection methods have better response time and small non-zero detection zone than passive detection time

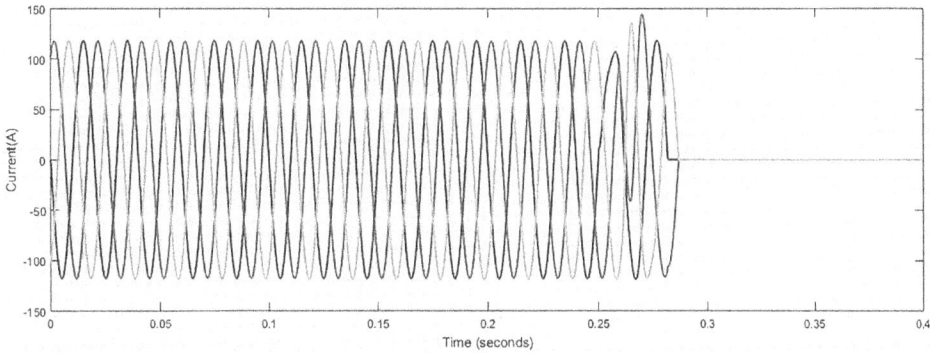

FIGURE 12.25 The waveform of grid current versus time.

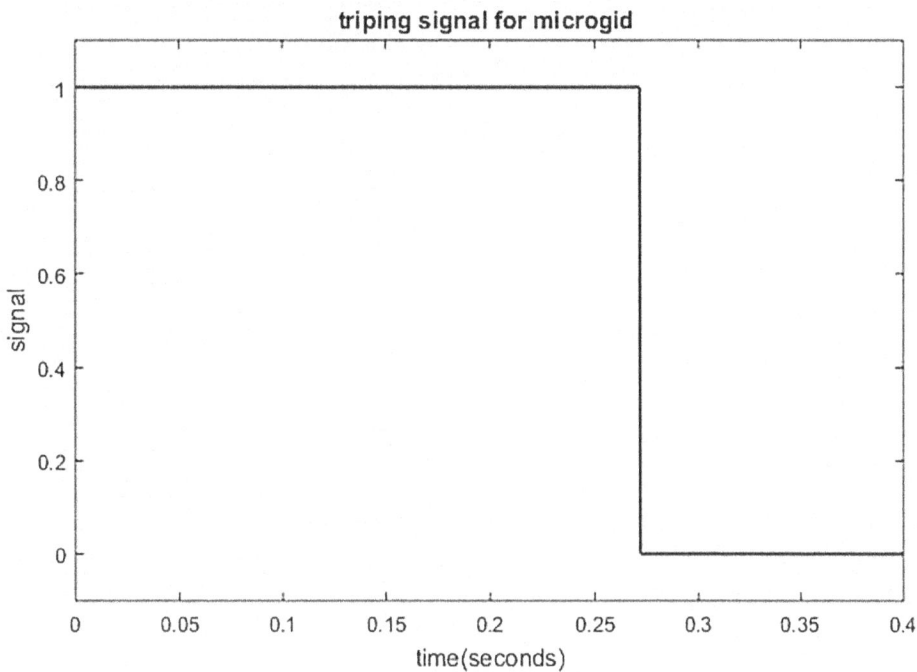

FIGURE 12.26 The waveform of tripping signal versus time for the microgrid.

but have little higher harmonics. Grid is decoupled at t = 0.25 seconds from the main network. After relay detection in two different methods, their detection of islanding for the microgrid is different. The active method detects the islanding condition at t = 0.279 seconds, and the passive method detects the islanding condition at t = 0.2806 seconds.

12.6.4 CLOSED-LOOP CONTROL OF A VSM-BASED INVERTER UNDER ISLANDING CONDITION

The DG inverter is initially linked to a grid-tied system, and the grid's islanding detection signal is zero. The islanding situation occurs at t = 0.25 sec owing to the power mismatch in Figure 12.28, although it is identified after 0.05 sec of real-time given by time delay. After 0.30 seconds, the true islanding state is identified. Figures 12.32 and 12.33 show the waveforms of grid voltage and grid

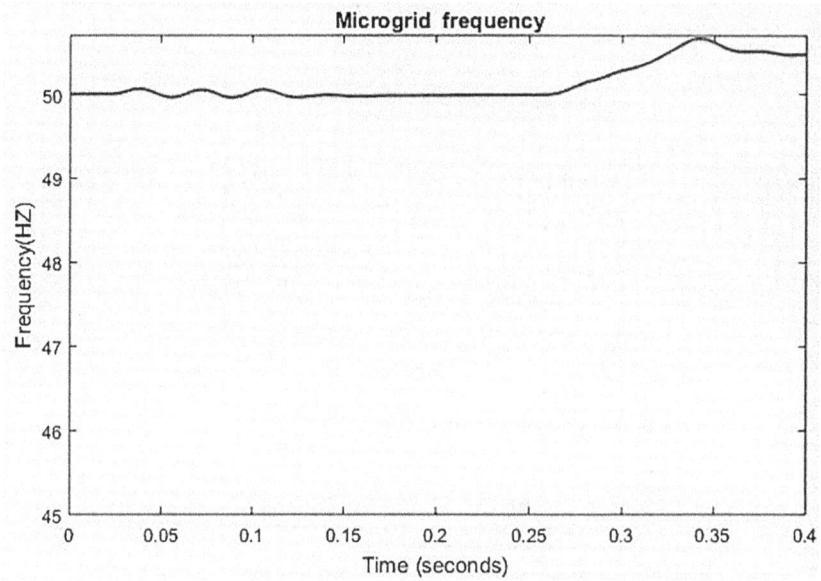

FIGURE 12.27 The waveform of microgrid frequency versus time for the microgrid.

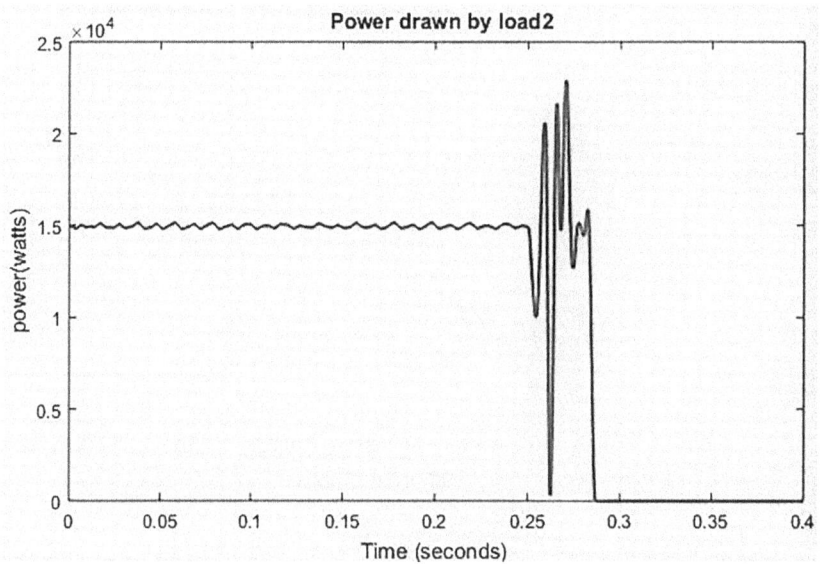

FIGURE 12.28 The waveform of power versus time drawn by load 2.

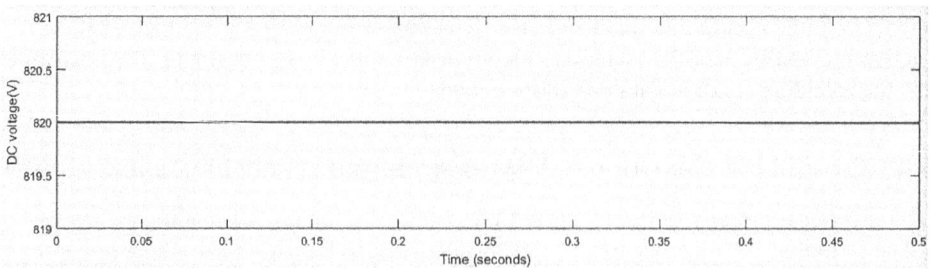

FIGURE 12.29 The waveform of DC voltage.

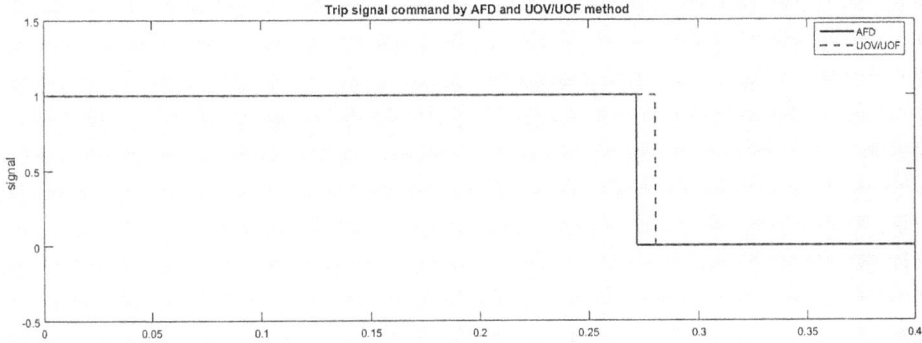

FIGURE 12.30 The waveform of comparison of UOV/UOF and AFD method.

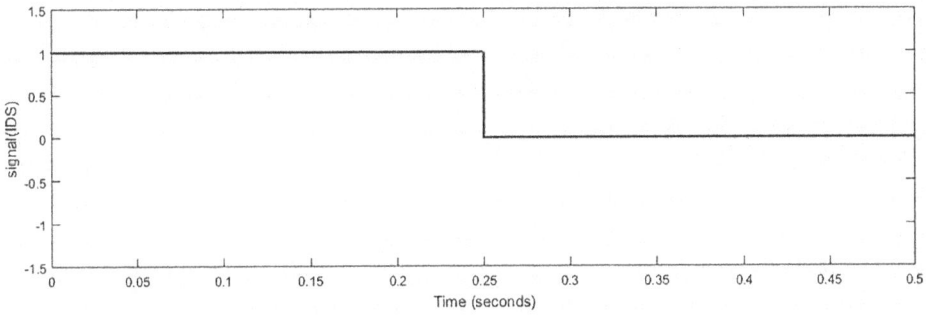

FIGURE 12.31 Waveform of islanding detection signal.

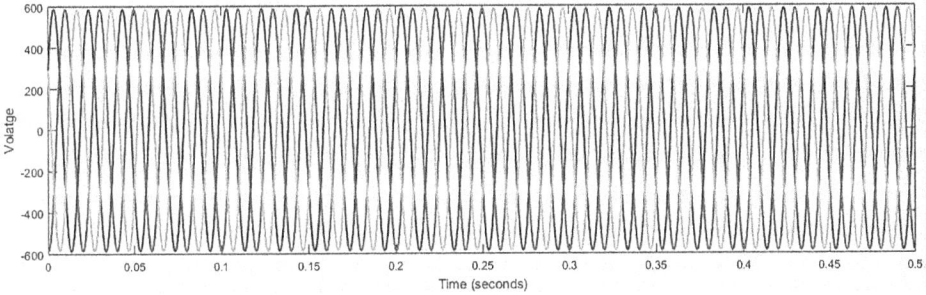

FIGURE 12.32 The waveform of grid voltage.

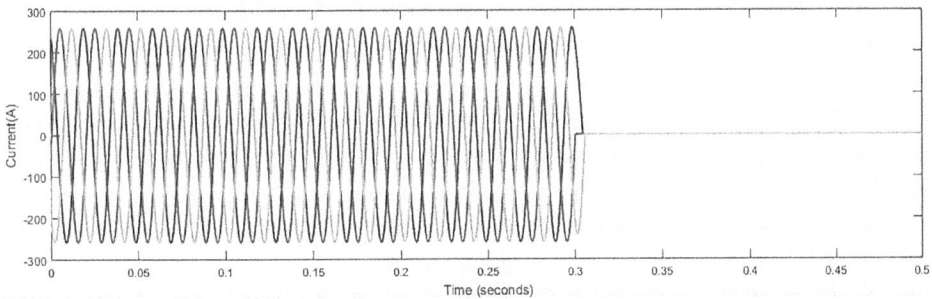

FIGURE 12.33 The waveform of grid current.

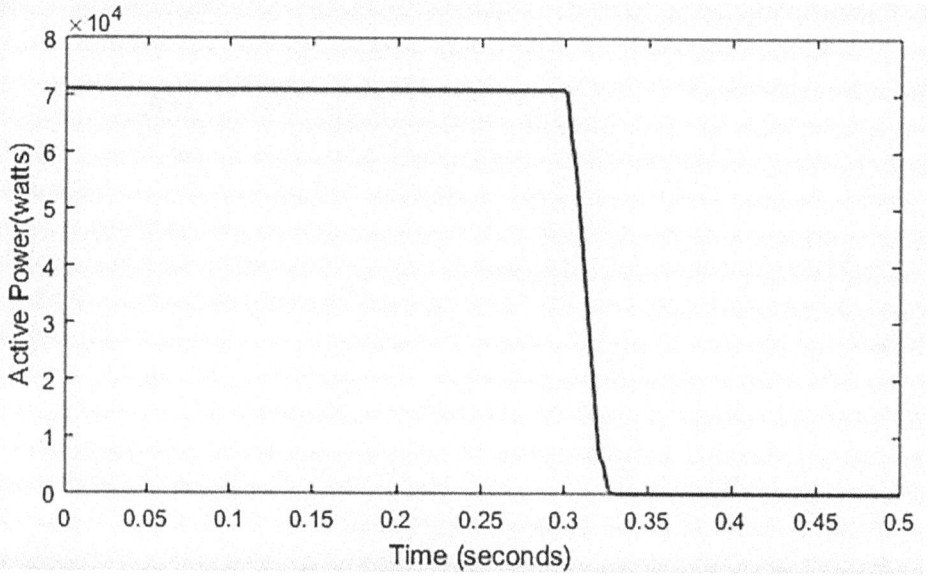

FIGURE 12.34   Waveform of active power exported from grid.

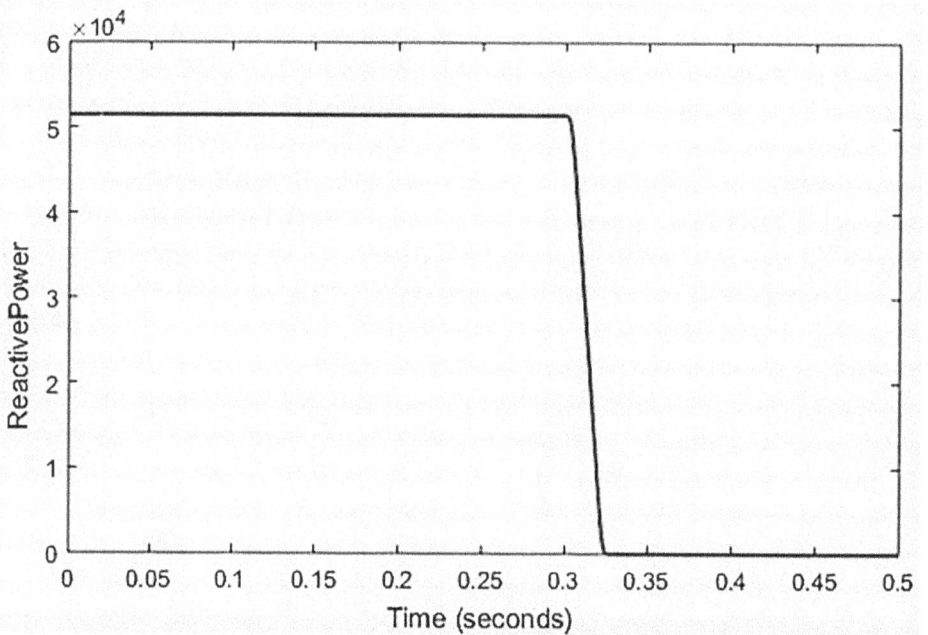

FIGURE 12.35   Waveform of reactive power (VAr) exported from grid.

current. The circuit breaker is unlocked to separate the inverter from the grid when the islanding condition is identified. Grid current is zero at t=0.30 sec in Figure 12.30. Real and reactive flowing from the grid is stopped feeding power to the load after 0.30 secs in Figure 12.34 and 12.35 respectively.

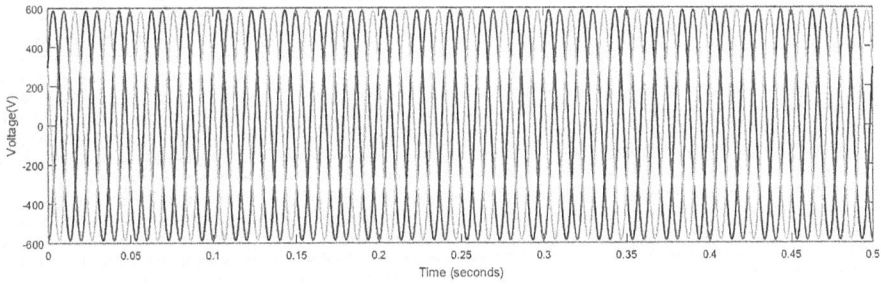

FIGURE 12.36   Waveform of inverter voltage.

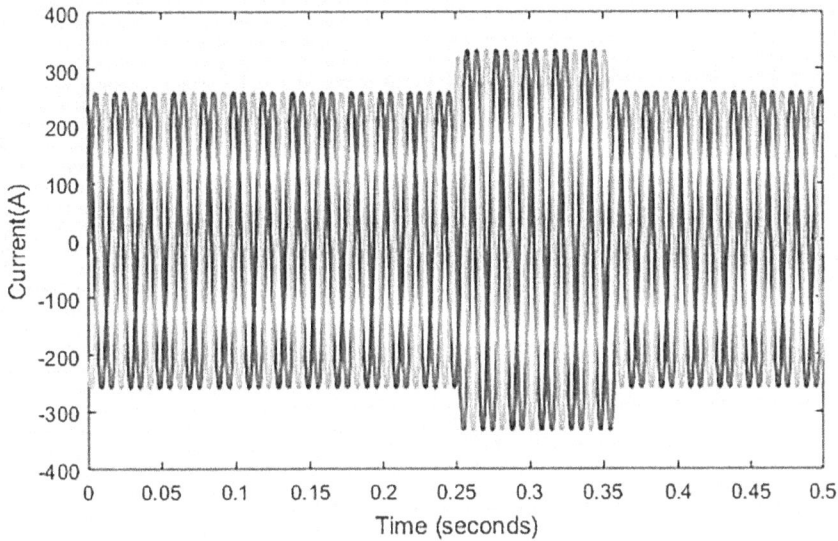

FIGURE 12.37   Waveform of inverter current.

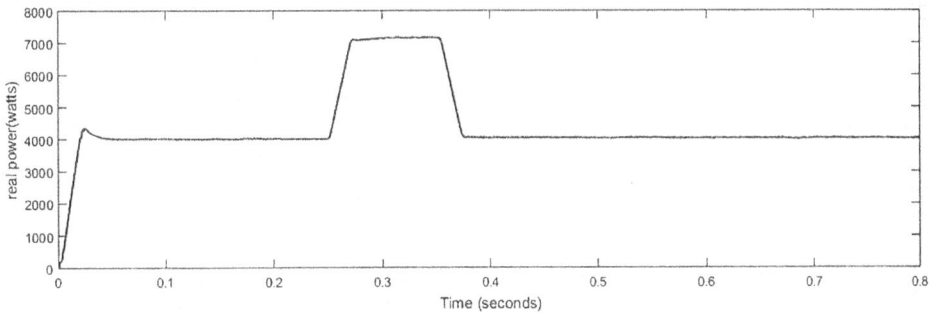

FIGURE 12.38   Waveform of real power from inverter.

When the circuit breaker is unlocked, the inverter voltage remains constant during the operation in Figure 12.36, but the inverter current rises from t = 0.25 to 0.35 sec in Figure 12.37 owing to a mismatch in load power. In Figures 12.38 and 12.39, the inverter continued to deliver real and reactive power to the critical load while remaining isolated from the grid. Due to real-power disturbances, the frequency loss is insignificant over a short period from t = 0.25 sec to 0.35 sec in Figure 12.40. In this, the DG system is considered as a DC equivalent voltage source in Figure 12.42.

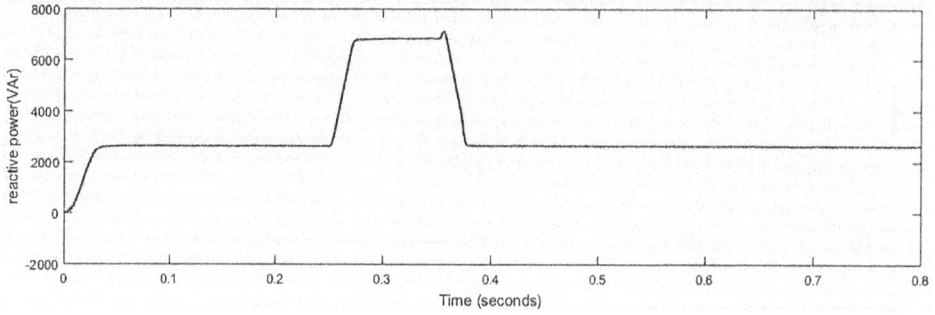

FIGURE 12.39 Waveform of reactive power from inverter.

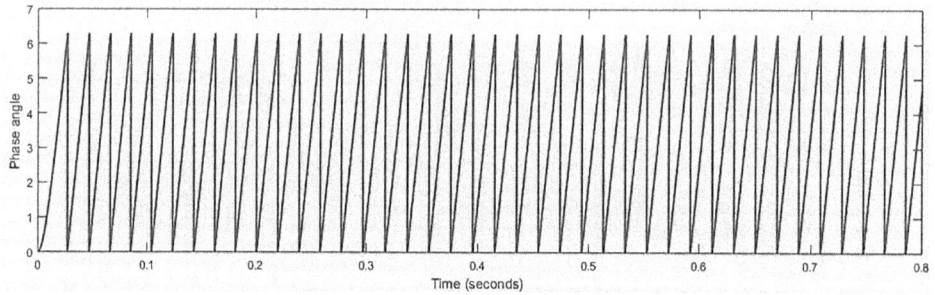

FIGURE 12.40 Waveform of phase angle.

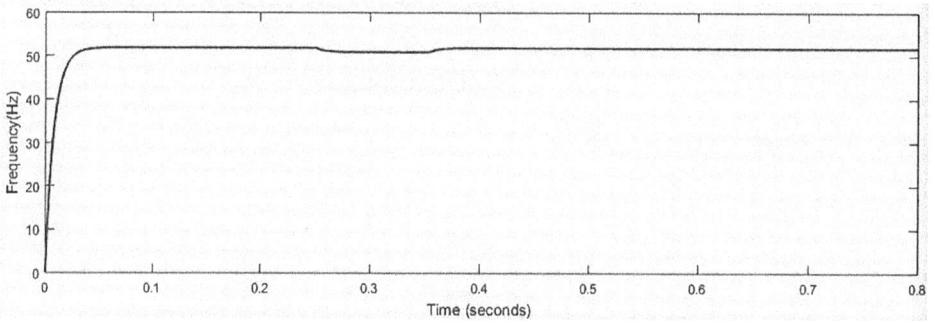

FIGURE 12.41 Waveform of frequency.

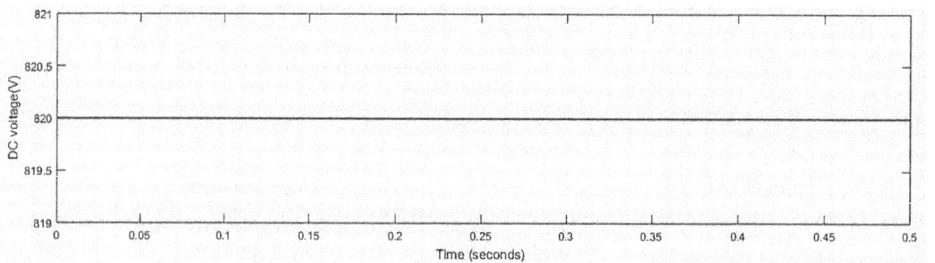

FIGURE 12.42 The waveform of DC voltage.

12.7   CONCLUSION

Islanding detection methods are necessary for distributed generators from a regulatory point of view based on safety concerns. Here two methods of islanding detection active and passive method are discussed and simulated in MATLAB Simulink. Under/over voltage (passive method) and AFD (Active Method) are compared and verified with simulation results. However passive method having a non-zero detection zone, not detect the islanding condition when power consumed by the load is equal to the power generated by the DG inverter and it's difficult to select threshold limits of the parameter like voltage and frequency, harmonics. In the active method, frequency drift is analysed and manipulated after injecting a disturbance signal to the inverter terminal. After simulation results, it is substantiated that the AFD method is faster in detecting the islanding condition and more reliable and lower power quality degrade. Finally, implemented a hybrid IDT based on the concept of VSM, which enables the grid integration of DG sources to the grid. This hybrid islanding detection technique has the benefit of being able to be included in the operation of DG inverters without requiring any additional control modifications. The results are verified using MATLAB Simulink.

ACKNOWLEDGEMENT

The idea of work is supported by DST project Scheme for Young Scientists and Technologists (SP/YO/2019/1349).

REFERENCES

[1] S. D'Arco, and J. A. Suul. Equivalence of virtual synchronous machines and frequency-droops for converter-based microgrids. *IEEE Transactions on Smart Grid,* vol. 5, no. 1, pp. 394–395, Dec. 2012.

[2] Q. C. Zhong, W. L. Ming, and Y. Zeng. Self-synchronized universal droop controller. *IEEE Access*, vol. 4, pp. 7145–7153, Oct. 2016.

[3] M. S. S. Chandra, L. V. Kumar, and S. Mohapatro. Voltage Control and Energy Management of Solar PV fed Stand-alone Low Voltage DC Microgrid for Rural Electrification. *IEEE National Power Systems Conference (NPSC)*, 2020, pp. 1–6.

[4] O. Raipala, A. Mäkinen, S. Repo, and P. Järventausta. An anti-islanding protection method based on reactive power injection and ROCOF. *IEEE Transactions on Power Delivery*, vol. 32, no. 1, pp. 401–410, 2016.

[5] Y. Si, Y. Liu, C. Liu, Z. Zhang, and Q. Lei. Reactive power injection and SOGI based active anti-islanding protection method. *2019 IEEE Energy Conversion Congress and Exposition (ECCE)*, 2019, pp. 2637–2642.

[6] A. G. Abokhalil, A. B. Awan, and A. R. Al-Qawasmi. Comparative study of passive and active islanding detection methods for PV grid-connected systems' " *Sustainability*, vol. 10, no. 6, p. 1798, Jun. 2018.

[7] Kim, M. S., Haider, R., Cho, G. J., Kim, C. H., Won, C. Y. and Chai, J. S., 2019. Comprehensive review of islanding detection methods for distributed generation systems. *Energies*, vol. 12, no. 5, p. 837. Jan 2019.

[8] A. Ghosh, S. Banerjee, M. K. Sarkar, and P. Dutta. Design and implementation of type-II and type-III controller for DC—DC switched mode boost converter by using K-factor approach and optimisation techniques. *IET Power Electronics*, vol. 9, no. 5, pp. 938–950, Apr. 2016.

[9] H. Tiwari, and A. Ghosh. Power Flow Control in Solar PV Fed DC Microgrid with Storage. *2020 IEEE 9th Power India International Conference (PIICON)*, 2020, pp. 1–6.

[10] J. Meher, and A. Ghosh. Comparative Study of DC/DC Bidirectional SEPIC Converter with Different Controllers. *2018 IEEE 8th Power India International Conference (PIICON)*, 2018, pp. 1–6.

[11] G. Vikash, D. Funde, and A. Ghosh. Implementation of the virtual synchronous machine in grid-connected and stand-alone mode. In *DC—DC Converters for Future Renewable Energy Systems*. Singapore: Springer, 2022, pp. 335–353.

[12] V. Gurugubelli, A. Ghosh, and A. K. Panda. Different oscillator-controlled parallel three-phase inverters in stand-alone microgrid. In *Sustainable Energy and Technological Advancements*, Singapore: Springer, 2022, pp. 67–79.

[13] V. Gurugubelli, A. Ghosh, and A. K. Panda. Droop controlled voltage source converter with different classical controllers in voltage control loop. In *2022 IEEE International Conference on Power Electronics, Smart Grid, and Renewable Energy (PESGRE)*, 2022, pp. 1–6, IEEE.

[14] S. Patel, A. Ghosh, and P. K. Ray. Adaptive power management in PV/Battery integrated hybrid microgrid system. In *2022 IEEE International Conference on Power Electronics, Smart Grid, and Renewable Energy (PESGRE)*, 2022, pp. 1–6.

[15] V. Gurugubelli, A. Ghosh, A. K. Panda, and S. Rudra. Implementation and comparison of droop control, virtual synchronous machine, and virtual oscillator control for parallel inverters in standalone microgrid. *International Transactions on Electrical Energy Systems*, vol. 31, no. 5, p. e12859, 2021.

[16] V. Gurugubelli, and A. Ghosh. Control of inverters in standalone andgrid-connected microgrid using different control strategies. *World Journal of Engineering*, July 2021.

[17] G. Vikash, A. Ghosh, and S. Rudra. Integration of Distributed Generation to Microgrid with Virtual Inertia. In *2020 IEEE 17th India Council International Conference (INDICON)*, 2020, pp. 1–6, IEEE.

[18] G. Vikash, and A. Ghosh. Parallel inverters control in standalone microgrid using different droop control methodologies and virtual oscillator control. *Journal of The Institution of Engineers (India): Series B*, pp. 1–9, 2021.

[19] V. Gurugubelli, A. Ghosh, and A. K. Panda. Comparison of deadzone and vanderpol oscillator controlled voltage source inverters in Islanded Microgrid. In *2021 IEEE 2nd International Conference on Smart Technologies for Power, Energy and Control (STPEC)*, 2021, pp. 1–6, IEEE.

[20] V. Gurugubelli, A. Ghosh, and A. K. Panda. Design of different classical controllers in the voltage control loop of a virtual synchronous machine in standalone mode. In *2022 IEEE International Conference on Power Electronics, Smart Grid, and Renewable Energy (PESGRE)*, 2022, pp. 1–6, IEEE.

[21] N. Priyadarshi, S. Padmanaban, J. B. Holm-Nielsen, F. Blaabjerg, and M. S. Bhaskar. An experimental estimation of hybrid ANFIS—PSO-based MPPT for PV grid integration under fluctuating sun irradiance. *IEEE Systems Journal*, vol. 14, no. 1, pp. 1218–1229, Nov. 2019.

[22] S. Padmanaban, N. Priyadarshi, J. B. Holm-Nielsen, M. S. Bhaskar, F. Azam, A. K. Sharma, and E. Hossain. A novel modified sine-cosine optimized MPPT algorithm for grid integrated PV system under real operating conditions. *IEEE Access*, vol. 7, pp. 10467–10477, 2019.

[23] S. Padmanaban, N. Priyadarshi, M. S. Bhaskar, J. B. Holm-Nielsen, E. Hossain, and F. Azam. A hybrid photovoltaic-fuel cell for grid integration with jaya-based maximum power point tracking: experimental performance evaluation. *IEEE Access*, vol. 7, pp. 82978–82990, 2019.

[24] N. Priyadarshi, V. K. Ramachandaramurthy, S. Padmanaban, and f. Azam. An ant colony optimized MPPT for standalone hybrid PV-wind power system with single Cuk converter. *Energies*, vol. 12, no. 1, p. 167, 2019.

[25] N. Priyadarshi, S. Padmanaban, L. Mihet-Popa, F. Blaabjerg, and F. Azam. Maximum power point tracking for brushless DC motor-driven photovoltaic pumping systems using a hybrid ANFIS-FLOWER pollination optimization algorithm. *Energies*, vol. 11, no. 5, p. 1067, 2018.

[26] N. Priyadarshi, A. K. Sharma, and F. Azam. A hybrid firefly-asymmetrical fuzzy logic controller based MPPT for PV-wind-fuel grid integration. *International Journal of Renewable Energy Research (IJRER)*, vol. 7, no. 4, pp. 1546–1560, 2017.

[27] N. Priyadarshi, A. Anand, A. Sharma, F. Azam, V. Singh, and R. Sinha. An experimental implementation and testing of GA based maximum power point tracking for PV system under varying ambient conditions using dSPACE DS 1104 controller. *International Journal of Renewable Energy Research (IJRER)*, vol. 7, no. 1, pp. 255–265, 2017.

[28] N. Priyadarshi, M. S. Bhaskar, S. Padmanaban, F. Blaabjerg, and F. Azam. New CUK—SEPIC converter based photovoltaic power system with hybrid GSA—PSO algorithm employing MPPT for water pumping applications. *IET Power Electronics*, vol. 13, no. 13, pp. 2824–2830, 2020.

[29] N. Priyadarshi, S. Padmanaban, M. S. Bhaskar, F. Blaabjerg, and A. Sharma. Fuzzy SVPWM-based inverter control realisation of grid integrated photovoltaic-wind system with fuzzy particle swarm optimisation maximum power point tracking algorithm for a grid-connected PV/wind power generation system: Hardware implementation. *IET Electric Power Applications*, vol. 12, no. 7, pp. 962–971, 2018.

[30] K. Kamalapathi, N. Priyadarshi, S. Padmanaban, J. B. Holm-Nielsen, F. Azam, C. Umayal, and V. K. Ramachandaramurthy. A hybrid moth-flame fuzzy logic controller based integrated cuk converter fed brushless DC motor for power factor correction. *Electronics*, vol. 7, no. 11, p. 288, 2018.

[31] N. Priyadarshi, S. Padmanaban, J. B. Holm-Nielsen, M. S. Bhaskar, and F. Azam. Internet of things augmented a novel PSO-employed modified zeta converter-based photovoltaic maximum power tracking system: hardware realisation. *IET Power Electronics*, vol. 13, no. 13, pp. 2775–2781, 2020.

[32] N. Priyadarshi, S. Padmanaban, D. M. Ionel, L. Mihet-Popa, and F. Azam. Hybrid PV-wind, micro-grid development using quasi-Z-source inverter modeling and control—experimental investigation. *Energies*, vol. 11, no. 9, p. 2277, 2018.

[33] F. Azam, S. K. Yadav, N. Priyadarshi, S. Padmanaban, and R. C. Bansal. A comprehensive review of authentication schemes in vehicular ad-hoc network. *IEEE Access*, vol. 9, pp. 31309–31321, 2021.

[34] F. Azam, N. Priyadarshi, H. Nagar, S. Kumar, and A. K. Bhoi. An overview of solar-powered electric vehicle charging in vehicular adhoc network. *Electric Vehicles*, pp. 95–102, 2021.

[35] N. Priyadarshi, F. Azam, A. K. Sharma, P. Chhawchharia, and P. R. Thakura. An Interleaved ZCS Supplied Switched Power Converter for Fuel Cell-Based Electric Vehicle Propulsion System. In *Advances in Smart Grid Automation and Industry 4.0*. Singapore: Springer, pp. 355–362, 2021.

[36] F. Azam, A. Biradar, N. Priyadarshi, S. Kumari, D. Almakhles, and S. Tangade. A framework for secured dissemination of messages in Internet of Vehicle (IoV) using blockchain approach. *2021 IEEE International Conference on Mobile Networks and Wireless Communications (ICMNWC)*, 2021, pp. 1–6.

[37] F. Azam, A. Biradar, N. Priyadarshi, S. Kumari, and S. Tangade. A review of blockchain based approach for secured communication in Internet of Vehicle (IoV) scenario. *2021 Second International Conference on Smart Technologies in Computing, Electrical and Electronics (ICSTCEE)*, 2021, pp. 1–6.

[38] N. Priyadarshi, M. S. Bhaskar, P. Sanjeevikumar, F. Azam, and B. Khan. High-power DC-DC converter with proposed HSFNA MPPT for photovoltaic based ultra-fast charging system of electric vehicles. *IET Renewable Power Generation*, 2022.

[39] V. Gurugubelli, A. Ghosh, and A. K. Panda. A new virtual oscillator control for synchronization of single-phase parallel inverters in islanded microgrid. *Energy Sources, Part A: Recovery, Utilization, and Environmental Effects*, vol. 44, no. 4, pp. 8842–8859, 2022.

[40] N. Priyadarshi, S. Padmanaban, M. S. Bhaskar, F. Azam, B. Khan, and M. G. Hussien. A novel hybrid grey wolf optimized fuzzy logic control based photovoltaic water pumping system. *IET Renewable Power Generation*, 2022.

[41] V. Gurugubelli, A. Ghosh, and A. K. Panda. Parallel inverter control using different conventional control methods and an improved virtual oscillator control method in a standalone microgrid. *Protection and Control of Modern Power Systems*, vol. 7, no. 1, pp. 1–13, 2022.

[42] N. Priyadarshi, P. Sanjeevikumar, M. S. Bhaskar, F. Azam, I. B. Taha, and M. G. Hussien. An adaptive TS-fuzzy model based RBF neural network learning for grid integrated photovoltaic applications. *IET Renewable Power Generation*, 2022.

[43] B. Sujith, A. Ghosh, and V. Gurugubelli. Design of PFC boost converter with stand-alone inverter for microgrid applications. In *2022 IEEE Delhi Section Conference (DELCON)*, IEEE, February 2022, pp. 1–5.

[44] J. K. Nayak, H. Thalla, and A. Ghosh. Efficient maximum power point tracking algorithms for photovoltaic systems with reduced number of sensors. *Process Integration and Optimization for Sustainability*, pp. 1–23, 2022.

[45] S. Patel, A. Ghosh, and P. K. Ray. Improved power flow management with proposed fuzzy integrated hybrid optimized fractional order cascaded proportional derivative filter (1+ proportional integral) controller in hybrid microgrid systems. *ISA Transactions*, 2022.

[46] D. Ravi, and A. Ghosh. Voltage mode control of buck converter using practical PID controller. In *2022 International Conference on Intelligent Controller and Computing for Smart Power (ICICCSP)*, IEEE, July 2022, pp. 1–6.

[47] S. Saurav, and A. Ghosh. Fourth order interleaved boost converter with PID, type II and type III controllers for smart grid applications. *Cyber-Physical Systems: Foundations and Techniques*, pp. 179–207, 2022.

[48] S. Joarder, and A. Ghosh. Design and implementation of dual active bridge converter for DC microgrid application. In *2022 IEEE Delhi Section Conference (DELCON)*, IEEE, February 2022, pp. 1–6.

[49] T. Barker, and A. Ghosh. Neural network-based PV powered electric vehicle charging station. In *2022 IEEE Delhi Section Conference (DELCON)*, IEEE, February 2022, pp. 1–6.

[50] A. Singh, and A. Ghosh. Comparison of quantitative feedback theory dependent controller with conventional PID and sliding mode controllers on DC-DC boost converter for microgrid applications. *Technology and Economics of Smart Grids and Sustainable Energy*, vol. 7, no. 1, pp. 1–12, 2022.

[51] H. Tiwari, A. Ghosh, P. K. Ray, B. Subudhi, G. Putrus, and M. Marzband. Direct power control of a three-phase AC-DC converter for grid-connected solar photovoltaic system. In *2021 International Symposium of Asian Control Association on Intelligent Robotics and Industrial Automation (IRIA)*, IEEE, September 2021, pp. 125–130.

13 Behavioral Analysis of Multi-Source DC to DC Converter for Integrating Renewable Energy and Sourcing to Residential Loads

R. Rajesh Kanna and R. Raja Singh

CONTENTS

Nomenclature

I_{ph}	Current in photovoltaic cell (A)	V_W	Wind velocity
I_{sc}	Short circuit current (A)	N_0	Number of fuel cell attached series in the stack
K_i	Equivalent of when cell at 25°C	I_{fc}	Current in fuel cell stack(A)
T	Operating temperature (K)	F	Faraday's constant (C/kmol)
I_{sr}	Solar irradiation (W/m²)	K_r	Constant of modelling (kmol/(sA)$^{-1}$)

DOI: 10.1201/9781003323471-13

I_{rs-pv}	PV module reverse saturation current	η_{act}	Function of oxygen concentration CO_2
$V_{oc\text{-}fc}$	Fuel cell open circuit voltage (V)	n_{ohmic}	Function of I_{fc}
Q	Charge of electron (C)	R_{int}	Stack internal resistance
N_S	Summation of cell joined in series	L_{fc}	Inductance of boost converter in fuel cell (H)
n	Ideality factor of the diode	L_{fc}	Resistance of boost converter in fuel cell (Ω)
K	Boltzmann's constant, J/K	D_f	Boost converter duty ratio
T_r	Nominal temperature (°C)		DC link capacitance (F)
E_g	Band gap energy of the semiconductor	V_{dc}	DC link voltage(V)
I_d	Diode current (A)	V_{eff}	Total efficiency of the MISO converter
N_p	Summation of cell joined in parallel	V_o	Output voltage of MISO converter
R_s	Resistance connected series (Ω)	d_{c1}, d_{c2}, d_{c3}	Duty cycle of MISO converter
R_p	Resistance connected parallel (Ω)	V_1, V_2, V_3	Input voltage of MISO converter
I_{sh}	Shunt current (A)	Δ_{iL}	Ripple in inductor current (%)
V_t	Diode thermal voltage (V)	Δ_{V_0}	Ripple in output voltage (%)
ρ	Air density	d_{c1}, d_{c2}, d_{c3}	Time period of MISO converter
λ	Dip speed ratio	$C_p(\lambda, \beta)$	Bentz limit
RES	Renewable energy sources	U	Hydrogen utilisation factor
PG	Generated power from RES (kW)	r	Radius of blade
PD	Power demand (kW)	$V_L(t)$	Voltage across inductor in MISO (V)
β	Pitch angle (θ)	$I_L(t)$	Inductor current in MISO (A)
ω_m	Generator speed in rad/sec	V_c	Voltage across capacitor (V)
MISO	Multi input single output	I_{fc-ref}	Reference current in fuel cell (A)

13.1 INTRODUCTION

Renewable energy sources (RESs) are envisioned as essential components of modern power grids due to their low generation costs and reduced pollution emissions [1]–[4]. Wind turbines and photovoltaic (PV) systems are unreliable energy sources because they are inconsistent and unpredictable. To address this problem, renewable resources are either blended each other. Otherwise, fuel cells (FCs) or energy storage devices are used [5], [6]. The isolated solar, wind and fuel cell hybrid energy system is widely explored as a typical standalone microgrid since fuel cell implemented into the system. This standalone power system can be employed not only in necessary services but also in remote, isolated regions where wind and solar energy are abundant [7], [8]. In a standalone wind, solar, and fuel cell hybrid power systems, the primary control objective is to preserve grid stability by keeping a balance between source and load under fluctuating of wind/ solar sources [9]. Multiple-input single output (MISO) converters have recently been presented as a way to incorporate a different renewable energy sources [10]. These converters have simple control topologies, excellent durability, centralised control, low production costs, and compact size. The isolation transformer provided galvanic isolation between the inputs and the load in the isolated converter system topologies in [11]–[13]. Non isolated power converters are easier to build and control than isolated power converters and when compared to replacing a capacitor in a non-isolated converter, the expense of replacing the transformer is substantial [14], [15]. One of the simplest methods to construct a nonisolated converter, as presented in [16] and [17], is to link some of the direct sources to the DC bus while others are coupled to the DC grid by means of bidirectional DC-DC converters; however, the DC-bus voltage, cannot be adjusted using this way. Furthermore, in [18], [19], it is recommended to use separate DC to DC converters for each input. Unlike the previous design, the multiple-converter configuration allows for output voltage management; nevertheless, it is an affluent strategy because it necessitates the use of many converters. MISO converter topologies have been described in the literature [20]–[22] to lower the cost of multiple-converter topologies. As a result, this study presents a unidirectional MISO converter using a single inductor held in common for all input sources. The converter is designed to work in buck/boost operation in vice versa according to the renewable energy sources output variations. It consists of an inductor, capacitor and operated by four IGBT switches and four diodes. The proposed converter is used for residential load application in isolated areas where the simultaneous power supply is achieved. DC grids are becoming more prominent for integrating renewable energy and house electrification. The DC grid and AC grid system designs are described in [23], [24]. The voltage standards for the DC grid are still being developed. As stated in [24] standard voltages are in the 48–400 V range. Commercial applications require 380–400 V, whereas domestic applications require 120V. The 48 V DC power supply is utilised in telecommunications usage. In India, 110 V is utilised for phone charging devices in electrical transmission systems. The grid voltage is set at 110 volts since it may be utilised for household lighting and electric traction. The main objectives of this work are,

(1) To develop a DC micro grid with PI controller for islanded residential load.
(2) To operate the multi-input single output converter in buck-boost mode therefore obtained voltage from renewable energy sources are supplied to residential load.

Adapting the energy storage increase the cost and maintenance. Therefore, the hybrid energy sources is the optimal choice for reliable and economical energy generation system.

This chapter is structured as follows: the detailed mathematical modelling of overall standalone hybrid system and operating principle of MISO converter is presented in Section 13.2. Followed by the integration of RES with MIISO converter and control strategy in Section 13.3. Finally, the simulation results are accomplished and presented in Section 13.4 and concluded in Section 13.5.

13.2 MODELLING OF STANDALONE HYBRID SYSTEM

This section contains the mathematical modelling of the hybrid energy system. The mathematical modelling of the hybrid energy system is provided [25], [26].

13.2.1 SOLAR PV

Figure 13.1 depicts an analogous arrangement of a PV cell. The cell photocurrent is represented by I_{ph}. Series and shunt resistances of the photovoltaic cell are denoted as R_s and R_{sh} respectively. In

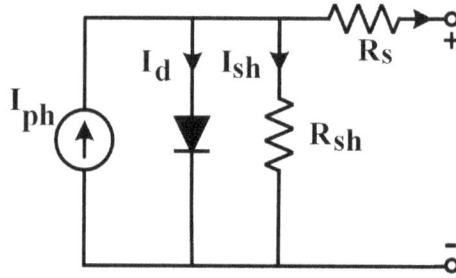

FIGURE 13.1   Equivalent circuit of PV cell.

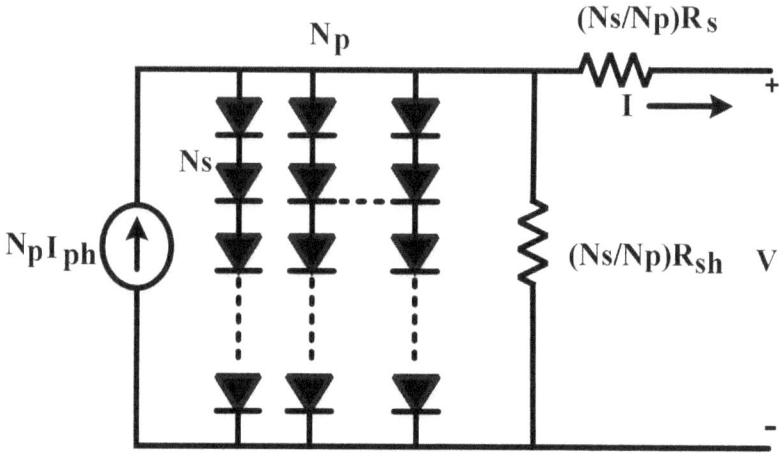

FIGURE 13.2   A solar array's circuit model.

general, R_s is quite little while R_{sh} is much higher. PV cells are grouped into superior units known as PV modules [27], [28]. As illustrated in Figure 13.2, these modules are attached in parallel or series to form PV arrays that are being used to generate energy in solar power plants.

13.2.1.1 Module Photo Current I_{ph}

The module's photocurrent could be calculated as follows:

$$I_{ph} = \left[I_{Sc} + P_j (T - 298) \right] * \frac{I_{Sr}}{1000} \tag{1}$$

Where I_{ph} represents current of photovoltaic cell (A), I_{sc} denotes short circuit current flowing through shunt resistance (A), P_j is equivalent I_{sc} at 25°C, T_k denotes the temperature of PV panel (K), I_{sr} signifies solar irradiation (W/m²).

Reverse saturation current of PV module (I_{rs-pv}) can be calculated by given eqn. (2):

$$I_{rs-pv} = \frac{I_{sc}}{\left[exp\left(\frac{qV_{oc}}{N_S knT} \right) - 1 \right]} \tag{2}$$

Where q represents electron charge, the open circuit voltage is denoted as V_{oc}, N_S is series connected cell, the diode's ideality factor is denoted n, k is Boltzmann's constant.

13.2.1.2   Module Saturation Current (I_o)

It differs with the temperature of photovoltaic cell, which is given by

$$I_o = I_{rs-pv} \left[\frac{T_k}{T_r} \right]^3 \exp \left[\frac{q*E_g}{nk} \left(\frac{1}{T_k} - \frac{1}{T_r} \right) \right] \tag{3}$$

Here, T_r is nominal temperature, E_g is the semiconductor's band gap energy.

The solar PV module's current output can be expressed as

$$I = \left\{ \left(I_{ph} N_p \right) - I_d - I_{sh} \right\} \tag{4}$$

Where I_d is diode current and I_{sh} is shunt current.

From eqn. (3), (4) the diode current can be derived as

$$I_d = N_p * I_o * \left[exp \left(\frac{\frac{V}{N_s} + I * \frac{R_s}{R_p}}{n * V_t} \right) - 1 \right] \tag{5}$$

Where V_t is the diode. Thermal voltage and shunt current is given by

$$V_t = \frac{kT}{q} \tag{6}$$

$$I_{sh} = \frac{V * \frac{N_p}{N_s} + I * R_s}{R_{sh}} \tag{7}$$

Where N_p is parallel connected PV modules.

MPPT algorithm is used to extract the most power from PV panels and transmit the supreme power from PV modules to loads [29], DC to DC converters are used, which is shown in Figure 13.3. A MPPT perturb and observation approach is established to run a stationary PV module at MPP.

13.2.2   WIND TURBINE

The wind's power is produced in the turbine blades, as represented by the following equation [30],

$$P_w = \frac{1}{2} * \rho \pi r^2 V_w^3 \tag{8}$$

Where ρ is constant value according to air density, V_w signifies wind velocity m/s, r represents blade radius m.

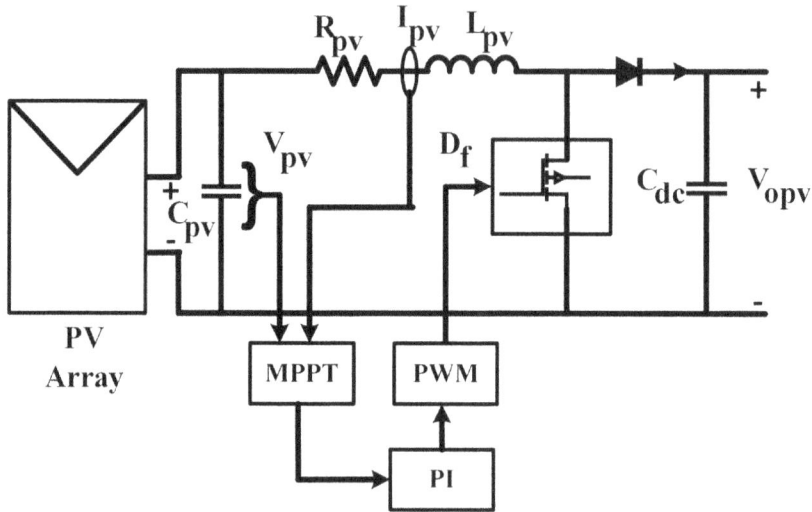

FIGURE 13.3   PV array with MPPT algorithm.

The power collected by a wind generator is determined by the bentz constraint $c_p(\lambda, \beta)$.

$$P_t = \frac{1}{2} * \rho \pi r^2 V_w^3 c_p(\lambda, \beta) \tag{9}$$

Where λ is the tip speed ratio and ß denotes the pitch angle.

$$\lambda = \frac{(\omega_m \cdot R)}{V_w} \tag{10}$$

Where ω_m denotes speed of generator (rad/sec), c_p is power coefficient.
Various power coefficient models have been established, mainly in [31], which incorporates C_p by

$$C_p(\lambda, \beta) = 0.517 \left(\frac{116}{\lambda_1} - 0.4\beta - 5 \right) e \left(-\frac{21}{\lambda_1} \right) + 0.006795 * \lambda_1 \tag{11}$$

Where λi value can be calculated as

$$\lambda_i = \frac{1}{\dfrac{1}{\lambda + 0.008 * \beta} - \dfrac{0.03}{\beta^3 + 1}} \tag{12}$$

The graph in Figure 13.4 illustrates the relation between produced output power (W) and wind velocity V_w (m/s). The wind energy conversion system runs across cut in to cut off speeds, which are 3.1 m/s and 25 m/s respectively [32]. To extract the most electricity from the wind, the wind

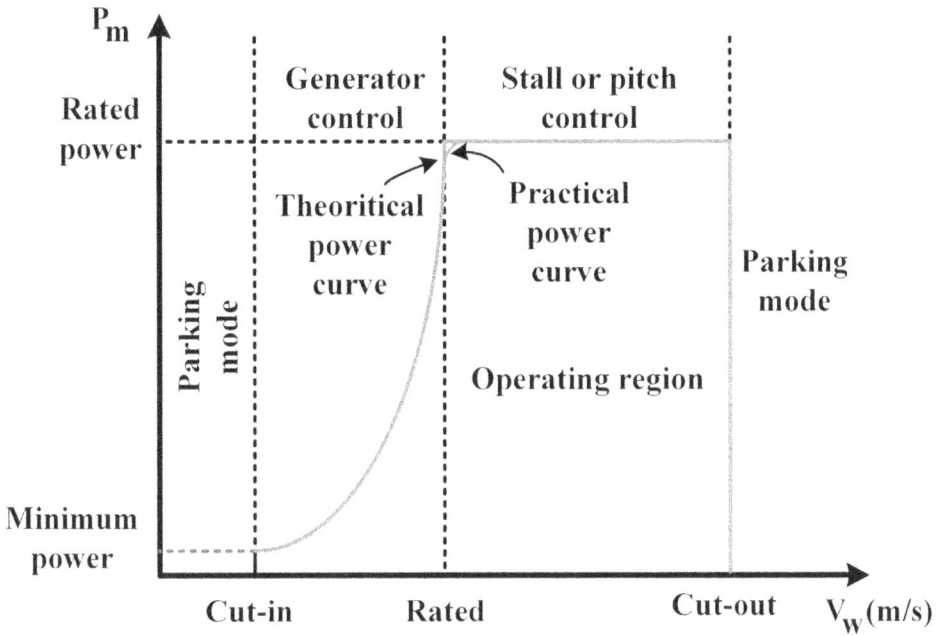

FIGURE 13.4 Output power versus wind speed relationship curve.

turbine should operate at optimal all of the time. The rotor speed of permanent magnet synchronous generator can be derived as

$$\omega_{opt} = \frac{\lambda_{opt} \cdot {*}V_w}{R}$$

(13)

Where λ_{opt} is optimum tip speed ratio, and the turbine's torque equation is derived as

$$T_t = \frac{P_t}{\omega_m}$$

(14)

Therefore, the torque equation will be

$$T_t = \frac{0.5 \cdot \rho \cdot \pi \cdot r^2 \cdot V_w^3 \cdot C_p(\lambda, \beta)}{\omega_m}$$

(15)

13.2.3 Fuel Cell

The following are some of the key relationships that explain the fuel cell modelling [33]. The current of fuel cell stack are linked with percentage of hydrogen flow as follows:

$$q_{h_2}^r = \frac{N_0 I_{fc}}{2F} = 2K_r I_{fc}$$

(16)

Where N_0 is the count of fuel cells in the stack coupled in series, I_{fc} denotes current in the fuel cell stack (A), and K_r denotes constant of modelling $\left(kmol/(SA)^{-1}\right)$. The hydrogen fuel cell polarisation curve provides the fuel cell voltage, which is stated mathematically as follows:

$$V_{cell} = E + \eta_{act} + \eta_{ohmic} \tag{17}$$

Where η_{act} denotes an oxygen concentration function CO_2 and I_{fc}, η_{ohmic} is an ohmic function related to I_{fc} and R_{int} represents internal resistance related to stack (Ω). Let's take constant O_2 absorption and temperature could be derived as,

$$V_{cell} = E - Bln\left(CI_{fc}\right) - R_{int}I_{fc} \tag{18}$$

$$B = 0.04777V, C = 0.0136A^{-1} \tag{19}$$

$$E = N_0 - Bln\left(CI_{fc}\right) - R_{int}I_{fc} \tag{20}$$

The Nernst voltage can be determined using gas molarities as follows [4]:

$$E = N_0\left[E_0 + \frac{RT}{2F}log\left[\frac{\rho h_2 \rho^{0.5}O_2}{\rho H_2O}\right]\right] \tag{21}$$

Where E_0 is open circuit voltage of fuel cell (V). The utilisation factor (U) of hydrogen is determined by fraction of H_2 processed inner side of the stack to H_2 supplied into the tank and ranges from 0 to 1,

$$U = \frac{q_{h_2}^r}{q_{h_2}^{in}} \tag{22}$$

Values greater than 0.9 indicate that hydrogen is being used too much, causing it to run out of fuel and reducing its performance and durability. On the other side, values under 0.8, show hydrogen underuse, indicating the presence of surplus hydrogen, resulting in a considerable spike in output voltage and a drop in efficacy of the system. Consequently, current in stack is limited by the following criterion for the best utilisation of H_2 in an FC stack,

$$\frac{0.8q_{h_2}^{in}}{2K_r} \leq I_{fc} \leq \frac{0.9q_{h_2}^{in}}{2K_r} \tag{23}$$

The current reference I_{fc-ref} obtained from the power management system provides the H_2 reference, which is represented by

$$q_{h_2}^{ref} = \frac{2K_r I_{fc-ref}}{U_{opt}} \tag{24}$$

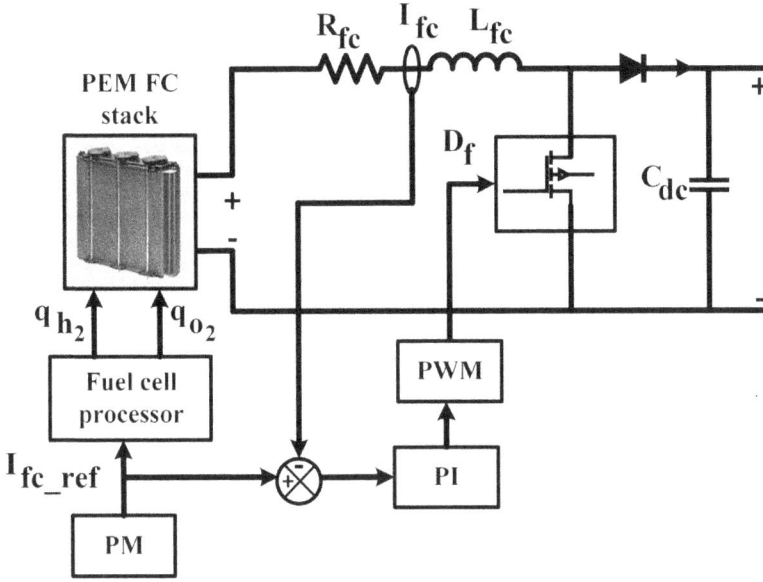

FIGURE 13.5   Fuel cell stack with boost converter.

The PEM (Proton Exchange Membrane) fuel cell connected to the DC bus by means of boost converters functioning with voltage control mode which is detailed in Figure 13.5. A standard proportional integral controller has been used to observe the reference voltage signal obtained from the DC link side and establish the duty cycle value to DC-DC boost converter in order to sustain the required voltage [34]. The transfer function current which is getting from inductor to duty cycle of DC-DC converter is represented as,

$$\frac{\hat{i}FC(s)}{\hat{d}_f(s)} = \frac{\dfrac{I_{fc}}{1-D_f}\left(1+\dfrac{V_{dc}C_{dc}s}{1-D_f}\right)}{S^2 L_{FC}C_{FC}+SR_{FC}C_{FC}+\left(1-D_f\right)^2} \tag{25}$$

Where the L_{fc} represents inductance of fuel cell, R_{fc} indicates resistance of fuel cell, I_{fc} denotes the current of boost converter, D_f denotes as the duty ratio of converter, and DC-link capacitance and voltage represented as C_{dc} and V_{dc} respectively.

13.2.4 MODELLING OF MISO CONVERTER

The MISO converter is modelled with the premise that the current which emerged from inductor is incessant in steady state. As measured by volt-sec balance equation, In steady state, the mean voltage across the inductor is zero [35].

Consider V_1, V_2 and V_3 are the voltage sources and dt_1, dt_2 and dt_3 are the time periods of the converter.

$$\int_0^{T_s} V_L = 0 \tag{26}$$

$$V_1 dt_1 + V_2\left(dt_2 + dt_1\right) + V_2\left(dt_3 + dt_2 + dt_3\right) = V_0\left(T - \left(dt_1 + dt_2 + dt_3\right)\right) \tag{27}$$

The switches S_1, S_2 and S_3 duty cycle are stated as d_{c1}, d_{c2}, d_{c3}. So the duty cycle of the switches can be derived as

$$dc_1 = \frac{dt_1}{T_s}, dc_2 = \frac{dt_1 + dt_2}{T_s}, dc_3 = \frac{dt_1 + dt_2 + dt_3}{T_s} \tag{28}$$

From (1) and (2),

$$V_0 = \frac{V_1 dc_1 + V_2 dc_2 + V_3 dc_3}{1 - dc_3} \tag{29}$$

The total efficiency is V_{eff} based on the output voltage V_o is

$$V_0 = \frac{V_{eff}}{1 - V_3} \tag{30}$$

Thus,

$$V_{eff} = V_1 dc_1 + V_2 dc_2 + V_3 dc_3 \tag{31}$$

The switches S_1, S_2 and S_3 turn on as well as turn off concurrently depending on the mode of operation. When all three input voltages are equivalent, the duty cycle expression is

$$dc_1 = dc_2 = dc_3 = D \tag{32}$$

$$V_1 + V_2 + V_3 = V \tag{33}$$

Thus, equation (5) is reduced to

$$V_0 = \frac{3VD}{1 - D} \tag{34}$$

$$\Delta i_L = \frac{DV}{Lf} \tag{35}$$

$$\Delta v_o = \frac{DV}{LCf} \tag{36}$$

$$V_L(t) = \begin{cases} V_1(t) + V_2(t) + V_3(t), & dt_0 < t < dt_1 \\ V_2(t) + V_3(t), & dt_1 < t < dt_2 \\ V_1(t) + V_2(t) + V_3(t) - V_o(t), & dt_3 < t < dt_4 \end{cases} \tag{37}$$

The output capacitor current based on the state can be derived as

$$I_L(t) = \begin{cases} -\dfrac{V_o}{R}, & 0 < t < dt_1 \\[2mm] -\dfrac{V_o}{R}, & dt_1 < t < dt_2 \\[2mm] I_L - \dfrac{V_o}{R}, & dt_3 < t < dt_4 \end{cases} \tag{38}$$

Thus, it's clearly showing the MISO converter provides buck as well as boost operation.

13.3 ANALYSIS OF PROPOSED CONVERTER

Figure 13.6 depicts the proposed multi input single output converter. In the proposed system, solar PV, wind, and fuel cell are considered as sources. A voltage source could be conceived as an entity or cell when paired with a switch which is connected in series and a diode which is connected in parallel. Since the units are linked in series, it is a series linked non-isolated converter. The input voltages V_1, V_2, and V_3 might come from the same source or from dissimilar sources with inconstant voltages.

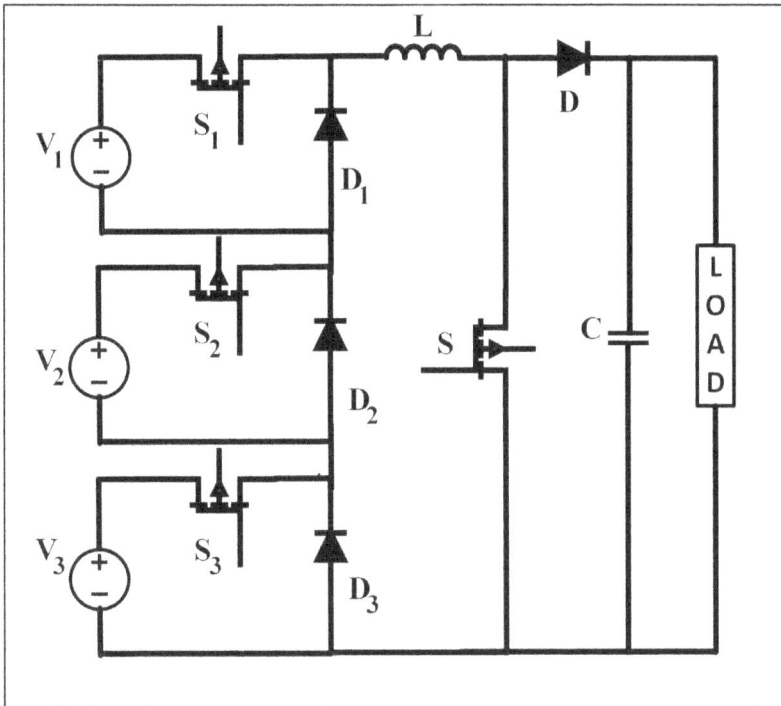

FIGURE 13.6 Multi-input single output converter topology.

13.3.1 OPERATING PRINCIPLE

In an entity, whichever the switch or the diode is turned on at the same period. The inductor current, I_L, is a replication of the MISO converter output current, I_{out}. It represents the overall current that flows between a switch and a diode in a single entity. The average potential difference across each diode such as D1, D2, and D3 fluctuates according to duty cycle value of the related switch. The mean current delivered from each source varies according to the mean voltage throughout the diode, switch internal resistance, and diode internal resistance. The excess current is carried by the diode in the cell during the switches OFF state. The circuit will still operate if one of the switches fails, but the inductor current must be carried by the diode. Therefore diode must be valued for the current in inductor. In case the diode gets defective, the circuit will not function effectively, and it must be replaced. Rising Edge Synchronization [36] is the PWM technique used for switches, in which switched on time of S_1, S_2, and S_3 are similar, but they have different turn off times.

The logical OR function is used to drive S_1, S_2 and S_3 switches in order to get the maximum ON periods S. If maximum turn ON time of switch S is considered as $(dt_3 + dt_2 + dt_1)$, then the switch S off period will be determined as $T-(dt_3 + dt_2 + dt_1)$ and the forward bias of diode D causes the absorbed power in the inductor to be delivered to the load. The MISO converter, which is linked to the residential load, delivers a constant voltage and a range of currents as required by the households. The inductor charges when switching pulses are turned on, and the ripple inductor current is computed by multiplying the inductor ripple I_L by the output current I_{out}. According to the source, the converter can be operated separately or concurrently. Figure 13.7 depicts the four operational stages of proposed MISO converter. When all sources of power are available at the same time, concurrent operation is feasible. When either source provides power, the converter will operate in an independent power supply state. When the energy requirement exceeds the supplied energy, the converter enters boost mode; when power demand is lesser than the input power, it operates in buck mode.

13.3.2 MISO CONVERTER TOPOLOGY AND OPERATION MODES

Mode 1 ($dt_0 < t < dt_1$): Switches S, S_1, S_2 and S_3 are operated in ON condition and diodes D, D_1, D_2 and D_3 functioned in reverse biased state, the inductor L is being charged by input voltages V_1, V_2 and V_3. The current is rising gradually, and the capacitor is ideally charged. In this mode power is not delivered to the load. The inductor current and capacitor voltage of mode 1 can be derived as follows,

$$\frac{dI_L}{dt} = \frac{V_1 + V_2 + V_3}{L} \tag{39}$$

$$C\frac{dV_C}{dt} = \frac{-V_0}{R} \tag{40}$$

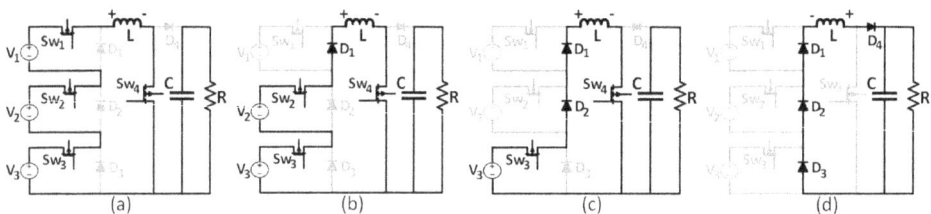

FIGURE 13.7 Modes of operation of MISO converter: (a) mode 1, (b) mode 2, (c) mode 3, (d) mode 4.

Mode 2 ($dt_1 < t < dt_2$): Switches S, S_2 and S_3 are functioned in ON state and S_1 is kept in OFF state, diodes D, D_2 and D_3 functioned in reverse biased, then D_1 act as forward-biased. The inductor L is being stored the energy by the input voltages V_2 and V_3. As a result, the current will increase gradually, and the capacitor will be in an ideal charged state. The power supply is not supplied to load. The inductor current and capacitor voltage of mode 2 can be derived as follows,

$$\frac{dI_L}{dt} = \frac{V_2 + V_3}{L} \tag{41}$$

$$C\frac{dV_C}{dt} = \frac{-V_0}{R} \tag{42}$$

Mode 3 ($dt_2 < t < dt_3$): S and S_3 switches are functioned in ON state, the rest switches S_2 and S_3 is in OFF state. D and D_3 diodes are operated in reverse-biased state whereas D_2 and D_4 is functioned in forward biased mode. So the inductor L is being stored the energy from the input voltages V_3. As a result, power is flowing inside the circuit rather than being transferred to the load. The inductor current and capacitor voltage of mode 3 can be derived as follows,

$$\frac{dI_L}{dt} = \frac{V_3}{L} \tag{43}$$

$$C\frac{dV_C}{dt} = \frac{-V_0}{R} \tag{44}$$

Mode 4 ($dt_3 < t < dt_4$): Switches S, S_1, S_2 and S_3 are in OFF state and diodes D, D_1, D_2 and D_3 functioned in forward-biased condition. The inductor L starts discharging power to capacitor and load. The inductor current and capacitor voltage of mode 4 can be derived as follows,

$$\frac{dI_L}{dt} = \frac{V_1 + V_2 + V_3 - V_0}{L} \tag{45}$$

$$C\frac{dV_C}{dt} = I_L - \frac{V_0}{R} \tag{46}$$

13.4 INTEGRATION OF RES WITH MISO CONVERTER FOR RESIDENTIAL LOAD APPLICATION

A multi-input single output converter takes 'n' renewable energy sources as input to produce a single DC output. For instance, the envisioned MISO converter integrates a hybrid PV panel, wind turbine, and fuel cell. The MPPT is utilised in solar PV systems to obtain the extreme power from the sun. Likewise, the wind turbine generates power based on the wind speed.

All these three sources are directly connected to the MISO to provide the supply to the residential load in an isolated area, which is detailed in Figure 13.9. Figure 13.8 depicts the switching arrangement as well as the logical waveforms of proposed multi-source converter. A developed MISO DC-DC converter is utilised to integrate the intermittent renewable energy sources. The operation of the converter is determined by the amount of electricity produced by renewable energy sources. According on the weather, electricity production may rise or decrease. When power generation from renewable sources exceeds the required power generation, the MISO converter operates as a buck converter. If renewable energy generation falls short of the expected power demand, the MISO converter will act as a boost converter.

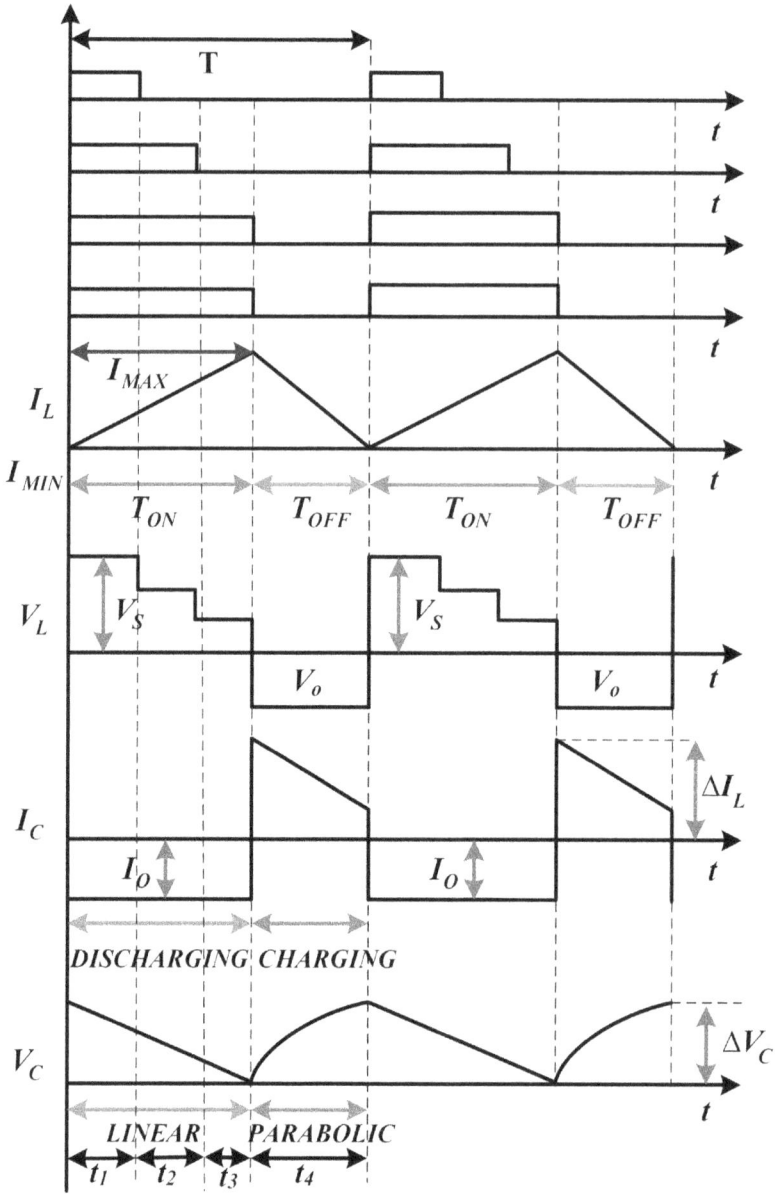

FIGURE 13.8   Waveform of multi-input single output converter.

The PI-controller regulates the functioning of the MISO converter, which creates PWM pulses based on proportional and integral values. The PI-controller gain value is designed using a trial-and-error strategy. On other side variable resistance is used as residential load where the loads are scheduled as per time. The overall performance of proposed system using PI controller and PWM pulse generation by MPPT for RES is described as a flowchart in Figure 13.10.

13.4.1   CONTROL STRATEGY

Figure 13.11 depicts the complete control framework. The wind MPPT algorithm generates the MPP current, which serves as the wind controller's reference current [37]. The control strategy of the recommended system is a two side control system, whilst sun, wind, and fuel cell control are source side controls.

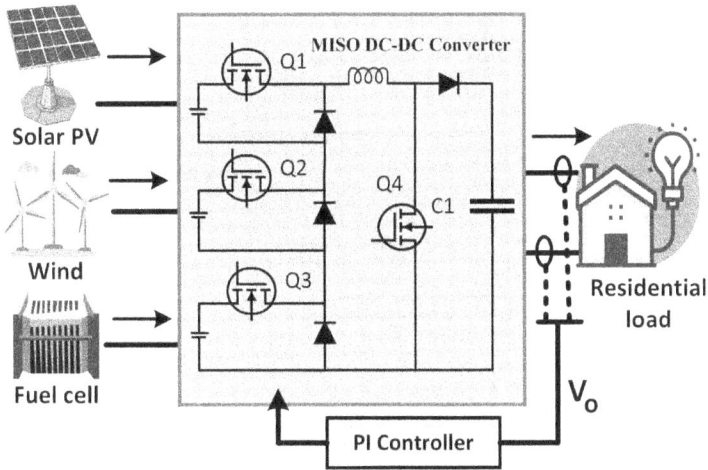

FIGURE 13.9 Integration of RES with MISO converter.

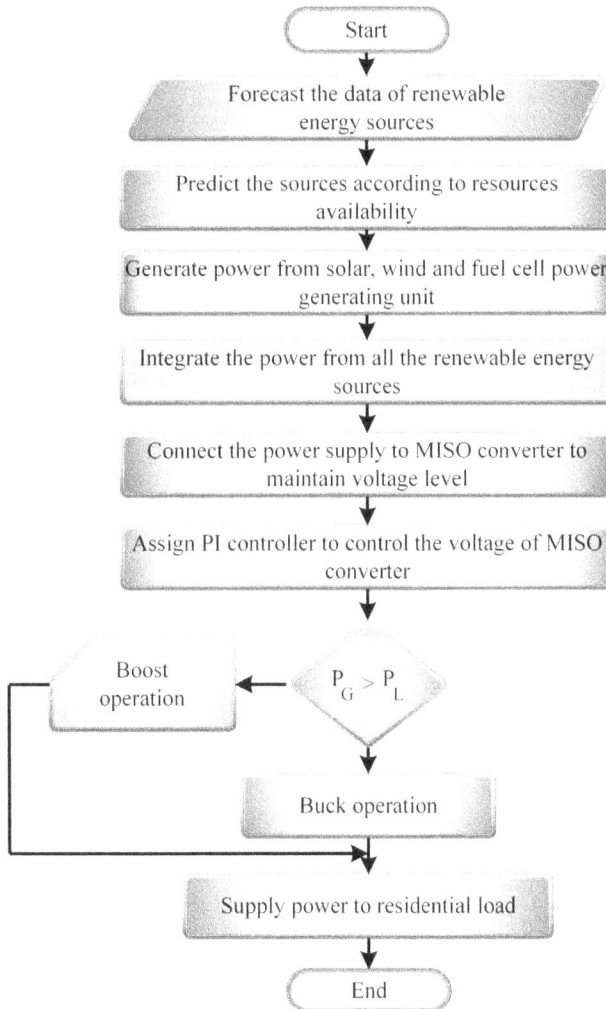

FIGURE 13.10 Overall performance of MISO converter.

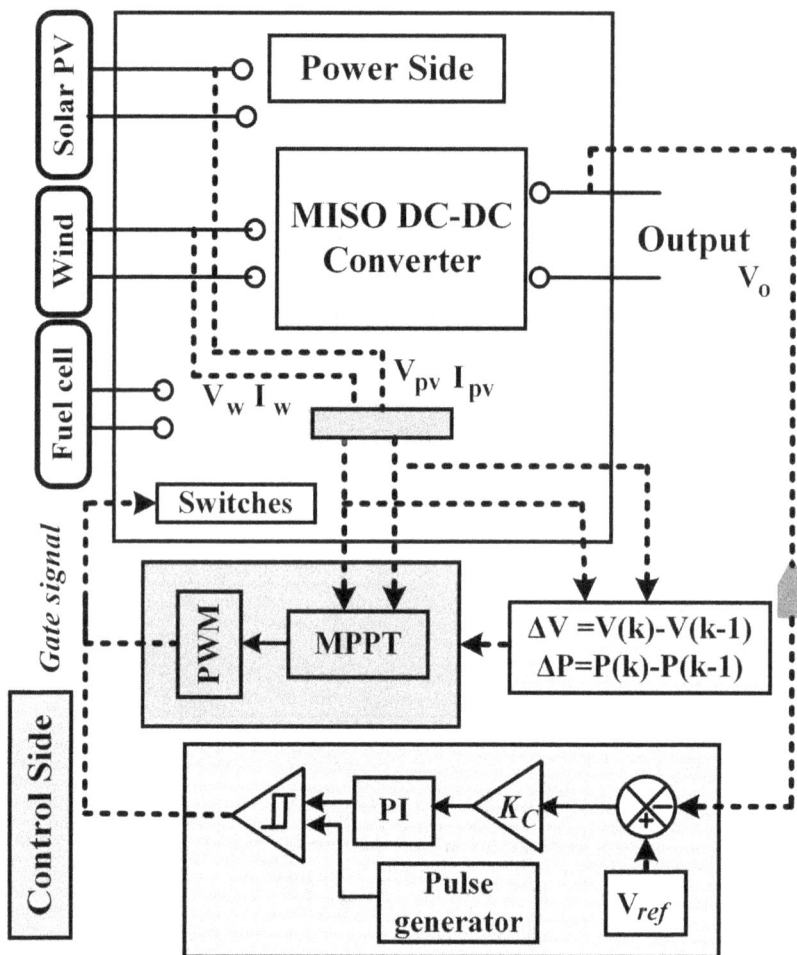

FIGURE 13.11 Overall control circuit.

The PI-controller can be used to produce PWM pulses for the multi-source DC-DC converter. The PWM controller's output is directed to IGBT switches S_1, S_2 and S_3 and then to switch S through a logical OR gate. The Rising Edge Synchronization PWM approach creates duty cycle dynamically in response to changes in load and reference voltage. The converter is not used at high or low duty ratios. The voltage controller's output is restricted between 0.1 and 0.85 and fed into the PWM controller to generate appropriate duty ratios. Converters in this series connected topology can operate either independently or concurrently. When all of the input voltage sources are joined together and energy is delivered concurrently, the options of the boost converter attaining its limit are reduced. This is a superiority over a parallel connection, which can only be used individually.

The factors of the PI-controller can be characterised as follows:

$$P = K_{pv} * e(t) \tag{47}$$

$$I = K_{iv} \int e(\tau) * d\tau \tag{48}$$

$$K_{pv} = 2\zeta\omega_n C_0 - \frac{1}{R} \tag{49}$$

TABLE 13.1
Design Parameters of MISO Converter

Parameters	Values
Solar PV input voltage	56 V
Wind turbine voltage	160 V
Fuel cell input	70 V
Inductor	5 mH
Capacitor	4700 μF
Switching frequency	10 kHz
Output voltage	110V
Proportional (P)	0.4
Integral (I)	10

$$K_{iv} = \omega_n^2 C_0 \qquad\qquad (50)$$

Where ζ represents the damping. The design specifications of MISO converter is given in Table 13.1.

13.4.1.1 Perdurb and Observation Method for MPPT

The power versus voltage characteristics of a solar PV array and power coefficient (Cp) λ curves are nonlinear [38] as seen in Figures 13.12 (a) and (b). The operating point is determined by the load impedance connected to the array as well as wind turbine terminals at any given time.

The spot of action on the PV curve and power coefficient (Cp) curves of a wind turbine is tracked using a DC-DC converter. There are several strategies for determining the maximum power point that have been reported in many works. The perturb and observe approach comprises altering the DC-DC converter's reference voltage or duty ratio in steps and evaluating the energy output. This MPPT strategy works by capturing both current and voltage of PV array to calculate output power of solar PV and change in output power. Figure 13.13 demonstrates the process for modifying the duty cycle of a DC-DC converter.

The real wind current (I_W) is matched to the reference current (I_{Wr}) and the difference is considered as an error, then supplied into the PI controller. The PI controller produces the necessary duty ratio to achieve zero percent error. The acquired duty cycle (D) is correlated to a large frequency ramp signal that defines the switching pulses of the source side DC-DC converter. While PV is accessible, MPPT algorithm creates V_{mpp} which is act as the reference voltage for PV voltage controller. The variation in PV voltages between real and reference is equated, and the difference is fed into the PI controller. The reference current (I_{pv}) is generated by the voltage controller to preserve output voltage of solar PV at V_{mpp}. Then, the real I_{pv} and reference current I_{pvr} are matched, and the difference is given to the PI controller, which is continued by the PWM generator to create switching pulses.

13.5 RESULT AND DISCUSSION

The studies were performed using MATLAB/Simulink to scrutinise the effectiveness of the proposed scheme under various operating situations. The specifications utilised in simulation performance are listed in Table 13.1. Variations in solar irradiation, wind speed, and load resistance are used to test the efficiency of the proposed framework. The proposed system's performance is tested under fundamental operating circumstances. These are (i) solar PV and wind power is available when fuel cell is connected, and (ii) both PV and wind are not available.

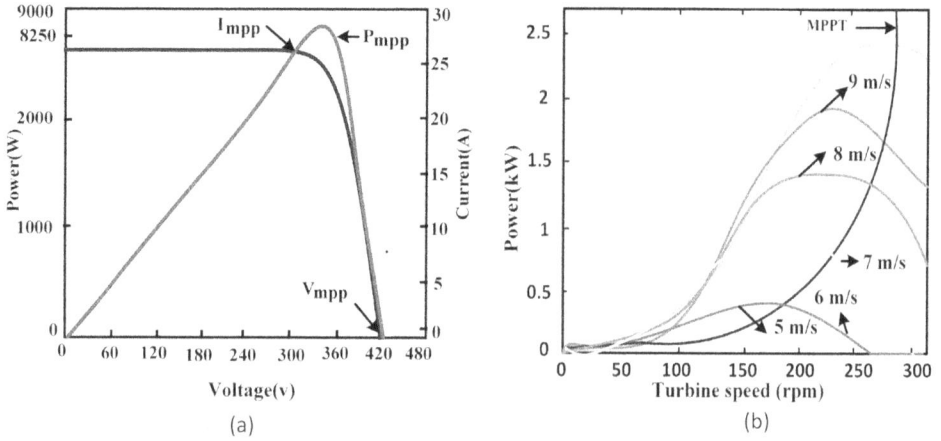

FIGURE 13.12   (a) Power versus voltage characteristics of solar, (b) power coefficient (Cp) λ curves of wind turbine.

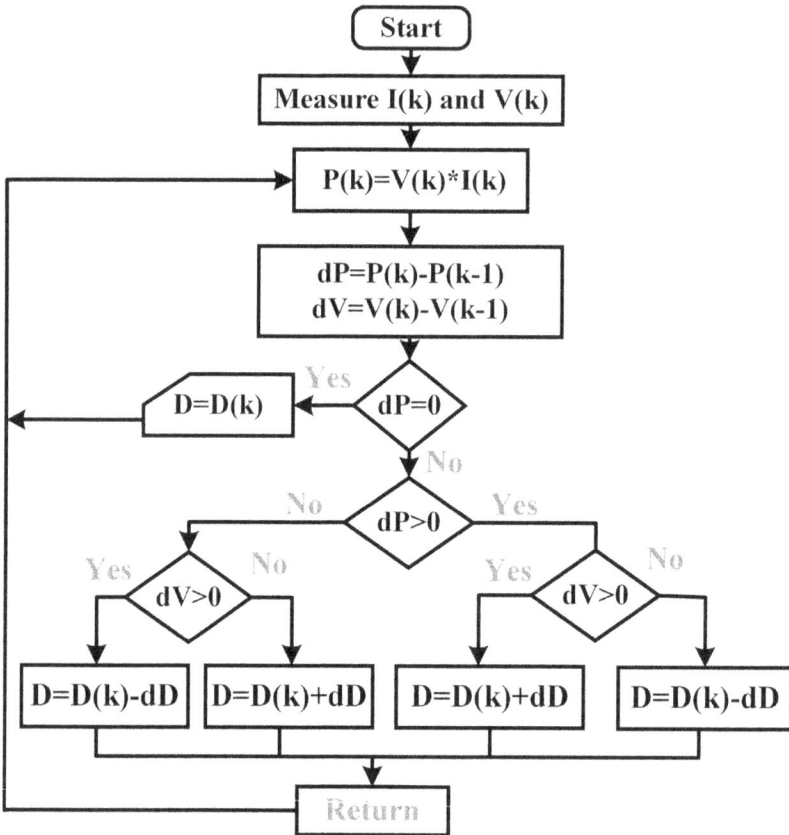

FIGURE 13.13   Flow chart for perturb and observe algorithm for MPPT.

13.5.1 Solar PV and Wind Energy Are Accessible

Figure 13.14 depicts the overall structure of proposed system under this functioning state. Since the data for solar PV, wind, and residential load is real-time data, their variation occur at periodic intervals under this working state. The data have been collected for 2 hours and retrieved from 2 days of data, with only 5 seconds taken into account while assessing the simulation. The balancing of output power is preserved in the system throughout all disturbances, as seen in Figure 13.15.

The output voltage of the MISO converter should always remain constant at 110 V, regardless of load variations. Figures 13.15 (a)–(d) shows the result about solar irradiance, voltage, current, and power profile respectively. Initially, the irradiance is captured 980 W/m^2 with varying temperature of 24°C–36°C. Then it is periodically increasing and decreasing according to the time variation, which is shown in Figure 13.15 (c) details about the output current from the solar PV.

Figures 13.15 (f)–(h) shows the specifics about wind input, voltage, current, and power profile respectively. The initial condition of the wind speed is noted as 11.5 (m/s), typically the wind speed is varying based on the weather. The minimum wind speed is noted as 9.4 (m/s), which is shown in Figure 13.15(e). The voltage from the wind maintained as 160 V at wind speed of 12.85 m/s and varies to 180 V due to change in wind speed above 13 m/s, which is shown in Figure 13.15(f). The current is varies from 0 to 5 A.

Figures 13.15(i)–(l) detail the utilisation of fuel, voltage, current, and power profile of a fuel cell. Since the two renewable sources available, the utilisation of hydrogen and oxygen level 35 and 60 percent respectively, which is shown in Figure 13.15(i). So the utilisation of hydrogen under 80 percent is indicating the presence of enormous hydrogen, resulting in a small spike in fuel cell output voltage, which is shown in Figure 13.15(j).

Figure 13.16(a) and (b) illustrate the inductor voltage and current when solar, wind, and fuel cell input voltages are 56 V, 160 V, and 70 V, respectively. Simulation result of load voltage, current and power is represented in Figure 13.17(a), (b), and (c) respectively.

Depending on the load, the load current is increased from 0.1 A to 1.4 A at 0.2 sec and then decreased from 1.4 A to 1 A at 4 sec. Even if a small fluctuation are detected in output voltage in solar as well as wind, the whole system is responding dynamically to settle to 110 V within a short period. It can also be discovered that the overall system reacts quicker when changes take place in source voltages as well as load current, as illustrated in Figure 13.17. The proposed system is evaluated by altering the input voltages; eventually, the system maintains the output voltage at 110 V.

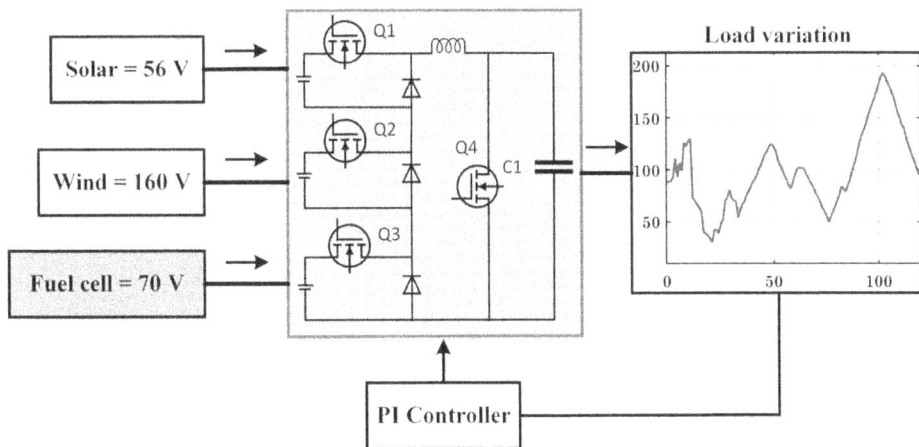

FIGURE 13.14 MISO converter with solar PV, wind, and fuel cell.

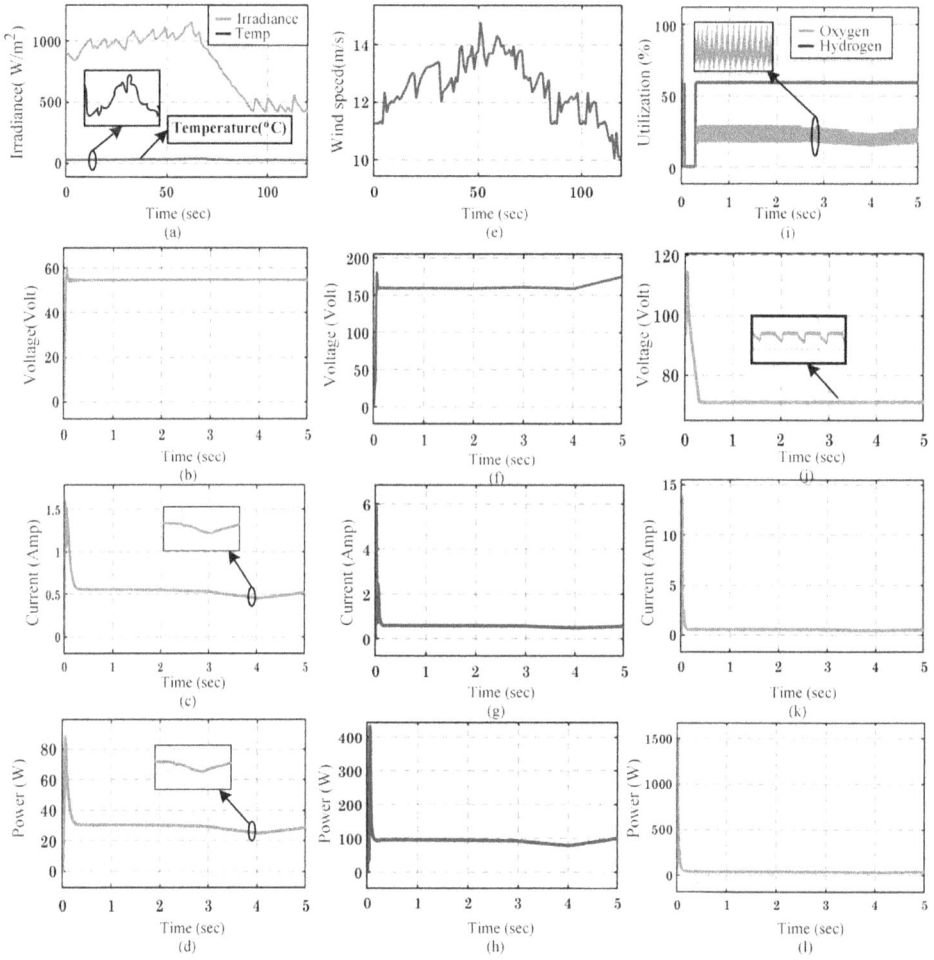

FIGURE 13.15   Input of MISO converter: (a) solar irradiance and temperature, (b) solar output voltage, (c) solar output current, (d) solar output power, (e) wind speed, (f) wind output voltage, (g) wind output current, (h) wind output power, (i) fuel consumption of fuel cell, (j) fuel cell output voltage, (k) fuel cell output current, (l) fuel cell output power.

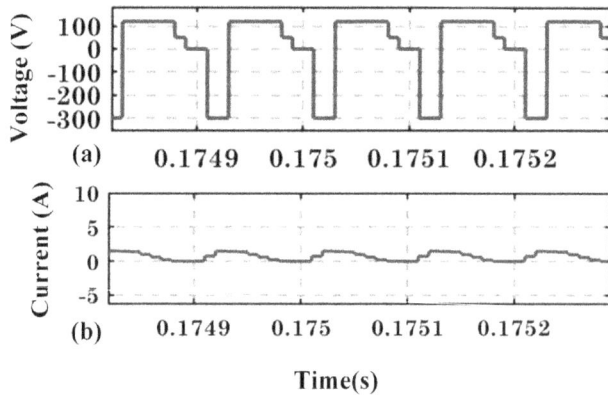

FIGURE 13.16   Inductor parameter: (a) inductor voltage and (b) inductor current.

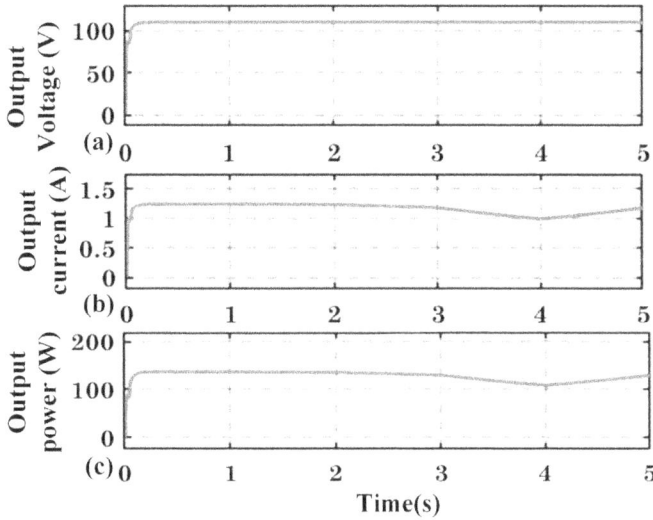

FIGURE 13.17 MISO converter output: (a) output voltage, (b) output current, (c) output power.

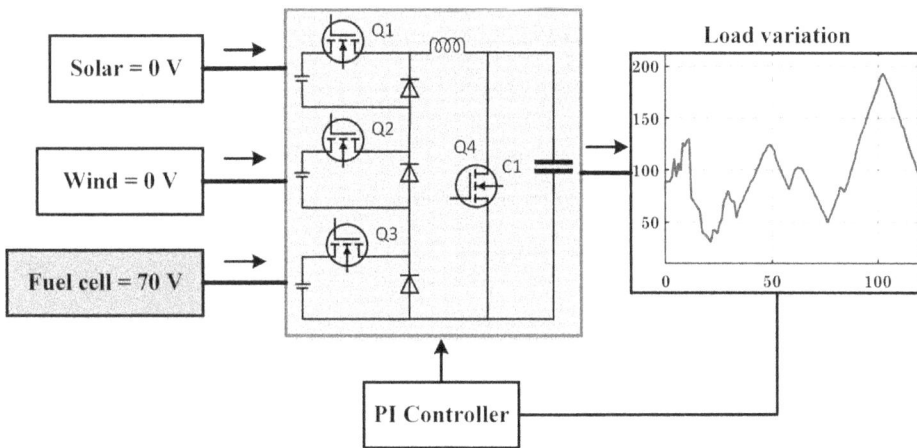

FIGURE 13.18 MISO converter when solar PV and wind is OFF.

13.5.2 SOLAR PV AND WIND ENERGY ARE INACCESSIBLE

The overall structure of the proposed system in this working state is depicted in Figure 13.18. Although solar PV as well as wind are both inaccessible, the whole demand power is delivered only by the fuel cell. So, from Figures 13.19 and 13.20 it has been discovered that during all the variation in load, oscillatory power is being supplied by fuel cell.

If one or both of the renewable energy input voltages are removed, the system is discovered to remain operational. But current from the other two sources is increased. Even if two of the sources like solar and wind are removed, the system maintains the stability voltage as 110 V within a few seconds, which is shown in Figures 13.20(a) and (b). Furthermore, the utilisation of hydrogen in fuel cell is initially hike to above 90 percent and after 3.2 sec it is maintain between 80 to 90 percent due to unavailability of solar PV and wind, which is shown in Figure 13.21. The hydrogen utilisation values between 80 and 90 percent indicate that hydrogen is being used efficiently and durability of

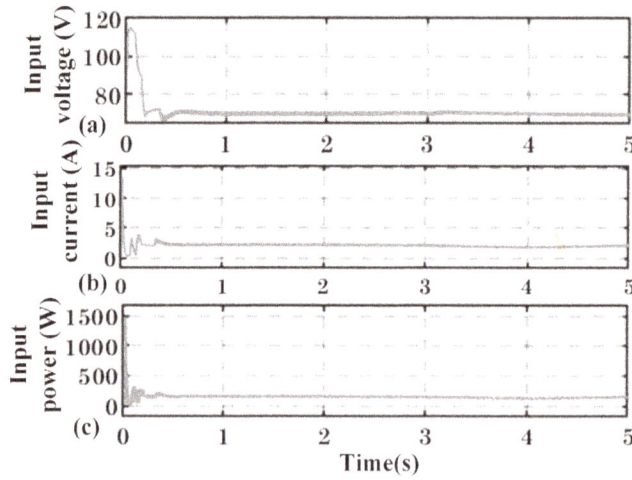

FIGURE 13.19   Fuel cell input: (a) input voltage, (b) input current, (c) input power.

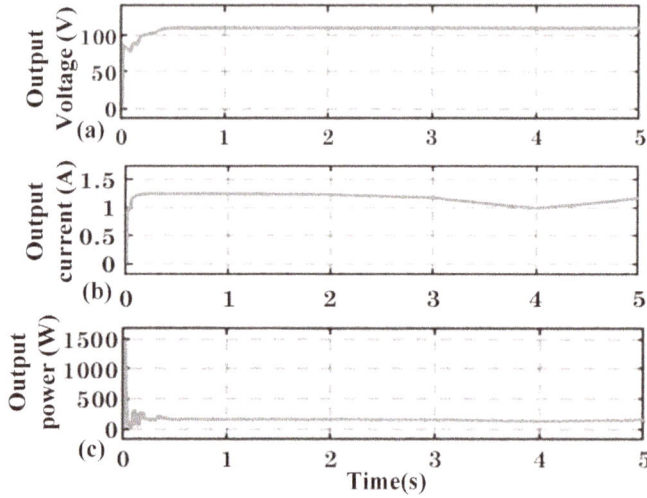

FIGURE 13.20   MISO converter output: (a) output voltage, (b) output current, (c) output power.

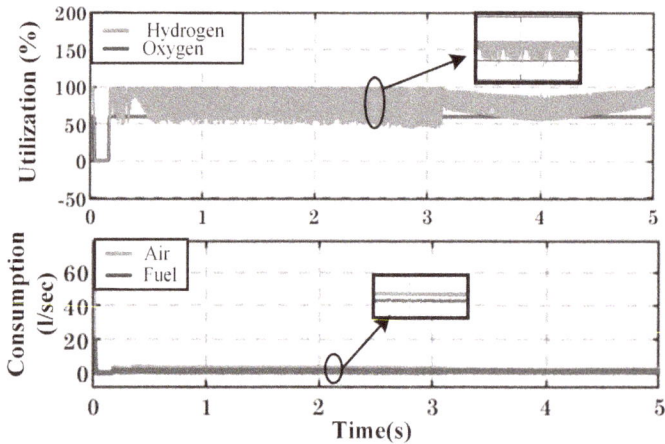

FIGURE 13.21   Fuel and stack consumption of fuel cell.

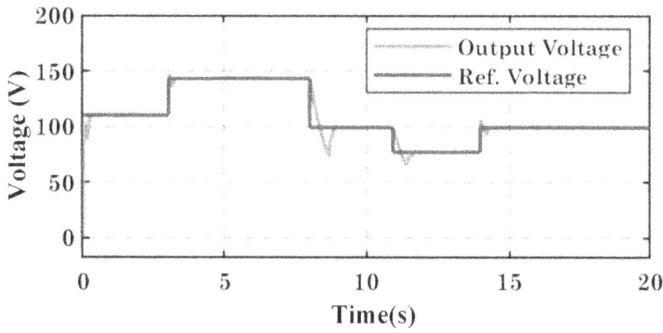

FIGURE 13.22 Output voltage for change in reference voltage.

the system. The reference voltage of the MISO converter is also examined. The reference voltage is changed between 80 V to 140 V while the current remains unchanged at 1 A. The load voltage is tracked by the reference voltage, as illustrated in Figure 13.22.

13.6 CONCLUSION

In this work, PI control based MISO converter along with RES is proposed for residential load for isolated area. The suggested converter has a single inductor and capacitor, resulting in less weight, lower loss, higher efficiency, and a more compact design. The MISO converter may supply electricity either individually or simultaneously according to the variation in RES. The RES have lower power losses than power grids, and they are more easily deployed in remote locations than power grid supplies. The converter is developed with computed values and executed with MATLAB/Simulink in this chapter. The output voltage is sustained by the PI-controller according to the estimated values. Even though unavailability of power in solar and wind sources, fuel cell alone has the ability to provide stability to the system within a few seconds.

REFERENCES

[1] N. Priyadarshi, S. Padmanaban, P. Kiran Maroti, and A. Sharma. An extensive practical investigation of FPSO-based MPPT for grid integrated PV system under variable operating conditions with anti-islanding protection. *IEEE Syst. J.*, vol. 13. pp. 1861–1871, 2019.
[2] R. R. Kanna, M. Baranidharan, R. Raja Singh, and V. Indragandhi. Solar energy application in indian irrigation system. *IOP Conf. Ser. Mater. Sci. Eng.*, vol. 937, no. 1, 2020.
[3] A. Banerji, K. Sharma, and R. R. Singh. Integrating Renewable energy and electric vehicle systems into power grid: benefits and challenges. *IPACT Conf.* pp. 1–6, 2022.
[4] A. F. Zobaa, M. A. M. Shaheen, H. M. Hasanien, R. A. Turky, M. Calasan, and S. H. E. A. Aleem. OPF of modern power systems comprising renewable energy sources using improved CHGS optimization algorithm. *J. Energies*. Vol. 14.14216962.2021.
[5] P. Roy, J. He, T. Zhao, and Y. V. Singh. Recent advances of wind-solar hybrid renewable energy systems for power generation : A Review. *IEEE. J. Industrial Electronics Society*. Vol. 3.2022.
[6] D. Market, Y. Yang, C. Qin, Y. Zeng, and C. Wang. Optimal Coordinated bidding strategy of wind and solar system with energy storage in day-ahead market. *IEEE. J. Modern Power Systems and Clean Energy*. vol. 10, no. 1, pp. 192–203, 2022.
[7] R. Alvarez. Fuel cell hybrid vehicles and their role in the decarbonisation of road transport. *Journal of Cleaner Production*. vol. 342.2022.
[8] A. S. Al-buraiki, and A. Al-sharafi. Hydrogen production via using excess electric energy of an off-grid hybrid solar/wind system based on a novel performance indicator. *J. Energy Conversion and Management*. Vol. 254.2022.
[9] J. Li, H. Wang, H. He, and S. Member. Battery optimal sizing under a synergistic framework with DQN-based power managements for the fuel cell hybrid powertrain. *J. Energy Conversion and Management*. vol. 8. pp. 36–47.2022.

[10] A. Almutairi, K. Sayed, N. Albagami, A. G. Abo-khalil, and H. Saleeb. Multi-Port PWM DC-DC power converter for renewable energy applications. *J. Energies*. Vol. 12.14123490.2021.

[11] B. N. Alajmi, M. I. Marei, S. Member, and I. Abdelsalam. A multiport DC—DC converter based on two-quadrant inverter topology for pv systems. *IEEE. Tran. Power Electron*. Vol. 36, no. 1, pp. 522–532, 2021.

[12] A. Avila, A. Garcia-bediaga, and I. Alzuguren. A modular multifunction power converter based on a multiwinding flyback transformer for EV spplication. *IEEE. Trans. Transportation Electrification*. vol. 8, no. 1, pp. 168–179, 2022.

[13] Q. Tian, S. Member, G. Zhou, S. Member, and H. Li. Symmetrical bipolar output isolated four-port converters based on center-tapped winding for bipolar DC bus applications. *IEEE. Tran. Power Electron*. vol. 37, no. 2, pp. 2338–2351, 2022.

[14] K. Suresh, C. Bharatiraja, S. Member, and B. Alamri. A multifunctional non-isolated dual input-dual output converter for electric vehicle application. *IEEE. Access*. Vol. 9. Pp. 64445–64460, 2021.

[15] W. Yi et al. Analysis and implementation of multi-port bidirectional converter for hybrid energy systems. *Energy Reports*, vol. 8, pp. 1538–1549, 2022.

[16] P. Xu, H. Wen, W. Hao, Y. Yang, J. Ma, and Y. Wang. Nonisolated switching-capacitor-integrated three-port converters with seamless PWM/PFM modulation. *Sol. Energy*, vol. 224. Pp. 160–174, 2021.

[17] M. B. Camara, H. Gualous, F. Gustin, A. Berthon, and B. Dakyo. DC/DC converter design for super-capacitor and battery power management in hybrid vehicle applications—polynomial control strategy. *IEEE. Tran*. Vol. 57, no. 2, pp. 587–597, 2010.

[18] S. K. Kollimalla, M. K. Mishra, and N. L. Narasamma. Design and analysis of novel control strategy for battery and supercapacitor storage system. *IEEE Trans. Sustain. Energy*, vol. 5, no. 4, pp. 1137–1144, 2014.

[19] B. Bendjedia, N. Rizoug, and M. Boukhnifer. Real time implementation of frequency separation management strategy of hybrid source for fuel cell electric vehicle applications. *IEEE 14th Int. Conf. Compat. Power Electron. Power Eng. CPE-Powereng 2020*, pp. 544–549, 2020.

[20] Q. Tian, G. Zhou, R. Liu, X. Zhang, and M. Leng. Topology synthesis of a family of integrated three-port converters for renewable energy. *IEEE Trans. Industrial Electronics*. Vol. 68, no. 7, pp. 5833–5846, 2021.

[21] D. Chen, S. Member, T. Zhao, L. Han, and Z. Feng. Single-stage multi-input buck type high-frequency power supply. *IEEE Trans. Power Electronics*. Vol. 37, no. 6, pp. 7411–7421, 2022.

[22] P. Yang, Y. Peng, Y. Xuan, X. Chen, S. Member, and X. Liu. Multi-input variable structure converter with optimal power extraction strategy for energy harvesting. *IEEE Journal on Emerging and Selected Topics in Circuits and Systems*. Vol. 12, no. 1, pp. 290–300, 2022.

[23] S. Chaturvedi, S. Member, D. Fulwani, and D. Patel. Dynamic virtual impedance-based second-order ripple regulation in DC microgrids. *IEEE Journal of Emerging and Selected Topics in Power Electronics*. Vol. 10, no. 1, pp. 1075–1083, 2022.

[24] O. Husev et al. Novel concept of solar converter with universal applicability for DC and AC microgrids. *IEEE Trans. Industrial Electronics*. vol. 69, no. 5, pp. 4329–4341, 2022.

[25] A. Abbassi, R. Ben, B. Touaiti, and L. Abualigah. Optik Parameterization of photovoltaic solar cell double-diode model based on improved arithmetic optimization algorithm. *Optik (Stuttg)*., vol. 253, no. p. 168600, 2022.

[26] N. Priyadarshi, S. Padmanaban, M. S. Bhaskar, F. Blaabjerg, and A. Sharma. Fuzzy SVPWM-based inverter control realisation of grid integrated photovoltaicwind system with fuzzy particle swarm optimisation maximum power point tracking algorithm for a grid-connected PV/wind power generation system: Hardware implementation. *IET Electr. Power Appl.*, vol. 12, no. 7, pp. 962–971, 2018.

[27] A. K. Singh, and R. R. Singh. An overview of factors influencing solar power efficiency and strategies for enhancing. *IPACT Conf*. pp. 1–6, 52855.2022.

[28] N. Priyadarshi, S. Padmanaban, M. S. Bhaskar, F. Azam, B. Khan, and M. G. Hussien. A novel hybrid grey wolf optimized fuzzy logic control based photovoltaic water pumping system. *Renewable Power Generation, IET*, 2022, doi: 10.1049/rpg2.12638.

[29] K. Patel, S. Borole, K. Ramaneti, A. Hejib, and R. Raja Singh. Design and implementation of sun tracking solar panel and smart wiping mechanism using Tinkercad. *IOP Conf. Ser. Mater. Sci. Eng.*, vol. 906, no. 1, 012030.2020.

[30] Y. Nie, F. Li, L. Wang, J. Li, and M. Sun. A mathematical model of vibration signal for multistage wind turbine gearboxes with transmission path effect analysis. *Mech. Mach. Theory*. vol. 167. p. 104428, 2022.

[31] B. Benyachou, B. Bahrar, A. Moufakkir, K. Gueraoui, and M. S. Hassani. Optimization & control strategy for offshore wind turbine based on a dual fed induction generator. *Mater. Today Proc*. 1016.2022.

[32] W. Torki, F. Grouz, and L. Sbita. Vector control of a PMSG direct-drive wind turbine. *Int. Conf. Green Energy Convers. Syst. GECS 2017*, 8066247.2017.

[33] M. Y. El-Sharkh, A. Rahman, M. S. Alam, P. C. Byrne, A. A. Sakla, and T. Thomas. A dynamic model for a stand-alone PEM fuel cell power plant for residential applications. *J. Power Sources*, vol. 138, no. 1–2, pp. 199–204, 2004.

[34] P. Hema Rani, S. Navasree, S. George, and S. Ashok. Fuzzy logic supervisory controller for multi-input non-isolated DC to DC converter connected to DC grid. *Int. J. Electr. Power Energy Syst.*, vol. 112. pp. 49–60, 2019.

[35] N. Priyadarshi, M. S. Bhaskar, P. Sanjeevikumar, F. Azam, and B. Khan. High-power DC-DC converter with proposed HSFNA MPPT for photovoltaic based ultra-fast charging system of electric vehicles. *Renewable Power Generation, IET*, 2022, doi: 10.1049/rpg2.12513.

[36] L. Kumar, and S. Jain. Multiple-input DC/DC converter topology for hybrid energy system. *IET Power Electron.*, vol. 6, no. 8. Pp. 1483–1501, 2013.

[37] B. R. Ravada, N. R. Tummuru, and B. N. L. Ande. Photovoltaic-wind and hybrid energy storage integrated multisource converter configuration-based grid-interactive microgrid. *IEEE Trans. Ind. Electron.*, vol. 68, no. 5, pp. 4004–4013, 2021.

[38] P. Manoharan *et al.* Improved perturb and observation maximum power point tracking technique for solar photovoltaic power generation systems. *IEEE Syst. J.*, vol. 15, no. 2, pp. 3024–3035, 2021.

14 Analysis, Modelling and Design of DSP TMS320F2812 Based Digital PI Controller for DC-DC Buck Converter

Vanshika Jindal and Dheeraj Joshi

CONTENTS

14.1 INTRODUCTION

Power electronics application are well established in continuous-time analysis, averaged modelling of switched-mode power converters, and analog control theory. With advancements in high power electronics and low frequency operations (power semiconductor devices), digital control offers various technical and economic advantages, such as low controller cost and low power dissipation. However, in low to medium power applications, switching frequency ranges from 100kHz to a few MHz, fast dynamic response with lesser power dissipation is required. And hence, analog controllers meet the challenges respectively [1]. Analog based control methods are used to compute real time operation with very high bandwidth. These are fixed and hardwired with more complexity and are more susceptible to noise and environmental changes.

Digital controllers are based on dedicated microprocessors, digital control processors and programmable logic gates. Digital control with multiple high resolution PWM and many Analog to digital (A/D) channels, allows software-based implementation with efficiency optimisation in switched mode power supplies. Also, these are relatively less sophisticated as it reduces space, simplifies the hardware circuitry, reduces sensitivity to process and temperature variation while increasing the speed of computation [2].

DOI: 10.1201/9781003323471-14

Traditionally, Linear PI controllers were being used for tracking of reference voltage with appropriate performance which were designed using analog and digital design approaches. In discrete data systems, sample data is obtained by sampling of analog signal in the form of pulse train or numerical coded with amplitude of pulses as its information. Hence, it improves the performance of control system. Digital control offers an ease while implementing non-linear controllers in software coding using code composer studio [3]. However, few disadvantages of digital control include finite word length of the processor, that is, limited signal resolution, sampling time delay and so on. Time delay is another concern in carrying out necessary computation while executing controller algorithm.

Common practice to simulate continuous data control system to digital equivalent is called as Digital redesign using z-domain [4]. In general, transfer function of digital controller can be realised by digital program by direct programming, which in turn determines the desired satisfactory performance of the system. The second approach describes the direct conversion of digital controller design into z-domain, termed as digital design approach. In this chapter the digital redesign approach, also called design by emulation, is discussed, that is, the controller is designed in continuous domain and then converted into discrete form. This chapter's objective is to design and implement digital controllers in switched-mode DC-DC converters while providing adequate steady and dynamic performances of final closed-loop system [5].

The objectives of this chapter are to analyse, model, design, and implement digital feedback loops from converter's transfer function to practical implementation of converter. In subsection 14.2.1, mathematical modelling of a buck converter is discussed using state space averaging [6]. The designing of the buck converter with suitable components ratings and specification are discussed in subsection 14.2.2. In subsection 14.2.3, the steps for designing of PI controller are discussed. The design specifications are performed in frequency design in terms of cut-off frequency (ω_c) and phase margin, ϕ_m. In subsection 14.2.4, the transformation of compensator transfer function from analog control to digital control using design by emulation is performed. Subsection 14.2.5 presents the digital implementation of PI controller with TMS320F2812 DSP. Sections 14.3 and 14.4 verifies the hardware and simulation results of closed loop PI controlled buck converter.

14.2   SYSTEM DESIGN METHODOLOGY

14.2.1   MATHEMATICAL MODELLING OF DC-DC BUCK CONVERTER

Figure 14.1 shows the basic circuit diagram of non-ideal dc-dc Buck converter [7]. The circuit consists of MOSFET switch S, capacitor C, diode D, inductor L and load resistance R. For accurate modelling of Buck converter, the different non-idealities such as equivalent series resistances (ESR) of inductor – r_L, ESR of capacitor – r_c, switch on resistance – r_{sw}, diode forward resistance – r_d, and diode forward voltage drop – V_F have been considered. The converter is operating at duty cycle d, and f is the switching frequency [1].

The output voltage to duty ratio transfer function for non-ideal buck converter is derived as follows.

$$\frac{V_{out}(s)}{d(s)} = C(sI - A)^{-1}B \tag{1}$$

It can also be written as:

$$\frac{V_{out}(s)}{d(s)} = \frac{\dfrac{R\left(V_g + V_F - \left(r_{sw} - r_d\right)I_L\right)}{LC\left(R + r_C\right)}\left(r_C Cs + 1\right)}{s^2 + \left(\dfrac{1}{L}\left(r_x + r_L + \dfrac{r_C R}{R + r_C}\right) + \dfrac{1}{C}\left(\dfrac{R}{R + r_C}\right)\right)s + \left(\dfrac{r_L + r_x + R}{LC\left(R + r_C\right)}\right)} \tag{2}$$

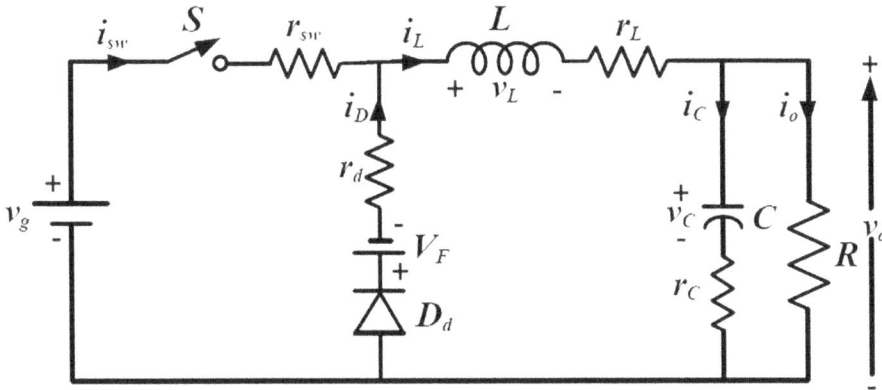

FIGURE 14.1   Buck Converter Topology.

TABLE 14.1

Parameters of Buck Converter

Parameters	Values
Input voltage, V_{in}	12–16V
Load resistance, R	2Ω
Output voltage, V_{out}	6V
Inductance, L/r_L	1.8mH / 0.18 Ω
Capacitance, C/r_C	47µF / 0.3 Ω
Frequency, f	20Khz
Switch- on resistance, r_{sw}	0.044Ω
Diode forward resistance, r_d	0.024Ω
Diode forward Voltage drop, V_d	0.7V

The dynamic model for switched-on and off stage is obtained and averaged over switching cycle as per state space averaging technique. The transfer function termed as G_{vd} depicts the effects of changes in duty cycle on the output voltage of a converter. G_{vd} is obtained by substituting the parameter values as in Table 14.1:

$$G_{vd}(s) = \frac{V_{out}(s)}{d(s)} = \frac{2417s + 1.714 \times 10^8}{s^2 + 9518s + 1.141 \times 10^7} \tag{3}$$

14.2.2 DESIGNING OF BUCK CONVERTER

The design parameters for the dc-dc buck converter are:

(1) Input voltage
(2) Maximum output current
(3) Regulated output voltage
(4) Switching frequency

14.2.2.1   Following Are the Specifications of DC-DC Buck Converter

14.2.2.1.1   Inductor Selection: The Selection of Inductor Is Dependent on the Current Ripple

$$L = \frac{V_O(1-D)}{f \Delta I_L} \tag{4}$$

ΔI_L = estimated inductor ripple current = 0.27% of I_L

D = Duty ratio

f = frequency

V_o = output voltage of converter

14.2.2.1.2 Output Capacitor Selection

$$C = \frac{\Delta I_L}{8f\Delta V_{out}}$$

(5)

ΔV_{OUT} = desired output voltage ripple = 0.2%

14.2.2.1.3 Efficiency of Buck Converter

Efficiency of the buck converter is affected by different losses, such as switching losses, conduction, and diode losses. Few of them are shown as follows:

1. **Switching losses, P_{sw}:** switching losses are generated when the transition to ON.

$$\frac{1}{2}V_{in} \times I_{out} \times \left(t_r + t_f\right) \times f_{sw}$$

P_{sw} = 6.66 mW

V_{in} = input voltage

I_{out} = output current

t_r = rise time

t_f = fall time

f_{sw} = switching frequency

2. **Conduction losses, P_c:** calculated from output current, R_{ON} and on state duty cycle.

$$D \times I_{orms} \times R_{on}$$

P_c = 3.825 W

D = duty ratio

Io_{rms} = output rms current

R_{on} = Mosfet on resistance

3. **Body diode losses/dead time losses, P_d:** calculated using dead time.

$$V_D \times I_{out} \times \left(t_{dr} + t_{df}\right) \times f_{sw}$$

P_d = 3.78 mW

t_{dr} = dead time at rise

t_{df} = dead time at fall

V_D = diode forward voltage drop

4. Inductor losses, P_L:

$$D \times I_L \times I_L \times r_L$$

$P_L = 0.81W$
I_L = inductor current
r_L = inductor resistance

5. **Efficiency** $= \dfrac{V_o \times I_o}{V_o \times I_o + P\ losses} = 79\%$

14.2.3 PI Controller Design of Buck Converter

An analytical procedure to identify the controller parameters from natural response to desired requirements is described in this section. Let $G_{vd}(s)$ be an Open loop transfer function, e(s) be the error between the reference y*(s) and feedback y(s), u(t) be the controller output and F(s), in equation (8) be a closed loop transfer function as shown in Figure 14.2. Let R(s) be the PI controller transfer function as in equation (7), where K_p and K_i being proportional and integral gains and L(s) in open loop can be defined as (6):

$$L(s) = \frac{Y(s)}{e(s)} = R(s)G(s) \tag{6}$$

$$R(s) = K_P + K_I / s \tag{7}$$

$$F(s) = \frac{y(s)}{y^*(s)} = \frac{L(s)}{1 + G(s)} \tag{8}$$

The following requirements for the desired response of final closed loop system are zero steady state error, settling time, that is, bandwidth (cut-off frequency ω_c) and robustness level, that is, phase margin ϕ_m. Based on these requirements the proportional and integral constants are computed, and tuning parameters (proportional and integral constants) are also calculated using the formulas in equations (9) and (10) [8].

$$K_P = \frac{\cos\left(\phi_m - \pi - \arg\left(G\left(j\omega_C\right)\right)\right)}{G\left(j\omega_C\right)} \tag{9}$$

$$K_I = \frac{-\omega_c \sin\left(\varnothing_m - \pi - \arg\left(G\left(j\omega_C\right)\right)\right)}{G\left(j\omega_C\right)} \tag{10}$$

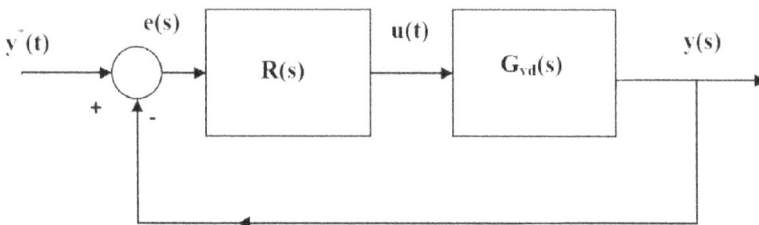

FIGURE 14.2   Closed loop control system.

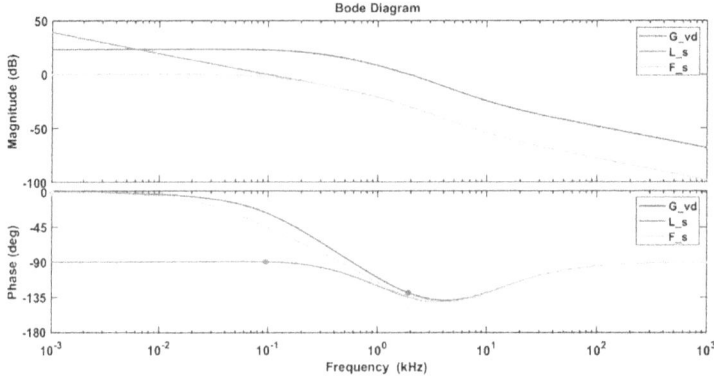

FIGURE 14.3 Bode plot of PI controller for uncompensated open loop gain and compensated buck converter.

In an uncompensated system, the second order denominator produces an overshoot and the controlled system is designed in such a way that the desired closed loop response is 20 times faster than open loop system [9]. The settling time, $T_{a,G}$ of a system must be greater than $5/\omega_g$, where ω_g is angular frequency of uncontrolled system. The desired controller is slowing down the natural response of the system by choosing $T_{a,F} = 23.9 \times T_{a,G} = 20 \times 0.00208$ comes out as 0.0083 μsec. Also, τ_F is defined as $T_{a,F}/5 = 0.0017$ sec and controller bandwidth, as $\omega_c = 1/\tau_F = 600$ rad/sec. By settling desired phase margin, ϕ_m as 90° and ω_c, proportional constants are computed, $K_p = 0.0328$ and $K_i = 38.8$ using [9] and [10]. Figure 14.3 shows the bode plot of PI controller for uncompensated buck converter, open loop gain and compensated buck converter with desired responses.

The bode plot of the G_{vd} transfer function is shown in Figure 14.3. This is common two pole low pass filter whose cut-off frequency ω_g is 1.2 rad/sec, settling time (T_{ag}) is 0.00278 sec, and phase margin as 50.4.

14.2.4 DIGITAL CONTROLLER DESIGN OF PI CONTROLLERS USING INDIRECT DESIGN APPROACH

A detailed block diagram describing the circuitry of a digital PI controller for dc-dc buck converter being interfaced with a TMS320F2812 DSP controller is shown in Figure 14.4. For realising analog PI controller into a digital controller appropriate numerical method such as Tustin approximation, Bilinear difference method, Backward Euler and Trapezoid are being used [13]. In this chapter, analog PI controller can be discretized by Backward Euler method, as given in equation (12), where T_s is the sampling time period defined as equal to the switching time period of the converter ($1/f_s = 50$μs).

$$S = \frac{1 - z^{-1}}{T_S} \tag{11}$$

$$R(z) = \frac{u(z)}{e(z)} = K_p + K_I T_s z / (z - 1) \tag{12}$$

The digital controller is also realised using Tustin approximations as in equation (in parallel form) (13).

$$S = \frac{2}{T_S} \frac{1 - z^{-1}}{1 + z^{-1}} \tag{13}$$

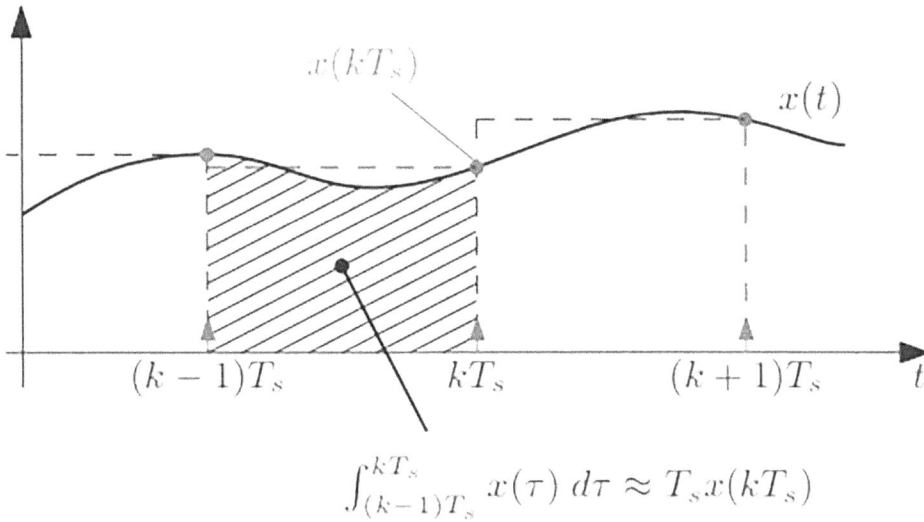

$$\int_{(k-1)T_s}^{kT_s} x(\tau)\, d\tau \approx T_s x(kT_s)$$

FIGURE 14.4 Backward Euler approximation.

$$R(z) = K_p + K_i \frac{T_s}{2}\frac{1+z^{-1}}{1-z^{-1}} + K_d \frac{2}{T_s}\frac{1-z^{-1}}{1+z^{-1}} \qquad (14)$$

where, u(k) is the control output is taken from k_{th} sample and e(k) is the error of the kth sample. The error e(k) is defined as V_{ref} −ADC(k), where ADC(k) is the digital converter value of the k_{th} sample of the sensed voltage and V_{ref} be the desired output voltage for the converter to operate. The controller is implemented using TMS320F2812 DSP in CCS window with instruction set using software programming. The closed loop dynamic performance of the prototype dc-dc converter is verified successfully with controller's implementation.

14.2.5 Loop Delays

Amplitude quantisation is performed by Analog to Digital converter with its resolution as $1/2^{12}$. Duty cycle quantisation (q_D) is also an essential quantity defined as the smallest variation in duty cycle, that is, T_{clk}/T_s -ratio of clock frequency and switching frequency. It is approximately equal to 0.16% for our converter with T_s as 20kHz and T_{clk} as 30MHz.

Digital control loop consists of A/D converter, digital compensator and digital pulse width modulator. Delays of various nature are most commonly observed in digital control loops and thus effects the frequency response of the system. Total loop delays (t_d) is the sum of control delay and modulation delay. Control delay is the time interval between when the sampling occurs and when the digital modulator generated the corresponding control command u[k] (calculated by the digital compensator). Modulation delay is defined as small-signal delay introduced by the digital modulator.

Therefore, for accounting sampling effects and delays in the feedback loop, total loop delays are included as referred here:

$$t_d = t_{cntrl} + t_{DPWM} = 600ns + \frac{1/20000}{2} = 25.6\mu s \qquad (15)$$

Where, t_{DPWM} is $DT_s \approx T_s/2$.

As the t_{cntrl} is increased, t_d is also increased and hence, the phase margin of the system, that is, the stability decreases.

The exact discrete model of the converter is obtained via discretization of the standard average modelling approach as shown in Figure 14.10. The magnitude and phase plots of the $G_{vd\,cont}$ is obtained from small signal modelling in s domain and the is discretized using Tustin approximation via MATLAB scripting as shown in G_{vdz}. The small departure at 100kHz is observed in between both plots. Therefore, the effective plant transfer function $G_{vds\,delay2}$ here provides the closest approximation with G_{vdz}.

$$Gvds\_delay2 = Gvd\_cont(s)e^{-std}$$

In this way the state space model correctly accounts for delays in the feedback loop in the discrete time modelling approach of the converter. The consideration of sampling effect on the converter discrete modelling can also be considered for more precise modelling.

14.2.6 IMPLEMENTATION OF DIGITAL PI CONTROLLERS

Power electronics applications demands high-speed data acquisition and control. Therefore, Digital Signal processor (DSP) are used to meet the processing requirements in the latest advances. TMS320 family has a fixed point and floating-point DSPs and targets a wide variety of digital control applications [12]. It also supports peripherals such as event manager (EVM), timers for PWM, dual 12-bit, 16 channel ultrafast ADC used for embedded control and communication. The overall program control and step by step implementation of PI controlled buck converter flowchart structure is shown in Figure 14.5.

The output voltage for a particular instant is sensed by voltage sensor circuit and is given input to DSP via ADC channel. The digital value of voltage output result from DSP, obtained after ADC conversion is compared to reference voltage. An interrupt is generated after AD conversion and controller implementation is initiated. The updated duty ratio inside ISR is computed and given to appropriate PWM compare register. The sampling frequency of the ADC is chosen as equal to switching frequency of the converter to reduce the sampling delay, $f_s = f_{sw} = 20$kHz. The sampling time of ADC is the total of acquisition time and conversion time. Acquisition time is sample and

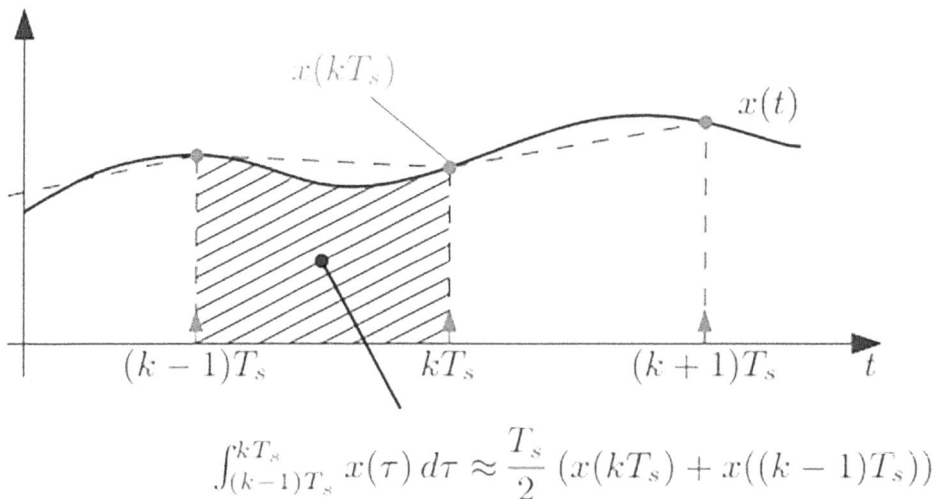

$$\int_{(k-1)T_s}^{kT_s} x(\tau)\,d\tau \approx \frac{T_s}{2}\left(x(kT_s) + x((k-1)T_s)\right)$$

FIGURE 14.5 Tustin approximation.

hold (SOH) cycles (set as 15) times the ADC clock time (1/3Mhz) and conversion time, which is 0.5ms for the ADC. The voltage reference was kept as 6V with scaling of 0.3. The resolution of 12bit ADC ($1/2^{12} \times 3$ V) is 0.73mV. The range of ADC is 0–3V. In the feedback circuitry, output voltage is scaled by gain of 0.3 using opamp 741 and is given to ADC of DSP using optocoupler.

14.3 RESULTS AND DISCUSSIONS

For this experiment, the prototype of buck converter of 12–16V input and 6V output was being developed with nominal duty ratio as 0.38. Load was varied from 4Ω to 2Ω as shown is Figure 14.6 and Figure 14.7. The transient response of converter is shown in Figure 14.10 and steady state output voltage, inductor current and switching pulses generated by controller are shown in Figure 14.9. The settling time as nominal 12V input voltage is 10ms. The results as shown in Table 14.2 verify our digital controlled buck converter at frequency = 20kHz and V_{in} = 12V. Figure 14.8 shows the hardware circuitry of the closed loop buck converter.

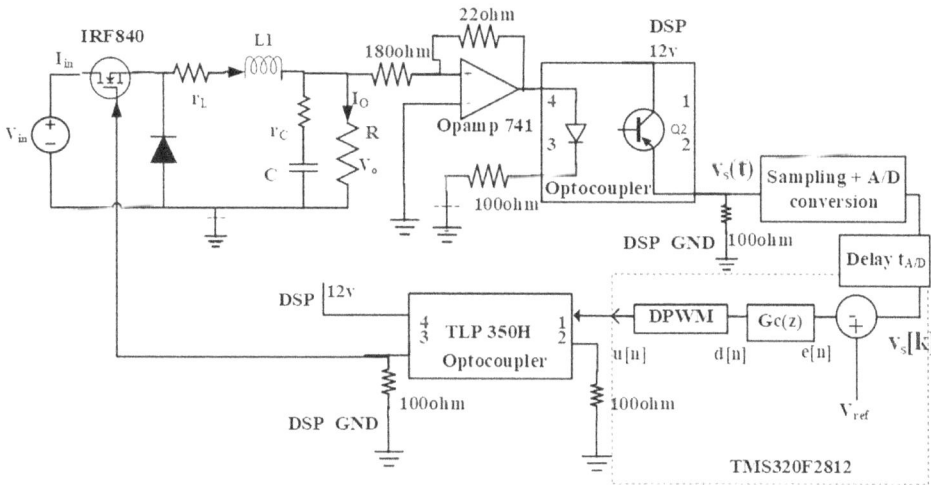

FIGURE 14.6   Discrete closed loop control system.

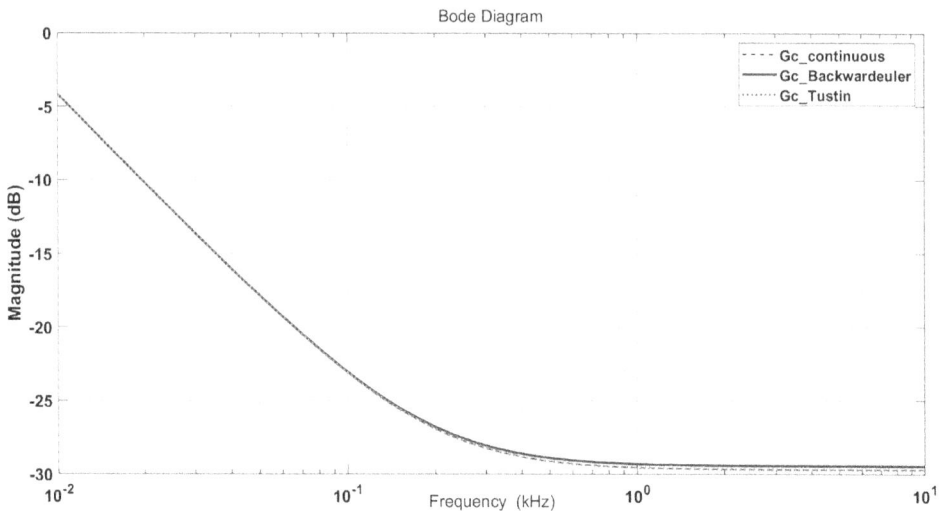

FIGURE 14.7   Comparison magnitude plot of controller for continuous vs discrete.

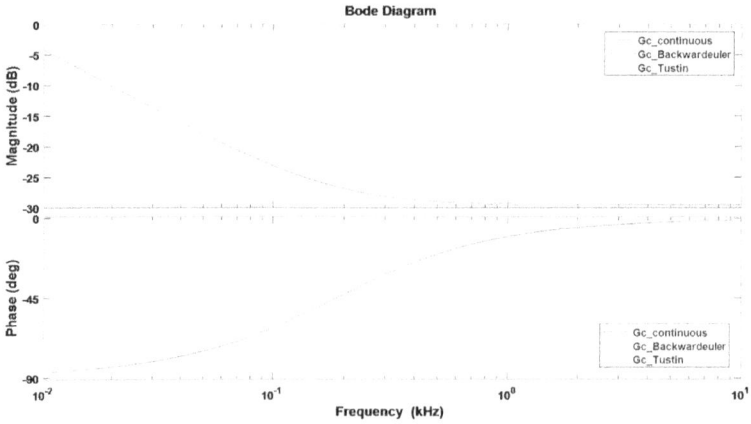

FIGURE 14.8   Comparison phase plot of controller for continuous vs discrete.

The figure compares the bode plots of the continuous-PI compensator design with Backward Euler and Tustin discretization. Here, the sampling process is synchronised with switching frequency of the converter, results in less severe and controllable aliasing effects. From the following bode plot (5) and (6), it is observed that the Tustin approximation is more accurate as compared with discretization methods.

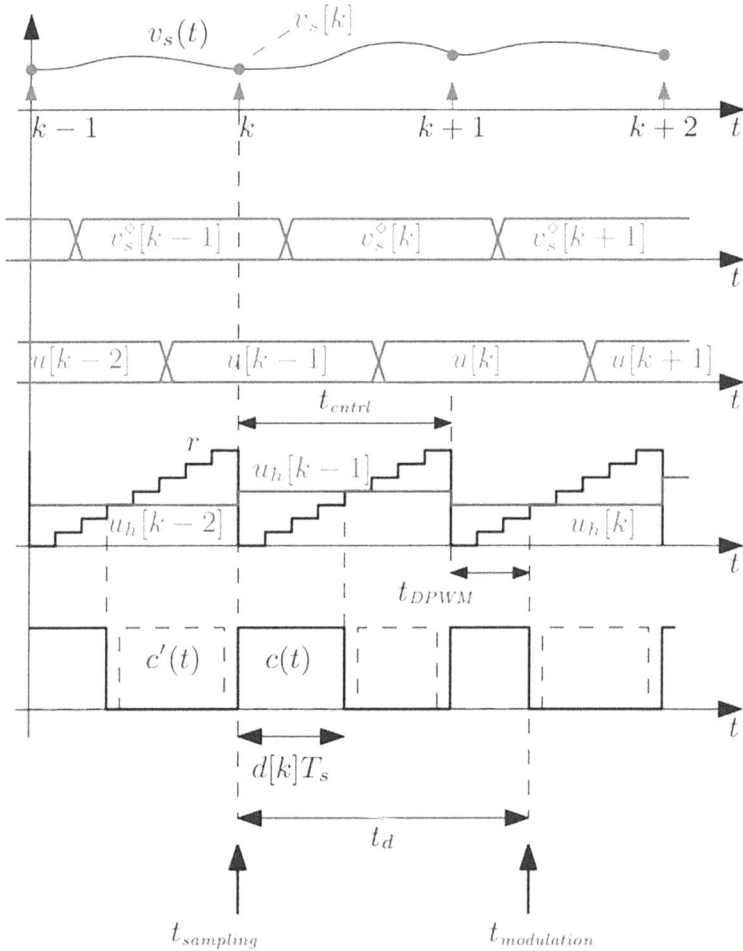

FIGURE 14.9   Timing diagram of DSP controller.

TABLE 14.2

Closed Loop Results (V_{in} = 12V, R = 2ohm)

V_{oref} (V)	V_o (measured) (V)	D (from DSP)	I_{out} (A)	$I_{inductor}$ (A)
3	2.96	0.23	1.50	1.5
5	4.60	0.38	2.30	2.4
6	5.70	0.48	2.90	3.0
8	7.20	0.61	3.75	4.0

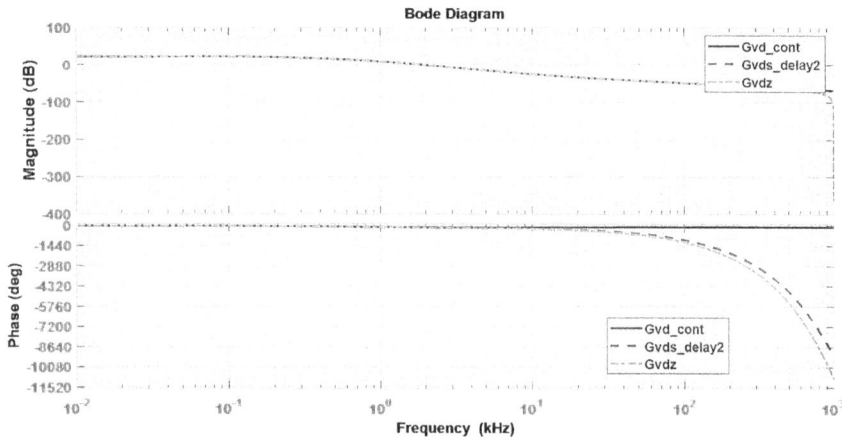

FIGURE 14.10   Synchronous buck converter comparsion.

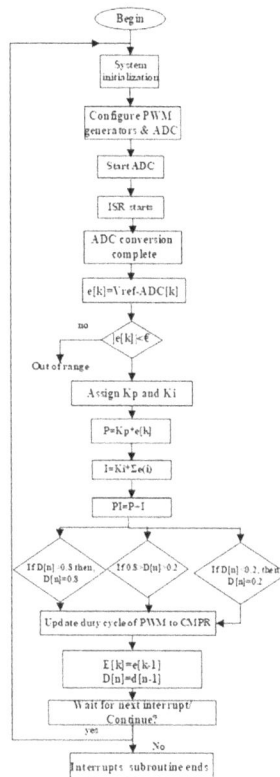

FIGURE 14.11   Flowchart for Digital controller implimentation.

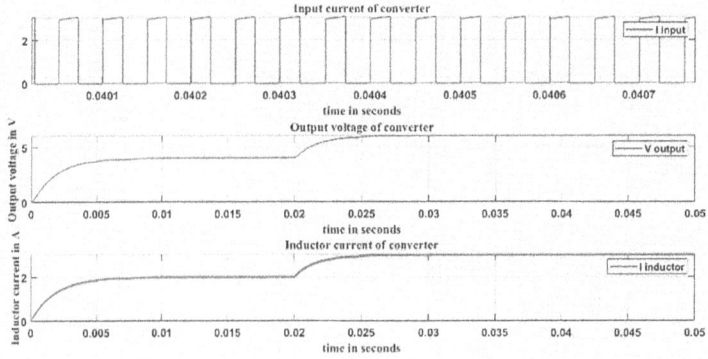

FIGURE 14.12   Transient and Load variation responses for buck converter.

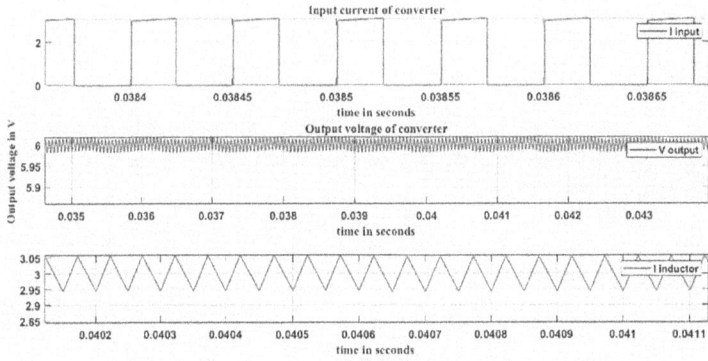

FIGURE 14.13   Steady state responses for buck converter.

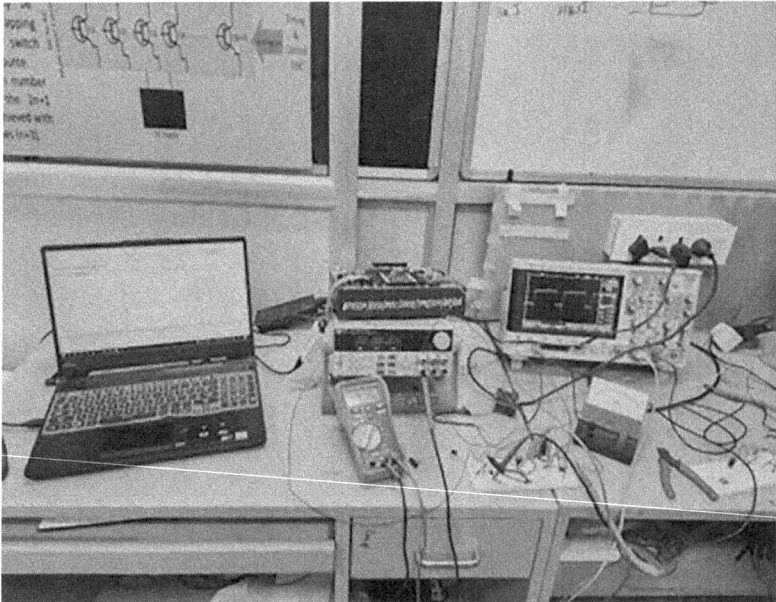

FIGURE 14.14   Buck Hardware circuit.

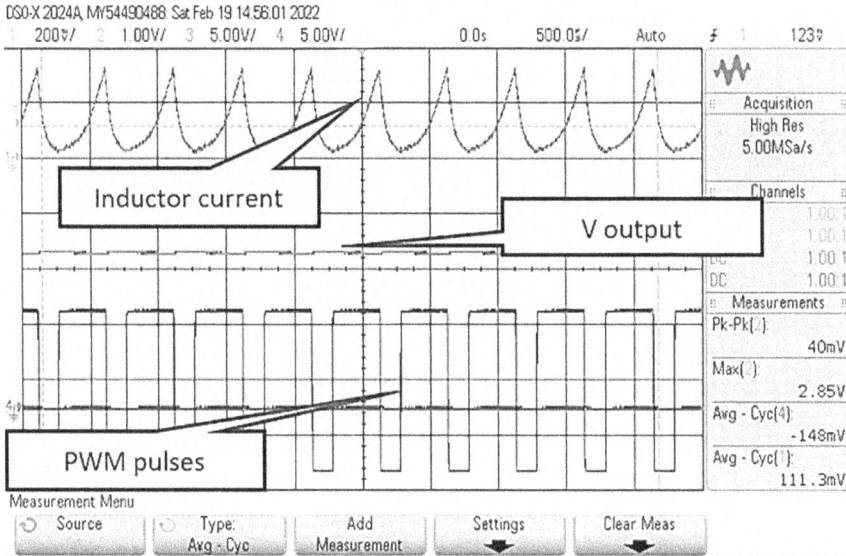

FIGURE 14.15 Steady state responses of closed loop buck converter.

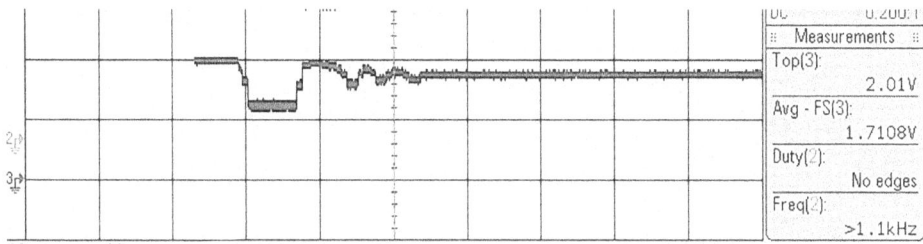

FIGURE 14.16 Transient response of closed loop buck converter.

14.4 CONCLUSION

This chapter discusses DSP based digital control loop design method for dc-dc buck converter. The converter was interfaced with a DSP controller of associated parameters to digital PI controller design. The design by emulation method allows to convert designed controller design to discrete controller. For time invariant buck converter topology, the exact discrete time model is also obtained via Tustin method with total loop delay taken into consideration for effective analysis. Then, the detailed implementation of controller design is presented using TMS320F2812. The MATLAB simulation results for digitally controlled converter are verified using hardware prototype of buck converter and satisfactory transient and steady state responses are received.

REFERENCES

[1] L. Corradini, D. Maksimovic, P. Mattavelli, R. Zane, *Digital Control of High-Frequency Switched-Mode Power Converters*. New York: Wiley-IEEE Press, 2015.
[2] V. Jindal, and D. Joshi. A comparative analysis of classical Tuning methods of PI controllers on non-ideal buck converter. *2022 2nd International Conference on Power Electronics & IoT Applications in Renewable Energy and its Control (PARC)*, 2022, pp. 1–5, doi: 10.1109/PARC52418.2022.9726251.
[3] M. F. N. b. Tajuddin, N. A. Rahim, and I. Daut. Design and implementation of a DSP based digital controller for a DC-DC converter. *2009 Second International Conference on Computer and Electrical Engineering*, 2009, pp. 209–213, doi: 10.1109/ICCEE.2009.217.

[4] M. M. Garg, Y. V. Hote, and M. K. Pathak. Design and performance analysis of a PWM dc—dc Buck Converter Using PI—Lead Compensator. *Arab. J. Sci. Eng.*, vol. 40, no. 12, pp. 3607–3626, Dec.2015.

[5] N. Mohan, T. M. Undeland, and W. P. *Robbins, Power Electronics Converters, Applications, and Design*, 3rd ed. Hoboken: John Wiley, 2003.

[6] R. W. Erickson, and D. *Maksimovic, Fundamentals of Power Electronics*, 2nd ed. London: Kluwer Academic Publishers, 2020.

[7] Choudhury, S. *Designing a TMS320F280x Based Digitally Controlled DC-DC Switching Power Supply*. Hoboken: John Wiley, 2005.

[8] S. Buso, and P. *Mattavelli, Digital Control in Power Electronics*, 1st ed. New York: Morgan & Claypool, 2006.

[9] Liping Guo. Implementation of digital PID controllers for DC-DC converters using digital signal processors. *2007 IEEE International Conference on Electro/Information Technology*, 2007, pp. 306–311, doi: 10.1109/EIT.2007.4374445.

[10] Rossi, M., Toscani, N., Mauri, M., Dezza, F. C. *Introduction to Microcontroller Programming for Power Electronics Control Applications: Coding with MATLAB® and Simulink®*, 1st ed. London: CRC Press, 2021.

[11] D. Ounnas, D. Guiza, Y. Soufi, R. Dhaouadi, and A. Bouden. Design and implementation of a digital PID controller for DC—DC buck converter. *2019 1st International Conference on Sustainable Renewable Energy Systems and Applications (ICSRESA)*, 2019, pp. 1–4, doi: 10.1109/ICSRESA49121.2019.9182430.

[12] D. Prarthana. Digital control of a closed loop buck converter. *International Journal of Engineering Applied Sciences and Technology*, vol. 5, no. 10, 2021, pp. 124–139.

[13] M. M. Garg, and M. Kumar Pathak. Performance comparison of non-ideal and ideal models of DC-DC buck converter. *2018 8th IEEE India International Conference on Power Electronics (IICPE)*, 2018, pp. 1–6, doi: 10.1109/IICPE.2018.8709450.

Index

For Product Safety Concerns and Information please contact our EU
representative GPSR@taylorandfrancis.com
Taylor & Francis Verlag GmbH, Kaufingerstraße 24, 80331 München, Germany

*9 7 8 1 0 3 2 3 4 7 1 5 8 *